The Linear Algebra a Beginning Graduate Student Ought to Know

Jonathan S. Golan

The Linear Algebra a Beginning Graduate Student Ought to Know

Third Edition

 Springer

Prof. Jonathan S. Golan
Dept. of Mathematics
University of Haifa
Haifa
Israel

ISBN 978-94-007-2635-2   e-ISBN 978-94-007-2636-9
DOI 10.1007/978-94-007-2636-9
Springer Dordrecht Heidelberg London New York

Library of Congress Control Number: 2012933373

Mathematics Subject Classification (2010): 15-XX, 16-XX, 17-XX, 65F30

Springer is part of Springer Science+Business Media (www.springer.com)

To the memory of Hemda, my wife of over 40 years:

בטח בה לב בעלה ושלל לא יחסר
(משלי, פרק ל"א)

And to my grandchildren: Shachar, Eitan, Sarel, Nachshon, Yarden, Itamar, Roni, and Naomi

תן לחכם ויחכם עוד
(משלי, פרק ט')

For Whom Is This Book Written?

> *Crow's Law: Do not think what you want to think until you know what you ought to know.*[1]

Linear algebra is a living, active branch of mathematical research which is central to almost all other areas of mathematics and which has important applications in all branches of the physical and social sciences and in engineering. However, in recent years the content of linear algebra courses required to complete an undergraduate degree in mathematics—and even more so in other areas—at all but the most dedicated universities, has been depleted to the extent that it falls far short of what is in fact needed for graduate study and research or for real-world application. This is true not only in the areas of theoretical work but also in the areas of computational matrix theory, which are becoming more and more important to the working researcher as personal computers become a common and powerful tool. Students are not only less able to formulate or even follow mathematical proofs, they are also less able to understand the underlying mathematics of the numerical algorithms they must use. The resulting knowledge gap has led to frustration and recrimination on the part of both students and faculty alike, with each silently—and sometimes not so silently—blaming the other for the resulting state of affairs. This book is written with the intention of bridging that gap. It was designed be used in one or more of several possible ways:

(1) As a self-study guide;
(2) As a textbook for a course in advanced linear algebra, either at the upper-class undergraduate level or at the first-year graduate level; or
(3) As a reference book.

It is also designed to be used to prepare for the linear algebra portion of prelim exams or Ph.D. qualifying exams.

This volume is self-contained to the extent that it does not assume any previous knowledge of formal linear algebra, though the reader is assumed to have been exposed, at least informally, to some basic ideas or techniques, such as matrix manipulation and the solution of a small system of linear equations. It does, however,

---

[1] This law, attributed to John Crow of King's College, London, is quoted by R.V. Jones in his book *Most Secret War*, Wordsworth, 1998 (ISBN 978-1853266997).

assume a seriousness of purpose, considerable motivation, and modicum of mathematical sophistication on the part of the reader.

The theoretical constructions presented here are illustrated with a large number of examples taken from various areas of pure and applied mathematics. As in any area of mathematics, theory and concrete examples must go hand in hand and need to be studied together. As the German philosopher Immanuel Kant famously remarked, concepts without precepts are empty, whereas precepts without concepts are blind.

The book also contains a large number of exercises, many of which are quite challenging, which I have come across or thought up in over thirty years of teaching. Many of these exercises have appeared in print before, in such journals as *American Mathematical Monthly*, *College Mathematics Journal*, *Mathematical Gazette*, or *Mathematics Magazine*, in various mathematics competitions or circulated problem collections, or even on the internet. Some were donated to me by colleagues and even students, and some originated in files of old exams at various universities which I have visited in the course of my career. Since, over the years, I did not keep track of their sources, all I can do is offer a collective acknowledgement to all those to whom it is due. Good problem formulators, like the God of the abbot of Citeaux, know their own. Deliberately, difficult exercises are not marked with an asterisk or other symbol. Solving exercises is an integral part of learning mathematics and the reader is definitely expected to do so, especially when the book is used for self-study. Try them all and remember the "grook" penned by the Danish genius Piet Hein: *Problems worthy of attack / Prove their worth by hitting back.*

Solving a problem using theoretical mathematics is often very different from solving it computationally, and so strong emphasis is placed on the interplay of theoretical and computational results. Real-life implementation of theoretical results is perpetually plagued by errors: errors in modeling, errors in data acquisition and recording, and errors in the computational process itself due to roundoff and truncation. There are further constraints imposed by limitations in time and memory available for computation. Thus the most elegant theoretical solution to a problem may not lead to the most efficient or useful method of solution in practice. While no reference is made to particular computer software, the concurrent use of a personal computer equipped symbolic-manipulation software such as MAPLE, MATHEMATICA, MATLAB, or MUPAD is definitely advised.

In order to show the "human face" of mathematics, the book also includes a large number of thumbnail photographs of researchers who have contributed to the development of the material presented in this volume.

Acknowledgements   Most of the first edition this book was written while I was a visitor at the University of Iowa in Iowa City and at the University of California in Berkeley. I would like to thank both institutions for providing the facilities and, more importantly, the mathematical atmosphere which allowed me to concentrate on writing. Subsequent, extensively revised editions, were prepared after I retired from teaching at the University of Haifa in April, 2004.

I have talked to many students and faculty members about my plans for this book and have obtained valuable insights from them. In particular, I would like to acknowledge the aid of the following colleagues and students who were kind enough to read the preliminary versions of this book and offer their comments and corrections: Prof. Daniel Anderson (University of Iowa), Prof. Adi Ben-Israel (Rutgers University), Prof. Robert Cacioppo (Truman State University), Prof. Joseph Felsenstein (University of Washington), Prof. Ryan Skip Garibaldi (Emory University), Mr. George Kirkup (University of California, Berkeley), Dr. Denis Sevee (John Albert College), Prof. Earl Taft (Rutgers University), Mr. Gil Vernik (University of Haifa).

Haifa, Israel   Jonathan S. Golan

Contents

Notation and Terminology

Sets will be denoted by braces, { }, between which we will either list the elements of the set or give a rule for determining whether something is an element of the set or not,[1] as in $\{x \mid p(x)\}$, which is read "the set of all x such that $p(x)$". If a is an element of a set A, we write $a \in A$; if it is not an element of A, we write $a \notin A$. When one enumerates the elements of a set, the order is not important. Thus $\{1, 2, 3, 4\}$ and $\{4, 1, 3, 2\}$ both denote the same set. However, we often do wish to impose an order on sets the elements of which we enumerate. Rather than introduce new and cumbersome notation to handle this, we will make the convention that when we enumerate the elements of a finite or countably-infinite set, we will assume an implied order, reading from left to right. Thus, the implied order on the set $\{1, 2, 3, \ldots\}$ is indeed the usual one, whereas $\{4, 1, 3, 2\}$ gives the first four positive integers, ordered alphabetically. The empty set, namely the set having no elements, is denoted by $\varnothing$. Sometimes we will use the word "collection" as a synonym for "set", generally to avoid talking about "sets of sets".

A finite or countably-infinite selection of elements of a set A is a *list*. Members of a list are assumed to be in a definite order, given by their indices or by the implied order of reading from left to right. Lists are usually written without brackets: $a_1, \ldots, a_n$, though, in certain contexts, it will be more convenient to write them as ordered n-tuples $(a_1, \ldots, a_n)$. Note that the elements of a list need not be distinct: 3, 1, 4, 1, 5, 9 is a list of six positive integers, the second and fourth elements of which are equal to 1. A countably-infinite list of elements of a set A is also often called a *sequence* of elements of A. The set of all distinct members of a list is called the *underlying subset* of the list.

If A and B are sets, then their *union* $A \cup B$ is the set of all elements that belong to either A or B, and their *intersection* $A \cap B$ is the set of all elements belonging both to A and to B. More generally, if $\{A_i \mid i \in \Omega\}$ is a (possibly-infinite) collection of

---

[1]Mathematically, these two ways of defining a set are equivalent, but philosophically and functionally they are not. Listing the elements of a set involves *denotation* whereas giving a rule for determining set membership involves *connotation*. This distinction becomes important when we attempt to use computers to manipulate sets.

J.S. Golan, *The Linear Algebra a Beginning Graduate Student Ought to Know*,
DOI 10.1007/978-94-007-2636-9_1, © Springer Science+Business Media B.V. 2012

sets, then $\bigcup_{i \in \Omega} A_i$ is the set of all elements that belong to at least one of the A_i and $\bigcap_{i \in \Omega} A_i$ is the set of all elements that belong to all of the A_i. If A and B are sets, then the *difference set* $A \smallsetminus B$ is the set of all elements of A which do not belong to B.

A *function* f from a nonempty set A to a nonempty set B is a rule which assigns to each element a of A a unique element $f(a)$ of B. The set A is called the *domain* of the function and the set B is called the *range* of the function. To denote that f is a function from A to B, we write $f : A \to B$. To denote that an element b of B is assigned to an element a of A by f, we write $f : a \mapsto b$. (Note the different form of the arrow!) This notation is particularly helpful in the case that the function f is defined by a formula. Thus, for example, if f is a function from the set of integers to the set of integers defined by $f : a \mapsto a^3$, then we know that f assigns to each integer its cube. The set of all functions from a nonempty set A to a nonempty set B is denoted by B^A. If $f \in B^A$ and if A' is a nonempty subset of A, then a function $f' \in B^{A'}$ is the *restriction* of f to A', and f is the *extension* of f' to A, if and only if $f' : a' \mapsto f(a')$ for all $a' \in A'$.

Functions f and g in B^A are *equal* if and only if $f(a) = g(a)$ for all $a \in A$. In this case, we write $f = g$. A function $f \in B^A$ is *monic* if and only if it assigns different elements of B to different elements of A, i.e., if and only if $f(a_1) \neq f(a_2)$ whenever $a_1 \neq a_2$ in A. A function $f \in B^A$ is *epic* if and only if every element of B is assigned by f to some element of A. A function which is both monic and epic is *bijective*. A bijective function from a set A to a set B determines a bijective correspondence between the elements of A and the elements of B. If $f : A \to B$ is a bijective function, then we can define the *inverse function* $f^{-1} : B \to A$ defined by the condition that $f^{-1}(b) = a$ if and only if $f(a) = b$. This inverse function is also bijective. A bijective function from a set A to itself is a *permutation* of A. Note that there is always at least one permutation of any nonempty set A, namely the identity function $a \mapsto a$.

The *Cartesian product* $A_1 \times A_2$ of nonempty sets A_1 and A_2 is the set of all ordered pairs (a_1, a_2), where $a_1 \in A_1$ and $a_2 \in A_2$. More generally, if $A_1, \ldots, A_n$ is a list of nonempty sets, then $A_1 \times \cdots \times A_n$ is the set of all ordered n-tuples $(a_1, \ldots, a_n)$ satisfying the condition that $a_i \in A_i$ for each $1 \leq i \leq n$. Note that each ordered n-tuple $(a_1, \ldots, a_n)$ uniquely defines a function $f : \{1, \ldots, n\} \to \bigcup_{i=1}^{n} A_i$ given by $f : i \mapsto a_i$ for each $1 \leq i \leq n$. Conversely, each function $f : \{1, \ldots, n\} \to \bigcup_{i=1}^{n} A_i$ satisfying the condition that $f(i) \in A_i$ for $1 \leq i \leq n$ defines such an ordered n-tuple, namely $(f(1), \ldots, f(n))$. This suggests a method for defining the Cartesian product of an arbitrary collection of nonempty sets. If $\{A_i \mid i \in \Omega\}$ is an arbitrary collection of nonempty sets, then the set $\prod_{i \in \Omega} A_i$ is defined to be the set of all those functions f from Ω to $\bigcup_{i \in \Omega} A_i$ satisfying the condition that $f(i) \in A_i$ for each $i \in \Omega$. The existence of such functions is guaranteed by a fundamental axiom of set theory, known as the *Axiom of Choice*. A certain amount of controversy surrounds this axiom, since it leads to some very counter-intuitive results. Thus, for example, in 1924 Polish mathematicians Stefan Banach and Alfred Tarski showed that if the Axiom of Choice is assumed then any solid sphere can be split into finitely-many pieces which can be reassembled to form two solid spheres of the same size as the original sphere. Therefore, there are mathematicians who prefer to make as little use of the Axiom of Choice as possible. In 1963, American mathematician P. J. Cohen showed that the Axiom of Choice is independent of the other axioms of Zermelo–Fraenkel set theory, and so one is—in principle—free to either assume it or its negation. Since we will need this axiom constantly throughout this book, we will always assume that it holds.

In the foregoing construction, we did not assume that the sets A_i were necessarily distinct. Indeed, it may very well happen that there exists a set A such that $A_i = A$ for all $i \in \Omega$. In that case, we see that $\prod_{i \in \Omega} A_i$ is just A^Ω. If the set Ω is finite, say $\Omega = \{1, \ldots, n\}$, then we write A^n instead of A^Ω. Thus, A^n is just the set of all ordered n-tuples $(a_1, \ldots, a_n)$ of elements of A.

Example The function $f_2 : \mathbb{N}^2 \to \mathbb{N}$ given by

$$f_2 : (i, j) \mapsto \frac{1}{2}(i^2 + j^2 + i + 2ij + 3j)$$

is bijective. For $k > 2$ we can define a bijective function $f_k : \mathbb{N}^k \to \mathbb{N}$ inductively by

$$f_k : (i_1, \ldots, i_k) \mapsto f_2(i_1, f_{k-1}(i_2, \ldots, i_k)).$$

We use the following standard notation for some common sets of numbers:

$\mathbb{N}$ the set of all nonnegative integers,
$\mathbb{Z}$ the set of all integers,
$\mathbb{Q}$ the set of all rational numbers,
$\mathbb{R}$ the set of all real numbers,
$\mathbb{C}$ the set of all complex numbers.

Other notion is introduced throughout the text, as is appropriate. See the Summary of Notation in Appendix A of the book.

Fields

The way of mathematical thought is twofold: the mathematician first proceeds inductively from the particular to the general and then deductively from the general to the particular. Moreover, throughout its development, mathematics has shown two aspects—the conceptual and the computational—the symphonic interleaving of which forms one of the major aspects of the subject's aesthetic.

Let us therefore begin with the first mathematical structure—numbers. By the Hellenistic times, mathematicians distinguished between two types of numbers: the *rational* numbers, namely those which could be written in the form $\frac{m}{n}$ for some integer m and some positive integer n, and those numbers representing the geometric magnitude of segments of the line, which today we call *real* numbers and which, in decimal notation, are written in the form $m.k_1k_2k_3\ldots$ where m is an integer and the k_i are digits. The fact that the set $\mathbb{Q}$ of rational numbers is not equal to the set $\mathbb{R}$ of real numbers was already noticed by the followers of the early Greek mathematician/mystic Pythagoras. On both sets of numbers we define operations of addition and multiplication which satisfy certain rules of manipulation. Isolating these rules as part of a formal system was a task first taken on in earnest by nineteenth-century British and German mathematicians. From their studies evolved the notion of a field, which will be basic to our considerations. However, since fields are not our primary object of study, we will delve only minimally into this fascinating notion. A serious consideration of field theory must be deferred to an advanced course in abstract algebra.

A nonempty set F together with two functions $F \times F \to F$, respectively called *addition* (as usual, denoted by $+$) and *multiplication* (as usual, denoted by $\cdot$ or by concatenation), is a *field* if the following conditions are satisfied:

(1) (*associativity of addition and multiplication*): $a + (b + c) = (a + b) + c$ and $a(bc) = (ab)c$ for all $a, b, c \in F$.
(2) (*commutativity of addition and multiplication*): $a + b = b + a$ and $ab = ba$ for all $a, b \in F$.
(3) (*distributivity of multiplication over addition*): $a(b + c) = ab + ac$ for all $a, b, c \in F$.

J.S. Golan, *The Linear Algebra a Beginning Graduate Student Ought to Know*, DOI 10.1007/978-94-007-2636-9_2, © Springer Science+Business Media B.V. 2012

(4) (*existence of identity elements for addition and multiplication*): There exist distinct elements of F, which we will denote by 0 and 1 respectively, satisfying $a + 0 = a$ and $a1 = a$ for all $a \in F$.

(5) (*existence of additive inverses*): For each $a \in F$ there exists an element of F, which we will denote by $-a$, satisfying $a + (-a) = 0$.

(6) (*existence of multiplicative inverses*): For each $0 \neq a \in F$ there exists an element of F, which we will denote by a^{-1}, satisfying $a^{-1}a = 1$.

With kind permission of the Archives of the Mathematisches Forschungsinstitut Oberwolfach (Weber, Dedekind, Kronecker and Steinitz).

The development of the abstract theory of fields is generally credited to the nineteenth-century German mathematician **Heinrich Weber**, based on earlier work by the German mathematicians **Richard Dedekind** and **Leopold Kronecker**. Another nineteenth-century mathematician, the British **Augustus De Morgan**, was among the first—along with French mathematician François Joseph Servois—to isolate the importance of such properties as associativity, distributivity, and so forth. The final axioms of a field are due to the twentieth-century German mathematician **Ernst Steinitz**.

Note that we did not assume that the elements $-a$ and a^{-1} are unique, though we will soon prove that in fact they are. If a and b are elements of a field F, we will follow the usual conventions by writing $a - b$ instead of $a + (-b)$ and $\frac{a}{b}$ instead of ab^{-1}. Moreover, if $0 \neq a \in F$ and if n is a positive integer, then na denotes the sum $a + \cdots + a$ (n summands) and a^n denotes the product $a \cdots a$ (n factors). If n is a negative integer, then na denotes $(-n)(-a)$ and a^n denotes $(a^{-1})^{-n}$. Finally, if $n = 0$ then na denotes the field element 0 and a^n denotes the field element 1. For $0 = a \in F$, we define $na = 0$ for all integers n and $a^n = 0$ for all positive integers n. The symbol 0^k is not defined for $k \leq 0$.

As an immediate consequence of the associativity and commutativity of addition, we see that the sum of any list $a_1, \ldots, a_n$ of elements of a field F is the same, no matter in which order we add them. We can therefore unambiguously write $a_1 + \cdots + a_n$. This sum is also often denoted by $\sum_{i=1}^{n} a_i$. Similarly, the product of these elements is the same, no matter in which order we multiply them. We can therefore unambiguously write $a_1 \cdots a_n$. This product is also often denoted by $\prod_{i=1}^{n} a_i$. Also, a simple inductive argument shows that multiplication distributes over arbitrary sums: if $a \in F$ and $b_1, \ldots, b_n$ is a list of elements of F then $a(\sum_{i=1}^{n} b_i) = \sum_{i=1}^{n} ab_i$.

We easily see that $\mathbb{Q}$ and $\mathbb{R}$, with the usual addition and multiplication, are fields.

A subset G of a field F is a *subfield* if and only if it contains 0 and 1, is closed under addition and multiplication, and contains the additive and multiplicative inverses of all of its nonzero elements. Thus, for example, $\mathbb{Q}$ is a subfield of $\mathbb{R}$. It is easy to verify[1] that the intersection of a collection of subfields of a field F is again a subfield of F.

We now want to look at several additional important examples of fields.

Example Let $\mathbb{C} = \mathbb{R}^2$ and define operations of addition and multiplication on $\mathbb{C}$ by setting $(a, b) + (c, d) = (a + c, b + d)$ and $(a, b) \cdot (c, d) = (ac - bd, ad + bc)$. These operations define the structure of a field on $\mathbb{C}$, in which the identity element for addition is $(0, 0)$, the identity element for multiplication is $(1, 0)$, the additive inverse of (a, b) is $(-a, -b)$, and

$$(a, b)^{-1} = \left(\frac{a}{a^2 + b^2}, \frac{-b}{a^2 + b^2} \right)$$

for all $(0, 0) \neq (a, b)$. This field is called the field of *complex numbers*. The set of all elements of $\mathbb{C}$ of the form $(a, 0)$ forms a subfield of $\mathbb{C}$, which we normally identify with $\mathbb{R}$ and therefore it is standard to consider $\mathbb{R}$ as a subfield of $\mathbb{C}$. In particular, we write a instead of $(a, 0)$ for any real number a. The element $(0, 1)$ of $\mathbb{C}$ is denoted by i. This element satisfies the condition that $i^2 = (-1, 0)$ and so it is often written as $\sqrt{-1}$. We also note that any element (a, b) of $\mathbb{C}$ can be written as $(a, 0) + b(0, 1) = a + bi$, and, indeed, that is the way complex numbers are usually written and how we will denote them from now on. If $z = a + bi$, then a is the *real part* of z, which is often denoted by $\mathrm{Re}(z)$, while bi is the *imaginary part* of z, which is often denoted by $\mathrm{Im}(z)$. The field of complex numbers is extremely important in mathematics. From a geometric point of view, if we identify $\mathbb{R}$ with the set of points on the Euclidean line, as one does in analytic geometry, then it is natural to identify $\mathbb{C}$ with the set of points in the Euclidean plane.

With kind permission of the Harvard Arts Museum (Descartes); With kind permission of ETH-Bibliothek Zurich, Image Archive (Euler); With kind permission of Bibliothèque nationale de France (Argand).

The term "imaginary" was coined by the seventeenth-century French philosopher and mathematician **René Descartes**. The use of i to denote $\sqrt{-1}$ was introduced by the eighteenth-century Swiss mathematician **Leonhard Euler**. The geometric representation of the complex numbers was first proposed at the end of the eighteenth century by the Norwegian surveyor Caspar Wessel, and later by the French accountant **Jean-Robert Argand**. It was studied in detail by the nineteenth-century Italian mathematician **Giusto Bellavitis**.

---

[1] When a mathematician says that something is "easy to see" or "trivial", it means that you are expected to take out a pencil and paper and spend some time—often considerable—checking it out by yourself.

If $z = a + bi \in \mathbb{C}$ then we denote the complex number $a - bi$, called the *complex conjugate* of z, by $\overline{z}$. It is easy to see that for all $z, z' \in \mathbb{C}$ we have $\overline{z + z'} = \overline{z} + \overline{z'}e$, $\overline{-z} = -\overline{z}$, $\overline{zz'} = \overline{z} \cdot \overline{z'}$, $\overline{z^{-1}} = (\overline{z})^{-1}$, and $\overline{\overline{z}} = z$. The number $z\overline{z}$ equals $a^2 + b^2$, which is a nonnegative real number and so has a square root in $\mathbb{R}$, which we will denote by $|z|$. Note that $|z|$ is nonzero whenever $z \neq 0$. From a geometric point of view, this number is just the distance from the number z, considered as a point in the Euclidean plane, to the origin, just as the usual absolute value $|a|$ of a real number a is the distance between a and 0 on the real line. It is easy to see that if y and z are complex numbers then $|yz| = |y| \cdot |z|$ and $|y + z| \leq |y| + |z|$. Moreover, if $z = a + bi$ then

$$z + \overline{z} = 2a \leq 2|a| = 2\sqrt{a^2} \leq 2\sqrt{a^2 + b^2} = 2|z|.$$

We also note, as a direct consequence of the definition, that $|z| = |\overline{z}|$ for every complex number z and so $z^{-1} = |z|^{-2}\overline{z}$ for all $0 \neq z \in \mathbb{C}$. In particular, if $|z| = 1$ then $z^{-1} = \overline{z}$.

Example The set $\mathbb{Q}^2$ is a subfield of the field $\mathbb{C}$ defined above. However, it is also possible to define field structures on $\mathbb{Q}^2$ in other ways. Indeed, let $F = \mathbb{Q}^2$ and let p be a fixed prime integer. Define addition and multiplication on F by setting $(a, b) + (c, d) = (a + c, b + d)$ and $(a, b) \cdot (c, d) = (ac + bdp, ad + bc)$.

Again, one can check that F is indeed a field and that, again, the set of all elements of F of the form $(a, 0)$ is a subfield, which we will identify with $\mathbb{Q}$. Moreover, the additive inverse of $(a, b) \in F$ is $(-a, -b)$ and the multiplicative inverse of $(0, 0) \neq (a, b) \in F$ is

$$\left(\frac{a}{a^2 - pb^2}, \frac{-b}{a^2 - pb^2} \right).$$

(We note that $a^2 - pb^2$ is the product of the nonzero real numbers $a + b\sqrt{p}$ and $a - b\sqrt{p}$ and so is nonzero.) The element $(0, 1)$ of F satisfies $(0, 1)^2 = (p, 0)$ and so one usually denotes it by $\sqrt{p}$ and, as before, any element of F can be written in the form $a + b\sqrt{p}$, where $a, b \in \mathbb{Q}$. The field F is usually denoted by $\mathbb{Q}(\sqrt{p})$. Since there are infinitely-many distinct prime integers, we see that there are infinitely-many ways of defining different field structures on $\mathbb{Q} \times \mathbb{Q}$, all having the same addition.

Example Fields do not have to be infinite. Let p be a positive integer and let $\mathbb{Z}/(p) = \{0, 1, \ldots, p - 1\}$. For each nonnegative integer n, let us, for the purposes of this example, denote the remainder after dividing n by p as $[n]_p$. Thus we note that $[n]_p \in \mathbb{Z}/(p)$ for each nonnegative integer n and that $[i]_p = i$ for all $i \in \mathbb{Z}/(p)$. We now define operations on $\mathbb{Z}/(p)$ by setting $[n]_p + [k]_p = [n + k]_p$ and $[n]_p \cdot [k]_p = [nk]_p$. It is easy to check that if the integer p is prime then $\mathbb{Z}/(p)$, together with these two operations, is again a field, known as the *Galois field* of order p. This field is usually denoted by $\mathrm{GF}(p)$. While Galois fields were first considered mathematical curiosities, they have since found important applications in coding theory, cryptography, and modeling of computer processes.

These are not the only possible finite fields. Indeed, it is possible to show that for each prime integer p and each positive integer n there exists an (essentially unique) field with p^n elements, usually denoted by GF(p^n).

The nineteenth-century French mathematical genius **Evariste Galois**, who died at the age of 21, was the first to consider such structures. The study of finite and infinite fields was unified in the 1890s by **Eliakim Hastings Moore**, the first American-born mathematician to achieve an international reputation.

Example Some important structures are "very nearly" fields. For example, let $\mathbb{R}_\infty = \mathbb{R} \cup \{\infty\}$, and define operations $\boxplus$ and $\boxdot$ on $\mathbb{R}_\infty$ by setting

$$a \boxplus b = \begin{cases} \min\{a, b\} & \text{if } a, b \in \mathbb{R}, \\ b & \text{if } a = \infty, \\ a & \text{if } b = \infty, \end{cases}$$

and

$$a \boxdot b = \begin{cases} a + b & \text{if } a, b \in \mathbb{R}, \\ \infty & \text{otherwise.} \end{cases}$$

This structure, called the *optimization algebra*, satisfies all of the conditions of a field *except* for the existence of additive inverses (such structures are known as *semifields*). As the name suggests, it has important applications in optimization theory and the analysis of discrete-event dynamical systems. There are several other semifields which have significant applications and which have been extensively studied.

Another possibility of generalizing the notion of a field is to consider an algebraic structure which satisfies all of the conditions of a field *except* for the existence of multiplicative inverses, and to replace that condition by the condition that if $a, b \neq 0$ then $ab \neq 0$. Such structures are known as *integral domains*. The set $\mathbb{Z}$ of all integers is the simplest example of an integral domain which is not a field. Algebras of polynomials over a field, which we will consider later, are also integral domains. In a course in abstract algebra, one proves that any integral domain can be embedded in a field.

In the field GF(p) which we defined above, one can easily see that the sum $1 + \cdots + 1$ (p summands) equals 0. On the other hand, in the field $\mathbb{Q}$, the sum of any number of copies of 1 is always nonzero. This is an important distinction which we will need to take into account in dealing with structures over fields. We therefore define the *characteristic* of a field F to be equal to the smallest positive integer p such that $1 + \cdots + 1$ (p summands) equals 0—if such an integer p exists—and to be

equal to 0 otherwise. We will not delve deeply into this concept, which is dealt with
in courses on field theory, except to note that the characteristic of a field, if nonzero,
always turns out to be a prime number, as we shall prove below.

In the definition of a field, we posited the existence of distinct identity elements
for addition and multiplication, but did not claim that these elements were unique.
It is, however, very easy to prove that fact.

Proposition 2.1 *Let F be a field.*
(1) *If e is an element of F satisfying $e + a = a$ for all $a \in F$ then $e = 0$;*
(2) *If u is an element of F satisfying $ua = a$ for all $a \in F$ then $u = 1$.*

Proof By definition, $e = e + 0 = 0$ and $u = u1 = 1$. $\square$

Similarly, we prove that additive and multiplicative inverses, when they exist, are
unique. Indeed, we can prove a stronger result.

Proposition 2.2 *If a and b are elements of a field F then:*
(1) *There exists a unique element c of F satisfying $a + c = b$.*
(2) *If $a \neq 0$ then there exists a unique element d of F satisfying $ad = b$.*

Proof (1) Choose $c = b - a$. Then

$$a + c = a + (b - a) = a + [b + (-a)]$$
$$= a + [(-a) + b] = [a + (-a)] + b = 0 + b = b.$$

Moreover, if $a + x = b$ then

$$x = 0 + x = [(-a) + a] + x$$
$$= (-a) + (a + x) = (-a) + b = b - a,$$

proving uniqueness.

(2) Choose $d = a^{-1}b$. Then $ad = a(a^{-1}b) = (aa^{-1})b = 1b = b$. Moreover, if
$ay = b$ then $y = 1y = (a^{-1}a)y = a^{-1}(ay) = a^{-1}b$, proving uniqueness. $\square$

We now summarize some of the elementary properties of fields, which are all we
will need for our discussion.

Proposition 2.3 *If a, b, and c are elements of a field F then:*
(1) $0a = 0$;
(2) $(-1)a = -a$;
(3) $a(-b) = -(ab) = (-a)b$;
(4) $-(-a) = a$;
(5) $(-a)(-b) = ab$;
(6) $-(a + b) = (-a) + (-b)$;
(7) $a(b - c) = ab - ac$;
(8) *If* $a \neq 0$ *then* $(a^{-1})^{-1} = a$;
(9) *If* $a, b \neq 0$ *then* $(ab)^{-1} = b^{-1}a^{-1}$;
(10) *If* $a + c = b + c$ *then* $a = b$;
(11) *If* $c \neq 0$ *and* $ac = bc$ *then* $a = b$;
(12) *If* $ab = 0$ *then* $a = b$ *or* $b = 0$.

Proof (1) Since $0a + 0a = (0 + 0)a = 0a$, we can add $-(0a)$ to both sides of the equation to obtain $0a = 0$.

(2) Since $(-1)a + a = (-1)a + 1a = [(-1) + 1]a = 0a = 0$ and also $(-a) + a = 0$, we see from Proposition 2.2 that $(-1)a = -a$.

(3) By (2) we have $a(-b) = a[(-1)b] = (-1)ab = -(ab)$ and similarly $(-a)b = -(ab)$.

(4) Since $a + (-a) = 0 = -(-a) + (-a)$, this follows from Proposition 2.2.

(5) From (3) and (4) it follows that $(-a)(-b) = a[-(-b)] = ab$.

(6) Since $(a + b) + [(-a) + (-b)] = a + b + (-a) + (-b) = 0$ and $(a + b) + [-(a + b)] = 0$, the result follows from Proposition 2.2.

(7) By (3) we have $a(b - c) = ab + a(-c) = ab + [-(ac)] = ab - ac$.

(8) Since $(a^{-1})^{-1}a^{-1} = 1 = aa^{-1}$, this follows from Proposition 2.2.

(9) Since $(a^{-1}b^{-1})(ba) = a^{-1}ab^{-1}b = 1 = (ab)^{-1}(ba)$, the result follows from Proposition 2.2.

(10) This is an immediate consequence of adding $-c$ to both sides of the equation.

(11) This is an immediate consequence of multiplying both sides of the equation by c^{-1}.

(12) If $b = 0$ we are done. If $b \neq 0$ then by (1) it follows that multiplying both sides of the equation by b^{-1} will yield $a = 0$.   □

The following two propositions are immediate consequences of Proposition 2.3.

Proposition 2.4 *Let a be a nonzero element of a finite field F having q elements. Then* $a^{-1} = a^{q-2}$.

Proof If $q = 2$ then $F = GF(2)$ and $a = 1$, so the result is immediate. Hence we can assume $q > 2$. Let $B = \{a_1, \ldots, a_{q-1}\}$ be the nonzero elements of F, written in some arbitrary order. Then $aa_i \neq aa_h$ for $i \neq h$ since, were they equal, we would have $a_i = a^{-1}(aa_i) = a^{-1}(aa_h) = a_h$. Therefore $B = \{aa_1, \ldots, aa_{q-1}\}$ and so $\prod_{i=1}^{q-1} a_i = \prod_{i=1}^{q-1}(aa_i) = a^{q-1}\big[\prod_{i=1}^{q-1} a_i\big]$. Moreover, this is a product of nonzero elements of F and so, by Proposition 2.3(12), is also nonzero. Therefore, by Proposition 2.3(11), $1 = a^{q-1}$, and so $aa^{-1} = 1 = a^{q-1} = a(a^{q-2})$, implying that $a^{-1} = a^{q-2}$. $\qquad\square$

Proposition 2.5 *If F is a field having characteristic $p > 0$, then p is prime.*

Proof Assume that p is not prime. Then $p = hk$, where $0 < h, k < p$. Therefore, $a = h1_F$ and $b = k1_F$ are nonzero elements of F. But $ab = (hk)1_F = p1_F = 0$, contradicting Proposition 2.3(12). $\qquad\square$

Of course, one can use Proposition 2.3 to prove many other identities among elements of a field. A typical example is the following

---

Proposition 2.6 (Hua's identity) *If a and b are nonzero elements of a field F satisfying $a \neq b^{-1}$ then*

$$a - aba = \big(a^{-1} + [b^{-1} - a]^{-1}\big)^{-1}.$$

Proof We note that

$$a^{-1} + (b^{-1} - a)^{-1} = a^{-1}\big[(b^{-1} - a) + a\big](b^{-1} - a)^{-1}$$
$$= a^{-1}b^{-1}(b^{-1} - a)^{-1},$$

so $(a^{-1} + [b^{-1} - a]^{-1})^{-1} = (b^{-1} - a)ba = a - aba$. $\qquad\square$

Loo-Keng Hua was a major twentieth-century Chinese mathematician.

Exercises

Exercise 1
Let F be a field and let $G = F \times F$. Define operations of addition and multiplication on G by setting $(a, b) + (c, d) = (a + c, b + d)$ and $(a, b) \cdot (c, d) = (ac, bd)$. Do these operations define the structure of a field on G?

Exercise 2
Let K be the set of the following four-tuples of elements of GF(3):

$$(0, 0, 0, 0), (1, 2, 1, 1), (2, 1, 2, 2), (1, 0, 0, 1), (2, 2, 1, 2),$$

$$(2, 0, 0, 2), (0, 1, 2, 0), (0, 2, 1, 0), (1, 1, 2, 1).$$

Define operations of addition and multiplication on K so that it becomes a field.

Exercise 3
Let $r \in \mathbb{R}$ and let $0 \neq s \in \mathbb{R}$. Define operations $\boxplus$ and $\boxdot$ on $\mathbb{R} \times \mathbb{R}$ by $(a, b) \boxplus (c, d) = (a + c, b + d)$ and $(a, b) \boxdot (c, d) = (ac - bd(r^2 + s^2), ad + bc + 2rbd)$. Do these operations, considered as addition and multiplication, respectively, define the structure of a field on $\mathbb{R} \times \mathbb{R}$?

Exercise 4
Define a new operation $\dagger$ on $\mathbb{R}$ by setting $a \dagger b = a^3 b$. Show that $\mathbb{R}$, on which we have the usual addition and this new operation as multiplication, satisfies all of the axioms of a field with the exception of one.

Exercise 5
Let $1 < t \in \mathbb{R}$ and let $F = \{a \in \mathbb{R} \mid a < 1\}$. Define operations $\oplus$ and $\odot$ on F as follows:
(1) $a \oplus b = a + b - ab$ for all $a, b \in F$;
(2) $a \odot b = 1 - t^{\log_t(1-a)\log_t(1-b)}$ for all $a, b \in F$.
For which values of t does F, together with these operations, form a field?

Exercise 6
Show that the set of all real numbers of the form $a + b\sqrt{2} + c\sqrt{3} + d\sqrt{6}$, where $a, b, c, d \in \mathbb{Q}$, forms a subfield of $\mathbb{R}$.

Exercise 7
Is $\{a + b\sqrt{15} \mid a, b \in \mathbb{Q}\}$ a subfield of $\mathbb{R}$?

Exercise 8
Show that the field $\mathbb{R}$ has infinitely-many distinct subfields.

Exercise 9
Let F be a field and define a new operation $*$ on F by setting $a * b = a + b + ab$. When is $(F, +, *)$ a field?

Exercise 10
Let F be a field and let G_n be the subset of F consisting of all elements which can be written as a sum of n squares of elements of F.
(1) Is the product of two elements of G_2 again an element of G_2?
(2) Is the product of two elements of G_4 again an element of G_4?

Exercise 11
Let $t = \sqrt[3]{2} \in \mathbb{R}$ and let S be the set of all real numbers of the form $a + bt + ct^2$, where $a, b, c \in \mathbb{Q}$. Is S a subfield of $\mathbb{R}$?

Exercise 12
Let F be a field. Show that the function $a \mapsto a^{-1}$ is a permutation of $F \setminus \{0_F\}$.

Exercise 13
Show that every $z \in \mathbb{C}$ satisfies

$$z^4 + 4 = (z - 1 - i)(z - i + i)(z + 1 + i)(z + 1 - i).$$

Exercise 14
In each of the following, find the set of all complex numbers $z = a + bi$ satisfying the given relation. Note that this set may be empty or may be all of $\mathbb{C}$. Justify your result in each case.
(a) $z^2 = \frac{1}{2}(1 + i\sqrt{3})$;
(b) $(\sqrt{2})|z| \geq |a| + |b|$;
(c) $|z| + z = 2 + i$;
(d) $z^4 = 2 - (\sqrt{12})i$;
(e) $z^4 = -4$.

Exercise 15
Let y be a complex number satisfying $|y| < 1$. Find the set of all complex numbers z satisfying $|z - y| \leq |1 - \bar{y}z|$.

Exercise 16
Let z_1, z_2, and z_3 be complex numbers satisfying the condition that $|z_i| = 1$ for $i = 1, 2, 3$. Show that $|z_1 z_2 + z_1 z_3 + z_2 z_3| = |z_1 + z_2 + z_3|$.

Exercise 17
For any $z_1, z_2 \in \mathbb{C}$, show that $|z_1|^2 + |z_2|^2 - z_1\bar{z_2} - \bar{z_1}z_2 = |z_1 - z_2|^2$.

Exercise 18
Show that $|z + 1| \leq |z + 1|^2 + |z|$ for all $z \in \mathbb{C}$.

Exercise 19
If $z \in \mathbb{C}$, find $w \in \mathbb{C}$ satisfying $w^2 = z$.

Exercise 20

Define new operations $\circ$ and $\diamond$ on $\mathbb{C}$ by setting $y \circ z = |y|z$ and

$$y \diamond z = \begin{cases} 0 & \text{if } y = 0, \\ \frac{1}{|y|} yz & \text{otherwise} \end{cases}$$

for all $y, z \in \mathbb{C}$. Is it true that $w \diamond (y \circ z) = (w \diamond y) \circ (w \diamond z)$ and $w \circ (y \diamond z) = (w \circ y) \diamond (w \circ z)$ for all $w, y, z \in \mathbb{C}$?

Exercise 21

Let $0 \neq z \in \mathbb{C}$. Show that there are infinitely-many complex numbers y satisfying the condition $y\bar{y} = z\bar{z}$.

Exercise 22

(Abel's inequality) Let $z_1, \ldots, z_n$ be a list of complex numbers and, for each $1 \leq k \leq n$, let $s_k = \sum_{i=1}^{k} z_i$. For real numbers $a_1, \ldots, a_n$ satisfying $a_1 \geq a_2 \geq \cdots \geq a_n \geq 0$, show that $|\sum_{i=1}^{n} a_i z_i| \leq a_1 (\max_{1 \leq k \leq n} |s_k|)$.

With kind permission of the Archives of the Mathematisches Forschungsinstitut Oberwolfach.

The nineteenth-century Norwegian mathematical genius **Niels Henrik Abel** died tragically at the age of 26.

Exercise 23

Let $0 \neq z_0 \in \mathbb{C}$ satisfy the condition $|z_0| < 2$. Show that there are precisely two complex numbers, z_1 and z_2, satisfying $|z_1| + |z_2| = 1$ and $z_1 + z_2 = z_0$.

Exercise 24

If p is a prime positive integer, find all subfields of $GF(p)$.

Exercise 25

Find 10^{-1} in $GF(33)$.

Exercise 26

Find elements $c, d \neq \pm 1$ in the field $\mathbb{Q}(\sqrt{5})$ satisfying $cd = 19$.

Exercise 27

Let F be the set of all real numbers of the form

$$a + b(\sqrt[3]{5}) + c(\sqrt[3]{5})^2,$$

where $a, b, c \in \mathbb{Q}$. Is F a subfield of $\mathbb{R}$?

Exercise 28

Let p be a prime positive integer and let $a \in \mathrm{GF}(p)$. Does there necessarily exist an element b of $\mathrm{GF}(p)$ satisfying $b^2 = a$?

Exercise 29

Let $F = \mathrm{GF}(11)$ and let $G = F \times F$. Define operations of addition and multiplication on G by setting $(a, b) + (c, d) = (a + c, b + d)$ and $(a, b) \cdot (c, d) = (ac + 7bd, ad + bc)$. Do these operations define the structure of a field on G?

Exercise 30

Let F be a field and let G be a finite subset of $F \smallsetminus \{0\}$ containing 1 and satisfying the condition that if $a, b \in F$ then $ab^{-1} \in G$. Show that there exists an element $c \in G$ such that $G = \{c^i \mid i \geq 0\}$.

Exercise 31

Let F be a field satisfying the condition that the function $a \mapsto a^2$ is a permutation of F. What is the characteristic of F?

Exercise 32

Is $\mathbb{Z}/(6)$ an integral domain?

Exercise 33

Let $F = \{a + b\sqrt{5} \in \mathbb{Q}(\sqrt{5}) \mid a, b \in \mathbb{Z}\}$. Is F an integral domain?

Exercise 34

Let F be an integral domain and let $a \in F$ satisfy $a^2 = a$. Show that $a = 0$ or $a = 1$.

Exercise 35

Let a be a nonzero element in an integral domain F. If $b \neq c$ are distinct elements of F, show that $ab \neq ac$.

Exercise 36

Let F be an integral domain and let G be a nonempty subset of F containing 0 and 1 and closed under the operations of addition and multiplication in F. Is G necessarily an integral domain?

Exercise 37

Let U be the set of all positive integers and let F be the set of all functions from U to $\mathbb{C}$. Define operations of addition and multiplication on F by setting

$f + g : k \mapsto f(k) + g(k)$ and $fg : k \mapsto \sum_{ij=k} f(i)g(j)$ for all $k \in U$. Is F, together with these operations, an integral domain? Is it a field?

Exercise 38
Let F be the set of all functions f from $\mathbb{R}$ to itself of the form $f : t \mapsto \sum_{k=1}^{n} [a_k \cos(kt) + b_k \sin(kt)]$, where the a_k and b_k are real numbers and n is some positive integer. Define addition and multiplication on F by setting $f + g : t \mapsto f(t) + g(t)$ and $fg : t \mapsto f(t)g(t)$ for all $t \in \mathbb{R}$. Is F, together with these operations, an integral domain? Is it a field?

Exercise 39
Show that every integral domain having only finitely-many elements is a field.

Exercise 40
Let F be a field of characteristic other than 2 in which there exist elements $a_1, \ldots, a_n$ satisfying $\sum_{i=1}^{n} a_i^2 = -1$. (This happens, for example, in the case $F = \mathbb{C}$.) Show that for any $c \in F$ there exist elements $b_1, \ldots, b_k$ of F satisfying $c = \sum_{i=1}^{k} b_i^2$.

Exercise 41
Let p be a prime integer. Show that for each $a \in GF(p)$ there exist elements b and c of $GF(p)$, not necessarily distinct, satisfying $a = b^2 + c^2$.

Exercise 42
Let F be a field in which we have elements a, b, and c (not necessarily distinct) satisfying $a^2 + b^2 + c^2 = -1$. Show that there exist (not necessarily distinct) elements d and e of F, satisfying $d^2 + e^2 = -1$.

Exercise 43
Is every nonzero element of the field $GF(5)$ in the form 2^i for some positive integer i? What happens in the case of the field $GF(7)$?

Exercise 44
Find the set of all fields F in which there exists an element a satisfying the condition that $a + b = a$ for all $b \in F \setminus \{a\}$.

Exercise 45
(Binomial formula) If a and b are elements of a field F, and if n is a positive integer, show that $(a + b)^n = \sum_{k=0}^{n} \binom{n}{k} a^k b^{n-k}$.

Exercise 46
Let F be a field of characteristic $p > 0$. Show that the function $\gamma : F \to F$ defined by $\gamma : a \mapsto a^p$ is monic.

Exercise 47
Let a and b be nonzero elements of a finite field F, and let m and n be positive integers satisfying $a^m = b^n = 1$. Show that there exists a nonzero element c of F satisfying $c^k = 1$, where k is the least common multiple of m and n.

Exercise 48
If a is a nonzero element of a field F, show that $(-a)^{-1} = -(a^{-1})$.

Exercise 49
Let $F = \mathrm{GF}(7)$ and let $K = F \times F$. Define addition and multiplication on K by setting $(a, b) + (c, d) = (a + b, c + d)$ and $(a, b) \cdot (c, d) = (ac - bd, ad + bc)$. Do these operations turn K into a field? What happens if $F = \mathrm{GF}(5)$?

Exercise 50
A field F is *orderable* if and only if there exists a subset P closed under addition and multiplication such that for each $a \in F$ precisely one of the following conditions holds: (i) $a = 0$; (ii) $a \in P$; (iii) $-a \in P$. Show that $\mathrm{GF}(5)$ is not orderable.

Exercise 51
Let F be a field and let K be the set of all functions $f \in F^{\mathbb{Z}}$ satisfying the condition that there exists an integer (perhaps negative) n_f such that $f(i) = 0$ for all $i < n_f$. Define operations of addition and multiplication on K by setting $f + g : i \mapsto f(i) + g(i)$ and $fg : i \mapsto \sum_{j+h=i} f(j)g(h)$. Show that K is a field, called the *field of formal Laurent series* over F.[2]

Exercise 52
Let F be a field. Find $A = \{(x, y) \in F^2 \mid x^2 + y^2 = 1\}$.

Exercise 53
Let F be a field having characteristic $p > 0$ and let $c \in F$. Show that there is at most one element b of F satisfying $b^p = c$.

Exercise 54
A *ternary ring* is a set R containing distinguished elements 0 and 1, together with a function $\theta : R^3 \to R$ satisfying the following conditions:
(1) $\theta(1, a, 0) = \theta(a, 1, 0) = a$ for all $a \in R$;
(2) $\theta(a, 0, c) = \theta(0, a, c) = c$ for all $c \in R$;
(3) If $a, b, c \in R$ then there is a unique element y of R satisfying $\theta(a, b, y) = c$;
(4) If $a, a', b, b' \in R$ with $a \neq a'$ then there is a unique element x of R satisfying $\theta(x, a, b) = \theta(x, a', b')$;
(5) If $a, a', b, b' \in R$ with $a \neq a'$ then there are unique elements x and y of R satisfying $\theta(a, x, y) = b$ and $\theta(a', x, y) = b'$.

---

[2]These series were first studied by the nineteenth-century French engineer and mathematician, **Pierre Alphonse Laurent**.

Such structures have applications in projective geometry. If F is a field, show that we can define a function $\theta : F^3 \to F$ in such a way that F becomes a tertiary ring (with 0 and 1 being the neutral elements of the field).

Exercise 55

For $h = 1, 2, 3$, let $z_h = a_h + b_h i$ be a complex number satisfying $|z_h| = 1$. Assume, moreover, that $\sum_{i=1}^{3} z_i = 0$. Show that the points (a_h, b_h) are the vertices of an equilateral triangle in the Euclidean plane.

Vector Spaces Over a Field

3

If $n > 1$ is an integer and if F is a field, it is natural to define addition on the set F^n componentwise:

$$(a_1, \ldots, a_n) + (b_1, \ldots, b_n) = (a_1 + b_1, \ldots, a_n + b_n).$$

More generally, if Ω is any nonempty set and if F^Ω is the set of all functions from Ω to the field F, we can define addition on F^Ω by setting $f + g : i \mapsto f(i) + g(i)$ for each $i \in \Omega$. Given these definitions, is it possible to define multiplication in such a manner that F^n or F^Ω will become a field naturally containing F as a subfield? We have seen that if $n = 2$ and if $F = \mathbb{R}$ or $F = \mathbb{Q}$, this is possible—and, indeed, in the latter case there are several different methods of doing it. If $F = \mathrm{GF}(p)$ then it is possible to define such a field structure on F^n for every integer $n > 1$. However, in general the answer is negative—as we will show in a later chapter for the specific case of $\mathbb{R}^k$, where $k > 2$ is an odd integer. Nonetheless, it is possible to construct another important and useful structure on these sets, and this structure will be the focus of our attention for the rest of this book. We will first give the formal definition, and then look at a large number of examples.

Let F be a field. A nonempty set V, together with a function $V \times V \to V$ called *vector addition* (denoted, as usual, by $+$) and a function $F \times V \to V$ called *scalar multiplication* (denoted, as a rule, by concatenation) is a *vector space* over F if the following conditions are satisfied:

(1) (*associativity of vector addition*): $v + (w + y) = (v + w) + y$ for all $v, w, y \in V$.
(2) (*commutativity of vector addition*): $v + w = w + v$ for all $v, w \in V$.
(3) (*existence of a identity element for vector addition*): There exists an element 0_V of V satisfying the condition that $v + 0_V = v$ for all $v \in V$.
(4) (*existence of additive inverses*): For each $v \in V$ there exists an element of V, which we will denote by $-v$, which satisfies $v + (-v) = 0_V$.
(5) (*distributivity of scalar multiplication over vector addition and of scalar multiplication over field addition*): $a(v + w) = av + aw$ and $(a + b)v = av + bv$ for all $a, b \in F$ and $v, w \in V$.

J.S. Golan, *The Linear Algebra a Beginning Graduate Student Ought to Know*,
DOI 10.1007/978-94-007-2636-9_3, © Springer Science+Business Media B.V. 2012

(6) *(associativity of scalar multiplication)*: $(ab)v = a(bv)$ for all $a, b \in F$ and
$v \in V$.
(7) *(existence of identity element for scalar multiplication)*: $1v = v$ for all $v \in V$.
The elements of V are called *vectors* and the elements of F are called *scalars*.

With kind permission of the Manuscripts & Archives, Yale University (Gibbs); © the estate of Oliver Heaviside.
Reproduced with kind permission of Alan Heather (Heaviside); With kind permission of Special collections,
Fine Arts Library, Harvard University (Maxwell).

The theory of vector spaces was developed in the 1880s by the American engineer and
physicist, **Josiah Willard Gibbs** and the British engineer **Oliver Heaviside**, based on the
work of the Scottish physicist **James Clerk Maxwell**, the German high-school teacher
Herman Grassmann, and the French engineer **Jean Claude Saint-Venant**.

Example Note that condition (7), apparently trivial, does not follow from the other
conditions. Indeed, if we take $V = F$ but define scalar multiplication by $av = 0_V$
for all $a \in F$ and $v \in V$, we would get a structure which satisfies conditions (1)–(6)
but not condition (7).

If $v, w \in V$ we again write $v - w$ instead of $v + (-w)$. As we noted when we
talked about fields, if $v_1, \ldots, v_n$ is a list of vectors in a vector space V over a field F,
the associativity of vector addition allows us to unambiguously write $v_1 + \cdots + v_n$,
and this sum is often denoted by $\sum_{i=1}^{n} v_i$. Moreover, if $a \in F$ is a scalar then we
surely have $a(\sum_{i=1}^{n} v_i) = (\sum_{i=1}^{n} av_i)$. Similarly, if $a_1, \ldots, a_n$ is a list of scalars and
if $v \in V$, then we have $(\sum_{i=1}^{n} a_i)v = \sum_{i=1}^{n} a_i v$. We will also adopt the convention
that the sum of an empty set of vectors is equal to 0_V.

Clearly, any field F is a vector space over itself, where we take the vector addition
to be the addition in F and scalar multiplication to be the multiplication in F.

We also note an extremely important construction. Let F be a field and let Ω be
a nonempty set. Assume that, for each $i \in \Omega$, we are given a vector space V_i over F,
the addition in which we will denote by $+_i$ (the vector spaces V_i need not, however,
be distinct from one another). Recall that $\prod_{i \in \Omega} V_i$ is the set of all those functions f
from Ω to $\bigcup_{i \in \Omega} V_i$ which satisfy the condition that $f(i) \in V_i$ for each $i \in \Omega$. We
now define the structure of a vector space on $\prod_{i \in \Omega} V_i$ as follows: if $f, g \in \prod_{i \in \Omega} V_i$
then $f + g$ is the function in $\prod_{i \in \Omega} V_i$ given by $f + g : i \mapsto f(i) +_i g(i)$ for each
$i \in \Omega$. Moreover, if $a \in F$ and $f \in \prod_{i \in \Omega} V_i$, then af is the function in $\prod_{i \in \Omega} V_i$
given by $af : i \mapsto a[f(i)]$ for each $i \in \Omega$. It is routine to verify that all of the
axioms of a vector space are satisfied in this case. For example, the identity element
for vector addition is just the function in $\prod_{i \in \Omega} V_i$ given by $i \mapsto 0_{V_i}$ for each $i \in \Omega$.

This vector space is called the *direct product* of the vector spaces V_i over F. If the set Ω is finite, say $\Omega = \{1, \ldots, n\}$, then we often write $V_1 \times \cdots \times V_n$ instead of $\prod_{i \in \Omega} V_i$. If all of the vector spaces V_i are equal to the same vector space V, then we write V^Ω instead of $\prod_{i \in \Omega} V_i$ and if $\Omega = \{1, \ldots, n\}$ we write V^n instead of V^Ω. Note that a function f from a finite set $\Omega = \{1, \ldots, n\}$ to a vector space V is totally defined by the list $f(1), f(2), \ldots, f(n)$ of its values. Conversely, any list $v_1, \ldots, v_n$ of elements of V uniquely defines such a function f given by $f : i \mapsto v_i$. Therefore, this notation agrees with our previous use of the symbol V^n to denote sets of n-tuples of elements of V. However, to emphasize the vector space structure here, we will write the elements of V^n as columns of the form $\begin{bmatrix} v_1 \\ \vdots \\ v_n \end{bmatrix}$, where the v_i are (not necessarily distinct) elements of V. Usually, we will consider the case $V = F$. Vector addition and scalar multiplication in V^n are then defined by the rules

$$\begin{bmatrix} v_1 \\ \vdots \\ v_n \end{bmatrix} + \begin{bmatrix} w_1 \\ \vdots \\ w_n \end{bmatrix} = \begin{bmatrix} v_1 + w_1 \\ \vdots \\ v_n + w_n \end{bmatrix} \text{ and } c \begin{bmatrix} v_1 \\ \vdots \\ v_n \end{bmatrix} = \begin{bmatrix} cv_1 \\ \vdots \\ cv_n \end{bmatrix}.$$

The "classical" study of vector spaces centers around the spaces $\mathbb{R}^n$, the vectors in which are identified with the points in n-dimensional Euclidean space. However, other vector spaces also have important applications. Vector spaces of the form $\mathbb{C}^n$ are needed for the study of functions of several complex variables. In algebraic coding theory, one is interested in spaces of the form F^n, where F is a finite field. The vectors in this space are *words* of length n and the field F is the *alphabet* in which these words are written. Thus, one choice for F is the Galois field $\mathrm{GF}(2^8)$, the 256 elements of which are identified with the 256 ASCII symbols.

© National Maritime Museum, Greenwich, London (Galilei); With kind permission of Frommann–Holzboog Publishers (Bolzano).

The first explicit statement of the geometric "parallelogram law" for adding geometric vectors was given by the sixteenth-century Pisan scientist **Galileo Galilei**. This idea was extended at the beginning of the nineteenth century by Bohemian priest **Bernard Bolzano**.

Let V be a vector space, let k and n be positive integers, and let $\Omega = \{(i, j) \mid 1 \le i \le k, 1 \le j \le n\}$. There exists a bijective correspondence between V^Ω and the set of all rectangular arrays of the form $\begin{bmatrix} v_{11} & \cdots & v_{1n} \\ \vdots & \ddots & \vdots \\ v_{k1} & \cdots & v_{kn} \end{bmatrix}$ in which the entries v_{ij} are elements of V. Such an array is called a $k \times n$ *matrix* over V. We will denote the set of all such matrices by $\mathcal{M}_{k \times n}(V)$. Addition in $\mathcal{M}_{k \times n}(V)$ is given by

$$\begin{bmatrix} v_{11} & \cdots & v_{1n} \\ \vdots & \ddots & \vdots \\ v_{k1} & \cdots & v_{kn} \end{bmatrix} + \begin{bmatrix} w_{11} & \cdots & w_{1n} \\ \vdots & \ddots & \vdots \\ w_{k1} & \cdots & w_{kn} \end{bmatrix} = \begin{bmatrix} v_{11} + w_{11} & \cdots & v_{1n} + w_{1n} \\ \vdots & \ddots & \vdots \\ v_{k1} + w_{k1} & \cdots & v_{kn} + w_{kn} \end{bmatrix}$$

and scalar multiplication in $\mathcal{M}_{k \times n}(V)$ is given by

$$c \begin{bmatrix} v_{11} & \cdots & v_{1n} \\ \vdots & \ddots & \vdots \\ v_{k1} & \cdots & v_{kn} \end{bmatrix} = \begin{bmatrix} cv_{11} & \cdots & cv_{1n} \\ \vdots & \ddots & \vdots \\ cv_{k1} & \cdots & cv_{kn} \end{bmatrix}.$$

The identity element for vector addition in $\mathcal{M}_{k \times n}(V)$ is the 0-*matrix* O, all entries of which are equal to 0_V. Note that $V^n = \mathcal{M}_{n \times 1}(V)$.

The term "matrix" was first coined by the nineteenth-century British mathematician **James Joseph Sylvester**, one of the major researchers in the theory of matrices and determinants.

If V is a vector space and if $\Omega = \mathbb{N}$, then the elements of V^{Ω} are infinite sequences $[v_0, v_1, \ldots]$ of elements of V. We will denote this vector space, which we will need later, by V^{∞}. Again, the space of particular interest will be F^{∞}.

Example If F is a subfield of a field K, then K is a vector space over F, with addition and scalar multiplication just being the corresponding operations in K. Thus, in particular, we can think of $\mathbb{C}$ as a vector space over $\mathbb{R}$ and of $\mathbb{R}$ as a vector space over $\mathbb{Q}$.

Example Let A be a nonempty set and let V be the collection of all subsets of A. Let us define addition of elements of V as follows: if B and C are elements of V then $B + C = (B \cup C) \smallsetminus (B \cap C)$. This operation is usually called the *symmetric difference* of B and C. This definition turns V into a vector space over GF(2), where scalar multiplication is defined by $0B = \varnothing$ and $1B = B$ for all $B \in V$. This is actually just a special case of what we have seen before. Indeed, we note that there is a bijective function from V to GF(2)^A which assigns to each subset B of A its *characteristic function*, namely the function χ_B defined by

$$\chi_B : a \mapsto \begin{cases} 1 & \text{if } a \in B, \\ 0 & \text{otherwise,} \end{cases}$$

and it is easy to see that $\chi_A + \chi_B = \chi_{A+B}$, while $\chi_A \chi_B = \chi_{A \cap B}$.

Proposition 3.1 *Let V be a vector space over a field F.*
(1) *If $z \in V$ satisfies $z + v = v$ for all $v \in V$ then $z = 0_V$.*
(2) *If $v, w \in V$ then there exists a unique element $y \in V$ satisfying $v + y = w$.*

Proof The proof is similar to the proofs of Proposition 2.1(1) and Proposition 2.2(1). $\qquad\qquad\square$

Proposition 3.2 *Let V be a vector space over a field F. If $v, w \in V$ and if $a \in F$, then:*
(1) $a0_V = 0_V$;
(2) $0v = 0_V$;
(3) $(-1)v = -v$;
(4) $(-a)v = -(av) = a(-v)$;
(5) $-(-v) = v$;
(6) $av = (-a)(-v)$;
(7) $-(v + w) = -v - w$;
(8) $a(v - w) = av - aw$;
(9) *If $av = 0_V$ then either $v = 0_V$ or $a = 0$.*

Proof The proof is similar to the proof of Proposition 2.3. $\qquad\qquad\square$

Let V be a vector space over a field F. A nonempty subset W of V is a *subspace* of V if and only if it is a vector space in its own right with respect to the addition and scalar multiplication defined on V. Thus, any vector space V is a subspace of itself, called the *improper* subspace; any other subspace is *proper*. Also, $\{0_V\}$ is surely a subspace of V, called the *trivial* subspace; any other subspace is *nontrivial*.

Note that the two conditions for a nonempty subset of a vector space to be a subspace are independent: the set of all vectors in $\mathbb{R}^3$ all entries of which are integers is closed under vector addition but not under scalar multiplication; the set of all vectors $\begin{bmatrix} a \\ b \\ c \end{bmatrix} \in \mathbb{R}^3$ satisfying $abc = 0$ is closed under scalar multiplication but not under vector addition.

Example Let V be a vector space over a field F and let Ω be a nonempty set. We have already seen that the set V^Ω of all functions from Ω to V is a vector space over F. If Λ is a subset of Ω then the set $\{f \in V^\Omega \mid f(i) = 0_V \text{ for all } i \in \Lambda\}$ is a subspace of V^Ω. In particular, if $k < n$ are positive integers, then we can think of V^k as being a subspace of V^n, by identifying it with

$$\left\{ \begin{bmatrix} v_1 \\ \vdots \\ v_n \end{bmatrix} \in V^n \,\middle|\, v_{k+1} = \cdots = v_n = 0_V \right\}.$$ Note that if $y \in V$, then $\{f \in V^{\Omega} \mid$

$f(i) = y$ for all $i \in \Lambda\}$ is not a subspace of V^{Ω} unless $y = 0_V$.

Example Let $\{V_i \mid i \in \Omega\}$ be a collection of vector spaces over a field F. The set of all functions $f \in \prod_{i \in \Omega} V_i$ satisfying the condition that $f(i) \neq 0_{V_i}$ for at most finitely-many elements i of Ω is a subspace of $\prod_{i \in \Omega} V_i$, called the *direct coproduct* of the spaces V_i and denoted by $\coprod_{i \in \Omega} V_i$. The direct coproduct is a proper subset of $\prod_{i \in \Omega} V_i$ when and only when the set Ω is infinite. If each of the spaces V_i is equal to a given vector space V, we write $V^{(\Omega)}$ instead of $\coprod_{i \in \Omega} V_i$.

Example If V is a vector space over a field F and if $v \in V$, then the set $Fv = \{av \mid a \in F\}$ is a subspace of V which is contained in any subspace of V containing v.

Example Let $\mathbb{R}$ be the field of real numbers and let Ω be either equal to $\mathbb{R}$, to some closed interval $[a, b]$ on the real line, or to a ray $[a, \infty)$ on the real line. We have already seen that the set $\mathbb{R}^{\Omega}$ of all functions from Ω to $\mathbb{R}$ is a vector space over $\mathbb{R}$. The set of all continuous functions from Ω to $\mathbb{R}$ is a subspace of this vector space, as are the set of all differentiable functions from Ω to $\mathbb{R}$, the set of all infinitely-differentiable functions from Ω to $\mathbb{R}$, and the set of all analytic functions from Ω to $\mathbb{R}$. If $a < b$ are real numbers, we will denote the space of all continuous functions from the closed interval $[a, b]$ to $\mathbb{R}$ by $C(a, b)$. If $a \in R$ we will denote the space of all continuous functions from $[a, \infty)$ to $\mathbb{R}$ by $C(a, \infty)$. These spaces will be very important to us later.

> **Proposition 3.3** *If V is a vector space over a field F, then a nonempty subset W of V is a subspace of V if and only if it is closed under addition and scalar multiplication.*

Proof If W is a subspace of V then it is surely closed under addition and scalar multiplication. Conversely, suppose that it is so closed. Then for any $w \in W$ we have $0_V = 0w \in W$ and $-w = (-1)w \in W$. The other conditions are satisfied in W since they are satisfied in V. $\qquad\qquad\qquad\qquad\qquad\qquad\qquad\qquad\qquad\qquad\qquad\square$

With kind permission of the Archives of the Mathematisches Forschungsinstitut Oberwolfach.

The first fundamental research in spaces of functions was done by the German mathematician **Erhard Schmidt**, a student of David Hilbert, whose work forms one of the bases of functional analysis.

> **Proposition 3.4** *If V is a vector space over a field F, and if $\{W_i \mid i \in \Omega\}$ is a collection of subspaces of V, then $\bigcap_{i \in \Omega} W_i$ is a subspace of V.*

Proof Set $W = \bigcap_{i \in \Omega} W_i$. If $w, y \in W$ then, for each $i \in \Omega$, we have $w, y \in W_i$ and so $w + y \in W_i$. Thus $w + y \in W$. Similarly, if $a \in F$ and $w \in W$ then $aw \in W_i$ for each $i \in \Omega$, and so $aw \in W$. $\qquad\qquad\qquad\square$

We will also set the convention that the intersection of an empty collection of subspaces of V is V itself. Subspaces W and W' are *disjoint* if and only if $W \cap W' = \{0_V\}$. More generally, a collection $\{W_i \mid i \in \Omega\}$ of subspaces of V is *pairwise disjoint* if and only if $W_i \cap W_j = \{0_V\}$ for $i \neq j$ in Ω. (Note that disjointness of subspaces of a given space is not the same as disjointness of subsets!)

Now let us look at a very important method of constructing subspaces of vector spaces. Let D be a nonempty set of elements of a vector space V over a field F. A vector $v \in V$ is a *linear combination* of elements of D over F if and only if there exist elements $v_1, \ldots, v_n$ of D and scalars $a_1, \ldots, a_n$ in F such that $v = \sum_{i=1}^{n} a_i v_i$. We will denote the set of all linear combinations of elements of D over F by FD. Note that if $v \in V$ then $F\{v\}$ is the set Fv which we defined earlier.

It is clear that if D is a nonempty set of elements of a vector space V over a field F then $D \subseteq FD$. Also, $0_V \in FD$ for any nonempty subset D of V, and it is the only vector belonging to each of the sets FD. To simplify notation, we will therefore define $F\varnothing$ to be $\{0_V\}$. If $D' \subseteq D$ then surely $FD' \subseteq FD$. We also note that $FD = F(D \cup \{0_V\})$ for any subset D of V.

Example If $D = \left\{ \begin{bmatrix} 1 \\ 0 \\ 0 \end{bmatrix}, \begin{bmatrix} 0 \\ 1 \\ 0 \end{bmatrix} \right\}$ and $D' = \left\{ \begin{bmatrix} 0 \\ 2 \\ 0 \end{bmatrix}, \begin{bmatrix} 3 \\ 3 \\ 0 \end{bmatrix} \right\}$ are subsets of $\mathbb{R}^3$, then

$$FD = FD' = \left\{ \begin{bmatrix} a \\ b \\ 0 \end{bmatrix} \,\middle|\, a, b \in \mathbb{R} \right\}. \text{ Indeed,}$$

$$\begin{bmatrix} a \\ b \\ 0 \end{bmatrix} = a \begin{bmatrix} 1 \\ 0 \\ 0 \end{bmatrix} + b \begin{bmatrix} 0 \\ 1 \\ 0 \end{bmatrix} = \left(\frac{b-a}{2} \right) \begin{bmatrix} 0 \\ 2 \\ 0 \end{bmatrix} + \frac{a}{3} \begin{bmatrix} 3 \\ 3 \\ 0 \end{bmatrix} \quad \text{for all } a, b \in \mathbb{R}.$$

Example If $D = \left\{ \begin{bmatrix} 0 \\ 0 \\ 4 \end{bmatrix}, \begin{bmatrix} 2 \\ 2 \\ 0 \end{bmatrix}, \begin{bmatrix} 2 \\ 0 \\ 0 \end{bmatrix}, \begin{bmatrix} 1 \\ 1 \\ 1 \end{bmatrix} \right\} \subseteq \mathbb{R}^3$ then

$$\begin{bmatrix} 4 \\ 2 \\ 4 \end{bmatrix} = 1 \begin{bmatrix} 0 \\ 0 \\ 4 \end{bmatrix} + 1 \begin{bmatrix} 2 \\ 2 \\ 0 \end{bmatrix} + 1 \begin{bmatrix} 2 \\ 0 \\ 0 \end{bmatrix}$$

$$= 1 \begin{bmatrix} 2 \\ 0 \\ 0 \end{bmatrix} + (-1) \begin{bmatrix} 2 \\ 2 \\ 0 \end{bmatrix} + 4 \begin{bmatrix} 1 \\ 1 \\ 1 \end{bmatrix}.$$

Thus we see that there may be several ways of representing a vector as a linear combination of elements of a given subset of a vector space.

Proposition 3.5 *Let D be a subset of a vector space V over a field F. Then:*
(1) *FD is a subspace of V;*
(2) *Every subspace of V containing D also contains FD;*
(3) *FD is the intersection of all subspaces of V containing D.*

Proof If $D = \varnothing$ then $FD = \{0_V\}$ and we are done. Thus we can assume that D is nonempty. It is an immediate consequence of the definitions that the sum of two linear combinations of elements of D over F is again a linear combination of elements of D over F, and that the product of a scalar and a linear combination of elements of D over F is again a linear combination of elements of D over F. This proves (1). Moreover, (2) is an immediate consequence of (1) and Proposition 3.3, while (3) follows directly from (2). $\square$

If D is a subset of a vector space V over a field F then the subspace FD of V is called the subspace *generated* or *spanned* by D, and the set D is called a *generating set* or *spanning set* for this subspace. In particular, we note that $\varnothing$ is a generating set for $\{0_V\}$.

Example Let F be a field. Then $A = \left\{ \begin{bmatrix} 1 \\ 0 \\ 0 \end{bmatrix}, \begin{bmatrix} 0 \\ 1 \\ 0 \end{bmatrix}, \begin{bmatrix} 0 \\ 0 \\ 1 \end{bmatrix} \right\}$ is a generating set for F^3 over F. The set $B = \left\{ \begin{bmatrix} 1 \\ 1 \\ 0 \end{bmatrix}, \begin{bmatrix} 1 \\ 0 \\ 1 \end{bmatrix}, \begin{bmatrix} 0 \\ 1 \\ 1 \end{bmatrix} \right\}$ is also a generating set for F^3 if the characteristic of F is other than 2, but not for $F = \text{GF}(2)$, since $\begin{bmatrix} 1 \\ 0 \\ 0 \end{bmatrix} \notin \text{GF}(2)B$.

The set $D = \left\{ \begin{bmatrix} 1 \\ 0 \\ 0 \end{bmatrix}, \begin{bmatrix} 0 \\ 1 \\ 0 \end{bmatrix}, \begin{bmatrix} 1 \\ 1 \\ 0 \end{bmatrix} \right\}$ is not a generating set for F^3 for any field F since $\begin{bmatrix} 0 \\ 0 \\ 1 \end{bmatrix} \notin FD$.

Often, in applications, we need to restrict ourselves to linear combinations of special type. For example, let V be a vector space over a field F and let D be a nonempty subset of V. An *affine combination* of elements of D is an element of V of the form $\sum_{i=1}^{n} a_i v_i$, where the v_i are elements of D and the a_i are scalars satisfying $\sum_{i=1}^{n} a_i = 1$. This is usually interpreted as a weighted average of the vectors v_i. The set of all affine combinations of elements of D is called the *affine hull* of D and is denoted by affh(D). In general, this is not a subspace of V. One can, however, easily verify that affh(affh(D)) = affh(D) for any set D.

Proposition 3.6 *Let V be a vector space over a field F and let D_1 and D_2 be subsets of V satisfying $D_1 \subseteq D_2 \subseteq FD_1$. Then $FD_1 = FD_2$.*

Proof Since FD_1 is a subspace of V containing D_2, we know by Proposition 3.5 that $FD_2 \subseteq FD_1$. Conversely, any linear combination of elements of D_1 over F is also a linear combination of elements of D_2 over F and so $FD_1 \subseteq FD_2$, thus establishing equality. □

In particular, we note that $FD = F(FD)$ for any subset D of V.

Proposition 3.7 (Exchange Property) *Let V be a vector space over a field F and let $v, w \in V$. Let D be a subset of V satisfying $v \in F(D \cup \{w\}) \smallsetminus FD$. Then $w \in F(D \cup \{v\})$.*

Proof Since $v \in F(D \cup \{w\})$ we know that there exist elements $v_1, \dots, v_n$ of D and scalars $a_1, \dots, a_n, b$ in F satisfying the condition that $v = \sum_{i=1}^{n} a_i v_i + bw$. Moreover, since $v \notin FD$, we know that $b \neq 0$ and so $w = b^{-1}v - \sum_{i=1}^{n} b^{-1}a_i v_i \in F(D \cup \{v\})$. □

A vector space V over a field F is *finitely generated* over F if it has a finite generating set. Finitely-generated vector spaces are often much easier to deal with by purely algebraic methods and therefore, in several situations, we will have to restrict our discussion to these spaces.

Example If F is a field and n is a positive integer, then one sees that
$$\left\{ \begin{bmatrix} 1 \\ 0 \\ 0 \\ \vdots \\ 0 \end{bmatrix}, \begin{bmatrix} 0 \\ 1 \\ 0 \\ \vdots \\ 0 \end{bmatrix}, \dots, \begin{bmatrix} 0 \\ 0 \\ 0 \\ \vdots \\ 1 \end{bmatrix} \right\}$$ is a finite generating set for F^n over F, and so F^n is finitely generated over F. More generally, if V is a vector space finitely generated over a field F, say $V = F\{v_1, \dots, v_k\}$, and if n is a positive integer, then

$$\left\{ \begin{bmatrix} v_1 \\ 0 \\ \vdots \\ 0 \end{bmatrix}, \begin{bmatrix} v_2 \\ 0 \\ \vdots \\ 0 \end{bmatrix}, \ldots, \begin{bmatrix} v_k \\ 0 \\ \vdots \\ 0 \end{bmatrix}, \begin{bmatrix} 0 \\ v_1 \\ \vdots \\ 0 \end{bmatrix}, \begin{bmatrix} 0 \\ v_2 \\ \vdots \\ 0 \end{bmatrix}, \ldots, \begin{bmatrix} 0 \\ v_k \\ \vdots \\ 0 \end{bmatrix}, \ldots, \right.$$

$$\left. \begin{bmatrix} 0 \\ 0 \\ \vdots \\ v_1 \end{bmatrix}, \begin{bmatrix} 0 \\ 0 \\ \vdots \\ v_2 \end{bmatrix}, \ldots, \begin{bmatrix} 0 \\ 0 \\ \vdots \\ v_k \end{bmatrix} \right\}$$

is a generating set for V^n over F having kn elements.

Example If F is a field and if k and n are positive integers, then the vector space $\mathcal{M}_{k \times n}(F)$ of all $k \times n$ matrices over F is finitely generated over F. Similarly, if V is a finitely-generated vector space over F, then the vector space $\mathcal{M}_{k \times n}(V)$ is also finitely generated over F.

Example For any field F, the vector space F^∞ is not finitely generated over F.

Example The field $\mathbb{R}$ is finitely generated as a vector space over itself, but is not finitely generated as a vector space over $\mathbb{Q}$.

Let V be a vector space over a field F. In Proposition 3.4, we saw that if $\{W_i \mid i \in \Omega\}$ is a collection of subspaces of V then $\bigcap_{i \in \Omega} W_i$ is a subspace of V. In the same way, we can define the subspace $\sum_{i \in \Omega} W_i$ of V to be the set of all vectors in V of the form $\sum_{j \in \Lambda} w_j$, where Λ is a finite nonempty subset of Ω and $w_j \in W_j$ for each $j \in \Lambda$. In other words, $\sum_{i \in \Omega} W_i = F(\bigcup_{i \in \Omega} W_i)$. Indeed, from the definition of this sum, we see something stronger: if D_i is a generating set for W_i for each $i \in \Omega$ then $\sum_{i \in \Omega} W_i = F(\bigcup_{i \in \Omega} D_i)$.

As a special case of the above, we see that if W_1 and W_2 are subspaces of V, then $W_1 + W_2$ equals the set of all vectors of the form $w_1 + w_2$, where $w_1 \in W_1$ and $w_2 \in W_2$. If both W_1 and W_2 are finitely generated then $W_1 + W_2$ is also finitely generated. By induction, we can then show that if $W_1, \ldots, W_n$ are finitely-generated subspaces of V, then $\sum_{i=1}^n W_i$ is also finitely generated.

Proposition 3.8 *If V is a vector space over a field F and if $\{W_i \mid i \in \Omega\}$ is a collection of subspaces of V, then:*
(1) *W_h is a subspace of $\sum_{i \in \Omega} W_i$ for all $h \in \Omega$;*
(2) *If Y is a subspace of V satisfying the condition that W_h is a subspace of Y for all $h \in \Omega$, then $\sum_{i \in \Omega} W_i$ is a subspace of Y.*

Proof (1) is clear from the definition. As for (2), if we have a subspace Y satisfying the given condition, if Λ is a finite subset of Ω, and if $w_j \in W_j$ for each $j \in \Lambda$, then $w_j \in Y$ for each j and so $\sum_{j \in \Lambda} w_j \in Y$. Thus $\sum_{i \in \Omega} W_i \subseteq Y$. $\square$

> **Proposition 3.9** *If V is a vector space over a field F and if W_1, W_2, and W_3 are subspaces of V, then:*
> (1) $(W_1 + W_2) + W_3 = W_1 + (W_2 + W_3)$;
> (2) $W_1 + W_2 = W_2 + W_1$;
> (3) $W_3 \cap [W_2 + (W_1 \cap W_3)] = (W_1 \cap W_3) + (W_2 \cap W_3)$;
> (4) **(Modular law for subspaces)**: *If $W_1 \subseteq W_3$ then*
>
> $$W_3 \cap (W_2 + W_1) = W_1 + (W_2 \cap W_3).$$

Proof Parts (1) and (2) follow immediately from the definition, while part (4) is a special case of (3). We are therefore left to prove (3). Indeed, if v belongs to $W_3 \cap [W_2 + (W_1 \cap W_3)]$, then we can write $v = w_2 + y$, where $w_2 \in W_2$ and $y \in W_1 \cap W_3$. Since $v, y \in W_3$, it follows that $w_2 = v - y \in W_3$, and so $v = y + w_2 \in (W_1 \cap W_3) + (W_2 \cap W_3)$. Thus we see that $W_3 \cap [W_2 + (W_1 \cap W_3)] \subseteq (W_1 \cap W_3) + (W_2 \cap W_3)$. Conversely, assume that $v \in (W_1 \cap W_3) + (W_2 \cap W_3)$. Then, in particular, $v \in W_3$ and we can write $v = w_1 + w_2$, where $w_1 \in W_1 \cap W_3$ and $w_2 \in W_2 \cap W_3$. Thus $v = w_1 + w_2 \in W_3 \cap W_2 + (W_1 \cap W_3)$. This shows that $(W_1 \cap W_3) + (W_2 \cap W_3) \subseteq W_3 \cap [W_2 + (W_1 \cap W_3)]$, and so we have the desired equality. $\square$

Exercises

Exercise 56
Is it possible to define on $V = \mathbb{Z}/(4)$ the structure of a vector space over GF(2) in such a way that the vector addition is the usual addition in $\mathbb{Z}/(4)$?

Exercise 57
Consider the set $\mathbb{Z}$ of integers, together with the usual addition. If $a \in \mathbb{Q}$ and $k \in \mathbb{Z}$, define $a \cdot k$ to be $\lfloor a \rfloor k$, where $\lfloor a \rfloor$ denotes the largest integer less than or equal to a. Using this as our definition of "scalar multiplication", have we turned $\mathbb{Z}$ into a vector space over $\mathbb{Q}$?

Exercise 58
Let $V = \{0, 1\}$ and let $F = \mathrm{GF}(2)$. Define vector addition and scalar multiplication by setting $v + v' = \max\{v, v'\}$, $0v = 0$, and $1v = v$ for all $v, v' \in V$. Does this define on V the structure of a vector space over F?

Exercise 59
Let $p > 2$ and let V be a vector space over GF(p). Show that $v \neq -v$ for all $0_V \neq v \in V$.

Exercise 60
Let $V = C(0, 1)$. Define an operation $\boxplus$ on V by setting $f \boxplus g : x \mapsto \max\{f(x), g(x)\}$. Does this operation of vector addition, together with the usual

operation of scalar multiplication, define on V the structure of a vector space over $\mathbb{R}$?

Exercise 61

Let V be a nontrivial vector space over $\mathbb{R}$. For each $v \in V$ and each complex number $a + bi$, let us define $(a + bi)v = av$. Does V, together with this new scalar multiplication, form a vector space over $\mathbb{C}$?

Exercise 62

Let I be the unit interval $[0, 1]$ on the real line and let $V = \mathbb{R} \times I$. Define operations of addition and scalar multiplication on V as follows: $(a, s) + (b, t) = (a + b, \min\{s, t\})$ and $c \cdot (a, s) = (ca, s)$. Is V a vector space over $\mathbb{R}$?

Exercise 63

Let $V = \{i \in \mathbb{Z} \mid 0 \le i < 2^n\}$ for some given positive integer n. Define operations of vector addition and scalar multiplication on V in such a way as to turn it into a vector space over the field GF(2).

Exercise 64

Let V be a vector space over a field F. Define a function from GF(3) $\times V$ to V by setting $(0, v) \mapsto 0_V$, $(1, v) \mapsto v$, and $(2, v) \mapsto -v$ for all $v \in V$. Does this function, together with the vector addition in V, define on V the structure of a vector space over GF(3)?

Exercise 65

Give an example of a vector space having exactly 125 elements.

Exercise 66

Let $V = \mathbb{Q}^2$, with the usual vector addition. If $a + b\sqrt{2} \in \mathbb{Q}(\sqrt{2})$ and if $\begin{bmatrix} c \\ d \end{bmatrix} \in \mathbb{Q}^2$, set $(a + b\sqrt{2}) \begin{bmatrix} c \\ d \end{bmatrix} = \begin{bmatrix} ac + 2bd \\ bc + ad \end{bmatrix}$. Do these operations turn $\mathbb{Q}^2$ into a vector space over $\mathbb{Q}(\sqrt{2})$?

Exercise 67

Let $V = \mathbb{R} \cup \{\infty\}$ and extend the usual addition of real numbers by defining $v + \infty = \infty + v = \infty$ for all $v \in V$. Is it possible to define an operation of scalar multiplication on V in such a manner as to turn it into a vector space over $\mathbb{R}$?

Exercise 68

Let $V = \mathbb{R}^2$. If $\begin{bmatrix} a \\ b \end{bmatrix}, \begin{bmatrix} a' \\ b' \end{bmatrix} \in V$ and $r \in \mathbb{R}$, set $\begin{bmatrix} a \\ b \end{bmatrix} + \begin{bmatrix} a' \\ b' \end{bmatrix} = \begin{bmatrix} a + a' + 1 \\ b + b' \end{bmatrix}$ and $r \begin{bmatrix} a \\ b \end{bmatrix} = \begin{bmatrix} ra + r - 1 \\ rb \end{bmatrix}$. Do these operations define on V the structure of a vector space over $\mathbb{R}$? If so, what is the identity element for vector addition in this space?

Exercise 69

Let $V = \mathbb{R}$ and let o be an operation on $\mathbb{R}$ defined by $a \circ b = a^3 b$. Is V, together with the usual addition and "scalar multiplication" given by o, a vector space over $\mathbb{R}$?

Exercise 70

Show that $\mathbb{Z}$ is not a vector space over any field.

Exercise 71

Let V be a vector space over the field GF(2). Show that $v = -v$ for all $v \in V$.

Exercise 72

In the definition of a vector space, show that the commutativity of vector addition is a consequence of the other conditions.

Exercise 73

Let W be the subset of $\mathbb{R}^5$ consisting of all vectors an odd number of the entries in which are equal to 0. Is W a subspace of $\mathbb{R}^5$?

Exercise 74

Let F be a field and fix $0 < k \in \mathbb{Z}$. Let W be the subset of $F^{\mathbb{Z}}$ consisting of all those functions f satisfying

$$f(i + k) = \sum_{j=0}^{k-1} f(i + j)$$

for each $i \in \mathbb{Z}$. Is W a subspace of $F^{\mathbb{Z}}$?

Exercise 75

Let W be the subset of $\mathbb{R}^3$ consisting of all vectors $\begin{bmatrix} a \\ b \\ c \end{bmatrix}$ satisfying $|a| + |b| = |c|$.
Is W a subspace of $\mathbb{R}^3$?

Exercise 76

Let $V = \mathbb{R}^{\mathbb{R}}$ and let W be the subset of V containing the constant function $x \mapsto 0$ and all of those functions $f \in V$ satisfying the condition that $f(a) = 0$ for at most finitely-many real numbers a. Is W a subspace of V?

Exercise 77

Let $V = \left\{ \begin{bmatrix} a_1 \\ \vdots \\ a_5 \end{bmatrix} \,\middle|\, 0 < a_i \in \mathbb{R} \right\}$. If $v = \begin{bmatrix} a_1 \\ \vdots \\ a_5 \end{bmatrix}$ and $w = \begin{bmatrix} b_1 \\ \vdots \\ b_5 \end{bmatrix}$ belong to V, and if $c \in \mathbb{R}$, set $v + w = \begin{bmatrix} a_1 b_1 \\ \vdots \\ a_5 b_5 \end{bmatrix}$ and $cv = \begin{bmatrix} a_1^c \\ \vdots \\ a_5^c \end{bmatrix}$. Do these operations turn V into

a vector space over $\mathbb{R}$?

Exercise 78
How many elements are there in the subspace of $\mathrm{GF}(3)^3$ generated by
$\left\{ \begin{bmatrix} 1 \\ 2 \\ 1 \end{bmatrix}, \begin{bmatrix} 2 \\ 2 \\ 1 \end{bmatrix} \right\}$?

Exercise 79
A function $f \in \mathbb{R}^{\mathbb{R}}$ is *piecewise constant* if and only if it is a constant function $x \mapsto c$ or there exist $a_1 < a_2 < \cdots < a_n$ and $c_0, \ldots, c_n$ in $\mathbb{R}$ such that

$$f : x \mapsto \begin{cases} c_0 & \text{if } x < a_1, \\ c_i & \text{if } a_i \leq x < a_{i+1} \text{ for } 1 \leq i < n, \\ c_n & \text{if } a_n \leq x. \end{cases}$$

Does the set of all piecewise constant functions form a subspace of the vector space $\mathbb{R}^{\mathbb{R}}$ over $\mathbb{R}$?

Exercise 80
Let V be the vector space of all continuous functions from $\mathbb{R}$ to itself and let W be the subset of all those functions $f \in V$ satisfying the condition that $|f(x)| \leq 1$ for all $-1 \leq x \leq 1$. Is W a subspace of V?

Exercise 81

Let W be the subspace of $V = \mathrm{GF}(2)^5$ consisting of all vectors $\begin{bmatrix} a_1 \\ \vdots \\ a_5 \end{bmatrix}$ satisfying

$\sum_{i=1}^{5} a_i = 0$. Is W a subspace of V?

Exercise 82
Let $V = \mathbb{R}^{\mathbb{R}}$ and let W be the subset of V consisting of all monotonically-increasing or monotonically-decreasing functions. Is W a subspace of V?

Exercise 83
Let $V = \mathbb{R}^{\mathbb{R}}$ and let W be the subset of V consisting of the constant function $a \mapsto 0$, and all epic functions. Is W a subspace of V?

Exercise 84
Let $V = \mathbb{R}^{\mathbb{R}}$ and let W be the subset of V containing the constant function $a \mapsto 0$ and all of those functions $f \in V$ satisfying the condition that $f(\pi) > f(-\pi)$. Is W a subspace of V?

Exercise 85

Let $V = \mathbb{R}^{\mathbb{R}}$ and let W be the subset of V consisting of all functions f satisfying the condition that there exists a real number c (which depends on f) such that $|f(a)| \leq c|a|$ for all $a \in \mathbb{R}$. Is W a subspace of V?

Exercise 86

Let $V = \mathbb{R}^{\mathbb{R}}$ and let W be the subset of V consisting of all functions f satisfying the condition that there exist real numbers a and b such that $|f(x)| \leq a|\sin(x)| + b|\cos(x)|$ for all $x \geq 0$. Is W a subspace of V?

Exercise 87

Let F be a field and let $V = F^F$, which is a vector space over F. Let W be the set of all functions $f \in V$ satisfying $f(1) = f(-1)$. Is W a subspace of V?

Exercise 88

For any real number $0 < t \leq 1$, let V_t be the set of all functions $f \in \mathbb{R}^{\mathbb{R}}$ satisfying the condition that if $a < b$ in $\mathbb{R}$ then there exists a real number $u(a,b)$ satisfying $|f(x) - f(y)| \leq u(a,b)|x - y|^t$ for all $a \leq x, y \leq b$. For which values of t is V_t a subspace of $\mathbb{R}^{\mathbb{R}}$?

Exercise 89

Let U be a nonempty subset of a vector space V. Show that U is a subspace of V if and only if $au + u' \in U$ for all $u, u' \in U$ and $a \in F$.

Exercise 90

Let V be a vector space over a field F and let v and w be distinct vectors in V. Set $U = \{(1 - t)v + tw \mid t \in F\}$. Show that there exists a vector $y \in V$ such that $\{u + y \mid u \in U\}$ is a subspace of V.

Exercise 91

Let V be a vector space over a field F and let W and Y be subspaces of V^2. Let U be the set of all vectors $\begin{bmatrix} v \\ v' \end{bmatrix} \in V^2$ satisfying the condition that there exists a vector $v'' \in V$ such that $\begin{bmatrix} v \\ v'' \end{bmatrix} \in W$ and $\begin{bmatrix} v'' \\ v' \end{bmatrix} \in Y$. Is U a subspace of V^2?

Exercise 92

Consider $\mathbb{R}$ as a vector space over $\mathbb{Q}$. Given a nonempty subset W of $\mathbb{R}$, let $\overline{W}$ be the set of all real numbers b for which there exists a sequence $a_1, a_2, \ldots$ of elements of W satisfying $\lim_{i \to \infty} a_i = b$. Show that $\overline{W}$ is a subspace of $\mathbb{R}$ whenever W is.

Exercise 93

Let V be a vector space over a field F and let P be the collection of all subsets of V, which we know is a vector space over GF(2). Is the collection of all subspaces of V a subspace of P?

Exercise 94

Let W be the set of all functions $f \in \mathbb{R}^{\mathbb{N}}$ satisfying the condition that if $f(i) \neq 0$ then $f(ji) \neq 0$ for all positive integers j. Is W a subspace of $\mathbb{R}^{\mathbb{N}}$?

Exercise 95

Let W be the set of all functions $f \in \mathbb{R}^{\mathbb{N}}$ satisfying the condition that if $f(i) = 0$ then $f(ji) = 0$ for all positive integers j. Is W a subspace of $\mathbb{R}^{\mathbb{N}}$?

Exercise 96

Let V be a vector space over a field F and let Y be the set of all matrices of the form $\begin{bmatrix} v_1 & v_2 & 0_V \\ 0_V & v_1 + v_2 & 0_V \\ 0_V & v_1 & v_2 \end{bmatrix}$ in $\mathcal{M}_{3 \times 3}(V)$. Is Y a subspace of $\mathcal{M}_{3 \times 3}(V)$?

Exercise 97

Let W be the set of all functions $f \in \mathbb{R}^{\mathbb{R}}$ satisfying the following conditions: there exist positive real numbers a and b such that for all $x \in \mathbb{R}$ satisfying $|x| \geq a$ we have $|f(x)| \leq b|x|$. Show that W is a subspace of $\mathbb{R}^{\mathbb{R}}$.

Exercise 98

Let W be a subspace of a vector space V over a field F. Is the set $(V \smallsetminus W) \cup \{0_V\}$ necessarily a subspace of V?

Exercise 99

Let V be a vector space over a field F and let f be a function from V to the unit interval $[0, 1]$ on the real line satisfying the condition that $f(au + bv) \geq \min\{f(u), f(v)\}$ for all $a, b \in F$ and all $u, v \in V$. Show that $f(0_V) \geq f(v)$ for all $v \in V$ and that if $0 \leq h \leq f(0_V)$ then $V_h = \{v \in V \mid f(v) \geq h\}$ is a subspace of V.

Exercise 100

Consider the elements f, g, h of $\mathbb{Q}^{\mathbb{Q}}$ defined by $f : t \mapsto t - 1$, $g : t \mapsto t + 1$, and $h : t \mapsto t^2 + 1$. Does the function $t \mapsto t^2$ belong to $\mathbb{Q}\{f, g, h\}$?

Exercise 101

Let $F = \mathrm{GF}(3)$ and let $D = \left\{ \begin{bmatrix} 1 \\ 1 \\ 0 \end{bmatrix}, \begin{bmatrix} 1 \\ 0 \\ 2 \end{bmatrix} \right\}$. For which scalars c is $\begin{bmatrix} 0 \\ 1 \\ c \end{bmatrix}$ a linear combination of elements of D?

Exercise 102

Find a real number c such that $\begin{bmatrix} 4 \\ 3 \\ 1 \end{bmatrix} \in \mathbb{R}^3$ is a linear combination of $\left\{ \begin{bmatrix} 3 \\ 1 \\ c \end{bmatrix}, \begin{bmatrix} -1 \\ 2 \\ 1 \end{bmatrix} \right\}$.

Exercise 103

Find subsets D and D' of $\mathbb{R}^3$ such that $\mathbb{R}(D \cap D') \neq \mathbb{R}D \cap \mathbb{R}D'$.

Exercise 104

Find subspaces W and Y of $\mathbb{R}^3$ having the property that $W \cup Y$ is not a subspace of $\mathbb{R}^3$.

Exercise 105

Let V be a vector space over a field F and let $0_V \neq w \in V$. Given a vector $v \in V \smallsetminus Fw$, find the set G of all scalars $a \in F$ satisfying $F\{v, w\} = F\{v, aw\}$.

Exercise 106

Let p be a prime integer and let V be a vector space over $F = \mathrm{GF}(p)$. Show that V is not the union of k subspaces, for any $k \leq p$.

Exercise 107

Let V be a vector space over a field F and let c and d be fixed elements of F. Define a new operation $\boxplus$ on V by setting $v \boxplus v' = cv + dv'$. Is V, with this new vector addition and the old scalar multiplication, still a vector space over F?

Exercise 108

Let I be the closed unit interval $[0, 1]$ on the real line. A function $(a, b) \mapsto a \circ b$ from $I \times I$ to I is a *triangular norm*[1] if and only if the following conditions hold for all $a, b, c \in I$:
(1) $a \circ 1 = a$;
(2) $a \leq c$ implies that $a \circ b \leq c \circ b$;
(3) $a \circ b = b \circ a$;
(4) $a \circ (b \circ c) = (a \circ b) \circ c$.
Given a vector space V over a field F, and given a triangular norm $\circ$ on I, a function $f : V \to I$ is a $\circ$-*fuzzy subspace* of V if and only if, for each $v, w \in V$ and each $d \in F$, we have $f(v + w) \geq f(v) \circ f(w)$ and $f(dv) \geq f(v)$. Find a condition that a $\circ$-fuzzy subspace f of V must satisfy for the set $\{v \in V \mid f(v) \geq a\}$ to be a subspace of V for any $a \in I$.

---

[1]Triangular norms play a very important part in the theory of probabilistic metric spaces and have important applications in statistics and in mathematical economics, as well as such areas as pattern recognition and capacity theory.

Exercise 109

Let V be a vector space over a field F and let D be a nonempty subset of V. A *zero-sum combination* of elements of D is an element of the form $\sum_{i=1}^{n} a_i v_i$, where the v_i are elements of D and the a_i are scalars satisfying $\sum_{i=1}^{n} a_i = 0$. The set $z(D)$ of all zero-sum combinations of elements of D is called the *zero-sum hull* of D. Is it true that $z(z(D)) = z(D)$? Is $z(D)$ necessarily a subspace of V?

Exercise 110

Let V be a vector space over a field F and let D be a nonempty subset of V. A *uniform combination* of elements of D is an element of the form $\sum_{i=1}^{n} a_i v_i$, where the v_i are elements of D and $a_1 = \cdots = a_n$. The set $u(D)$ of all uniform combinations of elements of D is called the *uniform hull* of D. Is it true that $u(u(D)) = u(D)$? Is $u(D)$ necessarily a subspace of V?

Exercise 111

If we identify $\mathbb{R}^2$ with the Euclidean plane in the usual way, and if $v \neq w$ are two vectors in $\mathbb{R}^2$, show that affh($\{v, w\}$) is the line passing through these two points.

Exercise 112

If we identify $\mathbb{R}^3$ with the three-dimensional Euclidean space in the usual way and if v, w, y are distinct vectors in $\mathbb{R}^3$ which are not collinear, show that affh($\{v, w, y\}$) is the plane determined by the three points.

Exercise 113

Let V be a vector space over a field F and let D be a subset of V containing 0_V. Show that affh(D) is a subspace of V.

Algebras Over a Field

In general, a vector space does not carry with it the notion of multiplying two vectors in the space to produce a third vector. However, sometimes such multiplication may be possible. A vector space K over a field F is an F-*algebra* if and only if there exists a function $(v, w) \mapsto v \bullet w$ from $K \times K$ to K such that

(1) $u \bullet (v + w) = u \bullet v + u \bullet w$;

(2) $(u + v) \bullet w = u \bullet w + v \bullet w$;

(3) $a(v \bullet w) = (av) \bullet w = v \bullet (aw)$

for all $u, v, w \in K$ and $a \in F$. As in the proof of Proposition 2.3(1), these conditions suffice to show that $0_K \bullet v = v \bullet 0_K = 0_K$ for all $v \in K$.

Note that the operation $\bullet$ need not be associative, nor need there exist an identity element for this operation. When the operation is associative, i.e. when it satisfies

(4) $v \bullet (w \bullet y) = (v \bullet w) \bullet y$

for all $v, w, y \in K$, then the algebra is called an *associative F-algebra*. If an identity element for $\bullet$ exists, that is to say, if there exists an element $0_K \neq e \in K$ satisfying $v \bullet e = v = e \bullet v$ for all $v \in K$, we say the F-algebra K is *unital*. In a unital F-algebra, as with the case of fields, the identity element must be unique. In this case, we can then identify F with the subset $\{ae \mid a \in F\}$ of K and we note that $a \bullet v = v \bullet a$ for all $v \in K$ and $a \in F$.

If v is an element of an associative F-algebra $(K, \bullet)$ and if n is a positive integer, we write v^n instead of $v \bullet \cdots \bullet v$ (n factors). If K is also unital and has a multiplicative identity e, we set $v^0 = e$ for all $0_K \neq v \in K$. The element $(0_K)^0$ is not defined.

If $v \bullet w = w \bullet v$ for all v and w in some F-algebra K, then the algebra is *commutative*. An F-algebra $(K, \bullet)$ satisfying the condition that $v \bullet w = -w \bullet v$ for all $v, w \in K$ is *anticommutative*. If the characteristic of F is other than 2, it is easy to see that this condition is equivalent to the condition that $v \bullet v = 0_K$ for all $v \in K$. Of course, in that case K cannot possibly be unital.

J.S. Golan, *The Linear Algebra a Beginning Graduate Student Ought to Know*, DOI 10.1007/978-94-007-2636-9_4, © Springer Science+Business Media B.V. 2012

The first systematic study of associative algebras was initiated by the nineteenth-century American mathematician **Benjamin Peirce** and continued by his son, the mathematician and logician **Charles Sanders Peirce**. Other major contributors at the beginning of the twentieth century were the American mathematician **Leonard Dickson**, the Scottish mathematician **Joseph Henry Wedderburn**, and the German mathematician **Emmy Noether**, generally known as "the mother of modern algebra".

If $(K, \bullet)$ is an associative unital F-algebra having a multiplicative identity e, and if $v \in K$ satisfies the condition that there exists an element $w \in K$ such that $v \bullet w = w \bullet v = e$, then we say that v is a *unit* of K. As with the case of fields, such an element w, if it exists, is unique and is usually denoted by v^{-1}. If v is a unit, then so is $-v$, for one immediately notes that $(-v)^{-1} = -(v^{-1})$. Also, it is easy to see that if v and w are units of K, then so is $v \bullet w$. Indeed,

$$(v \bullet w) \bullet \left(w^{-1} \bullet v^{-1}\right) = \left(v \bullet \left(w \bullet w^{-1}\right)\right) \bullet v^{-1}$$

$$= (v \bullet e) \bullet v^{-1} = v \bullet v^{-1} = e$$

and similarly $(w^{-1} \bullet v^{-1}) \bullet (v \bullet w) = e$, so $(v \bullet w)^{-1} = w^{-1} \bullet v^{-1}$. If $v \in K$ is a unit and if $n > 1$ is an integer, we write v^{-n} instead of $(v^{-1})^n$. Note that the Hua's identity (Proposition 2.6) in fact holds in any associative unital F-algebra in which the needed inverses exist, since the proof relies only on associativity of addition and multiplication and distributivity of multiplication over addition.

If $(K, \bullet)$ is an F-algebra, and $v, w \in K$, then (v, w) forms a *commuting pair* if and only if $v \bullet w = w \bullet v$. Of course, if the algebra K is commutative, all pairs of elements commute, but in general that will not be the case. Note that if (v, w) is a commuting pair in a unital associative F-algebra $(K, \bullet)$ and v^{-1} exists, then (v^{-1}, w) is also a commuting pair. Indeed, $(v^{-1} \bullet w) \bullet v = v^{-1} \bullet (w \bullet v) = v^{-1} \bullet (v \bullet w) = w$ so $w \bullet v^{-1} = [(v^{-1} \bullet w) \bullet v] \bullet v^{-1} = v^{-1} \bullet w$.

Example Any vector space V over a field F can be turned into an associative and commutative F-algebra which is not unital by setting $v \bullet w = 0_V$ for all $v, w \in V$.

Example If F is a subfield of a field K, then K has the structure of an associative F-algebra, with multiplication being the multiplication in K. Thus, $\mathbb{C}$ is an

$\mathbb{R}$-algebra and $\mathbb{Q}(\sqrt{p})$ is a $\mathbb{Q}$-algebra for every prime integer p. These algebras are, of course, unital.

Example Let F be a field, let $(K, \bullet)$ be an F-algebra, and let Ω be a nonempty set. Then the vector space K^{Ω} of all functions from Ω to K has the structure of an F-algebra with respect to the operation $\bullet$ defined by $f \bullet g : i \mapsto f(i) \bullet g(i)$ for all $i \in \Omega$. This F-algebra is associative if K is. If K is unital with multiplicative identity element e, then K^{Ω} is also unital, with identity element given by the constant function $i \mapsto e$. In particular, if F is a field and if Ω is a nonempty set then F^{Ω} is an associative unital F-algebra with respect to the operation $\bullet$ defined by $f \bullet g : i \mapsto f(i)g(i)$ for all $i \in \Omega$.

Example We have seen that the collection of all subsets of a given nonempty set A is a vector space over GF(2). It is in fact an associative and commutative unital GF(2)-algebra with respect to the operation $\cap$. The identity element with respect to this operation is A itself.

Example Define an operation $*$ on $C(0, \infty)$ by setting

$$f * g : t \mapsto \int_0^t f(t - u)g(u)\,du.$$

This turns $C(0, \infty)$ into an associative and commutative $\mathbb{R}$-algebra, known as the *convolution algebra* on $\mathbb{R}$.

Example Let K be the vector space over $\mathbb{R}$ consisting of all functions in $\mathbb{R}^{\mathbb{R}}$ which are infinitely differentiable, and define an operation $\bullet$ on K by setting $f \bullet g = (fg)'$ (where $'$ denotes differentiation). Then $(K, \bullet)$ is an algebra which is commutative but not associative.

The collection of all operations $\bullet$ on a vector space V over a field F which turn V into an F-algebra will be studied in more detail in Chap. 20.

Let F be a field. If $(K, \bullet)$ is an F-algebra, then a subspace L of K satisfying the condition that $w \bullet w' \in L$ for all $w, w' \in L$ is an F-*subalgebra* of K. If $(K, \bullet)$ is a unital F-algebra, then L is a *unital subalgebra* if it contains the multiplicative identity element of K.

Let F be a field. An anticommutative F-algebra $(K, \bullet)$ is a *Lie algebra* over F if and only if it satisfies the additional condition

$$(\textit{Jacobi identity}) \quad u \bullet (v \bullet w) + v \bullet (w \bullet u) + w \bullet (u \bullet v) = 0_K;$$

for all $u, v, w \in K$. This algebra is not associative unless $u \bullet v = 0_K$ for all $u, v \in K$.

Sophus Lie was a nineteenth-century Norwegian mathematician who developed mathematical concepts that provide the basic model for quantum theory and an important tool in differential geometry. They were independently defined by the nineteenth-century German teacher **Wilhelm Karl Joseph Killing**, in connection with his work on non-Euclidean geometry. Another pioneer in the study of noncommutative algebras because of their importance in physics was the twentieth-century British mathematician **Dudley Littlewood**.

Wilhelm Karl Joseph Killing   Dudley Littlewood

Example Let F be a field and let $(K, *)$ be an associative F-algebra. Define a new operation $\bullet$ on K by setting $v \bullet w = v * w - w * v$. Then $(K, \bullet)$ is a Lie algebra over F, which is usually denoted by K^-. The operation in K^- is known as the *Lie product* defined on the given F-algebra K. This example is very important because one can show that any Lie algebra over a field F can be considered as a subalgebra of a Lie algebra of the form K^- for some associative F-algebra K. (A proof of this result, known as the *Poincaré–Birkhoff–Witt Theorem*, is far beyond the scope of this book.) If $v, w \in K$, then $v \bullet w = 0_K$ precisely when $v * w = w * v$, in other words, precisely when (v, w) forms a commuting pair in $(K, *)$.

Lie algebras are of fundamental importance in the modeling problems in physics, and have many other applications; they are in the forefront of current mathematical research. One particular Lie algebra defined on $\mathbb{R}^3$ goes back to the work of Grassmann. Define the structure of an $\mathbb{R}$-algebra on $\mathbb{R}^3$ with multiplication $\times$ given by

$$
\begin{bmatrix} a_1 \\ a_2 \\ a_3 \end{bmatrix} \times \begin{bmatrix} b_1 \\ b_2 \\ b_3 \end{bmatrix} = \begin{bmatrix} a_2 b_3 - a_3 b_2 \\ a_3 b_1 - a_1 b_3 \\ a_1 b_2 - a_2 b_1 \end{bmatrix}.
$$

This operation, called the *cross product*, has very important applications in physics and engineering. It is easy to check that the algebra $(\mathbb{R}^3, \times)$ is a Lie algebra over $\mathbb{R}$. Note that if $v_1 = \begin{bmatrix} 1 \\ 0 \\ 0 \end{bmatrix}$, $v_2 = \begin{bmatrix} 0 \\ 1 \\ 0 \end{bmatrix}$, and $v_3 = \begin{bmatrix} 0 \\ 0 \\ 1 \end{bmatrix}$, then, surely, $v_1 \times v_2 = v_3$, $v_1 \times v_2 = v_3$, and $v_3 \times v_1 = v_2$. Moreover, the cross product is the only possible anticommutative product which can be defined on $\mathbb{R}^3$ and which satisfies this condition. Indeed, if $\bullet$ is any such product defined on $\mathbb{R}^3$ then

$$\begin{bmatrix} a_1 \\ a_2 \\ a_3 \end{bmatrix} \bullet \begin{bmatrix} b_1 \\ b_2 \\ b_3 \end{bmatrix} = \left(\sum_{i=1}^{3} a_i v_i \right) \bullet \left(\sum_{j=1}^{3} b_j v_j \right) = \sum_{i=1}^{3} \sum_{j=1}^{3} a_i b_j (v_i \bullet v_j)$$

$$= \begin{bmatrix} a_2 b_3 - a_3 b_2 \\ a_3 b_1 - a_1 b_3 \\ a_1 b_2 - a_2 b_1 \end{bmatrix} = \begin{bmatrix} a_1 \\ a_2 \\ a_3 \end{bmatrix} \times \begin{bmatrix} b_1 \\ b_2 \\ b_3 \end{bmatrix}.$$

Proposition 4.1 *If v and w are nonzero elements of $\mathbb{R}^3$, then $v \times w = \begin{bmatrix} 0 \\ 0 \\ 0 \end{bmatrix}$ if and only if $\mathbb{R}v = \mathbb{R}w$.*

Proof Suppose $v = \begin{bmatrix} a_1 \\ a_2 \\ a_3 \end{bmatrix}$ and $w = \begin{bmatrix} b_1 \\ b_2 \\ b_3 \end{bmatrix}$. These vectors are nonzero and so one of the entries in w is nonzero; without loss of generality, we can assume that $b_1 \neq 0$. Then $a_2 b_3 - a_3 b_2 = a_3 b_1 - a_1 b_3 = a_1 b_2 - a_2 b_1 = 0$ and so, if we define $c = a_1 b_1^{-1}$, we have $v = cw$. Hence $v \in \mathbb{R}w$. Moreover, $c \neq 0$ so $w = c^{-1}v \in \mathbb{R}v$, proving the desired equality. Conversely, if $\mathbb{R}v = \mathbb{R}w$ then there exists an $0 \neq d \in \mathbb{R}$ such that $w = dv$. Then $v \times w = d(v \times v) = \begin{bmatrix} 0 \\ 0 \\ 0 \end{bmatrix}$. $\qquad\square$

The cross product is very particular to the vector space $\mathbb{R}^3$, and does not generalize easily to spaces of the form $\mathbb{R}^n$ for $n > 3$, with the exception of $n = 7$, which we will see in a later chapter.

An important non-associative algebra is the following: let F be a field of characteristic other than 2, and let $(K, *)$ be an associative algebra. We can define a new operation $\bullet$ on K, called the *Jordan product*, by setting $v \bullet w = \frac{1}{2}(v * w + w * v)$. Then $(K, \bullet)$ is a commutative F-algebra, usually denoted by K^+, called the *Jordan algebra* defined by K. It is not associative in general, but does satisfy

(*Jordan identity*) $(v \bullet w) \bullet (v \bullet v) = v \bullet \left(w \bullet (v \bullet v) \right)$

for all $v, w \in K$. Jordan algebras have important applications in physics. Note that if $v * w = w * v$, then $v \bullet w = v * w$. This observation will have important consequences later. In particular, if $(K, *)$ is unital with multiplicative identity e, then $e \bullet v = v \bullet e = v$ for all $v \in K$, so K^+ is also unital.

Jordan algebras were developed by the twentieth-century German physicist **Pascual Jordan**, one of the fathers of quantum mechanics and quantum electrodynamics. The algebraic structure of Lie algebras and Jordan algebras was studied in detail by the twentieth-century American mathematicians **Nathan Jacobson** and **A. Adrian Albert**.

We now come to an extremely important algebra. Let F be a field and let X be an element not in F, which we will call an *indeterminate*. A *polynomial* in X with coefficients in F is a formal sum $f(X) = \sum_{i=0}^{\infty} a_i X^i$, in which the elements a_i belong to F, and no more than a finite number of these elements differ from 0. The elements a_i are called the *coefficients* of the polynomial. If all of the a_i equal 0, then the polynomial is called the *0-polynomial*. Otherwise, there exists a nonnegative integer n satisfying the condition that $a_n \neq 0$ and $a_i = 0$ for all $i > n$. The coefficient a_n is called the *leading coefficient* of the polynomial; the integer n is called the *degree* of the polynomial, and is denoted by $\deg(f)$. If the leading coefficient of a polynomial is 1, the polynomial is *monic*. The degree of the 0-polynomial is defined to be $-\infty$, where we assume that $-\infty < i$ for each integer i and $(-\infty) + i = -\infty$ for all integers i. If $f(X)$ is a polynomial of degree $n \neq -\infty$, we often write it as $\sum_{i=0}^{n} a_i X^i$. The set of all polynomials in X with coefficients in F is denoted by $F[X]$. We identify the elements of F with the polynomials of degree at most 0, and so can consider F as a subdomain of $F[X]$. We can associate the 0-polynomial with the identity element 0 of F for addition and the polynomials of degree 0 with the nonzero elements of F and so, without any problems, consider F as a subset of $F[X]$.

Example The polynomials $5X^3 + 2X^2 + 1$ and $5X^3 - X^2 + X + 4$ in $\mathbb{Q}[X]$ both have degree 3 and leading coefficient 5. Therefore, they are not monic. The polynomials $X^3 + 2X^2 + 1$ and $X^3 - X^2 + X + 4$ in $\mathbb{Q}[X]$ are both monic and have the same degree 3.

We define addition and multiplication of polynomials over a field as follows: if $f(X) = \sum_{i=0}^{\infty} a_i X^i$ and $g(X) = \sum_{i=0}^{\infty} b_i X^i$ are polynomials in $F[X]$, then $f(X) + g(X)$ is the polynomial $\sum_{i=0}^{\infty} c_i X^i$, where $c_i = a_i + b_i$ for all $i \geq 0$ and $f(X)g(X)$ is the polynomial $\sum_{i=0}^{\infty} d_i X^i$, where $d_i = \sum_{j=0}^{i} a_j b_{i-j}$ for all $i \geq 0$. It is easy to verify that these definitions turn $F[X]$ into an associative and commutative unital F-algebra with the 0-polynomial acting as the identity element for addition and the degree-0 polynomial 1 acting as the identity element for multiplication. This algebra is an integral domain, that is, the product of two nonzero elements of $F[X]$ is again nonzero. In general, an algebra having this property is said to be *entire*.

Thus, commutative, associative, entire, unital F-algebras are integral domains. The converse of this is not true: $\mathbb{Z}$ is an integral domain which is not an F-algebra for any field F. Not every commutative and associative unital $\mathbb{R}$-algebra is entire. Indeed, the functions $f : a \mapsto \max\{a, 0\}$ and $g : a \mapsto \max\{-a, 0\}$ are both nonzero elements of the $\mathbb{R}$-algebra $\mathbb{R}^{[-1,1]}$, but their product is the 0-function.

If $f(X) = \sum_{i=0}^{\infty} a_i X^i$ and $g(X) = \sum_{i=0}^{\infty} b_i X^i$ are polynomials in $F[X]$ then we define the polynomial $f(g(X))$ to be $\sum_{i=0}^{\infty} a_i g(X)^i$. Then, for any fixed $g(X)$, the set $F[g(X)] = \{f(g(X)) \mid f(X) \in F[X]\}$ is a unital subalgebra of $F[X]$.

Note that every polynomial in $F[X]$ is a linear combination of elements of the set $B = \{1, X, X^2, \ldots\}$ over F, so B is a set of generators of $F[X]$ over F. On the other hand, it is clear that no finite set of polynomials can be a generating set for $F[X]$ over F, and so $F[X]$ is not finitely generated as a vector space over F.

We should remark that the formal definition of multiplication of polynomials does not translate into the fastest method of carrying out such multiplication in practice on a computer, especially for polynomials of large degree. The problem of fast polynomial multiplication has been the subject of extensive research over the years, and many interesting algorithms to perform such multiplication have been devised. A typical such algorithm is *Karatsuba's algorithm*, which is easy to implement on a computer: let $f(X)$ and $g(X)$ be polynomials in $F[X]$, where F is a field. We can write these polynomials as $f(X) = \sum_{i=0}^{n} a_i X^i$ and $g(X) = \sum_{i=0}^{n} b_i X^i$, where n is a nonnegative power of 2 satisfying $n \geq \max\{\deg(f), \deg(g)\}$. (Of course, in this case a_n and b_n may equal 0.) We now calculate $f(X)g(X)$ as follows:

(1) If $n = 1$ then $f(X)g(X) = a_1 b_1 X^2 + (a_0 b_1 + a_1 b_0)X + a_0 b_0$.
(2) Otherwise, write $f(X) = f_1(X)X^{n/2} + f_0(X)$ and

$$g(X) = g_1(X)X^{n/2} + g_0(X),$$

where the polynomials $f_0(X)$, $f_1(X)$, $g_0(X)$, and $g_1(X)$ are all of degree at most $n/2$.
(3) Recursively, calculate $f_0(X)g_0(X)$, $f_1(X)g_1(X)$, and

$$(f_0 + f_1)(X)(g_0 + g_1)(X).$$

(4) Then

$$f(X)g(X) = X^n (f_1 g_1)(X) + X^{n/2}[(f_0 + f_1)(g_0 + g_1) - f_0 g_0 - f_1 g_1](X)$$
$$+ (f_0 g_0)(X).$$

Indeed, if the multiplication of two polynomials of degree at most n using the definition of polynomial multiplication takes an order of $2n^2$ arithmetic operations (i.e., additions and multiplications), it is possible to prove that there exists a fixed positive integer c such that the multiplication of two polynomials of degree at most n using Karatsuba's algorithm takes at most $cn^{1.59}$ arithmetic operations. If n is sufficiently large, the difference between these two bounds can be significant.

The main idea of Karatsuba's algorithm lies in the recursive reduction of the degrees of the polynomials involved. The method of recursive reduction has since been extended to fast algorithms in many other areas of mathematics. We will encounter it again when we consider the Strassen–Winograd algorithms for matrix multiplication.

With kind permission of Ekatherina Karatsuba.

Anatoli Alexeevich Karatsuba is a contemporary Russian mathematician whose research is primarily in number theory.

There are other highly-sophisticated algorithms for multiplying two polynomials of degree at most n in an order of $n \log(n)$ arithmetic operations.

Proposition 4.2 (Division Algorithm) *If F is a field and if $f(X)$ and $g(X) \neq 0$ are elements of $F[X]$, then there exist unique polynomials $u(X)$ and $v(X)$ in $F[X]$ satisfying $f(X) = g(X)u(X) + v(X)$ and $\deg(v) < \deg(g)$.*

Proof Assume that $f(X) = \sum_{i=0}^{\infty} a_i X^i$ and $g(X) = \sum_{i=0}^{\infty} b_i X^i$ are the given polynomials. If $f(X) = 0$ or if $\deg(f) < \deg(g)$, choose $u(X) = 0$ and $v(X) = f(X)$, and we are done. Thus we can assume that $n = \deg(f) \geq \deg(g) = k$, and will prove our result by induction on n. If $n = 0$ then $k = 0$, and therefore we can choose $u(X)$ to be $a_0 b_0^{-1}$, which is a polynomial of degree 0, and choose $v(X)$ to be the 0-polynomial. Now assume, inductively, that $n > 0$ and that the proposition has been established for all functions $f(X)$ of degree less than n. Set $h(X) = f(X) - a_n b_k^{-1} X^{n-k} g(X)$. If this is the 0-polynomial, choose $u(X) = a_n b_k^{-1} X^{n-k}$ and let $v(X)$ be the 0-polynomial. Otherwise, since $\deg(f) > \deg(h)$, we see by the induction hypothesis that there exist polynomials $v(X)$ and $w(X)$ in $F[X]$ satisfying $h(X) = g(X)w(X) + v(X)$, where $\deg(g) > \deg(v)$. Thus $f(X) = [a_n b_k^{-1} X^{n-k} + w(X)]g(X) + v(X)$, as required.

We are left to show uniqueness. Indeed, assume that $f(X)$ equals $g(X)u_1(X) + v_1(X)$ and $g(X)u_2(X) + v_2(X)$, where $\deg(v_1) < \deg(g)$ and $\deg(v_2) < \deg(g)$. Then $g(X)[u_1(X) - u_2(X)] + [v_1(X) - v_2(X)]g(X)[u_1(X) - u_2(X)] + [v_1(X) - v_2(X)]$ equals the 0-polynomial. If we have $u_1(X) = u_2(X)$ then $v_1(X) = v_2(X)$, and we are done. Therefore, assume that $u_1(X) \neq u_2(X)$. But then, since $\deg(g[u_1 - u_1]) > \deg(v_1 - v_2)$ and since $F[X]$ is entire, this is a contradiction. Thus we have established uniqueness. $\qquad \square$

Let us emphasize that the set $F[X]$ is composed of formal expressions and not functions. Every polynomial $f(X) = \sum_{i=0}^{\infty} a_i X^i \in F[X]$ defines a corresponding *polynomial function* in F^F given by $c \mapsto f(c) = \sum_{i=0}^{\infty} a_i c^i$, but the correspondence between polynomials and polynomial functions is not bijective. Indeed, it is possible for two distinct polynomials to define the same polynomial function. Thus, for example, if $F = \mathrm{GF}(2)$ then the distinct polynomials $X, X^2, X^3, \ldots$ all define the same function from F to itself, namely the function given by $0 \mapsto 0$ and $1 \mapsto 1$. The *degree* of a polynomial function is the least of the degrees of the (perhaps many) polynomials which define that function.

With kind permission of the Archives of the Mathematisches Forschungsinstitut Oberwolfach.

The first person to systematically consider the best methods of calculating $f(c)$ for a polynomial $f(X) \in F[X]$ and for $c \in F$, was the twentieth-century Russian mathematician **Alexander Markovich Ostrowski**.

Let $p_1(X)$ and $p_2(X)$ be polynomials in $F[X]$ and let $c \in F$. If we set $f(X) = p_1(X) + p_2(X)$ and $g(X) = p_1(X)p_2(X)$ then it is clear that $f(c) = p_1(c) + p_2(c)$ and $g(c) = p_1(c)p_2(c)$.

Proposition 4.3 *Let F be a field and let $p(X)$ be a polynomial in $F[X]$. Then an element c of F satisfies the condition that $p(c) = 0$ if and only if there exists a polynomial $u(X) \in F[X]$ satisfying $p(X) = (X - c)u(X)$.*

Proof By Proposition 4.2, we know that there exist polynomials $u(X)$ and $v(X)$ in $F[X]$ satisfying $p(X) = (X - c)u(X) + v(X)$, where $\deg(v) < \deg(X - c) = 1$. Therefore, $v(X) = b$ for some $b \in F$. If $b = 0$ then $p(c) = (c - c)u(c) = 0$. Conversely, if $p(c) = 0$ then $0 = p(c) = (c - c)u(c) + b = b$ and so $p(X) = (x - c)u(X)$. $\qquad\square$

As an immediate consequence of this result, we see that if F is a field and if $p(X) \in F[X]$, then the set of all elements c of F satisfying $p(c) = 0$ is finite and, indeed, cannot exceed the degree of $p(X)$.

Let F be a field. A polynomial $p(X) \in F[X]$ is *reducible* if and only if there exist polynomials $u(X)$ and $v(X)$ in $F[X]$, each of degree at least 1, satisfying $p(X) = u(X)v(X)$. Otherwise, the polynomial is *irreducible*. Many tests for the irreducibility of polynomials in $\mathbb{Q}[X]$ have been devised. One of the earliest and well-known is *Eisenstein's criterion*: if $p(X) = \sum_{i=0}^{n} a_i X^i \in \mathbb{Q}[X]$, where each a_i is an integer, and if there exists a prime integer q such that q does not divide a_n, q divides a_i for all $0 \le i \le n - 1$, and q^2 does not divide a_0, then $p(X)$ is irreducible. (A proof of this can be found in books on abstract algebra.) Thus, using this criterion, we see that $3X^3 + 7X^2 + 49X - 7$ is an irreducible polynomial in $\mathbb{Q}[X]$.

Gauss' brilliant student, **Ferdinand Eisenstein**, died of tuberculosis at the age of 29.

Example If $F = \text{GF}(5)$ then the polynomial $X^3 + X + 1 \in F[X]$ is irreducible, a fact which can be established, if necessary, by testing all possibilities. However, when $F = \text{GF}(3)$ it is easy to verify the factorization $X^3 + X + 1 = (X+2)(X^2+X+2)$, and thus see that the polynomial is reducible.

Example If $p(X) = u(X)v(X)$ in $F[X]$, then surely $p(X+c) = u(X+c)v(X+c)$ for any $c \in F$, and so to prove that a polynomial $p(X)$ is irreducible it suffices to prove that $p(X+c)$ is irreducible for some $c \in F$. For example, let q be a prime integer. The qth *cyclotomic polynomial* in $\mathbb{Q}[X]$ is defined to be $\Phi_q(X) = \sum_{i=0}^{q-1} X^i$. We claim that this polynomial is irreducible. To see that this is so, we observe that $\Phi_q(X+1) = X^{q-1} + \sum_{i=0}^{q-2} \binom{q}{q-1-i} X^i$, which is irreducible by Eisenstein's criterion.

It is known that the number of monic irreducible polynomials of positive degree m in $\text{GF}(p)$ equals $N(p) = \frac{1}{m} \sum \mu(d) p^{m/d}$, where the sum ranges over all integers d which divide m and the *Möbius function* $\mu(d)$ is defined by

$$\mu(d) = \begin{cases} 1 & \text{if } d = 1, \\ (-1)^k & \text{if } d \text{ is the product of } k \text{ distinct primes}, \\ 0 & \text{otherwise.} \end{cases}$$

This means that the probability of a randomly-selected monic polynomial of degree m in $\text{GF}(p)[X]$ being irreducible is $N(p)/p^m$, which is roughly $\frac{1}{m}$. In particular, we note that for every positive integer m there exists at least one monic irreducible polynomial of degree m in $\text{GF}(p)$.

Any polynomial in $F[X]$ can be written as a product of irreducible polynomials. How to find such a decomposition, especially in the case of polynomials over a finite field or over $\mathbb{Q}$, is a very difficult and important problem, which attracted such great mathematicians as Newton and which continues to attract many important mathematicians until this day. Indeed, the problem of factoring polynomials over finite fields into irreducible components has become even more important, since it is the basis for many current cryptographic schemes. There are algorithms, such as Berlekamp's algorithm, which factor a polynomial $f(X) \in F[X]$, where $F = \text{GF}(p^n)$, in a time polynomial in p, n, and $\deg(f)$. Moreover, under various assumptions, such as the Generalized Riemann Hypothesis, polynomials of special forms can be factored much more rapidly.

A polynomial $p(X) \in F[X]$ of positive degree is *completely reducible* if and only if it can be written as a product of polynomials in $F[X]$ of degree 1. Not every polynomial over every field is completely reducible. For example, the polynomial $X^2 + 1 \in \mathbb{Q}[X]$ is not completely reducible. The field F is *algebraically closed* if every polynomial of positive degree in $F[X]$ is completely reducible. The fields $\mathbb{Q}$ and $\mathbb{R}$ are not algebraically closed. The field $\mathbb{C}$ is algebraically closed, by a theorem known as the *Fundamental Theorem of Algebra*. This theorem is in fact analytic and not algebraic, and relies on various analytic properties of functions of a complex variable. Most of the great mathematicians of the eighteenth century—d'Alembert, Euler, Laplace, Lagrange, Argand, Cauchy, and others—tried in vain to prove this theorem. The first proof was given by Gauss in his doctoral thesis in 1799. His proof was basically topological and relied on the work of Euler. During his lifetime, Gauss published several proofs of this theorem.

With kind permission of the Archives of the Mathematisches Forschungsinstitut Oberwolfach.

A "nearly algebraic" proof was given by German/American mathematician **Hans Zassenhaus** in 1969. Most proofs of the Fundamental Theorem of Algebra are existence proofs and do not give a constructive method of finding the degree-one factors of a polynomial over an algebraically-closed field. The first constructive proof was given by the German mathematician **Helmut Kneser** in 1940.

Example The field $F = \mathrm{GF}(2)$ is not algebraically closed since the polynomial $X^2 + X + 1 \in F[X]$ is not completely reducible.

Note that if a field F is algebraically closed then every polynomial function $F \rightarrow F$ defined by a polynomial of positive degree is epic. Indeed, let $p(X) \in F[X]$ be a polynomial of positive degree and let $d \in F$. Then $q(X) = p(X) - d$ is a polynomial of positive degree in $F[X]$ and so there exists an element c of F such that $q(c) = 0$. In other words, $p(c) = d$.

It is easy to see that a polynomial $p(X) \in \mathbb{R}[X]$ of degree 1 is irreducible. If $p(X) = aX^2 + bX + c$ is of degree 2 then, considering it as an element of $\mathbb{C}[X]$, we have

$$p(X) = a\left(X + \frac{b}{2a} + \frac{1}{2a}\sqrt{b^2 - 4ac} \right)\left(X - \frac{b}{2a} + \frac{1}{2a}\sqrt{b^2 - 4ac} \right).$$

Then this factorization holds in $\mathbb{R}[X]$ as well if and only if $b^2 - 4ac \geq 0$, and so $p(X)$ is irreducible if and only if $b^2 - 4ac < 0$. From the following result we deduce immediately that there are no irreducible polynomials in $\mathbb{R}[X]$ of degree greater than 2.

Proposition 4.4 *A monic polynomial* $p(X) \in \mathbb{R}[X]$ *is irreducible if and only if it is of the form* $X - a$ *or* $(X - a)^2 + b^2$, *where* $a \in \mathbb{R}$ *and* $0 \neq b \in \mathbb{R}$.

Proof Clearly, every polynomial of the form $X - a$ is irreducible. Now assume that $f(X) = (X - a)^2 + b^2 = X^2 - 2aX + a^2 + b^2$. Were this polynomial reducible, we could find real numbers c and d satisfying

$$f(X) = (X - c)(X - d) = X^2 - (c + d)X + cd$$

and so $c + d = 2a$ and $cd = a^2 + b^2$. This implies that $c^2 - 2ac + a^2 + b^2 = 0$ and hence $c = \frac{1}{2}[2a \pm \sqrt{4a^2 - 4(a^2 + b^2)}] = a \pm \sqrt{-b^2}$, which contradicts the assumption that $c \in \mathbb{R}$ since b is assumed to be nonzero. Thus polynomials of both of the given forms are indeed irreducible.

Conversely, let $p(X) = \sum_{i=0}^{n} c_i X^i$ be a monic irreducible polynomial in $\mathbb{R}[X]$ that is not of the form $X - a$. By the Fundamental Theorem of Algebra, we know that there exists a complex number $z = a + bi$ satisfying $p(z) = 0$. Since the coefficients of $p(X)$ are real, this means that $p(\overline{z}) = 0$ as well, since $0 = \overline{0} = \overline{p(z)} = \sum_{i=0}^{n} c_i \overline{z}^i = p(\overline{z})$. Thus there exists a polynomial $u(X) \in \mathbb{R}[X]$ satisfying $p(X) = (X - z)(X - \overline{z})u(X)$, where $(X - z)(X - \overline{z}) = X^2 - (z + \overline{z})X + z\overline{z} = X^2 - 2aX + a^2 + b^2$. Since $p(X)$ was assumed irreducible, we conclude that $z \notin \mathbb{R}$ (i.e., $b \neq 0$) and that $p(X)$ equals $X^2 - 2aX + a^2 + b^2$, as desired.   □

An obvious generalization of the above construction is the following: Let F be a field and let $(K, \bullet)$ be an associative and commutative unital F-algebra. If X is an element not in K, we can define a *polynomial* with coefficients in K as a formal sum $f(X) = \sum_{i=0}^{\infty} a_i X^i$, in which the elements a_i belong to K and no more than a finite number of them differ from 0_K. The set of all such polynomials will be denoted by $K[X]$. As above, we define addition and multiplication in $K[X]$ as follows: if $f(X) = \sum_{i=0}^{\infty} a_i X^i$ and $g(X) = \sum_{i=0}^{\infty} b_i X^i$ belong to $K[X]$, then $f(X) + g(X)$ is the polynomial $\sum_{i=0}^{\infty} c_i X^i$, in which $c_i = a_i + b_i$ for each $0 \leq i \leq \infty$, and $f(X)g(X)$ is the polynomial $\sum_{i=0}^{\infty} d_i X^i$, in which $d_i = \sum_{j=0}^{i} a_j \bullet b_{i-j}$ for each $0 \leq i \leq \infty$. Again, it is easy to check that $K[X]$ is an F-algebra. Moreover, as a direct consequence of the definition of multiplication, we see that if K is entire then so is $K[X]$. This generalization allows us to consider algebras of *polynomials in several commuting indeterminates* with coefficients in K defined inductively by setting $K[X_1, \ldots, X_n] = K[X_1, \ldots, X_{n-1}][X_n]$ for each $n > 1$. Elements of this algebra are of the form

$$f(X_1, \ldots, X_n) = \sum a_{i_1, \ldots, i_n} X_1^{i_1} \cdots X_n^{i_n},$$

where the sum ranges over all n-tuples $(i_1, \ldots, i_n)$ of nonnegative integers and at most finitely-many of the coefficients $a_{i_1, \ldots, i_n} \in K$ are nonzero. The *degree* of $f(X_1, \ldots, X_n)$ is the maximal value of

$$\{i_1 + \cdots + i_n \mid a_{i_1, \ldots, i_n} \neq 0\}.$$

A polynomial $\sum a_{i_1,\dots,i_n} X_1^{i_1} \cdots X_n^{i_n}$ in $K[X_1, \dots, X_n]$ is *flat* if and only if $a_{i_1,\dots,i_n} \neq 0$ only when each i_j is either 0 or 1.

Proposition 4.5 *Let F be a field of characteristic other than 2 and let K be an associative and commutative unital entire F-algebra. Let $f(X_1, \dots, X_n) = \sum a_{i_1,\dots,i_n} X_1^{i_1} \cdots X_n^{i_n} \in K[X_1, \dots, X_n]$ be a flat polynomial of degree n. Then for each n-tuple $(c_1, \dots c_n)$ of nonzero elements of K there exist $e_1, \dots, e_n$ in K, each equal to 1_K or -1_K, such that $f(e_1 c_1, \dots, e_n c_n) \neq 0$.*

Proof We will prove this result by induction on n. If $n = 1$, then $f(X_1) = a_1 X_1 + a_0$, where $a_1 \neq 0$. If $0_K \neq c \in K$ then either $a_0 + a_1 c$ or $a_0 - a_1 c$ is nonzero, for otherwise we would have $2a_1 c = 0_K$, which is impossible since K is entire and the characteristic of F is not 2. Hence the case $n = 1$ has been established. Now assume that $n > 1$ and that the proposition has been established for flat polynomials in $F[X_1, \dots, X_{n-1}]$. We can write the polynomial $f(X_1, \dots, X_n)$ in the form $g(X_1, \dots, X_{n-1}) + h(X_1, \dots, X_{n-1}) X_n$, where $h(X_1, \dots, X_{n-1})$ is a flat polynomial in $K[X_1, \dots, X_{n-1}]$ of degree $n - 1$. If $(c_1, \dots, c_n)$ is an n-tuple of nonzero elements of K then, by the induction hypothesis, we can find $e_1, \dots, e_{n-1}$ in K, each equal to 1_K or -1_K, such that $h(e_1 c_1, \dots, e_{n-1} c_{n-1}) \neq 0$. But then we have $g(e_1 c_1, \dots, e_{n-1} c_{n-1}) + h(e_1 c_1, \dots, e_{n-1} c_{n-1}) X_n \in K[X_n]$ and so, by the case $n = 1$, we can find e_n equal to 1_K or -1_K such that $f(e_1 c_1, \dots, e_n c_n) \neq 0$. □

Exercises

Exercise 114
Let F be a field and let $(K, \bullet)$ and $(L, *)$ be F-algebras. Define an operation $\diamond$ on $K \times L$ by $\begin{bmatrix} a \\ b \end{bmatrix} \diamond \begin{bmatrix} a' \\ b' \end{bmatrix} = \begin{bmatrix} a \bullet a' \\ b * b' \end{bmatrix}$. Is $(K \times L, \diamond)$ an F-algebra?

Exercise 115
Let F be a field and let $(K, \bullet)$ be a unitary, associative, commutative, and entire F-algebra which, as a vector space, is finitely generated over F. Is K necessarily a field?

Exercise 116
Let F be a field and let $(K, \bullet)$ be an associative F-algebra which, as a vector space, is finitely generated over F. Given an element $a \in K$, do there necessarily exist elements $a_1, a_2 \in K$ satisfying $a_1 \bullet a_2 = a$?

Exercise 117
Define an operation $\bullet$ on $\mathbb{R}^2$ by setting $\begin{bmatrix} a \\ b \end{bmatrix} \bullet \begin{bmatrix} c \\ d \end{bmatrix} = \begin{bmatrix} 2ac - bd \\ ad + bc \end{bmatrix}$. Show that this operation turns $\mathbb{R}^2$ into an $\mathbb{R}$-algebra. Is this algebra associative?

Exercise 118

Let F be a field and let $(K, \bullet)$ be a unital F-algebra. Define an operation $\diamond$ on the vector space $L = K \times F$ by setting $\begin{bmatrix} v \\ a \end{bmatrix} \diamond \begin{bmatrix} w \\ b \end{bmatrix} = \begin{bmatrix} v \bullet w + bv + aw \\ ab \end{bmatrix}$ for all $v, w \in K$ and $a, b \in F$. Is L an F-algebra? Is it unital?

Exercise 119

Let F be a field. An F-algebra $(K, \bullet)$ is a *division algebra* if and only if for every $v \in K$ and for every $0_K \neq w \in K$ there exist unique vectors $x, y \in K$, not necessarily equal, satisfying $w \bullet x = v$ and $y \bullet w = v$. Is the algebra defined in the previous exercise a division algebra?

Exercise 120

Let K be the subset of $\mathcal{M}_{4 \times 4}(\mathbb{R})$ consisting of all matrices of the form $\begin{bmatrix} a & -b & -c & -d \\ b & a & -d & c \\ c & d & a & -b \\ d & -c & b & a \end{bmatrix}$ for $a, b, c, d \in \mathbb{R}$ and let L be the subset of $\mathcal{M}_{4 \times 4}(\mathbb{R})$

consisting of all matrices of the form $\begin{bmatrix} a & -b & -c & -d \\ b & a & d & -c \\ c & -d & a & b \\ d & c & -b & a \end{bmatrix}$. Are K and L uni-

tal subalgebras of $\mathcal{M}_{4 \times 4}(\mathbb{R})$? Are they division algebras?

Exercise 121

Let F be a field and let $(K, \bullet)$ be an associative unital F-algebra with multiplicative identity e. For units $v, w \in K$, show that:
(1) $v \bullet (v^{-1} + w^{-1}) = (v + w) \bullet w^{-1}$;
(2) $(v + w)^{-1} \bullet w = v^{-1} \bullet (v^{-1} + w^{-1})^{-1}$ whenever $v + w$ and $v^{-1} + w^{-1}$ are also a units;
(3) $v \bullet w^{-1} + e = v \bullet (v^{-1} + w^{-1})$.

Exercise 122

Let F be a field and let $(K, \bullet)$ be an associative F-algebra which, as a vector space, is finitely generated over F. Suppose that there exists an element $y \in K$ satisfying the condition that for each $v \in K$ there exists an element $v' \in K$ satisfying $v' \bullet y = v$. Show that each such element v' must be unique.

Exercise 123

Let F be an infinite field and let $(K, *)$ be an associative unital F-algebra. If $v, w \in K$, show that there are infinitely-many elements w' of K satisfying $v \bullet w = v \bullet w'$ in K.

Exercise 124

Let F be a field and let $(K, \bullet)$ be an associative unital F-algebra. If A and B are subsets of K, we let $A \bullet B$ be the set of all elements of K of the form $a \bullet b$, with

$a \in A$ and $b \in B$ (in particular, $\varnothing \bullet B = A \bullet \varnothing = \varnothing$). We know that the set V of all subsets of K is a vector space over GF(2). Is $(V, \bullet)$ a GF(2)-algebra? If so, is it associative? Is it unital?

Exercise 125
Let F be a field and let $(K, \bullet)$ be an associative F-algebra. If V and W are subspaces of K, we let $V \bullet W$ be the set of all finite sums of the form $\sum_{i=1}^{n} v_i \bullet w_i$, with $v_i \in V$ and $w_i \in W$. Is $V \bullet W$ necessarily a subspace of K?

Exercise 126
Let $(K, \bullet)$ be an associative F-algebra and let $v \in K$. If there exists an element y of K satisfying $v \bullet y \bullet v = v$, show that there also exists an element w of K satisfying $v \bullet w \bullet v = v$ and $w \bullet v \bullet w = w$.

Exercise 127
For $v, w \in \mathbb{R}^3$, simplify the expression $(v + w) \times (v - w)$.

Exercise 128
For $u, v, w \in \mathbb{R}^3$, simplify the expression $(u + v + w) \times (v + w)$.

Exercise 129
Let F be a field and let $(K, \bullet)$ be an F-algebra satisfying the Jacobi identity. Show that K is a Lie algebra if and only if $v \bullet v = 0_K$ for all $v \in K$.

Exercise 130
Let F be a field and let $(K, *)$ be an associative F-algebra. For each $0_F \neq c \in F$ and define an operation $\bullet_c$ on K by setting $v \bullet_c w = c(v * w + w * v)$. For which values of c is $(K, \bullet)$ a Jordan algebra over F?

Exercise 131
Let F be a field and let $(K, \bullet)$ be a unitary F-algebra. For each $v \in K$, let $S(v)$ be the set of all $a \in F$ satisfying the condition that $v - a1_K$ does not have an inverse with respect to the operation $\bullet$. If $v \in K$ has a multiplicative inverse v^{-1} with respect to this operation, show that either $S(v) = \varnothing = S(v^{-1})$ or $S(v) \neq \varnothing$ and $S(v^{-1}) = \{a^{-1} \mid a \in S(v)\}$.

Exercise 132
Let F be a field and let L be the set of all polynomials $f(X) \in F[X]$ satisfying the condition that $f(-a) = -f(a)$ for all $a \in F$. Is L a subspace of $F[X]$?

Exercise 133
Let F be a field and let L be the set of all polynomials $f(X) \in F[X]$ satisfying the condition that $\deg(f)$ is even. Is L a subspace of $F[X]$?

Exercise 134
Let F be a field and let $f(X), g(X) \in F[X]$. Show that $\deg(fg) = \deg(f) + \deg(g)$.

Exercise 135
Let F be a field and let $f(X), g(X) \in F[X]$. Show that $\deg(f + g) \leq \max\{\deg(f), \deg(g)\}$, and give an example in which we do not have equality.

Exercise 136
Find polynomials $u(X), v(X) \in \mathbb{Q}[X]$ satisfying

$$X^4 + 3X^3 = (X^2 + X + 1)u(X) + v(X).$$

Exercise 137
Let $F = \text{GF}(2)$. Find polynomials $u(X), v(X) \in F[X]$ satisfying

$$X^5 + X^2 = (X^3 + X + 1)u(X) + v(X).$$

Exercise 138
Let $F = \text{GF}(7)$. Find a nonzero polynomial $p(X) \in F[X]$ such that the polynomial function defined by p is the 0-function.

Exercise 139
Is the polynomial $6X^4 + 3X^3 + 6X^2 + 2X + 5 \in \text{GF}(7)[X]$ irreducible?

Exercise 140
Is the polynomial $X^7 + X^4 + 1 \in \mathbb{Q}[X]$ irreducible?

Exercise 141
Find $t \in \mathbb{R}$ such that there exist $a, b \in \mathbb{R}$ satisfying $a + b = 1$ and $2a^3 - a^2 - 7a + t = 0 = 2b^3 - b^2 - 7b + t$.

Exercise 142
For a field F, compare the subsets $F[X^2]$ and $F[X^2 + 1]$ of $F[X]$.

Exercise 143
Let $F = \text{GF}(p)$, where p is a prime integer, and let g be an arbitrary function from F to itself. Show that there exists a polynomial $p(X) \in F[X]$ of degree less than p satisfying the condition that $g(c) = p(c)$ for all $c \in F$.

Exercise 144
Let c be a nonzero element of a field F and let $n > 1$ be an integer. Show that there exists a polynomial $p(X) \in F[X]$ satisfying $c^n + c^{-n} = p(c + c^{-1})$.

Exercise 145

Let F be a field. Find the set of all polynomials $0 \neq p(X) \in F[X]$ satisfying $p(X^2) = p(X)^2$.

Exercise 146

Let $p_k(X) = \frac{1}{k!}X(X-1)\cdots(X-k+1) \in \mathbb{Q}[X]$ for some positive integer k. Show that $p_k(n) \in \mathbb{Z}$ for every nonnegative integer n.

Exercise 147

Let $p_n(X) = nX^{n+1} - (n+1)X^n + 1 \in \mathbb{Q}[X]$ for any positive integer n. Show that there exists a polynomial $q_n(X) \in \mathbb{Q}[X]$ satisfying $p_n(X) = (X-1)^2 q_n(X)$.

Exercise 148

Let F be a field and let W be a nontrivial subspace of the vector space $F[X]$ over F. Let $p(X) \in F[X]$ be a given monic polynomial and let $p(X)W = \{p(X)f(X) \mid f(X) \in W\}$. Show that $p(X)W$ is a subspace of $F[X]$ and find a necessary and sufficient condition for it to equal W.

Exercise 149

Let p be a prime integer and let n be a positive integer. Does there necessarily exist an irreducible monic polynomial in $GF(p)[X]$ of degree n?

Exercise 150

Let p be a prime integer and let n be a positive integer. Show that the product of all irreducible monic polynomials in $GF(p)[X]$ of degree dividing n is equal to $X^{p^n} - X$.

Exercise 151

Let $n > 1$ be an integer. Is the polynomial $p(X) = 1 + \sum_{h=1}^{n} \frac{1}{h!}X^h \in \mathbb{Q}[X]$ necessarily irreducible?

Exercise 152

Show that the polynomial $X^4 + 1$ is irreducible in $\mathbb{Q}[X]$ but reducible in $GF(p)[X]$ for every prime p.

Exercise 153

Show that $X^4 + 2(1-c)X^2 + (1+c)^2 \in \mathbb{Q}[X]$ is irreducible for every $c \in \mathbb{Q}$ satisfying $\sqrt{c} \notin \mathbb{Q}$.

Exercise 154

Let F be a field and let $K = F^{\mathbb{N}}$. Define operations $+$ and $\bullet$ on K by setting $f + g : i \mapsto f(i) + g(i)$ and $f \bullet g : i \mapsto \sum_{j+k=i} f(j)g(k)$. Show that K is an associative and commutative unital F-algebra. Is it entire?

Exercise 155

Let k be a positive integer and let $a < b$ be real numbers. A function $f \in \mathbb{R}^{[a,b]}$ is a *spline function* of degree k if and only if there exist real numbers $a = a_0 < \cdots < a_n = b$ and polynomials $p_0(X), \ldots, p_{n-1}(X)$ of degree k in $\mathbb{R}[X]$ satisfying the condition that $f : x \mapsto p_i(x)$ for all $a_i \leq x \leq a_{i+1}$ and all $0 \leq i \leq n - 1$. Spline functions play an important part in interpolation theory and in numerical procedures for solving differential equations. Is the set of all spline functions of fixed degree k a subspace of the vector space $\mathbb{R}^{[a,b]}$?

With kind permission of the Archives of the Mathematisches Forschungsinstitut Oberwolfach.

Spline functions were first defined and studied by the twentieth-century Romanian/American mathematician **Isaac Jacob Schoenberg**.

Exercise 156

Let F be a finite field, let $k > 1$ be an integer, and let V be the vector space over F consisting of all polynomials in $F[X]$ having degree less than k. Let $a_1, \ldots, a_n$ be distinct elements of F and let W be the subset of F^n consisting of all vectors of the form $\begin{bmatrix} p(a_1) \\ \vdots \\ p(a_n) \end{bmatrix}$ for some $p \in V$. Is W a subspace of F_n?

Exercise 157

A *trigonometric polynomial* in $\mathbb{R}^{\mathbb{R}}$ is a function of the form $t \mapsto a_0 + \sum_{h=1}^{k}[a_h \cos(ht) + b_h \sin(ht)]$, where $a_0, \ldots, a_k, b_1, \ldots, b_k \in \mathbb{R}$. Show that the subset of $\mathbb{R}^{\mathbb{R}}$ consisting of all trigonometric polynomials is an entire $\mathbb{R}$-algebra.

Linear Independence and Dimension

5

In this chapter, we will see how a restricted collection of vectors in a vector space over a field can dictate the structure of the entire space, and we will deduce far-ranging conclusions from this. Let V be a vector space over a field F. A nonempty subset D of V is *linearly dependent* if and only if there exist distinct vectors $v_1, \ldots, v_n$ in D and scalars $a_1, \ldots, a_n$ in F, not all of which are equal to 0, satisfying $\sum_{i=1}^{n} a_i v_i = 0_V$. A list of elements of V is linearly dependent if it has two equal members or if its underlying subset is linearly dependent. Clearly, any set of vectors containing 0_V is linearly dependent. A nonempty set of vectors which is not linearly dependent is *linearly independent*. That is to say, D is linearly independent if and only if $D = \varnothing$ or $D \neq \varnothing$ and we have $\sum_{i=1}^{n} a_i v_i = 0_V$ with the a_i in F and the v_i in V, when and only when $a_i = 0$ for all $1 \leq i \leq n$. As a consequence of this definition, we see that an infinite set of vectors is linearly dependent if and only if it has a finite linearly-dependent subset, and an infinite set of vectors is linearly independent if and only if each of its finite subsets is linearly independent. It is also clear that any set of vectors containing a linearly-dependent subset is linearly dependent and that any subset of a linearly-independent set of vectors is linearly independent.

With kind permission of The Shelby White and Leon Levy Archives Center, USA.

The notion of linear independence of vectors was introduced by Grassmann; it was extensively generalized to other mathematical contexts the by the twentieth-century American mathematician **Hassler Whitney**.

Example The subset $\left\{ \begin{bmatrix} 1 \\ 2 \\ 1 \end{bmatrix}, \begin{bmatrix} -1 \\ 3 \\ 4 \end{bmatrix}, \begin{bmatrix} -4 \\ 7 \\ 11 \end{bmatrix} \right\}$ of $\mathbb{Q}^3$ is linearly dependent

J.S. Golan, *The Linear Algebra a Beginning Graduate Student Ought to Know*,
DOI 10.1007/978-94-007-2636-9_5, © Springer Science+Business Media B.V. 2012

since $\begin{bmatrix} 0 \\ 0 \\ 0 \end{bmatrix} = (-1) \begin{bmatrix} 1 \\ 2 \\ 1 \end{bmatrix} + 3 \begin{bmatrix} -1 \\ 3 \\ 4 \end{bmatrix} + (-1) \begin{bmatrix} -4 \\ 7 \\ 11 \end{bmatrix}$. Similarly, the subset

$\left\{ \begin{bmatrix} 1 \\ 0 \\ 0 \end{bmatrix}, \begin{bmatrix} 1 \\ 1 \\ 0 \end{bmatrix}, \begin{bmatrix} 1 \\ 1 \\ 1 \end{bmatrix} \right\}$ of $\mathbb{Q}^3$ is linearly independent, since if

$$\begin{bmatrix} 0 \\ 0 \\ 0 \end{bmatrix} = a \begin{bmatrix} 1 \\ 0 \\ 0 \end{bmatrix} + b \begin{bmatrix} 1 \\ 1 \\ 0 \end{bmatrix} + c \begin{bmatrix} 1 \\ 1 \\ 1 \end{bmatrix},$$

then $\begin{bmatrix} 0 \\ 0 \\ 0 \end{bmatrix} = \begin{bmatrix} a+b+c \\ b+c \\ c \end{bmatrix}$ and this implies that $a = b = c = 0$.

Example The subset $\left\{ \begin{bmatrix} 1 \\ 1 \\ 0 \\ 1 \\ 1 \\ 0 \\ 0 \end{bmatrix}, \begin{bmatrix} 1 \\ 0 \\ 1 \\ 1 \\ 0 \\ 1 \\ 0 \end{bmatrix}, \begin{bmatrix} 0 \\ 1 \\ 1 \\ 1 \\ 0 \\ 0 \\ 1 \end{bmatrix} \right\}$ of $GF(2)^7$ is linearly independent and

generates a subspace of V composed of eight vectors:

$$\begin{bmatrix} 0 \\ 0 \\ 0 \\ 0 \\ 0 \\ 0 \\ 0 \end{bmatrix}, \begin{bmatrix} 1 \\ 1 \\ 0 \\ 1 \\ 1 \\ 0 \\ 0 \end{bmatrix}, \begin{bmatrix} 1 \\ 0 \\ 1 \\ 1 \\ 0 \\ 1 \\ 0 \end{bmatrix}, \begin{bmatrix} 0 \\ 1 \\ 1 \\ 1 \\ 0 \\ 0 \\ 1 \end{bmatrix}, \begin{bmatrix} 0 \\ 0 \\ 0 \\ 1 \\ 1 \\ 1 \\ 1 \end{bmatrix}, \begin{bmatrix} 1 \\ 1 \\ 0 \\ 0 \\ 0 \\ 1 \\ 1 \end{bmatrix}, \begin{bmatrix} 1 \\ 0 \\ 1 \\ 0 \\ 1 \\ 0 \\ 1 \end{bmatrix}, \text{ and } \begin{bmatrix} 0 \\ 1 \\ 1 \\ 0 \\ 1 \\ 1 \\ 0 \end{bmatrix}.$$

Note that in every element of V other than its identity element for addition, a majority of the entries are nonzero. This property makes this subspace of V important in algebraic coding theory.

Example Let $b > 1$ be a real number, let $\{p_1, p_2, \ldots\}$ be the set of prime integers and, for each i, let $u_i = \log_b(p_i)$. We claim that $D = \{u_1, u_2, \ldots\}$ is a linearly-independent subset of $\mathbb{R}$ when it is considered as a vector space over $\mathbb{Q}$. Indeed, assume that this is not the case. Then there are a positive integer n and rational numbers $a_1, \ldots, a_n$, not all equal to 0, satisfying $\sum_{i=1}^{n} a_i u_i = 0$. If we multiply both sides by the product of the denominators of the a_i, we can assume that the a_i

are integers. Then

$$1 = b^0 = b^{\sum a_i u_i} = \prod_{i=1}^{n} b^{a_i u_i} = \prod_{i=1}^{n} (b^{u_i})^{a_i} = \prod_{i=1}^{n} p_i^{a_i},$$

and this is a contradiction. Therefore, D must be linearly independent.

Example Let F be a field and let Ω be a nonempty set. Let V_i be a vector space over F for each $i \in \Omega$, and set $V = \prod_{i \in \Omega} V_i$. We have already seen that the identity for addition in this vector space is the function $g_0 : \Omega \to \bigcup_{i \in \Omega} V_i$ given by $g_0 : i \mapsto 0_{V_i}$. For each $i \in \Omega$, let $f_i : \Omega \to \bigcup_{i \in \Omega} V_i$ be a function satisfying the condition that $f_i(i) \neq g_0(i)$ but $f_i(h) = g_0(h)$ for all $h \in \Omega \smallsetminus \{i\}$. We claim that the subset $\{f_i \mid i \in \Omega\}$ of V is linearly independent. To see this, assume that there exists a finite subset Λ of Ω and a family of scalars $\{c_h \mid h \in \Lambda\}$ such that $\sum_{h \in \Lambda} c_h f_h = g_0$. Then for each $k \in \Lambda$ we have $g_0(k) = (\sum_{h \in \Lambda} c_h f_h)(k) = \sum_{h \in \Lambda} c_h f_h(k) = c_k f_k(k)$ and since, by definition, $f_k(k) \neq g_0(k)$, we must have $c_k = 0$.

Example If F is a field, the subset $\{1, X, X^2, \ldots\}$ of $F[X]$ is surely linearly independent, since $\sum_{i=0}^{n} a_i X^i = 0$ if and only if each of the coefficients a_i equals 0.

Example Let $V = \mathbb{R}^{\mathbb{R}}$ be the vector space, over $\mathbb{R}$, of all functions from $\mathbb{R}$ to itself. Let D be the set of all functions of the form $x \mapsto e^{ax}$ for some real number a. We claim that D is linearly independent. Indeed, assume that there are distinct real numbers $a_1, \ldots, a_n$ and real numbers $c_1, \ldots, c_n$ such that the function $x \mapsto \sum_{i=1}^{n} c_i e^{a_i x}$ equals the 0-function $f_0 : x \mapsto 0$, which is the identity element of V for addition. We need to show that each of the c_i equals 0, and this we will do by induction on n.

If $n = 1$ then we must have $c_1 = 0$ since the function $x \mapsto e^{ax}$ is different from f_0 for each $a \in \mathbb{R}$. Assume therefore that $n > 1$ and that every subset of D having no more than $n - 1$ elements is linearly independent. For each $1 \leq i \leq n$, set $b_i = a_i - a_n$. Then

$$f_0 = e^{-a_n x} \sum_{i=1}^{n} c_i e^{a_i x} = \left[\sum_{i=1}^{n-1} c_i e^{b_i x} \right] + c_n$$

and if we differentiate both sides of the equation, we see that $f_0 = \sum_{i=1}^{n-1} b_i c_i e^{b_i x}$. By the induction hypothesis and the choice of the scalars a_i as being distinct, it follows that $b_i c_i = 0 \neq b_i$ for each $1 \leq i \leq n - 1$ and so $c_i = 0$ for all $1 \leq i \leq n - 1$. This in turn implies that $c_n = 0$ as well.

Similarly, let G be the subset of V consisting of all of the functions of the form $g_i : x \mapsto x^{i-1} 2^{x-1}$. We claim that this set too is linearly independent. Indeed, assume otherwise. Then there exists a positive integer n and there exist real numbers $c_1, \ldots, c_n$, such that $\sum_{i=1}^{n} c_i g_i = f_0$. But this implies that $2^{x-1} (\sum_{i=1}^{n} c_i x^{i-1}) = 0$ for each real number x. Since $2^{x-1} \neq 0$ for each $x \in \mathbb{R}$, we conclude that $\sum_{i=1}^{n} c_i x^{i-1} = 0$ for all x. But the polynomial function $x \mapsto \sum_{i=1}^{n} c_i x^{i-1}$ from $\mathbb{R}$ to itself has infinitely-many roots if and only if $c_i = 0$ for all i, proving linear independence.

Note that if $\{v, w\}$ is a linearly-dependent set of vectors in an anticommutative algebra $(K, \bullet)$ over a field of characteristic other than 2, then there exist scalars a and b, not both equal to 0, such that $av + bw = 0_K$. Relabeling if necessary, we can assume that $b \neq 0$. Then $0_K = a(v \bullet v) + b(v \bullet w) = b(v \bullet w)$ and so $v \bullet w = 0_K$. A simple induction argument shows that if D is a linearly-dependent subset of K then $v_1 \bullet \cdots \bullet v_k = 0_K$ for any finite subset $\{v_1, \ldots, v_k\}$ of D.

Note too that Proposition 3.7 can be easily iterated to get the more general result that if D is a nonempty subset of a vector space V over a field F and if B is a finite linearly-independent subset of FD having k elements, then there exists a subset D' of D also having k elements satisfying the condition that $F((D \setminus D') \cup B) = FD$. Moreover, if D is linearly independent, so is $(D \setminus D') \cup B$. This result is sometimes known as the *Steinitz Replacement Property*.

Proposition 5.1 *Let V be a vector space over a field F. A nonempty subset D of V is linearly dependent if and only if some element of D is a linear combination of the others over F.*

Proof Assume D is linearly dependent. Then there exists a finite subset $\{v_1, \ldots, v_n\}$ of D and scalars $a_1, \ldots, a_n$, not all of which equal 0, satisfying $\sum_{i=1}^{n} a_i v_i = 0_V$. Say $a_h \neq 0$. Then $v_h = -a_h^{-1} \sum_{i \neq h} a_i v_i$ and so we see that v_h is a linear combination of the other elements of D over F. Conversely, assume that there is some element of D is a linear combination of the others over F. That is to say, there is an element v_1 of D, elements $v_2, \ldots, v_n$ of $D \setminus \{v_1\}$ and scalars $a_2, \ldots, a_n$ in F satisfying $v_1 = \sum_{i=2}^{n} a_i v_i$. If we set $a_1 = -1$, we see that $\sum_{i=1}^{n} a_i v_i = 0_V$ and so D is linearly dependent. $\qquad \square$

Example For every real number a, let f_a be the function in $\mathbb{R}^{\mathbb{R}}$ defined by $f_a : x \mapsto |x - a|$. We claim that the subset $D = \{f_a \mid a \in \mathbb{R}\}$ of $\mathbb{R}^{\mathbb{R}}$ is linearly independent. Indeed, assume that this is not the case. Then there exists a real number b such that f_b is a linear combination of other members of D. In other words, there exist a finite subset E of $\mathbb{R} \setminus \{b\}$ and scalars c_a for each $a \in E$ such that $f_b = \sum_{a \in E} c_a f_a$. But the function on the right-hand side of this equation is differentiable at b, while the function on the left-hand side is not. From this contradiction, we see that D is linearly independent.

If A is a nonempty set, then a relation $\preccurlyeq$ between elements of A is called a *partial order relation* if and only if the following conditions are satisfied:
(1) $a \preccurlyeq a$ for all $a \in A$;
(2) If $a \preccurlyeq b$ and $b \preccurlyeq a$ then $a = b$;
(3) If $a \preccurlyeq b$ and $b \preccurlyeq c$ then $a \preccurlyeq c$.
The term "partial" comes from the fact that, given elements a and b of A, it may happen that neither $a \preccurlyeq b$ nor $b \preccurlyeq a$. A set on which a partial order has been defined is a *partially-ordered set*. A partially-ordered set A satisfying the condition that for

all $a, b \in A$ we have either $a \preccurlyeq b$ or $b \preccurlyeq a$ is called a *chain*. A nonempty subset B of a partially-ordered set A is itself partially-ordered relative to the partial order relation defined on A; it is a *chain subset* if it is a chain relative to the partial order defined on A.

If A is a nonempty set on which we have a partial order relation $\preccurlyeq$ defined, then an element a_0 of A is *maximal* in A if and only if $a_0 \preccurlyeq a$ when and only when $a = a_0$. An element a_1 is *minimal* if and only if $a \preccurlyeq a_1$ when and only when $a = a_1$. Maximal and minimal elements need not exist or, if they exist, need not be unique. The *Well Ordering Principle*, one of the fundamental axioms of number theory, says that any nonempty subset of $\mathbb{N}$, ordered with the usual partial order, has a minimal element. This principle is equivalent to the principle of mathematical induction.

Partial order relations are ubiquitous in mathematics, and often play a very important, though not usually highlighted, part in the analysis of mathematical structures.

Example Let A be a nonempty set and let P be the collection of all subsets of A. Define a relation $\preccurlyeq$ between elements of P by setting $B \preccurlyeq B'$ if and only if $B \subseteq B'$. It is easy to verify that this is indeed a partial order relation. Moreover, P has a unique maximal element, namely A, and a unique minimal element, namely $\varnothing$. The set P is not a chain whenever A has more than one element since, if a and b are distinct elements of A, then $\{a\} \not\subseteq \{b\}$ and $\{b\} \not\subseteq \{a\}$.

Example Let $A = \{1, 2, 3\}$ and let P be the collection of all subsets of A having one or two elements. Thus P has six elements: $\{1\}$, $\{2\}$, $\{3\}$, $\{1, 2\}$, $\{1, 3\}$, and $\{2, 3\}$. Again, the relation $\preccurlyeq$ between elements of P defined by setting $B \preccurlyeq B'$ if and only if $B \subseteq B'$ is a partial order relation. Moreover, P has three minimal elements: $\{1\}$, $\{2\}$, and $\{3\}$; it also has three maximal elements: $\{1, 2\}$, $\{1, 3\}$, and $\{2, 3\}$.

In general, if we have a collection of subsets of a given set, the collection is partially-ordered by setting $B \preccurlyeq B'$ if and only if $B \subseteq B'$. Therefore, it makes sense for us to talk about "a minimal generating set" of a vector space V—namely a minimal element in the partially-ordered collection of all generating sets of V— and about "a maximal linearly-independent subset" of a vector space V—namely a maximal element of the partially-ordered collection of all linearly-independent subsets of V. However, we have no a priori guarantee that such minimal or maximal elements in fact exist.

Example Consider the set A of all integers greater than 1, and define a relation $\preccurlyeq$ on A by setting $k \preccurlyeq n$ if and only if there is a positive integer t satisfying $n = tk$. This is a partial order relation on A. Moreover, A has infinitely-many minimal elements, since each prime integer is a minimal element of A, while it has no maximal elements, since $n \preccurlyeq 2n$ for each $n \in A$.

Proposition 5.2 *Let V be a vector space over a field F. Then the following conditions on a subset D of V are equivalent:*
(1) *D is a minimal set of generators of V;*
(2) *D is a maximal linearly-independent subset of V;*
(3) *D is a linearly-independent set of generators of V.*

Proof (1) $\Rightarrow$ (2): Let D be a minimal set of generators of V, and assume that D is linearly dependent. By Proposition 5.1, there exists an element $v_0 \in D$ which is a linear combination of elements of the set $E = D \smallsetminus \{v_0\}$ over F. Say $v_0 = \sum_{i=1}^n a_i u_i$, where the u_i belong to E and the a_i are scalars in F. If v is arbitrary element of V then, since D is a set of generators of V, there exists elements $v_1, \ldots, v_n$ of E and scalars $b_0, b_1, \ldots, b_n$ such that $v = \sum_{j=0}^n b_j v_j$. But this then implies that $v = b_0 v_0 + \sum_{j=1}^n b_j v_j = \sum_{i=1}^n b_0 a_i u_i + \sum_{j=1}^n b_j v_j$ and so E is also a set of generators of V, contradicting the minimality of D. This establishes the claim that D is linearly independent. If $v \in V \smallsetminus D$, the set $D \cup \{v\}$ is linearly dependent since v is a linear combination of elements of D. Thus D is a maximal linearly-independent set.

(2) $\Rightarrow$ (3): Assume that D is a maximal linearly-independent subset of V.

Consider a vector v_0 in $V \smallsetminus D$. By (2), we know that the set $D \cup \{v_0\}$ is linearly dependent, and so $0_V \in F(D \cup \{v_0\}) \smallsetminus FD$ by Proposition 3.7, this implies $v_0 \in F(D \cup \{0_V\}) = FD$, which proves that D is a set of generators of V.

(3) $\Rightarrow$ (1): Assume that D is a linearly-independent set of generators of V and that E is a proper subset of D which is also a set of generators for V. Let $v_0 \in D \smallsetminus E$. Then there exist elements $v_1, \ldots, v_n$ of E and scalars $a_1, \ldots, a_n$ such that $v_0 = \sum_{i=1}^n a_i v_i$. But, by Proposition 5.1, this implies that the set D is linearly dependent, contradicting (3). Therefore, no such E exists and so D is a minimal set of generators of V. $\square$

Proposition 5.3 *Let V be a vector space over a field F and let D be a linearly-independent subset of V. If $v_0 \in V \smallsetminus FD$ then the set $D \cup \{v_0\}$ is linearly independent.*

Proof Assume that this set is linearly dependent. Then there exist elements $v_1, \ldots, v_n$ of D and scalars $a_0, a_1, \ldots, a_n$, not all equal to 0, such that $\sum_{i=0}^n a_i v_i = 0_V$. The scalar a_0 must be different from 0, for otherwise D would be linearly dependent, which is a contradiction. Therefore, $v_0 = \sum_{i=1}^n -a_0^{-1} a_i v_i \in FD$, which contradicts the choice of v_0. Thus $D \cup \{v_0\}$ must be linearly independent. $\square$

Proposition 5.3 has important implications. For example, let V be a vector space over a field F which is not finitely generated and let $D = \{v_1, \ldots, v_n\}$ be a linearly-independent subset of V. Then $FD \neq V$, since V is not finitely generated, and so there exists a vector $v_{n+1} \in V \smallsetminus FD$ such that $\{v_1, \ldots, v_{n+1}\}$ is linearly independent. Thus we see that a vector space which is not finitely generated has linearly-independent finite subsets of arbitrarily-large size.

A generating set for a vector space V over a field F which is also linearly independent, is called a *basis* of V over F. In Proposition 5.2, we gave some equivalent conditions for determining of a subset of a vector space is a basis. However, we have not yet proven that every (or, indeed, any) vector space must have a basis.

Example Clearly, $\left\{ \begin{bmatrix} 1 \\ 0 \\ 0 \end{bmatrix}, \begin{bmatrix} 0 \\ 1 \\ 0 \end{bmatrix}, \begin{bmatrix} 0 \\ 0 \\ 1 \end{bmatrix} \right\}$ and $\left\{ \begin{bmatrix} 1 \\ 0 \\ 0 \end{bmatrix}, \begin{bmatrix} 1 \\ 1 \\ 0 \end{bmatrix}, \begin{bmatrix} 1 \\ 1 \\ 1 \end{bmatrix} \right\}$ are bases

of F^3 for any field F. If the characteristic of F is other than 2, then

$\left\{ \begin{bmatrix} 1 \\ 1 \\ 0 \end{bmatrix}, \begin{bmatrix} 1 \\ 0 \\ 1 \end{bmatrix}, \begin{bmatrix} 0 \\ 1 \\ 1 \end{bmatrix} \right\}$ is a basis of F^3, but if the field F has characteristic 2,

then the set is linearly dependent, since $\begin{bmatrix} 1 \\ 1 \\ 0 \end{bmatrix} + \begin{bmatrix} 1 \\ 0 \\ 1 \end{bmatrix} + \begin{bmatrix} 0 \\ 1 \\ 1 \end{bmatrix} = \begin{bmatrix} 0 \\ 0 \\ 0 \end{bmatrix}$.

Example Let F be a field and let both k and n be positive integers. For each $1 \le s \le k$ and each $1 \le t \le n$, let H_{st} be the matrix $[a_{ij}]$ in $\mathcal{M}_{k \times n}(F)$ defined by

$$a_{ij} = \begin{cases} 1 & \text{if } (i, j) = (s, t), \\ 0 & \text{otherwise.} \end{cases}$$

Then $\{H_{st} \mid 1 \le s \le k \text{ and } 1 \le t \le n\}$ is a basis of $\mathcal{M}_{k \times n}(F)$.

Example If F is a field, then we have already seen that the subset $\{1, X, X^2, \ldots\}$ of $F[X]$ is a linearly-independent generating set for $F[X]$ as a vector space over F, and so is a basis of this space. The same is true for the subset $\{1, X + 1, X^2 + X + 1, \ldots\}$ of $F[X]$. More generally, if $\{p_0(X), p_1(X), \ldots\}$ is a subset of $F[X]$ satisfying the condition that $\deg(p_i(X)) = i$ for all $i \ge 0$, then it is a basis of $F[X]$ as a vector space over F.

Since every element of a vector space V over a field F has a unique representation as a linear combination of elements of a basis, if one wants to define a structure of an F-algebra on V it suffices to define the product of any pair of basis elements, and then extend the definition by distributivity and associativity. This is illustrated by the following example, and we will come back to it again in Proposition 5.5.

Example We have already noted that if F is a field then $F[X]$ is an associative F-algebra. Let us generalize this construction. Let H be a nonempty set on which we have defined an associative operation $*$. Thus, for example, H could be the set of nonnegative integers with the operation of addition or multiplication. Let V be the vector space over F with basis $\{v_h \mid h \in H\}$ and define an operation $\bullet$ on V as follows: if $v = \sum_{g \in H} a_g v_g$ and $w = \sum_{h \in H} b_h v_h$ are elements of V (where at most finitely-many of the a_g and the b_h are nonzero), then set $v \bullet w = \sum_{g \in H} \sum_{h \in H} a_g b_h v_{g*h}$. This turns V into an associative F-algebra. In the case $H = \{X^i \mid i \geq 0\}$, we get $F[X]$. Such constructions are very important in advanced applications of linear algebra.

Note that a vector space may have (and usually does have) many bases and so the problem arises as to whether there is a preferred basis among all of these. For vector spaces of the form F^n, there are reasons to prefer the basis

$$\left\{ \begin{bmatrix} 1 \\ 0 \\ 0 \\ \vdots \\ 0 \end{bmatrix}, \begin{bmatrix} 0 \\ 1 \\ 0 \\ \vdots \\ 0 \end{bmatrix}, \dots, \begin{bmatrix} 0 \\ 0 \\ 0 \\ \vdots \\ 1 \end{bmatrix} \right\} ; \text{ for vector spaces of the form } \mathcal{M}_{k \times n}(F) \text{ there are rea-}$$

sons to prefer the basis $\{H_{st} \mid 1 \leq s \leq k \text{ and } 1 \leq t \leq n\}$ defined above; and for vector spaces of the form $F[X]$ there are reasons to prefer the basis $\{1, X, X^2, \dots\}$. These bases are called the *canonical bases* of their respective spaces. However, in various applications—especially those involving large calculations—it is often convenient and sometimes extremely important to pick other bases which fit the problem under consideration. Indeed, in applications many considerations arise in choosing a basis D for a given vector space V. For example, we would like representation of elements of V as linear combinations of elements of the basis to be stable under perturbations of the coefficients. That is to say, if $v = \sum_{i=1}^{n} a_i v_i$, where the v_i are elements of D, and if a_i' is a scalar near a_i for each $1 \leq i \leq n$, then we would like $\sum_{i=1}^{n} a_i' v_i$ to be, in some sense, near v. (What "near" means here depends on notions of distance arising from the particular situation under consideration.) This is especially important if our data is based on observation or measurement which is not assumed to be entirely accurate. For instance, we might want to choose the basis taking into account the fact that the coefficient of v_h is much more dubious than the coefficients of the other basis elements, or choose it so that all of the coefficients a_i be of the same numerical order of magnitude for those vectors v in which we are really interested and for which we will have to do extensive calculation.

It is also important to emphasize another point. When we defined the notation for this book, we stressed that when a set is defined by listing its elements, the set comes with an implicit order defined by that listing. When we deal with bases, and especially finite bases, the order in which the elements of the basis are written often plays a critical role, and one should never lose track of this.

Proposition 5.4 *Let V be a vector space over a field F and let D be a nonempty subset of V. Then D is a basis of V if and only if every vector in V can be written as a linear combination of elements of D over F in precisely one way.*

Proof First, let us assume that D is a basis of V and that there exists an element v of V which can be written as a linear combination of elements of D over F in two different ways. That is to say, that there exists a finite subset $\{v_1, \ldots, v_n\}$ of D and there exist scalars $a_1, \ldots, a_n, b_1, \ldots, b_n$ in F such that $v = \sum_{i=1}^n a_i v_i = \sum_{i=1}^n b_i v_i$, where $a_h \neq b_h$ for at least one index h. Then $0_V = v - v = (\sum_{i=1}^n a_i v_i) - (\sum_{i=1}^n b_i v_i) = \sum_{i=1}^n [a_i - b_i] v_i$, where at least one of the scalars $a_i - b_i$ is nonzero. This contradicts the assumption that D is a basis and hence linearly independent. Therefore, every vector in V can be written as a linear combination of elements of D over F in precisely one way.

Conversely, assume that every vector in V can be written as a linear combination of elements of D over F in precisely one way. That certainly implies that D is a generating set for V over F. If $\{v_1, \ldots, v_n\}$ is a subset of D and if $a_1, \ldots, a_n$ are scalars satisfying $\sum_{i=1}^n a_i v_i = 0_V$, then we have $\sum_{i=1}^n a_i v_i = \sum_{i=1}^n 0 v_i$ and so, by uniqueness of representation, we have $a_i = 0$ for each $1 \leq i \leq n$. This shows that D is linearly independent and so a basis. $\square$

We can look at Proposition 5.4 from another point of view. Let D be a nonempty subset of a vector space V over a field F, and define a function $\theta : F^{(D)} \to V$ by setting $\theta : f \mapsto \sum_{u \in D} f(u) u$. (This sum is well-defined since only finitely-many of the summands are nonzero.) Then:

(1) The function θ is monic if and only if D is linearly independent;
(2) The function θ is epic if and only if D is a generating set;
(3) The function θ is bijective if and only if D is a basis.

Proposition 5.5 *Let D be a basis for a vector space V over a field F. Then any function $f : D \times D \to V$ can be extended in a unique manner to a function $V \times V \to V$ which defines on V the structure of an F-algebra. Moreover, all F-algebra structures on V arise in this manner.*

Proof Let $D = \{y_i \mid i \in \Omega\}$. Suppose that we are given a function $f : D \times D \to V$. We define an operation $\bullet$ on V as follows: if $v, w \in V$, then, by Proposition 5.4, we know that we can write $v = \sum_{i \in \Omega} a_i y_i$ and $w = \sum_{j \in \Omega} b_j y_j$ in a unique manner, where the a_i and b_j are scalars, only a finite number of which are nonzero; then set $v \bullet w = \sum_{i \in \Omega} \sum_{j \in \Omega} a_i b_j f(y_i, y_j)$. It is straightforward to show that this defines the structure of an F-algebra on V. Conversely, if $(V, \bullet)$ is an F-algebra, define the function $f : D \times D \to V$ by $f : (y_i, y_j) \mapsto y_i \bullet y_j$. $\square$

The function f in Proposition 5.5 is the *multiplication table* of the vector multiplication operation $\bullet$ with respect to the basis D.

Example Let F be a field and let $a, b \in F$. Let $B = \{v_1, v_2, v_3, v_4\}$ be the canonical basis for F^4 over F. Define an operation $\bullet$ on B according to the multiplication table:

$\bullet$	v_1	v_2	v_3	v_4
v_1	v_1	v_1	v_3	v_4
v_2	v_2	av_1	v_4	av_3
v_3	v_3	$-v_4$	bv_1	$-bv_2$
v_4	v_4	$-av_3$	bv_2	$-abv_1$

and extend this operation to F^4 by setting

$$\left(\sum_{i=1}^{4} a_i v_i\right) \bullet \left(\sum_{j=1}^{4} b_j v_j\right) = \sum_{i=1}^{4}\sum_{j=1}^{4} a_i b_j (v_i \bullet v_j).$$

Then F^4, together with this operation, is a unital associative algebra known as a *quaternion algebra* over F, in which v_1 is the identity element of for multiplication. In the special case of $F = \mathbb{R}$ and $a = b = -1$, we get the algebra of *real quaternions*, which is denoted by $\mathbb{H}$. The algebra of real quaternions was first defined by Hamilton in 1844 as a generalization of the field of complex numbers (and earlier studied by Gauss, who did not publish his results). It is a division algebra over $\mathbb{R}$ since every nonzero quaternion is a unit of $\mathbb{H}$. These were subsequently generalized by Clifford and used in his study of non-Euclidean spaces. Lately, they have also been used in computer graphics and in signal analysis. If F is a field having characteristic $p > 0$, quaternion algebras over F are not even entire. However, they arise naturally in the theory of elliptic curves, and so are of great importance in cryptography. If $p > 2$, then no quaternion algebras over F are commutative.

With kind permission of the Special collections, Fine Arts Library, Harvard University (Tait); With kind permission of the London Mathematical Society (Clifford).

Sir William Rowan Hamilton, a nineteenth-century Irish mathematician and physicist, helped create matrix theory in its modern formulation, together with Cayley and Sylvester. Hamilton was the first to use the terms "vector" and "scalar" in an algebraic context. His championship of quaternions as an alternative to vectors in physics was later taken up by Scottish mathematician **Peter Guthrie Tait**. The nineteenth-century British mathematician **William Kingdon Clifford** was one of the first to argue that energy and matter were just different types of curvature of space.

We now show that any vector space over a field F has a basis. Indeed, the following two propositions show somewhat stronger than that.

Proposition 5.6 *If V is a vector space finitely generated over a field F then every finite generating set of V over F contains a basis of V.*

Proof Let V be a vector space finitely generated over a field F and let D be a finite generating set for V over F. If D is minimal among all generating sets for V, then we know by Proposition 5.2 that it is a basis of V. If not, it properly contains other generating sets for V over F, one of which, say E, has the fewest elements. Then E cannot properly contain any other generating set for V over F, and so it must be a basis of V.   □

Proposition 5.7 *If V is a vector space finitely generated over a field F then every linearly-independent subset B of V is contained in a basis of V over F.*

Proof By assumption, there exists a finite generating set $\{v_1, \ldots, v_n\}$ for V over F. Let B be a linearly-independent subset of V. If $v_i \in FB$ for each $1 \leq i \leq n$, then $FB = V$, and B is itself a basis of V. Otherwise, let $h = \min\{i \mid v_i \notin FB\}$. By Proposition 5.3, the set $D = B \cup \{v_h\}$ is linearly independent. If it is a generating set for V, then it is a basis and we are done. If not, let $k = \min\{i \mid v_i \notin FD\}$, and replace D by $B \cup \{v_h, v_k\}$. Continuing in this manner, we see that after finitely-many steps we obtain a basis of V.   □

With kind permission of the Department of Mathematics, University of Torino, Italy.

The Italian mathematician **Giuseppe Peano**, best known for his axiomatization of the natural numbers, was the first to prove that every finitely-generated vector space has a basis at the end of the nineteenth century. He also gave the final form for the definition of a vector space, which we used above.

We now want to extend this result to vector spaces which are not finitely generated, and to do so we have to make use of an axiom of set theory known variously as the *Hausdorff Maximum Principle* or *Zorn's Lemma*. To state this principle, we need another concept about partially-ordered sets. Let A be a set on which we have defined a partial order $\preccurlyeq$. A subset B of A is *bounded* if and only if there exists an element $a_0 \in A$ satisfying $b \preccurlyeq a_0$ for all $b \in B$. Note that we do not require that a_0 belong to B. The Hausdorff maximum principle then says that if A is a partially-ordered set in which every chain subset is bounded, then A has a maximal element. Again, this is not really a "principle" or a "lemma"; it is an axiom of set theory which has been shown to be independent of the other (Zermelo–Fraenkel) axioms one usually assumes. Indeed, it is logically equivalent to the Axiom of Choice, which we mentioned in Chap. 1 as being somewhat controversial among those mathematicians dealing with the foundations of mathematics. However, in this book, we will assume that it holds. Given that assumption, we can now extend Proposition 5.7.

With kind permission of the Hausdorff Research Institute for Mathematics (Hausdorff); © Jens Zorn (Zorn).

Felix Hausdorff, one of the leading mathematicians of the early twentieth century and one of the founders of topology, died in a German concentration camp in 1942. **Max Zorn**, a German mathematician who emigrated to the United States, made skillful use of the Hausdorff Maximum Principle in his research, turning it into an important mathematical tool.

Proposition 5.8 *If V is a vector space over a field F then every linearly-independent subset B of V is contained in a basis of V.*

Proof Let B be a linearly-independent subset of V and let P be the collection of all linearly-independent subsets of V which contain B, which is partially-ordered by inclusion, as usual. Then P is nonempty since $B \in P$. Let Q be a chain subset of P. We want to prove that Q is bounded in P. That is to say, we want to find a linearly independent subset E of V which contains every element of Q. Indeed, let us take E to be the union of all of the elements of Q. To show that E is linearly independent, it suffices to show that every finite subset of E is linearly independent. Indeed, let $\{v_1, \ldots, v_n\}$ be a finite subset of E. Then for each $1 \leq i \leq n$, there exists an element D_i of Q containing v_i. Since Q is a chain, there exists an index h such that $D_i \subseteq D_h$ for all $1 \leq i \leq n$ and so $v_i \in D_h$ for all $1 \leq i \leq n$. Therefore, this set is a subset of a linearly-independent set and so is linearly independent. Thus we have shown that every chain subset of P is bounded and so, by the Hausdorff maximum principle, the set P has a maximal element. In other words, there exists a maximal linearly-independent subset of V containing B, and this, as we know, is a basis of V over F. $\qquad\square$

Taking the special case of $B = \varnothing$ in Proposition 5.8, we see that every vector space has a basis. In the above proof we used the Axiom of Choice to prove this statement. In fact, one can show something considerably stronger: in the presence of the other generally-accepted axioms of set theory, the Axiom of Choice is equivalent, in the sense of formal logic, to the statement that every vector space over any field has a basis.

© A. Blass.

The above result is due to the contemporary American mathematician, **Andreas Blass**.

Example Consider the field $\mathbb{R}$ as a vector space over its subfield $\mathbb{Q}$. A basis for this space is known as a *Hamel basis*. By Proposition 5.8, we know that Hamel bases exist, but nobody has been able to come up with a method of specifically constructing one. The subset C of $\mathbb{R}$ consisting of all real numbers which can be represented in the form $\sum_{i>0} u_i 3^{-i}$, where each u_i is either 0 or 2, is called the *Cantor set*, and it can be shown to be "sparse" (in a technical sense of the word we won't go into here) in the unit interval $[0, 1]$ in $\mathbb{R}$. It is possible to show that there is a Hamel basis of $\mathbb{R}$ contained in C.

The existence of Hamel bases leads to some very interesting results, as the following shows. Indeed, let H be a Hamel basis of $\mathbb{R}$. If $r \in \mathbb{R}$ then we can write $r = \sum_{a \in H} q_a(r)a$, where $q_a(r) \in \mathbb{Q}$ and there are only finitely-many elements $a \in H$ for which $q_a(r) \neq 0$. Since such a representation is unique, we see that

$$q_a(b) = \begin{cases} 1 & \text{if } a = b, \\ 0 & \text{otherwise} \end{cases}$$

for $a, b \in H$. Moreover, if $r, s \in \mathbb{R}$ and $a \in H$ then $q_a(r+s) = q_a(r) + q_a(s)$ so, if $a \neq b$ are elements of H then for any $r \in \mathbb{R}$ we have $q_a(r+b) = q_a(r) + q_a(b) = q_a(r)$. Thus we see that the function $q_a \in \mathbb{R}^{\mathbb{R}}$ is *periodic*, with period b for any $b \in H \smallsetminus \{a\}$, and its image is contained in $\mathbb{Q}$. Moreover, if we pick two distinct elements c and d of H, we see that for each $r \in \mathbb{R}$, we have $r = f(r) + g(r)$, where $f, g \in \mathbb{R}^{\mathbb{R}}$ are defined by $f : r \mapsto q_c(r)c$ and $g : r \mapsto \sum_{a \in H \smallsetminus \{c\}} q_a(r)a$. By our previous comments, f is periodic with period d and g is periodic with period c. We conclude that the identity function in $\mathbb{R}^{\mathbb{R}}$ is the sum of two periodic functions. A somewhat more sophisticated argument along the same lines shows that any polynomial function in $\mathbb{R}^{\mathbb{R}}$ of degree n is the sum of $n+1$ periodic functions. Of course, since we cannot specify H, there is no way of finding these periodic functions explicitly.

© Professor Richard von Mises.

The twentieth-century German mathematician **Georg Hamel** was a student of Hilbert who worked primarily in function theory. In his later years, he became notorious for his pro-Nazi views and activities.

We have seen that a vector space over a field can have many bases. We want to show next that if the vector space is finitely generated, then all of these bases are finite and have the same number of elements. First, however, we must prove a preliminary result.

Proposition 5.9 *Let V be a vector space over a field F which is generated by a finite set $B = \{v_1, \dots, v_n\}$ and let D be a linearly independent set of vectors in V. Then the number of elements in D is at most n.*

Proof Suppose that D has a subset $E = \{w_1, \dots, w_{n+1}\}$ having more than n elements. Since this set must also be linearly independent, we know that none of the w_i equals 0_V. For each $1 \leq k \leq n$, set $D_k = \{w_1, \dots, w_k, v_{k+1}, \dots, v_n\}$.

Since B is a generating set for V, we can find scalars $a_1, \dots, a_n$, not all equal to 0, such that $w_1 = \sum_{i=1}^{n} a_i v_i$. In order to simplify our notation, we will renumber the elements of B if necessary so that $a_1 \neq 0$. Then $v_1 = a_1^{-1} w_1 - \sum_{i=2}^{n} a_1^{-1} a_i v_i$ and so $D \subseteq FD_1$. But $D_1 \subseteq V = FD$ and so $V = FD_1$, by Proposition 3.6. Now assume that $1 \leq k < n$ and that we have already shown that $V = FD_k$. Then there exist scalars $b_1, \dots, b_n$, not all equal to 0, such that $w_{k+1} = \sum_{i=1}^{k} b_i w_i + \sum_{i=k+1}^{n} b_i v_i$. If the scalars $b_{k+1}, \dots b_n$ are all equal to 0, then we have shown that D is linearly dependent, which is not the case. Therefore, at least one of them is nonzero and, by renumbering if necessary, we can assume that $b_{k+1} \neq 0$. Thus $v_{k+1} = b_{k+1}^{-1} w_{k+1} - \sum_{i=1}^{k} b_{k+1}^{-1} b_i w_i - \sum_{i=k+2}^{n} b_{k+1}^{-1} b_i v_i$ and so, using the above reasoning, we get $V = FD_{k+1}$. Continuing in this manner, we see that after n steps we obtain $V = FD_n = F\{w_1, \dots, w_n\}$. But then $w_{n+1} \in F\{w_1, \dots, w_n\}$ and so E is linearly dependent, contrary to our assumption. This proves that D can have at most n elements. $\square$

Proposition 5.10 *Let V be a vector space finitely generated over a field F. Then any two bases of V have the same number of elements.*

Proof By hypothesis, there exists a finite generating set for V over F having, say, n elements. If B is a basis of V then, by Proposition 5.9, we know that B has at most n elements and so, in particular, is finite. Suppose B and B' are two bases for V having h and k elements, respectively. Since B is linearly independent and B' is a generating set, we know that $h \leq k$. But, on the other hand, B' is linearly independent and B is a generating set, so $k \leq h$. Thus $h = k$. $\square$

We should remark at this point that the assertion for linearly-dependent sets corresponding to Proposition 5.10 is not true. That is to say, a finite linearly-dependent set of vectors may have two minimal linearly-dependent subsets with different numbers of elements. Indeed, there is no efficient algorithm to find such subsets of a given linearly-dependent set. Minimal linearly-dependent sets of vectors are often called *circuits* because of applications to graph theory. We should also note that Proposition 5.9 is a special case of a more general theorem: If V is a vector space (not necessarily finitely generated) over a field F then there exists a bijective function between any two bases of V. The proof of this result makes use of techniques from advanced set theory, such as transfinite induction.

If V is a vector space finitely generated over a field F then V is *finite dimensional* and the number of elements in a basis of V is called the *dimension* of V over F. If V is not finite dimensional, it is *infinite dimensional*. (In choosing this latter terminology, we are deliberately skipping over the subject of various transfinite dimensions, since the reader is not assumed to be familiar with the arithmetic of transfinite cardinals. In certain mathematical contexts, distinction between infinite dimensions—for example the distinction between spaces of countably-infinite and uncountably-infinite dimension—can be very significant. We will not, however, need it in this book.) We denote the dimension of V over F by $\dim(V)$, or by $\dim_F(V)$ when it is important to emphasize the field of scalars.

With kind permission of the Archives of the Mathematisches Forschungsinstitut Oberwolfach.

The notion of dimension was implicit in the work of Peano, but was redefined and studied in a comprehensive manner by the twentieth-century German mathematician **Hermann Weyl**.

Notice that the proof of Proposition 5.9, which is in turn critical in proving Proposition 5.10, uses the fact that every nonzero element of F has a multiplicative inverse, and this cannot be avoided. If we try to weaken the notion of a vector space by allowing scalars to be, say, only integers, it may happen that such a space would have two bases of different sizes and so we could no longer define the notion of dimension in an obvious manner. We did not use, in an unavoidable manner, the commutativity of scalar multiplication and so we could weaken our notion of a vector space to allow scalars which do not commute among themselves, such as scalars coming from $\mathbb{H}$. However, the generality thus gained does not seem to outweigh the bother it causes, and so we will refrain from doing so. Thus, for us, the fact that scalars always come from a field is critical in the development of our theory.

Example If F is a field then $\dim(F^n) = n$ for every positive integer n, since the canonical basis of F^n has n elements. Similarly, if k and n are positive integers then $\dim_F(\mathcal{M}_{k \times n}(F)) = kn$, since the canonical basis of $\mathcal{M}_{k \times n}(F)$ has kn elements. The dimension of the space $F[X]$ is infinite since the canonical basis of $F[X]$ has infinitely-many elements.

Example If F is a field and n is a positive integer, then the set W of all polynomials in $F[X]$ having degree at most n is a subspace of $F[X]$ having dimension $n + 1$, since $\{1, X, \ldots, X^n\}$ is a basis of W having $n + 1$ elements.

Example Let V be a vector space over $\mathbb{R}$, Then $Y = V^2$ is a vector space over $\mathbb{R}$, but it also has the structure of a vector space over $\mathbb{C}$ with the same addition and with scalar multiplication given by $(a + bi) \begin{bmatrix} v_1 \\ v_2 \end{bmatrix} = \begin{bmatrix} av_1 - bv_2 \\ bv_1 + av_2 \end{bmatrix}$. This space is called the *complexification* of V. If B is a basis for V over $\mathbb{R}$ then it is easy to check that $\left\{ \begin{bmatrix} v \\ 0_V \end{bmatrix} \,\middle|\, v \in B \right\}$ is a basis for Y over $\mathbb{C}$. Thus, in particular, if V is finitely generated over $\mathbb{R}$ then Y is finitely generated over $\mathbb{C}$ and $\dim_{\mathbb{R}}(V) = \dim_{\mathbb{C}}(Y)$.

With kind permission of UC Berkeley.

Complexification of real vector spaces was first used extensively by the twentieth-century American mathematician **Angus Taylor**.

Example The dimension of $\mathbb{R}$ over itself is 1. Since $\{1, i\}$ is a basis of $\mathbb{C}$ as a vector space over $\mathbb{R}$, we see that $\dim_{\mathbb{R}}(\mathbb{C}) = 2$ and so there cannot be a proper subfield F of $\mathbb{C}$ properly containing $\mathbb{R}$. Indeed, if there were such a field, its dimension over $\mathbb{R}$ would have to be greater than 1 and less than 2 (else it would be equal to $\mathbb{C}$), which is impossible. Clearly, $\dim_{\mathbb{R}}(\mathbb{H}) = 4$. It turns out that the only possible dimensions of division $\mathbb{R}$-algebras are 1, 2, 4, and 8. The dimension 8 case is realized by a (non-associative) Cayley algebra over $\mathbb{R}$, as defined in Chap. 15. There are no associative division algebras of dimension 8 over $\mathbb{R}$.

With kind permission of the Archives of the Mathematisches Forschungsinstitut Oberwolfach.

The twentieth-century German mathematician **Heinz Hopf** used algebraic topology to prove that the only possible dimensions of division R-algebras were powers of 2, and the final result was obtained by the twentieth-century American mathematician **Raoul Bott** and contemporary American mathematician **John Milnor**, again using non-algebraic tools.

Example Let F be a field, let $(K, \bullet)$ be an associative unital F-algebra, and let $v \in K$. If $p(X) = \sum_{i=0}^{\infty} a_i X^i \in F[X]$, then $p(v) = \sum_{i=0}^{\infty} a_i v^i$ is an element of K and the set of all elements of K of this form is an F-subalgebra of K, which is in fact commutative, even though K itself may not be. We will denote this algebra by $F[v]$. If the dimension of $F[v]$, considered as a vector space over F, is finite, we know that there must exist a polynomial $p(X) \in F[X]$ of positive degree satisfying $p(v) = 0_K$. In that case, we say that v is *algebraic* over F. Otherwise, if the dimension of $F[v]$

is infinite, we say v is *transcendental* over F. Thus, for example, the real numbers π and e (the base of the natural logarithms) are transcendental over $\mathbb{Q}$. If F is a subfield of a field K then the set L of all elements of K which are algebraic over F is a subfield of K. Moreover, if K is algebraically closed, so is L, and in fact L is the smallest algebraically-closed subfield of K containing F. In particular, we can consider the field of all complex numbers algebraic over $\mathbb{Q}$. This is a proper subfield of $\mathbb{C}$, known as the *field of algebraic numbers*.

With kind permission of the Archives of the Mathematisches Forschungsinstitut Oberwolfach.

The transcendence of π was proven by German mathematician **Ferdinand von Lindemann** in 1882. The transcendence of e was proven by French mathematician **Charles Hermite** in 1873. As we shall see later, Hermite made many important contributions to linear algebra.

From the definition of dimension we see that if V is a vector space of finite dimension n over a field F then:
(1) Every subset of V having more than n elements must be linearly dependent;
(2) There exists a linearly-independent subset B of V having precisely n elements;
(3) If B is as in (2) then B is also a generating set of V over F.

Proposition 5.11 *Let V be a vector space finitely generated over a field F and let W be a subspace of V. Then:*
(1) W is finitely generated over F;
(2) Every basis of W can be extended to a basis of V;
(3) $\dim(W) \leq \dim(V)$, with equality when and only when $W = V$.

Proof Let $n = \dim(V)$.

(1) If W is not finitely generated, then, as we remarked after Proposition 5.3, W has a linearly-independent subset B having $n + 1$ elements. But B is also a subset of V, contradicting the assumption that $\dim(V) = n$.

(2) Let B be a basis of W. Then B is as linearly-independent set of elements of V and so, by Proposition 5.7, can be extended to a basis of V.

(3) By (2), we see that the number of elements of a basis of W can be no greater than the number of elements of a basis of V, and so $\dim(W) \leq \dim(V)$. Moreover, if we have equality then any basis B of W is also a basis of V, and so $W = FB = V$. $\square$

We now want to extend the notion of linear independence. Let U and W be subspaces of a vector space V over a field F. Any vector $v \in U + W$ can be written in the form $u + w$, where $u \in U$ and $w \in W$, but there is no reason for this representation to be unique. It will be unique, however, if U and W are disjoint. Indeed, if this condition holds and if $u, u' \in U$ and $w, w' \in W$ satisfy $u + w = u' + w'$, then $u - u' = w' - w \in U \cap W$ and so $u - u' = 0_V = w - w'$, which in turn implies that $u = u'$ and $w = w'$. To emphasize the importance of this situation, we will introduce new notation: if U and W are disjoint subspaces of a vector space V over a field F, we will write $U \oplus W$ instead of $U + W$. The subspace $U \oplus W$ is called the *direct sum* of U and W. We note that, by this definition, $U \oplus \{0_V\} = U$ for every subspace U of V.

Example It is easy to see that $\mathbb{R}^2 = \mathbb{R}\begin{bmatrix} 1 \\ 0 \end{bmatrix} \oplus \mathbb{R}\begin{bmatrix} 0 \\ 1 \end{bmatrix}$.

Of course, we would like to extend the notion of direct sum to cover more than two subspaces. In general, if V is a vector space over a field F, then a collection $\{W_h \mid h \in \Omega\}$ of subspaces of V is *independent* if and only if it satisfies the following condition: If Λ is a finite subset of Ω and if we choose elements $w_h \in W_h$ for all $h \in \Lambda$, then $\sum_{h \in \Lambda} w_h = 0_V$ when and only when $w_h = 0_V$ for each $h \in \Lambda$. Thus we see that an infinite collection of subspaces is independent if and only if every finite nonempty subcollection is independent. Clearly, a subset D of a vector space V over a field F is linearly independent if and only if the collection of subspaces $\{Fv \mid v \in D\}$ is independent.

Proposition 5.12 *Let V be a vector space over a field F and let $W_1, \ldots, W_n$ be distinct subspaces of V. Then the following conditions are equivalent:*

(1) $\{W_1, \ldots, W_n\}$ *is independent;*

(2) *Every vector $w \in \sum_{i=1}^{n} W_i$ can be written as $w_1 + \cdots + w_n$, with $w_i \in W_i$ for each $1 \leq i \leq n$, in exactly one way;*

(3) W_h *and $\sum_{i \neq h} W_i$ are disjoint, for each $1 \leq h \leq n$.*

Proof (1) $\Rightarrow$ (2): Let $w \in \sum_{i=1}^{n} W_i$ and assume that we can write $w = w_1 + \cdots + w_n = y_1 + \cdots + y_n$, where $w_i, y_i \in W_i$ for each $1 \leq i \leq n$. Then $\sum_{i=1}^{n} (w_i - y_i) = 0_V$ and so, by (i), it follows that $w_i - y_i = 0_V$ for each $1 \leq i \leq n$, proving (2).

(2) $\Rightarrow$ (3): Assume that $0_V \neq w_h \in W_h \cap \sum_{i \neq h} W_i$. Then for each $i \neq h$ there exists an element $w_i \in W_i$ satisfying $w_h = \sum_{i \neq h} w_i$, contradicting (2).

(3) $\Rightarrow$ (1): Suppose we can write $w_1 + \cdots + w_n = 0_V$, where $w_i \in W_i$ for each $1 \leq i \leq n$, and where $w_h \neq 0_V$ for some h. Then $w_h = -\sum_{i \neq h} w_i \in W_h \cap \sum_{i \neq h} W_i$, and this contradicts (3). Thus (1) must hold. $\qquad \square$

If V is a vector space over a field F and if $\{W_i \mid i \in \Omega\}$ is an independent collection of subspaces of V, we write $\bigoplus_{i \in \Omega} W_i$ instead of $\sum_{i \in \Omega} W_i$. If $\Omega = \{1, \ldots, n\}$, we will also write this sum as $W_1 \oplus \cdots \oplus W_n$. If $V = \bigoplus_{i \in \Omega} W_i$, then we say that V has a *direct-sum decomposition* relative to the subspaces W_i.

Example If B is a basis of a vector space V over a field F then $V = \bigoplus_{v \in B} Fv$.

The importance of direct-sum decompositions is illustrated by the following result.

Proposition 5.13 *Let V be a vector space over a field F, let $\{W_i \mid i \in \Omega\}$ be a pairwise disjoint collection of subspaces of V and, for each $i \in \Omega$, let B_i be a basis of W_i. Then $V = \bigoplus_{i \in \Omega} W_i$ if and only if $B = \bigcup_{i \in \Omega} B_i$ is a basis of V.*

Proof Assume $V = \bigoplus_{i \in \Omega} W_i$ and let $v \in V$. Then there exists a finite subset Λ of Ω such that $v \in \bigoplus_{i \in \Lambda} W_i$, and so for each $i \in \Lambda$ there is an element $w_i \in W_i$ satisfying $v = \sum_{i \in \Lambda} w_i$. Moreover, each w_i is a linear combination of elements of B_i. Thus v is a linear combination of elements of B, and so B is a generating set for V. We are left to show that B is linearly independent. If this is not the case, then there exist an element h of Ω, vectors $y_1, \ldots, y_t$ in B_h, and scalars $a_1, \ldots, a_t$ in F, not all of which equal to 0, such that $\sum_{j=1}^{t} a_j v_j + u = 0_V$, where u is a linear combination of elements of $\bigcup_{i \neq h} B_i$. But then $\sum_{j=1}^{t} a_j v_j \in W_h \cap \sum_{i \neq h} W_i$, contradicting our initial assumption. Thus $B = \bigcup_{i \in \Omega} B_i$.

Conversely, if $B = \bigcup_{i \in \Omega} B_i$, it then follows that every element of V can be written in a unique way as $\sum_{i \in \Lambda} w_i$, where Λ is some finite subset of Ω, which suffices to prove that $V = \bigoplus_{i \in \Omega} W_i$. $\qquad\square$

Let W be a subspace of a vector space V over a field F. A subspace Y of V is a *complement* of W in V if and only if $V = W \oplus Y$. We immediately note that if Y is a complement of W in V then W is a complement of Y in V. In general, a subspace of a vector space can have many complements.

Example Each of the following subspaces of $\mathbb{R}^2$ is a complement of each of the others in $\mathbb{R}^2$:

$$W_1 = \left\{ \begin{bmatrix} a \\ 0 \end{bmatrix} \,\middle|\, a \in \mathbb{R} \right\}; \qquad W_2 = \left\{ \begin{bmatrix} 0 \\ b \end{bmatrix} \,\middle|\, b \in \mathbb{R} \right\};$$

$$W_3 = \left\{ \begin{bmatrix} c \\ c \end{bmatrix} \,\middle|\, c \in \mathbb{R} \right\}; \quad \text{and} \quad W_4 = \left\{ \begin{bmatrix} d \\ 2d \end{bmatrix} \,\middle|\, d \in \mathbb{R} \right\}.$$

Proposition 5.14 *Every subspace W of a vector space V over a field F has at least one complement in V.*

Proof If W is improper, then $\{0_V\}$ is a complement of W in V. Similarly, V is a complement of $\{0_V\}$ in V. Otherwise, let B be a basis of W. By Proposition 5.8, we know that there exists a linearly-independent subset D of V such that $B \cup D$ is a basis of V. Then FD is a complement of W in V. $\qquad\square$

Example Let F be a field of characteristic other than 2, let n be a positive integer, and let V be a vector space over F. Let $W = \mathcal{M}_{n \times n}(V)$, which is also a vector space over F. Let W_1 be the set of all those matrices $A = [v_{ij}]$ in W satisfying $v_{ij} = v_{ji}$ for all $1 \leq i, j \leq n$, and let W_2 be the set of all those matrices $A = [v_{ij}]$ in W satisfying $v_{ij} = -v_{ji}$ for all $1 \leq i, j \leq n$. These two subspaces are disjoint. If $A = [v_{ij}]$ is an arbitrary matrix in W, then we can write $A = B + C$, where $B = [y_{ij}]$ is the matrix defined by $y_{ij} = \frac{1}{2}(v_{ij} + v_{ji})$ for all $1 \leq i, j \leq n$, and $C = [z_{ij}]$ is the matrix defined by $z_{ij} = \frac{1}{2}(v_{ij} - v_{ji})$ for all $1 \leq i, j \leq n$. Note that $A \in W_1$ and $B \in W_2$. Thus $V = W_1 \oplus W_2$.

Example A function $f \in \mathbb{R}^{\mathbb{R}}$ is *even* if and only if $f(a) = f(-a)$ for all $a \in \mathbb{R}$; it is *odd* if and only if $f(a) = -f(-a)$ for all $a \in \mathbb{R}$. The set W of all even functions is clearly a subspace of $\mathbb{R}^{\mathbb{R}}$, as is the set Y of all odd functions, and these two subspaces are disjoint. Moreover, if $f \in \mathbb{R}^{\mathbb{R}}$ then $f = f_1 + f_2$, where the function $f_1 : x \mapsto \frac{1}{2}[f(x) + f(-x)]$ is in W and the function $f_2 : x \mapsto \frac{1}{2}[f(x) - f(-x)]$ is in Y. Thus Y is a complement of W in $\mathbb{R}^{\mathbb{R}}$.

Proposition 5.15 *Let F be a field which is not finite and let V be a vector space over F having dimension at least 2. Then every proper nontrivial subspace W of V has infinitely-many complements in V.*

Proof By Proposition 5.14, we know that W has at least one complement U in V. Choose a basis B for U. If $0_V \neq w \in W$, then by Proposition 3.2(9) and the fact that F is infinite, we know that Fw is an infinite subset of W. Thus we know that the set W is infinite. For each $w \in W$, let $Y_w = F\{u + w \mid u \in B\}$. We claim that each of these spaces is a complement of W in V. Indeed, assume that $v \in W \cap Y_w$. Then there exist elements $u_1, \ldots, u_n$ of B and scalars $c_1, \ldots, c_n$ in F satisfying $v = \sum_{i=1}^{n} c_i(u_i + w)$. But then $\sum_{i=1}^{n} c_i u_i = v - (\sum_{i=1}^{n} c_i)w \in W \cap U = \{0_V\}$ and since the set $\{u_1, \ldots, u_n\}$ is linearly independent, we see that $c_i = 0$ for all i. This shows that $v = 0_V$, and we have thus shown that W and Y_w are disjoint. If v is an arbitrary element of V, let us write $v = x + (\sum_{i=1}^{n} c_i u_i)$, where $x \in W$, the vectors $u_1, \ldots, u_n$ belong to B, and the scalars $c_1, \ldots, c_n$ belong to F. Then $v = [x - (\sum_{i=1}^{n} c_i)w] + \sum_{i=1}^{n} c_i(u_i + w) \in W + Y_w$ and thus we have shown that $V = W + Y_w$ and so Y_w is a complement of W in V.

We are left to show that all of these complements are indeed different from each other. Indeed, assume that $w \neq x$ are elements of W satisfying $Y_w = Y_x$. If $u \in B$ then there exist elements $u_1, \ldots, u_n$ of B and scalars $c_1, \ldots, c_n$ such that $u + w = \sum_{i=1}^{n} c_i(u_i + x)$. From this it follows that $u - \sum_{i=1}^{n} c_i u_i = (\sum_{i=1}^{n} c_i)x - w$ and this belongs to $W \cap Y_w = \{0_V\}$. But B is a linearly-independent set and so u has to equal to one of the u_h for some $1 \leq h \leq n$, and we must have $c_i = 0$ for $i \neq h$ and

$c_h = 1$. Hence $x - w = 0_V$, namely $x = w$. This is a contradiction, and so the Y_w must all be distinct.   □

Proposition 5.16 (Grassmann's Theorem) *Let V be a vector space over a field F and let W and Y be subspaces of V satisfying the condition that $W + Y$ has finite dimension. Then* $\dim(W + Y) = \dim(W) + \dim(Y) - \dim(W \cap Y)$.

Proof Let $U_0 = W \cap Y$, which is a subspace both of W and of Y. In particular, U_0 has a complement U_1 in W and a complement U_2 in Y. Then $W + Y = U_0 + U_1 + U_2$. We claim that in fact $W + Y = U_0 \oplus U_1 \oplus U_2$. Indeed, assume that $u_0 + u_1 + u_2 = 0_V$, where $u_j \in U_j$ for $j = 0, 1, 2$. Then $u_1 = -u_2 - u_0 \in W \cap Y = U_0$. But U_0 and U_1 are disjoint and so $u_1 = 0_V$. Therefore, $u_0 = -u_2 \in U_0 \cap U_2 = \{0_V\}$. Therefore, $u_0 = 0_V$ and $u_2 = 0_V$ as well. Thus we see that the set $\{U_0, U_1, U_2\}$ is independent. Therefore, from the definition of the complement, we have

$$\dim(W + Y) = \dim(U_0) + \dim(U_1) + \dim(U_2) = \dim(W) + \dim(U_2)$$

and this equals $\dim(W) + \dim(Y) - \dim(W \cap Y)$ since $Y = U_2 \oplus (W \cap Y)$.   □

Example Consider the subspaces

$$W_1 = \mathbb{R}\left\{ \begin{bmatrix} 1 \\ 0 \\ 2 \end{bmatrix}, \begin{bmatrix} 1 \\ 2 \\ 2 \end{bmatrix} \right\} \quad \text{and} \quad W_2 = \mathbb{R}\left\{ \begin{bmatrix} 1 \\ 1 \\ 0 \end{bmatrix}, \begin{bmatrix} 0 \\ 1 \\ 1 \end{bmatrix} \right\}$$

of $\mathbb{R}^3$. Each one of these subspaces has dimension 2, and so we see that $2 \leq \dim(W_1 + W_2) \leq 3$. By Proposition 5.16, we see that, as a result of this, we have $1 \leq \dim(W_1 \cap W_2) \leq 2$. In order to ascertain the exact dimension of $W_1 \cap W_2$, we must find a basis for it. If $v \in W_1 \cap W_2$ then there exist scalars a, b, c, d satisfying $a \begin{bmatrix} 1 \\ 0 \\ 2 \end{bmatrix} + b \begin{bmatrix} 1 \\ 2 \\ 2 \end{bmatrix} = c \begin{bmatrix} 1 \\ 1 \\ 0 \end{bmatrix} + d \begin{bmatrix} 0 \\ 1 \\ 1 \end{bmatrix}$, and so $a + b = c$, $2b = c + d$, and $2a + 2b = d$, from which we conclude that $b = -3a$, $c = -2a$, and $d = -4a$. Thus v has to be of the form $(-2a) \begin{bmatrix} 1 \\ 1 \\ 0 \end{bmatrix} + (-4a) \begin{bmatrix} 0 \\ 1 \\ 1 \end{bmatrix} = a \begin{bmatrix} -2 \\ -6 \\ -4 \end{bmatrix}$, which shows that

$$W_1 \cap W_2 = \mathbb{R} \begin{bmatrix} -2 \\ -6 \\ -4 \end{bmatrix}, \text{ and so it has dimension 1.}$$

Very often, we can reduce our computations by passing to complements. A good example of this is given by the following proposition.

> **Proposition 5.17** *Let V be a vector space over a field F and let W be a subspace of V having a complement Y in V. Let $\{v_1, \dots, v_n\}$ be a subset of V and, for each $1 \le i \le n$, let $v_i = w_i + y_i$, where $w_i \in W$ and $y_i \in Y$. If the vectors $w_1, \dots, w_n$ are distinct and the set $\{w_1, \dots, w_n\}$ is linearly independent, then so is the set $\{v_1, \dots, v_n\}$.*

Proof Assume that there exist scalars $a_1, \dots, a_n$ satisfying $\sum_{i=1}^{n} a_i v_i = 0_V$. Then $\sum_{i=1}^{n} a_i w_i + \sum_{i=1}^{n} a_i y_i = 0_V$, and so $\sum_{i=1}^{n} a_i w_i = \sum_{i=1}^{n} a_i y_i = 0_V$. Since the vectors $w_1, \dots, w_n$ are distinct and $\{w_1, \dots, w_n\}$ is linearly independent, we must have $a_1 = \cdots = a_n = 0$, and so $\{v_1, \dots, v_n\}$ is linearly independent as well. $\square$

Exercises

Exercise 158
Let v_1, v_2, and v_3 be distinct elements of a vector space V over a field F and let $c_1, c_2, c_3 \in F$. Under what conditions is the subset $\{c_2 v_3 - c_3 v_2, c_1 v_2 - c_2 v_1, c_3 v_1 - c_1 v_3\}$ of V linearly dependent?

Exercise 159
For which values of the real number t is the subset

$$\left\{ \begin{bmatrix} \cos(t) + i\sin(t) \\ 1 \end{bmatrix}, \begin{bmatrix} 1 \\ \cos(t) - i\sin(t) \end{bmatrix} \right\}$$

of $\mathbb{C}^2$ linearly dependent?

Exercise 160
Let F be a field and let V be the subspace of $F[X]$ consisting of all those polynomials of degree at most 4. Let $p_1(X), \dots, p_5(X)$ be distinct polynomials in V satisfying the condition that $p_i(0) = 1$ for each $1 \le i \le 5$. Is the set $\{p_1(X), \dots, p_5(X)\}$ necessarily linearly dependent?

Exercise 161
Consider the functions $f : x \mapsto 5^x$ and $g : x \mapsto 5^{2x}$. Is $\{f, g\}$ a linearly-dependent subset of $\mathbb{R}^{\mathbb{R}}$?

Exercise 162
Find $a, b \in \mathbb{Q}$ such that the subset $\left\{ \begin{bmatrix} 2 \\ a - b \\ 1 \end{bmatrix}, \begin{bmatrix} a \\ b \\ 3 \end{bmatrix} \right\}$ of $\mathbb{Q}^3$ is linearly dependent.

Exercise 163

Let $F = \mathbb{Q}$. Is the subset $\left\{ \begin{bmatrix} 4 \\ 2 \\ 1 \end{bmatrix}, \begin{bmatrix} 1 \\ 0 \\ 0 \end{bmatrix}, \begin{bmatrix} 1 \\ 3 \\ 4 \end{bmatrix} \right\}$ of F^3 linearly independent?
What happens if $F = \mathrm{GF}(5)$?

Exercise 164

Let V be a vector space over a field F and let $n > 1$ be an integer. Let Y be the
set of all vectors $\begin{bmatrix} v_1 \\ \vdots \\ v_n \end{bmatrix} \in V^n$ satisfying the condition that the set $\{v_1, \ldots, v_n\}$ is
linearly dependent. Is Y necessarily a subspace of V^n?

Exercise 165

Is the subset $\left\{ \begin{bmatrix} 1+i \\ 3+8i \\ 5+7i \end{bmatrix}, \begin{bmatrix} 1-i \\ 5 \\ 2+i \end{bmatrix}, \begin{bmatrix} 1+i \\ 3+2i \\ 4-i \end{bmatrix} \right\}$ of $\mathbb{C}^3$ linearly independent
when we consider $\mathbb{C}^3$ as a vector space over $\mathbb{C}$? Is it linearly independent when
we consider $\mathbb{C}^3$ as a vector space over $\mathbb{R}$?

Exercise 166

For each nonnegative integer n, let $f_n \in \mathbb{R}^{\mathbb{R}}$ be the function defined by $f_n : x \mapsto \sin^n(x)$. Is the subset $\{f_n \mid n \geq 0\}$ of $\mathbb{R}^{\mathbb{R}}$ linearly independent?

Exercise 167

Let $V = C(-1, 1)$, which is a vector space over $\mathbb{R}$. Let $f, g \in V$ be the functions
defined by $f : x \mapsto x^2$ and $g : x \mapsto |x|x$. Is $\{f, g\}$ linearly independent?

Exercise 168

Let V be a vector space over $\mathrm{GF}(5)$ and let $v_1, v_2, v_3 \in V$. Is the subset
$\{v_1 + v_2, v_1 - v_2 + v_3, 2v_2 + v_3, v_2 + v_3\}$ of V linearly independent?

Exercise 169

Let F be a field of characteristic different from 2 and let V be a vector space
over F containing a linearly-independent subset $\{v_1, v_2, v_3\}$. Show that the set
$\{v_1 + v_2, v_2 + v_3, v_1 + v_3\}$ is also linearly independent.

Exercise 170

Is the subset $\left\{ \begin{bmatrix} 1 \\ 1 \\ 2 \\ 2 \end{bmatrix}, \begin{bmatrix} 1 \\ 2 \\ 1 \\ 2 \end{bmatrix}, \begin{bmatrix} 1 \\ 1 \\ 1 \\ 2 \end{bmatrix}, \begin{bmatrix} 0 \\ 2 \\ 2 \\ 0 \end{bmatrix} \right\}$ of $\mathrm{GF}(3)^4$ linearly independent?

Exercise 171

Let $t \leq n$ be positive integers and, for all $1 \leq i \leq t$, let $v_i = \begin{bmatrix} a_{i1} \\ \vdots \\ a_{in} \end{bmatrix}$ be a vector in $\mathbb{R}^n$ chosen so that $2|a_{jj}| > \sum_{i=1}^{t} |a_{ij}|$ for all $1 \leq j \leq n$. Show that $\{v_1, \ldots, v_t\}$ is linearly independent.

Exercise 172

If $\{v_1, v_2, v_3, v_4\}$ is a linearly-independent subset of a vector space V over the field $\mathbb{Q}$, is the set

$$\{3v_1 + 2v_2 + v_3 + v_4, 2v_1 + 5v_2, 3v_3 + 2v_4, 3v_1 + 4v_2 + 2v_3 + 3v_4\}$$

linearly independent as well?

Exercise 173

Let A be a subset of $\mathbb{R}$ having at least three elements and let $f_1, f_2, f_3 \in \mathbb{R}^A$ be the functions defined by $f_i : x \mapsto x^{i-1}2^{x-1}$. Is the set $\{f_1, f_2, f_3\}$ linearly independent?

Exercise 174

Let $F = GF(5)$ and let $V = F^F$, which is a vector space over F. Let $f : x \mapsto x^2$ and $g : x \mapsto x^3$ be elements of V. Find an element h of V such that $\{f, g, h\}$ is linearly independent.

Exercise 175

Consider $\mathbb{R}$ as a vector space over $\mathbb{Q}$. Is the subset $\{(a - \pi)^{-1} \mid a \in \mathbb{Q}\}$ of this space linearly independent?

Exercise 176

In the vector space $V = \mathbb{R}^{\mathbb{R}}$ over $\mathbb{R}$, consider the functions $f_1 : x \mapsto \ln((x^2 + 1)^3(x^4 + 7)^{-1})$, $f_2 : x \mapsto \ln(\sqrt{x^2 + 1})$, and $f_3 : x \mapsto \ln(x^4 + 7)$. Is the subset $\{f_1, f_2, f_3\}$ of V linearly independent?

Exercise 177

Show that the subset $\left\{ \begin{bmatrix} 1 \\ 2 \\ 0 \end{bmatrix}, \begin{bmatrix} 0 \\ 1 \\ 2 \end{bmatrix}, \begin{bmatrix} 2 \\ 0 \\ 1 \end{bmatrix} \right\}$ of $GF(p)^3$ is linearly independent if and only if $p \neq 3$.

Exercise 178

Let F be a subfield of a field K and let n be a positive integer. Show that a nonempty linearly-independent subset D of F^n remains linearly independent when considered as a subset of K^n.

Exercise 179
Let $F = \mathrm{GF}(5)$ and let $V = F^F$. For $4 \le k \le 7$, let $f_k \in V$ be defined by $f_k :$ $a \mapsto a^k$. Is the subset $\{f_k \mid 4 \le k \le 7\}$ of V linearly independent?

Exercise 180
Let V be a vector space over $\mathbb{R}$. For vectors $v \ne w$ in V, let $K(v, w)$ be the set of all vectors in V of the form $(1 - a)v + aw$, where $0 \le a \le 1$. Given vectors $v, w, y \in V$ satisfying the condition that the set $\{w - v, y - v\}$ is linearly independent (and so, in particular, its elements are distinct), show that the set

$$K\left(v, \frac{1}{2}(w + y)\right) \cap K\left(w, \frac{1}{2}(v + y)\right) \cap K\left(y, \frac{1}{2}(v + w)\right)$$

is nonempty, and determine how many elements it can have.

Exercise 181
Let V be a vector space finitely generated over a field F and let $B = \{v_1, \ldots, v_n\}$ be a basis for V. Let $y \in V \smallsetminus B$. Show that the set $\{v_1, \ldots, v_n, y\}$ has a unique minimal linearly-dependent subset.

Exercise 182
Find all of the minimal linearly-dependent subsets of the subset

$$\left\{ \begin{bmatrix} 1 \\ 0 \end{bmatrix}, \begin{bmatrix} 0 \\ 4 \end{bmatrix}, \begin{bmatrix} 0 \\ 0 \end{bmatrix}, \begin{bmatrix} 2 \\ 0 \end{bmatrix}, \begin{bmatrix} 2 \\ 2 \end{bmatrix} \right\}$$

of $\mathbb{Q}^2$.

Exercise 183
Let V be a vector space over a field F and let D and D' be distinct finite minimal linearly-dependent subsets of V which are not disjoint. If $v \in D \cap D'$, show that $(D \cup D') \smallsetminus \{v\}$ is linearly dependent.

Exercise 184
Let F be a field of characteristic other than 2. Let V be the subspace of $F[X]$ consisting of all polynomials of degree at most 3. Is $\{X + 2, X^2 + 1, X^3 + X^2, X^3 - X^2\}$ a basis of V?

Exercise 185
Let $\{v_1, \ldots, v_n\}$ be a basis for a vector space V over a field F. Is the set $\{v_1 + v_2, v_2 + v_3, \ldots, v_{n-1} + v_n, v_n + v_1\}$ necessarily also a basis for V over F?

Exercise 186
Is $\{1 + 2\sqrt{5}, -3 + \sqrt{5}\}$ a basis for $\mathbb{Q}(\sqrt{5})$ as a vector space over $\mathbb{Q}$?

Exercise 187
For which values of $a \in \mathbb{R}$ is the set

$$\left\{ \begin{bmatrix} a & 2a \\ 2 & 3a \end{bmatrix}, \begin{bmatrix} 1 & 2 \\ 2a & 3 \end{bmatrix}, \begin{bmatrix} 1 & 2a \\ a+1 & a+2 \end{bmatrix}, \begin{bmatrix} 1 & a+1 \\ 2 & 2a+1 \end{bmatrix} \right\}$$

a basis for $\mathcal{M}_{2\times2}(\mathbb{R})$ as a vector space over $\mathbb{R}$?

Exercise 188
Let F be an algebraically-closed field and let $(K, \bullet)$ be an associative F-algebra having a basis $\{v_1, v_2\}$ as a vector space over F. Show that $v_2^2 = v_2$ or $v_2^2 = 0_K$.

Exercise 189
Let V be a vector space over a field F. A nonempty subset U of V is *nearly linearly independent* if and only if U is linearly dependent but $U \smallsetminus \{u\}$ is linearly independent for every $u \in U$. Find an example of a set of three vectors in $\mathbb{R}^3$ which is nearly linearly independent. Does there exist a nearly linearly independent subset of $\mathbb{R}^3$ having four elements?

Exercise 190
Find a basis for the subspace W of $\mathbb{R}^4$ generated by

$$\left\{ \begin{bmatrix} 4 \\ 2 \\ 6 \\ -2 \end{bmatrix}, \begin{bmatrix} 1 \\ -1 \\ 3 \\ -1 \end{bmatrix}, \begin{bmatrix} 1 \\ 2 \\ 0 \\ 0 \end{bmatrix}, \begin{bmatrix} 1 \\ 5 \\ -3 \\ 1 \end{bmatrix} \right\}.$$

Exercise 191
For each real number a, let $f_a \in \mathbb{R}^\mathbb{R}$ be defined by

$$f_a : r \mapsto \begin{cases} 1 & \text{if } r = a, \\ 0 & \text{otherwise.} \end{cases}$$

Is $\{f_a \mid a \in \mathbb{R}\}$ a basis for $\mathbb{R}^\mathbb{R}$ over $\mathbb{R}$?

Exercise 192
Let A be a nonempty finite set and let V be the collection of all subsets of A, which is a vector space over $GF(2)$. For each $a \in A$, let $v_a = \{a\}$. Is $\{v_a \mid a \in A\}$ a basis for V?

Exercise 193
Show that $\left\{ \begin{bmatrix} 1 & 0 \\ 0 & 1 \end{bmatrix}, \begin{bmatrix} 0 & 1 \\ 1 & 0 \end{bmatrix}, \begin{bmatrix} 0 & -i \\ i & 0 \end{bmatrix}, \begin{bmatrix} 1 & 0 \\ 0 & -1 \end{bmatrix} \right\}$ is a basis for the vector space $\mathcal{M}_{2\times2}(\mathbb{C})$ over $\mathbb{C}$. (The last three of these matrices are known as the *Pauli matrices* and play a very important part in the formulation of quantum physics.

Exercise 194
Let F be a field and let $a, b, c \in F$. Determine whether

$$\left\{ \begin{bmatrix} 1 \\ a \\ b \end{bmatrix}, \begin{bmatrix} 0 \\ 1 \\ c \end{bmatrix}, \begin{bmatrix} 0 \\ 0 \\ a+b+c \end{bmatrix} \right\}$$

is a basis for F^3.

Exercise 195
Let V be a vector space finitely generated over a field F having a basis $\{v_1, \ldots, v_n\}$. Is $\{v_1, \sum_{i=1}^{2} v_i, \ldots, \sum_{i=1}^{n} v_i\}$ necessarily a basis for V?

Exercise 196
Let $F = \mathrm{GF}(p)$ for some prime integer p, let n be a positive integer, and let V be a vector space of dimension n over F. In how many ways can we choose a basis for V?

Exercise 197
Let V be a three-dimensional vector space over a field F, with basis $\{v_1, v_2, v_3\}$. Is $\{v_1 + v_2, v_2 + v_3, v_1 - v_3\}$ a basis for V?

Exercise 198
Let V be a vector space of finite dimension n over $\mathbb{C}$ having basis $\{v_1, \ldots, v_n\}$. Show that $\{v_1, \ldots, v_n, iv_1, \ldots, iv_n\}$ is a basis for V, considered as a vector space over $\mathbb{R}$.

Exercise 199
Let V be a vector space of finite dimension $n > 0$ over $\mathbb{R}$ and, for each positive integer i, let U_i be a proper subspace of V. Show that $V \neq \bigcup_{i=1}^{\infty} U_i$.

Exercise 200
Let V be a vector space over a field F which is not finite dimensional, and let W be a proper subspace of V. Show that there exists an infinite collection $\{Y_1, Y_2, \ldots\}$ of subspaces of V satisfying $\bigcap_{i=1}^{\infty} Y_i \subseteq W$ but $\bigcap_{i=1}^{n} Y_i \nsubseteq W$ for all $n \geq 1$.

Exercise 201
Let V be the subspace of $\mathbb{R}[X]$ consisting of all polynomials of degree at most 5, and let $A = \{X^5 + X^4, X^5 - 7X^3, X^5 - 1, X^5 + 3X\}$. Show that this subset of V is linearly independent and extend it to a basis of V.

Exercise 202
Let V be a vector space of finite dimension n over a field F, and let W be a subspace of V of dimension $n - 1$. If U is a subspace of V not contained in W, show that $\dim(W \cap U) = \dim(U) - 1$.

Exercise 203
Let a, b, c, d be rational numbers such that $\{a + c\sqrt{3}, b + d\sqrt{3}\}$ is a basis for $\mathbb{Q}(\sqrt{3})$ as a vector space over $\mathbb{Q}$. Is $\{c + a\sqrt{3}, d + b\sqrt{3}\}$ a basis for $\mathbb{Q}(\sqrt{3})$ as a vector space over $\mathbb{Q}$? Is $\{a + c\sqrt{5}, b + d\sqrt{5}\}$ a basis for $\mathbb{Q}(\sqrt{5})$ as a vector space over $\mathbb{Q}$?

Exercise 204
Find a real number a such that

$$\dim\left(\mathbb{R}\left\{\begin{bmatrix} -9 \\ a \\ -1 \\ -5 \\ -14 \end{bmatrix}, \begin{bmatrix} 2 \\ -5 \\ 3 \\ 0 \\ 2 \end{bmatrix}, \begin{bmatrix} 1 \\ 4 \\ -1 \\ 1 \\ 2 \end{bmatrix}, \begin{bmatrix} 3 \\ -1 \\ 2 \\ 1 \\ 4 \end{bmatrix}, \begin{bmatrix} -1 \\ 9 \\ -4 \\ 1 \\ 0 \end{bmatrix}\right\}\right) = 2.$$

Exercise 205
Let $W = \mathbb{R}\left\{\begin{bmatrix} 2 \\ 1 \\ 3 \\ 1 \end{bmatrix}, \begin{bmatrix} 1 \\ 2 \\ 0 \\ 1 \end{bmatrix}, \begin{bmatrix} -1 \\ 1 \\ -3 \\ 0 \end{bmatrix}\right\} \subseteq \mathbb{R}^4$. Determine the dimension of W and find a basis for it.

Exercise 206
Consider the vectors

$$v_1 = \begin{bmatrix} 0 \\ 1 \\ 0 \\ 1 \\ 0 \end{bmatrix}, \quad v_2 = \begin{bmatrix} 7 \\ 4 \\ 1 \\ 8 \\ 3 \end{bmatrix}, \quad v_3 = \begin{bmatrix} 0 \\ 3 \\ 0 \\ 4 \\ 0 \end{bmatrix}, \quad v_4 = \begin{bmatrix} 1 \\ 9 \\ 5 \\ 7 \\ 1 \end{bmatrix}, \quad \text{and} \quad v_5 = \begin{bmatrix} 0 \\ 1 \\ 0 \\ 5 \\ 0 \end{bmatrix}$$

in the vector space $\mathbb{Q}^5$. Do there exist rational numbers a_{ij}, for $1 \le i, j \le 5$, such that the subset $\{\sum_{j=1}^5 a_{ij} v_j \mid 1 \le i \le 5\}$ of $\mathbb{Q}^5$ is linearly independent?

Exercise 207
Let F be a subfield of a field K satisfying the condition that K is finitely generated as a vector space over F. For each $c \in K$, show that there exists a nonzero polynomial $p(X) \in F[X]$ satisfying $p(c) = 0$.

Exercise 208
Let $W = \mathbb{R}\left\{\begin{bmatrix} 1 \\ 2 \\ 1 \\ 0 \end{bmatrix}, \begin{bmatrix} -1 \\ 1 \\ 1 \\ 1 \end{bmatrix}\right\} \subseteq \mathbb{R}^4$ and let $V = \mathbb{R}\left\{\begin{bmatrix} 2 \\ -1 \\ 0 \\ 1 \end{bmatrix}, \begin{bmatrix} -5 \\ 6 \\ 3 \\ 0 \end{bmatrix}\right\}$. Compute $\dim(W + V)$ and $\dim(W \cap V)$.

Exercise 209
Let F be a subfield of a field K satisfying the condition that the dimension of K as a vector space over F is finite and equal to r. Let V be a vector space of finite dimension $n > 0$ over K. Find the dimension of V as a vector space over F.

Exercise 210
Let V be a vector space over a field F having infinite dimension over F. Show that there exists an infinite sequence $W_1, W_2, \ldots$ of proper subspaces of V, satisfying $\bigcup_{i=1}^{\infty} W_i = V$.

Exercise 211
Let $F = \mathrm{GF}(p)$, where p is a prime integer, and let V be a vector space over F having finite dimension n. How many subspaces of dimension 1 does V have?

Exercise 212
Let W be the subset of $\mathbb{R}^{\mathbb{R}}$ consisting of all functions of the form $x \mapsto a \cdot \cos(x - b)$, for real numbers a and b. Show that W is a subspace of $\mathbb{R}^{\mathbb{R}}$ and find its dimension.

Exercise 213
$$\text{Let } W = \mathbb{R} \left\{ \begin{bmatrix} 4 \\ 3 \\ 2 \\ 1 \end{bmatrix}, \begin{bmatrix} 6 \\ 2 \\ 2 \\ 2 \end{bmatrix}, \begin{bmatrix} 1 \\ 1 \\ 1 \\ 2 \end{bmatrix} \right\} \subseteq \mathbb{R}^4 \text{ and let } Y = \mathbb{R} \left\{ \begin{bmatrix} 4 \\ -2 \\ 0 \\ -2 \end{bmatrix}, \begin{bmatrix} 1 \\ 0 \\ 3 \\ 2 \end{bmatrix} \right\} \subseteq \mathbb{R}^4.$$
Find $\dim(W + Y)$ and $\dim(W \cap Y)$.

Exercise 214
Let V be a vector space of finite dimension n over a field F and let W and Y be distinct subspaces of V, each of dimension $n - 1$. What is $\dim(W \cap Y)$?

Exercise 215
Let V be a finite-dimensional vector space over a field F and let B be a basis of V such that $\left\{ \begin{bmatrix} w \\ w \end{bmatrix} \;\middle|\; w \in B \right\}$ is a basis for V^2. What is the dimension of V?

Exercise 216
Let F be a field and let V be the subspace of $F[X]$ consisting of all polynomials of degree at most 4. Find a complement for V in $F[X]$.

Exercise 217
Let F be a field and let V be the subspace of $F[X]$ consisting of all polynomials of the form $(X^3 + X + 1)p(X)$ for some $p(X) \in F[X]$. Find a complement for V in $F[X]$.

Exercise 218

Let B be a nonempty proper subset of a set A. Let F be a field and let $V = F^A$. Let W be the subspace of V consisting of all those functions $f \in V$ satisfying $f(b) = 0$ for all $b \in B$. Find a complement of W in V.

Exercise 219

Let F be a field of characteristic other than 2, let V be a vector space over F, and

let $U = \left\{ \left[\begin{array}{c} v \\ v' \\ v+v' \end{array} \right] \;\middle|\; v, v' \in V \right\} \subseteq V^3$. Is $Y = \left\{ \left[\begin{array}{c} v \\ v \\ v \end{array} \right] \;\middle|\; v \in V \right\}$ a complement

of U in V?

Exercise 220

Let F be a field and let $p(X) \in F[X]$ have positive degree k. Let W be the subspace of $F[X]$ composed of all polynomials of the form $p(X)g(X)$ for some $g(X) \in F[X]$. Show that W has a complement in $F[X]$ of dimension k.

Exercise 221

Let V be a vector space over a field F which is not finite dimensional, and let $V \supset W_1 \supset W_2 \supset \cdots$ be a chain of subspaces of V, each properly contained in the one before it. Is the subspace $\bigcap_{i=1}^{\infty} W_i$ of V necessarily finite-dimensional?

Exercise 222

Let V be a vector space finitely generated over a field F. Let W and Y be subspaces of V and assume that there is a function $f \in F^V$ satisfying the condition that $f(w) < f(y)$ for all $0_V \neq w \in W$ and $0_V \neq y \in Y$. Show that $\dim(W) + \dim(Y) \leq \dim(V)$.

Exercise 223

Let $(K, \bullet)$ be a division algebra of dimension 2 over $\mathbb{R}$ containing an element v_1 which satisfies the condition that $v_1 \bullet v = v = v \bullet v_1$ for all $v \in V$. Show that $(K, +, \bullet)$ is a field.

Exercise 224

For each $a \in \mathbb{R}$, the set $\mathbb{Q}[a] = \{p(a) \mid p(X) \in \mathbb{Q}[X]\}$ is a subspace of $\mathbb{R}$, considered as a vector space over $\mathbb{Q}$. Find all pairs (a, b) of real numbers $a \neq b$ satisfying the condition that the set $\{\mathbb{Q}[a], \mathbb{Q}[b]\}$ is independent over $\mathbb{Q}$.

Exercise 225

Let V be a vector space over a field F. Find a necessary and sufficient condition for there to exist subspaces W and W' of V such that $\{\{0_V\}, W, W'\}$ is independent.

Exercise 226

Let $(K, \bullet)$ be a unital $\mathbb{R}$-algebra (not necessarily associative) with multiplicative identity e, and let $\{v_i \mid i \in \Omega\}$ be a basis for K over $\mathbb{R}$ containing e (which is equal to v_t for some $t \in \Omega$). If $v = \sum_{i \in \Omega} c_i v_i \in K$, set $\overline{v} = c_t v_t - \sum_{i \neq t} c_i v_i$. If $v \in K$, is it true that $v \bullet \overline{v} = \overline{v} \bullet v$ and $\overline{\overline{v}} = v$? (Note that this construction generalizes the notion of the conjugate of a complex number.)

Exercise 227
For each nonnegative integer n, define the subsets P_n, A_n, and F_n of $\mathbb{R}$ as follows:
(1) $P_0 = \varnothing$, $A_0 = \{1\}$, and $F_0 = \mathbb{Q}$;
(2) If $n > 0$, then P_n is the set of the first n prime integers, A_n consists of 1 and the set of square roots of products of distinct elements of P_n, and $F_n = \mathbb{Q}A_n$.
Show that each A_n is a linearly-independent subset of $\mathbb{R}$, considered as a vector space over $\mathbb{Q}$, and that F_n is a subfield of $\mathbb{R}$, having the property that every element of F_n the square of which belongs to $\mathbb{Q}$ must belong to $\mathbb{Q}a$, for some $a \in A_n$.

Exercise 228
Find all $a \in \mathbb{R}$ (if any exist) satisfying the condition that the dimension of
$\mathbb{R} \left\{ \begin{bmatrix} -1 \\ 2a \\ -2 \end{bmatrix}, \begin{bmatrix} 2 \\ 2 \\ -1 \end{bmatrix}, \begin{bmatrix} 1 \\ 1 \\ 0 \end{bmatrix} \right\}$ is at most 2.

Exercise 229
Give an example of a vector space V finitely generated over a field F, together with nonempty subsets B_1, B_2, and B_3 of V satisfying the following conditions:
(1) Each B_i is linearly independent;
(2) For each $1 \leq i \neq j \leq 3$ there exists a basis of V containing $B_i \cup B_j$;
(3) There is no basis of V containing $B_1 \cup B_2 \cup B_3$.

Exercise 230
Let V be a vector space over $\mathbb{R}$. A *fuzzification* of V is a function μ from V to the unit interval $\mathbb{I}$ of real numbers, satisfying the condition that $\mu(av + bw) \geq \min\{\mu(v), \mu(w)\}$ for all $a, b \in \mathbb{R}$ and all $v, w \in V$. A finite nonempty linearly-independent subset $\{v_1, \ldots, v_n\}$ of V is μ-*linearly independent* if and only if it satisfies the additional condition that $\mu(\sum_{i=1}^n a_i v_i) = \min\{a_1 v_1, \ldots, a_n v_n\}$.
(1) Show that the function $\mu : \mathbb{R}^2 \rightarrow \mathbb{I}$ defined by

$$\mu : \begin{bmatrix} a \\ b \end{bmatrix} \mapsto \begin{cases} 1 & \text{if } a = b = 0, \\ \frac{1}{2} & \text{if } a = 0 \text{ and } b \neq 0, \\ \frac{1}{4} & \text{otherwise} \end{cases}$$

is a fuzzification on V.
(2) Is the linearly-independent subset $\left\{ \begin{bmatrix} 1 \\ 0 \end{bmatrix}, \begin{bmatrix} -1 \\ 1 \end{bmatrix} \right\}$ of $\mathbb{R}$ also μ-linearly independent?

Exercise 231

Let V be a vector space over a field F and let B be a fixed basis of V. We then know that each element $v \in V$ can be written in a unique way as $v = \sum_{w \in B} c_w w$, where the c_w are scalars, only finitely-many of which are nonzero. Let $n(v)$ be the number of nonzero scalars c_w in this representation. (Note that $n(v) = 0$ if and only if $v = 0_V$.) Define a relation $\preceq$ on V by setting $v_1 \preceq v_2$ if and only if $n(v_1) \leq n(v_2)$. Is this a partial order relation on V?

Exercise 232

Let V be a vector space over a field F and let D be a finite minimal linearly-dependent subset of V. Find $\dim(FD)$.

Linear Transformations

Let V and W be vector spaces over a field F. A function $\alpha : V \to W$ is a *linear transformation* or *homomorphism* if and only if for all $v_1, v_2 \in V$ and $a \in F$ we have $\alpha(v_1 + v_2) = \alpha(v_1) + \alpha(v_2)$ and $\alpha(av_1) = a\alpha(v_1)$. We note that, as a consequence of the second condition, we have $\alpha(0_V) = \alpha(00_V) = 0\alpha(0_V) = 0_W$. If $(K, \bullet)$ and $(L, *)$ are F-algebras, then a linear transformation $\alpha : K \to L$ is a *homomorphism of F-algebras* if it is a linear transformation and, in addition, satisfies $\alpha(v_1 \bullet v_2) = \alpha(v_1) * \alpha(v_2)$ for all $v_1, v_2 \in K$. If both K and L are unital, then it is a *homomorphism of unital F-algebras* if it also sends the identity element of K for $\bullet$ to the identity element of L for $*$.

With kind permission of the Archives of the Mathematisches Forschungsinstitut Oberwolfach.

Linear transformations between finite-dimensional vector spaces were studied by Peano. Linear transformations between infinite-dimensional spaces were first considered in the late nineteenth century by Italian mathematician **Salvadore Pincherle**.

Example Let V be a vector space over a field F. Every scalar $c \in F$ defines a linear transformation $\sigma_c : V \to V$ given by $\sigma_c : v \mapsto cv$. In particular, σ_1 is the identity function $v \mapsto v$ and σ_0 is the 0-function $v \mapsto 0_V$.

Example Let F be a field and let $a_1, \ldots, a_6$ be scalars in F. The function $\alpha : F^2 \to F^3$ defined by $\alpha : \begin{bmatrix} c_1 \\ c_2 \end{bmatrix} \mapsto \begin{bmatrix} a_1c_1 + a_2c_2 \\ a_3c_1 + a_4c_2 \\ a_5c_1 + a_6c_2 \end{bmatrix}$ is a linear transformation.

The previous example can be generalized in an extremely significant manner. Let k and n be positive integers and let F be a field. Every matrix $A = [a_{ij}] \in \mathcal{M}_{k \times n}(F)$

J.S. Golan, *The Linear Algebra a Beginning Graduate Student Ought to Know*,
DOI 10.1007/978-94-007-2636-9_6, © Springer Science+Business Media B.V. 2012

defines a linear transformation from F^n to F^k given by

$$\begin{bmatrix} c_1 \\ c_2 \\ \vdots \\ c_n \end{bmatrix} \mapsto \begin{bmatrix} a_{11}c_1 + \cdots + a_{1n}c_n \\ a_{21}c_1 + \cdots + a_{2n}c_n \\ \vdots \\ a_{k1}c_1 + \cdots + a_{kn}c_n \end{bmatrix}.$$

In what follows, we will show that every linear transformation from F^n to F^k can be defined in this manner.

Example Let F be a field of characteristic 0. Then there are linear transformations α and β from $F[X]$ to itself defined by

$$\alpha : \sum_{i=0}^{\infty} a_i X^i \mapsto \sum_{i=0}^{\infty} i a_i X^i \quad \text{and} \quad \beta : \sum_{i=0}^{\infty} a_i X^i \mapsto \sum_{i=0}^{\infty} (1+i)^{-1} a_i X^{i+1}.$$

(By $1+i$, we mean the sum of $1+i$ copies of the identity element for multiplication of F; since the characteristic of F is 0, we know that this element is nonzero, and so is a unit in F.)

Example Let V and W be vector spaces over a field F and let k and n be positive integers. For all $1 \le i \le k$ and $1 \le j \le n$, let $\alpha_{ij} : V \to W$ be a linear transformation. Then there is a linear transformation from $\mathcal{M}_{k \times n}(V)$ to $\mathcal{M}_{k \times n}(W)$ defined by

$$\begin{bmatrix} v_{11} & \cdots & v_{1n} \\ \vdots & \ddots & \vdots \\ v_{k1} & \cdots & v_{kn} \end{bmatrix} \mapsto \begin{bmatrix} \alpha_{11}(v_{11}) & \cdots & \alpha_{1n}(v_{1n}) \\ \vdots & \ddots & \vdots \\ \alpha_{k1}(v_{k1}) & \cdots & \alpha_{kn}(v_{kn}) \end{bmatrix}.$$

Example Let V be the subspace over $\mathbb{R}^{\mathbb{R}}$ consisting of all differentiable functions. For each $f \in V$, we define a function $Df : \mathbb{R} \times \mathbb{R} \to \mathbb{R}$, called the *differential* of f, by setting $Df : (a, b) \mapsto f'(a)b$, where f' is the derivative of f. Then the function $D : V \to \mathbb{R}^{\mathbb{R} \times \mathbb{R}}$ given by $f \mapsto Df$ is a linear transformation. Such linear transformations play an important part in differential geometry.

Example Sometimes linear transformations between F-algebras which are not homomorphisms of F-algebras play an important role. Let $(K, \bullet)$ be an associative algebra over a field F and let $c \in F$. Then K is a *Baxter algebra* over F of *weight* c if and only if there exists a linear transformation $\alpha : K \to K$ satisfying the condition that $\alpha(x) \bullet \alpha(y) = \alpha(\alpha(x) \bullet y) + \alpha(x \bullet \alpha(y)) + c\alpha(x \bullet y)$ for all $x, y \in K$. Thus, for example, if K is the $\mathbb{R}$-algebra of all continuous functions from $\mathbb{R}$ to itself, the linear transformation $\alpha : K \to K$ given by $\alpha(f) : t \mapsto \int_0^t f(s) \, ds$ defines on K the structure of a Baxter algebra of weight 0. If F is any field and if $K = F^{\mathbb{P}}$ with componentwise addition and multiplication, then the function $\alpha : K \to K$ given by $\alpha : [a_1, a_2, \ldots] \mapsto [a_1, a_1 + a_2, a_1 + a_2 + a_3, \ldots]$ defines on K the structure of a Baxter algebra of weight 1.

Example Linear transformations are considered nice from an algebraic point of view, but may be less so from an analytic point of view. Let B be a Hamel basis of $\mathbb{R}$ over $\mathbb{Q}$. Then for each real number r there exists a unique finite subset $\{u_1(r), \ldots, u_{n(r)}(r)\}$ of B and scalars $a_1(r), \ldots, a_{n(r)}(r)$ in $\mathbb{Q}$ satisfying $r = \sum_{j=1}^{n(r)} a_j(r) u_j(r)$. The function from $\mathbb{R}$ to $\mathbb{R}$ defined by $r \mapsto \sum_{j=1}^{n(r)} a_j(r)$ is a linear transformation, but is not continuous at any $r \in \mathbb{R}$.

Let V and W be vector spaces over a field F. To any function $f : V \to W$ we can associate the subset $\mathrm{gr}(f) = \left\{ \begin{bmatrix} v \\ f(v) \end{bmatrix} \,\middle|\, v \in V \right\}$ of $V \times W$, called the *graph* of f. We can use the notion of graph to characterize linear transformations in terms of subspaces.

Proposition 6.1 *Let V and W be vector spaces over a field F and let $\alpha : V \to W$ be a function. Then α is a linear transformation if and only if $\mathrm{gr}(\alpha)$ is a subspace of $V \times W$.*

Proof Assume that α is a linear transformation. If $v, v' \in V$ and $c \in F$ then in $V \times W$ we have $\begin{bmatrix} v \\ \alpha(v) \end{bmatrix} + \begin{bmatrix} v' \\ \alpha(v') \end{bmatrix} = \begin{bmatrix} v + v' \\ \alpha(v) + \alpha(v') \end{bmatrix} = \begin{bmatrix} v + v' \\ \alpha(v + v') \end{bmatrix} \in \mathrm{gr}(\alpha)$ and $c \begin{bmatrix} v \\ \alpha(v) \end{bmatrix} = \begin{bmatrix} cv \\ c\alpha(v) \end{bmatrix} = \begin{bmatrix} cv \\ \alpha(cv) \end{bmatrix} \in \mathrm{gr}(\alpha)$, showing that $\mathrm{gr}(\alpha)$ is closed under taking sums and scalar multiples, and so is a subspace of $V \times W$. Conversely, if it is such a subspace then for $v, v' \in V$ and $c \in F$ we note that $\begin{bmatrix} v \\ \alpha(v) \end{bmatrix} + \begin{bmatrix} v' \\ \alpha(v') \end{bmatrix} = \begin{bmatrix} v + v' \\ \alpha(v) + \alpha(v') \end{bmatrix} \in \mathrm{gr}(\alpha)$, and so we must have $\alpha(v) + \alpha(v') = \alpha(v + v')$. Similarly, $c \begin{bmatrix} v \\ \alpha(v) \end{bmatrix} = \begin{bmatrix} cv \\ c\alpha(v) \end{bmatrix} \in \mathrm{gr}(\alpha)$, and so we must have $c\alpha(v) = \alpha(cv)$. Thus α is a linear transformation. $\square$

Let V and W be vector spaces over a field F. If α and β are linear transformations from V to W, they are, in particular, functions in W^V, and so the function $\alpha + \beta : V \to W$ is defined by $\alpha + \beta : v \mapsto \alpha(v) + \beta(v)$ for all $v \in V$. For all $v, v' \in V$ and all $c \in F$, we have

$$(\alpha + \beta)(v + v') = \alpha(v + v') + \beta(v + v')$$
$$= \alpha(v) + \alpha(v') + \beta(v) + \beta(v')$$
$$= (\alpha + \beta)(v) + (\alpha + \beta)(v')$$

and $(\alpha + \beta)(cv) = \alpha(cv) + \beta(cv) = c\alpha(v) + c\beta(v) = c[\alpha(v) + \beta(v)] = c(\alpha + \beta)(v)$.

Thus we see that $\alpha + \beta$ is a linear transformation from V to W. If $c \in F$ is a scalar then the function $c\alpha$ from V to W is defined by $c\alpha : v \mapsto c\alpha(v)$ and this, again, is a linear transformation from V to W. It is easy to check that the set of all linear transformations from V to W is a subspace of W^V, which we will denote by $\mathrm{Hom}(V, W)$, or $\mathrm{Hom}_F(V, W)$ in case the field needs to be emphasized.

Example Let V and W be vector spaces over $\mathbb{R}$ and let having complexifications U and Y, respectively. If $\alpha \in \mathrm{Hom}_{\mathbb{R}}(V, W)$ then the function $\begin{bmatrix} v_1 \\ v_2 \end{bmatrix} \mapsto \begin{bmatrix} \alpha(v_1) \\ \alpha(v_2) \end{bmatrix}$ belongs to $\mathrm{Hom}_{\mathbb{C}}(U, Y)$.

Since $\mathrm{Hom}(V, W)$ is a vector space, we can apply concepts we have already considered for vectors to linear transformations. For example, we can talk about a linearly-dependent or linearly-independent set of linear transformations from a vector space V over a field F to a vector space W over F. However, we must be very careful to remember that when we are doing so, we are working in the space $\mathrm{Hom}(V, W)$, and not in either V or W. The following example illustrates the pitfalls one can encounter.

Example Let V and W be vector spaces over the same field F. A nonempty subset $D = \{\alpha_1, \ldots, \alpha_n\}$ of $\mathrm{Hom}(V, W)$ is *locally linearly dependent* if and only if the subset $\{\alpha_1(v), \ldots, \alpha_n(v)\}$ of W is linearly dependent for every $v \in V$. If D is a linearly-dependent subset of $\mathrm{Hom}(V, W)$, then there exist scalars $c_1, \ldots, c_n$, not all of which are equal to 0, such that $\sum_{i=1}^n c_i \alpha_i$ is the 0-function. In particular, for each $v \in V$ we see that $\sum_{i=1}^n c_i \alpha_i(v) = 0_W$ and so D is locally linearly dependent. The converse, however, is false. It may be possible for D to be linearly independent and still locally linearly dependent. To see this, take $V = W = F^2$ and let $D = \{\alpha_1, \alpha_2\} \subseteq \mathrm{Hom}(F^2, F^2)$, where we define $\alpha_1 : \begin{bmatrix} a \\ b \end{bmatrix} \mapsto \begin{bmatrix} a \\ 0 \end{bmatrix}$ and $\alpha_2 : \begin{bmatrix} a \\ b \end{bmatrix} \mapsto \begin{bmatrix} b \\ 0 \end{bmatrix}$. If $v \in F^2$, then $\{\alpha_1(v), \alpha_2(v)\}$ is a subset of the one-dimensional subspace $F \begin{bmatrix} 1 \\ 0 \end{bmatrix}$ of F^2 and so cannot be linearly independent. On the other hand, D is linearly independent since if there exist scalars c and d satisfying the condition that $c\alpha_1 + d\alpha_2$ is the 0-function, then

$$\begin{bmatrix} 0 \\ 0 \end{bmatrix} = c\alpha_1 \left(\begin{bmatrix} 1 \\ 0 \end{bmatrix} \right) + d\alpha_2 \left(\begin{bmatrix} 1 \\ 0 \end{bmatrix} \right) = \begin{bmatrix} c \\ 0 \end{bmatrix} + \begin{bmatrix} 0 \\ 0 \end{bmatrix} = \begin{bmatrix} c \\ 0 \end{bmatrix},$$

which implies that $c = 0$. Similarly,

$$\begin{bmatrix} 0 \\ 0 \end{bmatrix} = c\alpha_1 \left(\begin{bmatrix} 0 \\ 1 \end{bmatrix} \right) + d\alpha_2 \left(\begin{bmatrix} 0 \\ 1 \end{bmatrix} \right) = \begin{bmatrix} 0 \\ 0 \end{bmatrix} + \begin{bmatrix} d \\ 0 \end{bmatrix} = \begin{bmatrix} d \\ 0 \end{bmatrix},$$

which implies that $d = 0$ as well.

The following proposition shows that the operation of a linear transformation is entirely determined by its action on elements of a basis. This result is extremely important, especially if the vector spaces involved are finitely generated.

Proposition 6.2 *Let V and W be vector spaces over a field F, and let B be a basis of V. If $f \in W^B$ then there is a unique linear transformation $\alpha \in \mathrm{Hom}(V, W)$ satisfying the condition that $\alpha(u) = f(u)$ for all $u \in B$.*

Proof Since B is a basis of V, we know that each vector $v \in V$ can be written as a linear combination $v = \sum_{i=1}^{n} a_i u_i$ of elements of B in a unique way. We now define the function $\alpha : V \to W$ by $\alpha : v \mapsto \sum_{i=1}^{n} a_i f(u_i)$. This function is well defined as a result of the uniqueness of representation of v, as was shown in Proposition 5.4. Moreover, it is clear that α is a linear transformation. If $\beta : V \to W$ is a linear transformation satisfying the condition that $\beta(u) = f(u)$ for all $u \in B$ then $\beta(v) = \beta(\sum_{i=1}^{n} a_i u_i) = \sum_{i=1}^{n} a_i \beta(u_i) = \sum_{i=1}^{n} a_i f(u_i) = \alpha(v)$, and so $\beta = \alpha$. Thus α is unique. $\square$

Example Let F be a field and let $c_0, c_1, \ldots$ be a sequence of elements of F. Then we have a linear transformation $\alpha : F[X] \to F$ defined by $\alpha : \sum_{i=0}^{n} a_i X^i \mapsto \sum_{i=0}^{n} a_i c_i$.

Example We can use Proposition 6.2 to show how uncommon linear transformations really are. Let $F = \mathrm{GF}(3)$ and let $V = F^4$. Then V has $3^4 = 81$ elements and so the number of functions from V to itself is 81^{81}. On the other hand, a basis B for V over F has 4 elements and so, since every linear transformation from V to itself is totally determined by its action on B and that any function from B to V defines such a linear transformation, we see that the number of linear transformations from V to itself is 81^4. Therefore, the probability that a randomly-selected function from V to itself be a linear transformation is $81^4/81^{81} = 81^{-77}$, which is roughly 0.11134×10^{-146}.

Proposition 6.3 *Let V, W, and Y be vector spaces over a field F and let $\alpha : V \to W$ and $\beta : W \to Y$ be linear transformations. Then $\beta\alpha : V \to Y$ is a linear transformation.*

Proof If $v_1, v_2 \in V$ and if $a \in F$ then

$$(\beta\alpha)(v_1 + v_2) = \beta\big(\alpha(v_1 + v_2)\big) = \beta\big(\alpha(v_1) + \alpha(v_2)\big)$$

$$= \beta\big(\alpha(v_1)\big) + \beta\big(\alpha(v_2)\big) = (\beta\alpha)(v_1) + (\beta\alpha)(v_2)$$

and $(\beta\alpha)(cv_1) = \beta(\alpha(cv_1)) = \beta(c\alpha(v_1)) = c\beta(\alpha(v_1)) = c(\beta\alpha)(v_1)$, which proves the proposition. $\square$

Example It is often important and insightful to write a linear transformation as a composite of linear transformations of predetermined types. Consider the following situation: Let $a < b$ be real numbers and let V be the vector space over $\mathbb{R}$ consisting of all functions from the closed interval $[a, b]$ to $\mathbb{R}$. Let W be the subspace of V

consisting of all differentiable functions, and let $\delta : W \to V$ be the function which assigns to each function $f \in W$ its derivative. For each real number $a < c < b$, let $\varepsilon_c : V \to \mathbb{R}$ be the linear transformation defined by $\varepsilon_c : g \mapsto g(c)$. Then the Intermediate Value Theorem from calculus says that the linear transformation $\beta : W \to \mathbb{R}$ defined by $\beta : f \mapsto [f(b) - f(a)](b - a)^{-1}$ is of the form $\varepsilon_c \delta$ for some c.

Let V and W be vector spaces over a field F and let $\alpha : V \to W$ be a linear transformation. For $w \in W$, we denote $\{v \in V \mid \alpha(v) = w\}$ by $\alpha^{-1}(w)$. Note that this set may be empty. In particular, we will be interested in $\alpha^{-1}(0_W) = \{v \in V \mid \alpha(v) = 0_W\}$. This set is called the *kernel* of α and is denoted by $\ker(\alpha)$. Then $\ker(\alpha)$ is never empty, since it always contains 0_V. If U is a nonempty subset of W, set $\alpha^{-1}(U) = \{\alpha^{-1}(u) \mid u \in U\}$. It is easy to verify that $\alpha^{-1}(U)$ is a subspace of V whenever U is a subspace of W.

Example Let F be a field and let $\alpha \in \mathrm{Hom}(F^3, F^4)$ be the linear transformation defined by $\alpha : \begin{bmatrix} a \\ b \\ c \end{bmatrix} \mapsto \begin{bmatrix} a - b \\ 0 \\ c \\ c \end{bmatrix}$. Then $\ker(\alpha) = \left\{ \begin{bmatrix} a \\ a \\ 0 \end{bmatrix} \middle| \, a \in F \right\}$.

Proposition 6.4 *Let V and W be vector spaces over a field F and let $\alpha \in \mathrm{Hom}(V, W)$. Then $\ker(\alpha)$ is a subspace of V, which is trivial if and only if α is monic.*

Proof Let $v_1, v_2 \in \ker(\alpha)$ and let $a \in F$. Then $\alpha(v_1 + v_2) = \alpha(v_1) + \alpha(v_2) = 0_W + 0_W = 0_W$, and so $v_1 + v_2 \in \ker(\alpha)$. Similarly, $\alpha(av_1) = a\alpha(v_1) = a0_W = 0_W$ and so $av_1 \in \ker(\alpha)$. This proves that $\ker(\alpha)$ is a subspace of V.

If α is monic then $\alpha^{-1}(w)$ can have at most one element for each $w \in W$, and so, in particular, $\ker(\alpha) = \{0_V\}$. Conversely, suppose that $\ker(\alpha)$ is trivial and that there exist elements $v_1 \neq v_2$ of V satisfying $\alpha(v_1) = \alpha(v_2)$. Then $\alpha(v_1 - v_2) = \alpha(v_1) - \alpha(v_2) = 0_W$ and so $v_1 - v_2 \in \ker(\alpha)$. Thus $v_1 - v_2 = 0_V$ and so $v_1 = v_2$, which is a contradiction. Hence α must be monic. $\qquad\square$

Let V and W be vector spaces over a field and let $\alpha : V \to W$ be a linear transformation. The *image* of α is the subset $\mathrm{im}(\alpha) = \{\alpha(v) \mid v \in V\}$ of W. This set is nonempty since $0_W = \alpha(0_V) \in \mathrm{im}(\alpha)$. Note that $w \in \mathrm{im}(\alpha)$ if and only if $\alpha^{-1}(w) \neq \varnothing$. If U is a nonempty subset of V, we denote the subset $\{\alpha(u) \mid u \in U\}$ of W by $\alpha(U)$. Thus $\alpha(V) = \mathrm{im}(\alpha)$.

Proposition 6.5 *Let V and W be vector spaces over a field F and let $\alpha \in \mathrm{Hom}(V, W)$. Then $\mathrm{im}(\alpha)$ is a subspace of W, which is improper if and only if α is epic.*

Proof If $\alpha(v_1)$ and $\alpha(v_2)$ are in $\operatorname{im}(\alpha)$ and if $a \in F$, then $\alpha(v_1) + \alpha(v_2) = \alpha(v_1 + v_2) \in \operatorname{im}(\alpha)$ and similarly $a\alpha(v_1) = \alpha(av_1) \in \operatorname{im}(\alpha)$, proving that $\operatorname{im}(\alpha)$ is a subspace of W. The second part follows immediately from the definition of an epic function. $\qquad \square$

A monic linear transformation between vector spaces over a field F is called a *monomorphism*; an epic linear transformation between vector spaces is called an *epimorphism*. A bijective linear transformation between vector spaces is called an *isomorphism*. If both spaces are also F-algebras, then a bijective homomorphism of F-algebras is called an *isomorphism of F-algebras*. Similarly, a bijective homomorphism of unital F-algebras is an *isomorphism of unital F-algebras*.

Example Let F be a field and let k and n be positive integers. For each matrix $A = [a_{ij}] \in \mathcal{M}_{k \times n}(F)$, we can define the *transpose* of A to be the matrix $A^T \in \mathcal{M}_{n \times k}(F)$ obtained from A by interchanging its rows and columns. In other words, $A^T = \begin{bmatrix} a_{11} & \cdots & a_{k1} \\ \vdots & \ddots & \vdots \\ a_{1n} & \cdots & a_{kn} \end{bmatrix}$. It is easy to check that the function $A \mapsto A^T$ is an isomorphism from $\mathcal{M}_{k \times n}(F)$ to $\mathcal{M}_{n \times k}(F)$.

Example Let K and L be F-algebras. It is possible for a linear transformation $\alpha : K \to L$ to be an isomorphism of vector spaces without being an isomorphism of F-algebras. This is the case, for example, with the linear transformation $\alpha : \mathbb{Q}(\sqrt{2}) \to \mathbb{Q}(\sqrt{5})$ given by $\alpha : a + b\sqrt{2} \mapsto a + b\sqrt{5}$.

Example Let V be a vector space over a field F. Any linear transformation $\alpha : V \to F$ other than the 0-function is an epimorphism. Indeed, if α is a nonzero linear transformation and if $v_0 \in V$ satisfies the condition that $\alpha(v_0) = c \neq 0$, then for any $a \in F$ we have $a = (ac^{-1})c = (ac^{-1})\alpha(v_0) = \alpha((ac^{-1})v_0) \in \operatorname{im}(\alpha)$.

Example Let F be a field and let $\alpha : F^{(\infty)} \to F[X]$ be the function defined by $\alpha : f \mapsto \sum_{i=0}^{\infty} f(i)X^i$, which is well-defined since only finitely-many of the $f(i)$ are nonzero. This is easily checked to be an isomorphism of vector spaces.

We have already seen that if D is a basis of a vector space V over a field F then there exists a bijective function $\theta : F^{(D)} \to V$, and it is easy to verify that this is in fact an isomorphism of vector spaces. This leads us to the very important observation that for any nontrivial vector space V over a field F there exists a nonempty set Ω and an isomorphism $F^{(\Omega)} \to V$.

Let V and W be vector spaces over a field F and let B be a basis of V. Then we can define a function $\varphi : \operatorname{Hom}(V, W) \to W^B$ by restriction: $\varphi(\alpha) : u \mapsto \alpha(u)$ for all $u \in B$. It is straightforward to check that φ is a linear transformation of vector spaces over F. Moreover, by Proposition 6.2, we see that any function $f \in W^B$ is of the form $\varphi(\alpha)$ for a unique element α of $\operatorname{Hom}(V, W)$. Therefore, φ is an isomorphism.

Let V and W be vector spaces over a field F. If $\alpha : V \to W$ is a linear transformation and $0_W \neq w \in W$ then $\alpha^{-1}(w)$ is not a subspace of V. However, the next result shows that, if it is nonempty, it is close to being a subspace.

Proposition 6.6 *Let $\alpha : V \to W$ be a linear transformation of vector spaces over a field F and let $w \in \mathrm{im}(\alpha)$. For any $v_0 \in \alpha^{-1}(w)$ we have $\alpha^{-1}(w) = \{v + v_0 \mid v \in \ker(\alpha)\}$.*

Proof If $v \in \ker(\alpha)$ then $\alpha(v + v_0) = \alpha(v) + \alpha(v_0) = 0_W + w = w$ and so $v + v_0 \in \alpha^{-1}(w)$. Conversely, if $v_1 \in \alpha^{-1}(w)$ then $v_1 = (v_1 - v_0) + v_0$, where $v_1 - v_0 \in \ker(\alpha)$ since $\alpha(v_1 - v_0) = \alpha(v_1) - \alpha(v_0) = w - w = 0_W$. $\square$

Note that if $w \neq 0_W$ then $\alpha^{-1}(w)$ is not a subspace of V but rather the result of "shifting" a subspace by adding a fixed nonzero vector to each of its elements. Such a subset of a vector space is called an *affine subset*, or *linear variety* of a vector space. Let V and W be vector spaces over a field F. An *affine transformation* $\zeta : V \to W$ is a function of the form $v \mapsto \alpha(v) + y$, for some fixed $\alpha \in \mathrm{Hom}(V, W)$ and $y \in W$. It is clear that the sum of two affine transformations is again an affine transformation, as is the product of an affine transformation by a scalar, so that the set $\mathrm{Aff}(V, W)$ of all affine transformations from V to W is also a subspace of W^V which in turn contains $\mathrm{Hom}(V, W)$ as a subspace. Indeed, $\mathrm{Aff}(V, W) = F(\mathrm{Hom}(V, W) \cup K)$, where K is the set of all constant functions from V to W.

Moreover, if $\zeta : V \to W$ is the affine transformation defined by $v \mapsto \alpha(v) + y$ and if $w \in W$, then $\zeta^{-1}(w) = \alpha^{-1}(w - y)$ and so is an affine subset of V.

Analysis of computational procedures in linear algebra often hinges on the fact that when we think we are computing the effect of some linear transformation $\alpha \in \mathrm{Hom}(V, W)$, we are in fact computing that of an affine transformation $v \mapsto \alpha(v) + y$ where y is a vector arising from computational or random errors which, hopefully, is "very small" (in some sense) relative to $\alpha(v)$. Similarly, in linear models in statistics one must allow for such an affine transformation, where y is a random error vector, assumed to have expectation 0.

Example Let $V = C(0, 1)$ and let W be the subspace of V composed of all differentiable functions having a continuous derivative. Let $\delta : W \to V$ be the linear transformation which assigns to each function $f \in W$ its derivative. Then $\ker(\delta)$ consists of all constant functions. If $g \in \mathrm{im}(\delta)$ then $g = \delta(f)$, where f is the function $f : x \mapsto \int_0^x g(t)\, dt$. Thus $\delta^{-1}(g)$ consists of all functions of the form $f : x \mapsto \int_0^x g(t)\, dt + c$, where $c \in \mathbb{R}$.

Proposition 6.7 *If $\alpha : V \to W$ is an isomorphism of vector spaces over a field F then there exists an isomorphism $\beta : W \to V$ satisfying $\beta\alpha(v) = v$ and $\alpha\beta(w) = w$ for all $v \in V$ and all $w \in W$.*

Proof Define the function β by $\beta(w) = v$ if and only if $w = \alpha(v)$. This function is well-defined since every element w is of the form $\alpha(v)$ for a unique element $v \in V$. It is easy to check that the function β is an isomorphism which satisfies the stated conditions.   □

The function β defined in Proposition 6.7 is denoted by α^{-1}.

Let V and W be vector spaces over a field F. If there exists an isomorphism from V to W, we say that V and W are *isomorphic* and write $V \cong W$. It is easy to see that if V, W, and Y are vector spaces over F then:

(1) $V \cong V$;

(2) If $V \cong W$ then $W \cong V$;

(3) If $V \cong W$ and $W \cong Y$ then $V \cong Y$.

It is also clear that if $\alpha : V \to W$ is an isomorphism between vector spaces over F and if B is a basis of V then $\{\alpha(u) \mid u \in B\}$ is a basis of W. As an immediate consequence of this, we see that if $V \cong W$ then the dimensions of V and W are the same. The converse is true if V and W are finitely generated, as we shall now see.

Proposition 6.8 *Let V and W be vector spaces over a field F having bases B and D, respectively, and assume that there exists a bijective function $f : B \to D$. Then $V \cong W$.*

Proof By Proposition 6.2, we know that there exists a linear transformation $\alpha \in \mathrm{Hom}(V, W)$ satisfying the condition $\alpha(v) = f(v)$ for all $v \in B$. This linear transformation is epic since $\mathrm{im}(\alpha)$ contains a basis of W. If $v' = \sum_{v \in B} a_v v$ (where only finitely-many of the coefficients a_v are nonzero) belongs to $\ker(\alpha)$ then $0_W = \alpha(v') = \alpha(\sum_{v \in B} a_v v) = \sum_{v \in B} a_v \alpha(v) = \sum_{v \in B} a_v f(v)$ and so $a_v = 0$ for all $v \in B$, since D is linearly independent. Therefore, $\ker(\alpha)$ is trivial, and this shows that α is monic and hence an isomorphism.   □

In particular, if V and W are vector spaces of the same finite dimension n over a field F, then $V \cong W$.

Proposition 6.9 *If V and W are vector spaces finitely generated over a field F, then*

(1) *There exists a monomorphism from V to W if and only if $\dim(V) \leq \dim(W)$;*

(2) *There exists an epimorphism from V to W if and only if $\dim(V) \geq \dim(W)$.*

Proof (1) If there exists a monomorphism α from V to W then $V \cong \mathrm{im}(\alpha)$ and so $\dim(W) \geq \dim(\mathrm{im}(\alpha)) = \dim(V)$. Conversely, assume that $\dim(V) \leq \dim(W)$. Then there exists a basis $B = \{v_1, \ldots, v_n\}$ of V and there exists a basis $D =$

$\{w_1, \ldots, w_t\}$ of W, where $n \leq t$. The function from B to W given by $v_i \mapsto w_i$ for all $1 \leq i \leq n$ can be extended to a linear transformation $\alpha : V \to W$, which is monic and so is a monomorphism.

(2) If there exists an epimorphism α from V to W and if $\{v_1, \ldots, v_n\}$ is a basis of V, then $\{\alpha(v_i) \mid 1 \leq i \leq n\}$ is a generating set of W and so the dimension of W is at most $n = \dim(V)$. Conversely, if $n = \dim(V) \geq \dim(W) = t$, pick a basis $\{w_1, \ldots, w_t\}$ of W and a basis $B = \{v_1, \ldots, v_n\}$ of V. Define a function $f : B \to W$ by

$$f : v_i \mapsto \begin{cases} w_i & \text{for } 1 \leq i \leq t, \\ w_t & \text{for } t \leq i \leq n. \end{cases}$$

From Proposition 6.2, it follows that there exists a linear transformation $\alpha : V \to W$ satisfying $\alpha(v_i) = f(v_i)$ for all $1 \leq i \leq n$, and this is the desired epimorphism. $\qquad \square$

Proposition 6.10 *Let V and W be vector spaces over a field F, where V is finitely generated. Then $\dim(V) = \dim(\text{im}(\alpha)) + \dim(\ker(\alpha))$ for any linear transformation $\alpha \in \text{Hom}(V, W)$.*

Proof Let $\alpha \in \text{Hom}(V, W)$. Set $V_1 = \ker(\alpha)$ and let V_2 be a complement of V_1 in V. By Proposition 5.16, we see that $\dim(V) = \dim(V_1) + \dim(V_2)$ and so it suffices for us to show that $V_2 \cong \text{im}(\alpha)$. Let α_2 be the restriction of α to V_2. Then $\alpha_2 \in \text{Hom}(V_2, \text{im}(\alpha))$. If $v_2 \in \ker(\alpha_2)$ then $v_2 \in V_2 \cap V_1 = \{0_V\}$. Thus α_2 is a monomorphism. If $w \in \text{im}(\alpha)$ then there exists an element v of V satisfying $\alpha(v) = w$. Moreover, $v = v_1 + v_2$ for some $v_1 \in V_1$ and $v_2 \in V_2$ so $w = \alpha(v) = \alpha(v_1) + \alpha(v_2) = 0_W + \alpha(v_2) = \alpha(v_2) = \alpha_2(v_2)$. Therefore, $\text{im}(\alpha_2) = \text{im}(\alpha)$, showing that α_2 is also an epimorphism and hence the desired isomorphism. $\qquad \square$

Let V and W be vector spaces over a field F. If $\alpha \in \text{Hom}(V, W)$ then we define the *rank* $\text{rk}(\alpha)$ of α to be $\dim(\text{im}(\alpha))$ and define is the *nullity* $\text{null}(\alpha)$ of α to be $\dim(\ker(\alpha))$. Thus, Proposition 6.10 says that V has finite dimension n then both the rank and nullity of α are finite and their sum is n. The converse is also clearly true: if the rank and nullity of α are both finite, then the dimension of V is finite. Let us give bounds on the rank and nullity of compositions of linear transformations.

Proposition 6.11 (Sylvester's Theorem) *Let V, W, and Y be vector spaces finitely-generated over a field F and let $\alpha : V \to W$ and $\beta : W \to Y$ be linear transformations. Then*
(1) $\text{null}(\beta\alpha) \leq \text{null}(\alpha) + \text{null}(\beta)$;
(2) $\text{rk}(\alpha) + \text{rk}(\beta) - \dim(W) \leq \text{rk}(\beta\alpha) \leq \min\{\text{rk}(\alpha), \text{rk}(\beta)\}$.

Proof (1) Let β_1 be the restriction of β to $\mathrm{im}(\alpha)$. Then $\ker(\beta_1)$ is a subspace of $\ker(\beta)$. By Proposition 6.10, we have

$$\mathrm{null}(\beta\alpha) = \dim(V) - \mathrm{rk}(\beta\alpha) = \big[\dim(V) - \mathrm{rk}(\alpha)\big] + \big[\mathrm{rk}(\alpha) - \mathrm{rk}(\beta\alpha)\big]$$
$$= \mathrm{null}(\alpha) + \mathrm{null}(\beta_1) \le \mathrm{null}(\alpha) + \mathrm{null}(\beta).$$

(2) Clearly, $\mathrm{im}(\beta\alpha)$ is a subspace of $\mathrm{im}(\beta)$ and so its dimension is no greater than that of $\mathrm{im}(\beta)$. Moreover, $\mathrm{im}(\beta\alpha) = \mathrm{im}(\beta_1)$ and so $\mathrm{rk}(\beta\alpha) \le \mathrm{rk}(\alpha)$. Thus $\mathrm{rk}(\beta\alpha) \le \min\{\mathrm{rk}(\alpha), \mathrm{rk}(\beta)\}$. Moreover, from (1) we see that

$$\dim(V) - \mathrm{null}(\beta\alpha) \ge \dim(V) - \mathrm{null}(\alpha) + \dim(W) - \mathrm{null}(\beta) - \dim(W)$$
$$= \mathrm{rk}(\alpha) + \mathrm{rk}(\beta) - \dim(W),$$

and this proves that $\mathrm{rk}(\alpha) + \mathrm{rk}(\beta) - \dim(W) \le \mathrm{rk}(\beta\alpha)$. □

Exercises

Exercise 233
Which of the following statements are true for all vector spaces V and W over a field F and all $\alpha \in \mathrm{Hom}(V, W)$?
(1) $\alpha(A \cup B) = \alpha(A) \cup \alpha(B)$ for all nonempty subsets A and B of V;
(2) $\alpha(A \cap B) = \alpha(A) \cap \alpha(B)$ for all nonempty subsets A and B of V;
(3) $\alpha^{-1}(C \cup D) = \alpha^{-1}(C) \cup \alpha^{-1}(D)$ for all nonempty subsets C and D of W;
(4) $\alpha^{-1}(C \cap D) = \alpha^{-1}(C) \cap \alpha^{-1}(D)$ for all nonempty subsets C and D of W.

Exercise 234

Let $\alpha : \mathbb{R}^3 \to \mathbb{R}^3$ be a linear transformation satisfying $\alpha\left(\begin{bmatrix} 1 \\ 0 \\ 1 \end{bmatrix}\right) = \begin{bmatrix} -1 \\ 3 \\ 4 \end{bmatrix}$,

$\alpha\left(\begin{bmatrix} 1 \\ -1 \\ 1 \end{bmatrix}\right) = \begin{bmatrix} 0 \\ 1 \\ 0 \end{bmatrix}$, and $\alpha\left(\begin{bmatrix} 1 \\ 2 \\ -1 \end{bmatrix}\right) = \begin{bmatrix} 3 \\ 1 \\ 4 \end{bmatrix}$. What is $\alpha\left(\begin{bmatrix} 1 \\ 0 \\ 0 \end{bmatrix}\right)$?

Exercise 235

Let $\alpha : \mathbb{R}^3 \to \mathbb{R}^3$ be a linear transformation satisfying $\alpha\left(\begin{bmatrix} 1 \\ 1 \\ 0 \end{bmatrix}\right) = \begin{bmatrix} 1 \\ 2 \\ -1 \end{bmatrix}$,

$\alpha\left(\begin{bmatrix} 1 \\ 0 \\ -1 \end{bmatrix}\right) = \begin{bmatrix} 0 \\ 1 \\ 1 \end{bmatrix}$, and $\alpha\left(\begin{bmatrix} 0 \\ -1 \\ 1 \end{bmatrix}\right) = \begin{bmatrix} 3 \\ 3 \\ 3 \end{bmatrix}$. Find a vector $v \in \mathbb{R}^3$ for which

$\alpha(v) = \begin{bmatrix} 1 \\ 0 \\ 0 \end{bmatrix}$.

Exercise 236
Let F be a field and let V be the subspace of $F[X]$ consisting of all polynomials of degree at most 2. Let $\alpha : V \to F[X]$ be a linear transformation satisfying $\alpha(1) = X$, $\alpha(X + 1) = X^5 + X^3$, and $\alpha(X^2 + X + 1) = X^4 - X^2 + 1$. What is $\alpha(X^2 - X)$?

Exercise 237
For each $d \in \mathbb{R}$, let $\alpha_d : \mathbb{R}^2 \to \mathbb{R}^2$ be the function defined by

$$\alpha_d : \begin{bmatrix} a \\ b \end{bmatrix} \mapsto \begin{bmatrix} a + b + d^2 + 1 \\ a \end{bmatrix}.$$

Is there a number d having the property that α_d is a linear transformation? What if we consider α_d as a function from $GF(5)^2$ to itself?

Exercise 238
For each $d \in \mathbb{R}$, let $\alpha_d : \mathbb{R}^2 \to \mathbb{R}^2$ be the function defined by

$$\alpha_d : \begin{bmatrix} a \\ b \end{bmatrix} \mapsto \begin{bmatrix} 5da - db \\ 8d^2 - 8d - 6 \end{bmatrix}.$$

Is there a number d having the property that α_d is a linear transformation?

Exercise 239
Let V and W be vector spaces over $\mathbb{Q}$ and let $\alpha : V \to W$ be a function satisfying $\alpha(v + v') = \alpha(v) + \alpha(v')$ for all $v, v' \in V$. Is α necessarily a linear transformation?

Exercise 240
Let $\alpha : \mathbb{R} \to \mathbb{R}$ be a continuous function which satisfies $\alpha(a + b) = \alpha(a) + \alpha(b)$ for all $a, b \in \mathbb{R}$. Show that α is a linear transformation.

Exercise 241
Let W and W' be subspaces of a vector space V over a field F and assume that we have linear transformations $\alpha : W \to V$ and $\beta : W' \to V$ satisfying the condition that $\alpha(v) = \beta(v)$ for all $v \in W \cap W'$. Find a linear transformation $\theta : W + W' \to V$, the restriction of which to W equals α and the restriction of which to W' equals β, or show why no such linear transformation exists.

Exercise 242
Let $F = GF(3)$ and let $\theta \in F^F$ be the function defined by $\theta(0) = 0$, $\theta(1) = 2$, and $\theta(2) = 1$. Let n be a positive integer and let $\alpha : F^n \to F^n$ be the function defined by $\alpha : \begin{bmatrix} a_1 \\ \vdots \\ a_n \end{bmatrix} \mapsto \begin{bmatrix} \theta(a_1) \\ \vdots \\ \theta(a_n) \end{bmatrix}$. Is α a linear transformation?

Exercise 243
Does there exist a linear transformation $\alpha : \mathbb{Q}^4 \to \mathbb{Q}[X]$ satisfying

$$\alpha \left(\begin{bmatrix} 1 \\ 2 \\ 0 \\ -1 \end{bmatrix} \right) = 2, \quad \alpha \left(\begin{bmatrix} -1 \\ 1 \\ 1 \\ 1 \end{bmatrix} \right) = X, \quad \text{and} \quad \alpha \left(\begin{bmatrix} -1 \\ 4 \\ 2 \\ 1 \end{bmatrix} \right) = X + 1?$$

Exercise 244
Let B be a Hamel basis for $\mathbb{R}$ as a vector space over $\mathbb{Q}$ and let $1 \neq a \in \mathbb{R}$. Show that there exists an element $y \in B$ satisfying $ay \notin B$.

Exercise 245
For which nonnegative integers h is the function α from $GF(3)^3$ to itself defined

by $\alpha : \begin{bmatrix} a \\ b \\ c \end{bmatrix} \mapsto \begin{bmatrix} a^h \\ b \\ c^h \end{bmatrix}$ a linear transformation?

Exercise 246
For any field F, let $\theta : F \to F$ be the function defined by

$$\theta : a \mapsto \begin{cases} 0 & \text{if } a = 0, \\ a^{-1} & \text{otherwise.} \end{cases}$$

This is clearly a linear transformation when $F = GF(2)$. Does there exist a field other than $GF(2)$ for which θ is a linear transformation?

Exercise 247
Let $V = F^\infty$ and let $\alpha : V \to V$ be the function that assigns to each sequence $[a_1, a_2, \ldots] \in V$ its sequence of partial sums, namely $[a_1, a_2, \ldots] \mapsto [a_1, \sum_{i=1}^2 a_i, \sum_{i=1}^3 a_i, \ldots]$. Is α a linear transformation?

Exercise 248
Let $Y = \mathbb{R}^{\mathbb{R}} \times \mathbb{R}$. Is the function $\alpha : Y \to \mathbb{R}$ defined by $\alpha : (f, a) \mapsto f(a)$ a linear transformation?

Exercise 249
Let F be a field and let b and c be nonzero elements of F. Let $\alpha : F^\infty \to F^\infty$ be the linear transformation defined by

$$\alpha : [a_1, a_2, \ldots] \mapsto [a_3 + ba_2 + ca_1, a_4 + ba_3 + ca_2, \ldots].$$

Let $y \in \ker(\alpha)$ be a vector satisfying the condition that two successive entries in y equal 0. Show that $y = [0, 0, \ldots]$.

Exercise 250

Consider the field $F = \mathbb{Q}(\sqrt{2})$ as a $\mathbb{Q}$-algebra. Show that the only homomorphisms of $\mathbb{Q}$-algebras from F to itself are the identity function and the function $a + b\sqrt{2} \mapsto a - b\sqrt{2}$.

Exercise 251

Let V, W, and Y be vector spaces finitely-generated over a field F and let $\alpha : V \to W$ be a linear transformation. Show that the set of all linear transformations $\beta : W \to Y$ satisfying the condition that $\beta\alpha$ is the 0-transformation is a subspace of $\mathrm{Hom}(W, Y)$, and calculate its dimension.

Exercise 252

Let V and W be vector spaces over a field F and let V' be a proper subspace of V. Are $\{\alpha \in \mathrm{Hom}(V, W) \mid \ker(\alpha) \subseteq V'\}$ and $\{\alpha \in \mathrm{Hom}(V, W) \mid \ker(\alpha) \supseteq V'\}$ subspaces of $\mathrm{Hom}(V, W)$?

Exercise 253

Let V and W be vector spaces over a field F and assume that there are subspaces V_1 and V_2 of V, both of positive dimension, satisfying $V = V_1 \oplus V_2$. For $i = 1, 2$, let $U_i = \{\alpha \in \mathrm{Hom}(V, W) \mid \ker(\alpha) \supseteq V_i\}$. Show that $\{U_1, U_2\}$ is an independent set of subspaces of $\mathrm{Hom}(V, W)$. Is it necessarily true that $\mathrm{Hom}(V, W) = U_1 \oplus U_2$?

Exercise 254

Let F be a field, and let $\alpha : \mathcal{M}_{2\times2}(F) \to \mathcal{M}_{n\times n}(F)$ be a homomorphism of F-algebras for some $n > 1$. Show that $\alpha\left(\begin{bmatrix} 0 & 0 \\ 1 & 0 \end{bmatrix}\right) \neq I$.

Exercise 255

Let V and W be vector spaces over a field F and let $\alpha, \beta : V \to W$ be linear transformations satisfying the condition that for each $v \in V$ there exists a scalar $c_v \in F$ (depending on v) satisfying $\beta(v) = c_v\alpha(v)$. Show that there exists a scalar c satisfying $\beta = c\alpha$.

Exercise 256

Let V and W be vector spaces over a field F. Define a function $\varphi : \mathrm{Hom}(V, W) \to \mathrm{Hom}(V \times W, V \times W)$ by setting $\varphi(\alpha) : \begin{bmatrix} v \\ w \end{bmatrix} \mapsto \begin{bmatrix} 0_V \\ \alpha(v) \end{bmatrix}$. Is φ a linear transformation of vector spaces over F? Is it a monomorphism?

Exercise 257

Find the kernel of the linear transformation $\alpha : \mathbb{R}^5 \to \mathbb{R}^3$ defined by

$$\alpha : \begin{bmatrix} a \\ b \\ c \\ d \\ e \end{bmatrix} \mapsto \begin{bmatrix} b + c - 2d + e \\ a + 2b + 3c - 4d \\ 2a + 2c - 2e \end{bmatrix}.$$

Exercise 258

Let $\alpha : \mathbb{R}^3 \to \mathbb{R}^3$ be the linear transformation defined by

$$\alpha : \begin{bmatrix} a \\ b \\ c \end{bmatrix} \mapsto \begin{bmatrix} 2a + 4b - c \\ 0 \\ 3c + 2b - a \end{bmatrix}.$$

Are $\mathrm{im}(\alpha)$ and $\ker(\alpha)$ disjoint?

Exercise 259

Let W be the subspace $\mathbb{Q} \left\{ \begin{bmatrix} 2 \\ -1 \\ 0 \\ 1 \end{bmatrix}, \begin{bmatrix} 1 \\ 0 \\ 0 \\ 1 \end{bmatrix}, \begin{bmatrix} 1 \\ 1 \\ 0 \\ 1 \end{bmatrix} \right\}$ of $\mathbb{Q}^4$ and let $\alpha : W \to \mathbb{Q}^2$ be

the linear transformation defined by setting $\alpha : \begin{bmatrix} a \\ b \\ c \\ d \end{bmatrix} \mapsto \begin{bmatrix} a + 2b + c \\ -a - 2b - c \end{bmatrix}$. Find a

basis for $\ker(\alpha)$.

Exercise 260

Let $F = \mathrm{GF}(3)$ and let $\alpha : F^3 \to F^3$ be the linear transformation defined by
$\alpha : \begin{bmatrix} a \\ b \\ c \end{bmatrix} \mapsto \begin{bmatrix} a + b \\ 2b + c \\ 0 \end{bmatrix}$. Find the kernel of α.

Exercise 261

Let $\alpha : \mathbb{R}^4 \to \mathbb{R}^3$ be the linear transformation defined by

$$\alpha : \begin{bmatrix} a \\ b \\ c \\ d \end{bmatrix} \mapsto \begin{bmatrix} 2a + 4b + c - d \\ 3a + b - 2c \\ a + 5c + 4d \end{bmatrix}.$$

Do there exist $a, b, d \in \mathbb{Z}$ such that $\begin{bmatrix} a \\ b \\ 7 \\ d \end{bmatrix} \in \ker(\alpha)$?

Exercise 262

Let V and W be vector spaces over a field F. Let $\alpha \in \text{Hom}(V, W)$ and $\beta \in \text{Hom}(W, V)$ satisfy the condition that $\alpha\beta\alpha = \alpha$. If $w \in \text{im}(\alpha)$, show that $\alpha^{-1}(w) = \{\beta(w) + v - \beta\alpha(v) \mid v \in V\}$.

Exercise 263

Let V, W, and Y be vector spaces over a field F and let $\alpha \in \text{Hom}(V, W)$ and $\beta \in \text{Hom}(W, Y)$ satisfy the condition that $\text{im}(\alpha)$ has a finitely-generated complement in W and $\text{im}(\beta)$ has a finitely-generated complement in Y. Does $\text{im}(\beta\alpha)$ necessarily have a finitely-generated complement in Y?

Exercise 264

Let $\alpha : \mathcal{M}_{3\times 3}(\mathbb{R}) \to \mathbb{R}$ be defined by $\alpha : [a_{ij}] \mapsto \sum_{i=1}^{3} \sum_{j=1}^{3} a_{ij}$. Show that α is a linear transformation and find a basis for $\ker(\alpha)$.

Exercise 265

Let $F = \text{GF}(2)$ and let $n > 2$ be an integer. Let W be the set of all vectors $\begin{bmatrix} a_1 \\ \vdots \\ a_n \end{bmatrix}$ in F^n having an even number of nonzero entries. Show that W is a subspace of F^n by showing that it is the kernel of some linear transformation.

Exercise 266

Let A and B be nonempty sets. Let V be the collection of all subsets of A and let W be the collection of all subsets of B, both of which are vector spaces over $\text{GF}(2)$. Any function $f : A \to B$ defines a function $\alpha_f : W \to V$ by setting $\alpha_f : D \mapsto \{a \in A \mid f(a) \in D\}$. Show that each such function α_f is a linear transformation, and find its kernel.

Exercise 267

Let V be a vector space over a field F and let $\alpha : V^3 \to V$ be the function defined by $\alpha : \begin{bmatrix} v_1 \\ v_2 \\ v_3 \end{bmatrix} \mapsto v_1 + v_2 + v_3$. Show that α is a linear transformation and find its kernel.

Exercise 268

Let n be a positive integer and let V be the subspace of $\mathbb{R}[X]$ composed of all polynomials of degree at most n. Let $\alpha : V \to V$ be the linear transformation given by $\alpha : p(X) \mapsto p(X + 1) - p(X)$. Find $\ker(\alpha)$ and $\text{im}(\alpha)$.

Exercise 269
Let $\alpha : \mathbb{R}^3 \to \mathbb{R}^3$ be the linear transformation given by

$$\alpha : \begin{bmatrix} a \\ b \\ c \end{bmatrix} \mapsto \begin{bmatrix} a+b+c \\ -a-c \\ b \end{bmatrix}.$$

Find $\ker(\alpha)$ and $\text{im}(\alpha)$.

Exercise 270
Find the kernel of the linear transformation $\alpha : \mathbb{Q}[X] \to \mathbb{R}$ defined by $\alpha : p(X) \mapsto p(\sqrt{3})$.

Exercise 271
Let $V = C(0, 1)$. For each positive integer n, we define the nth *Bernstein function* $\beta_n : V \mapsto \mathbb{R}[X]$ by

$$\beta_n : f \mapsto \sum_{k=0}^{n} \frac{n!}{k!(n-k)!} f\left(\frac{k}{n}\right) X^k (1-X)^{n-k}.$$

Show that each β_n is a linear transformation and find $\bigcap_{n=1}^{\infty} \ker(\beta_n)$. (Note: the Bernstein functions are used in building polynomial approximations to continuous functions.)

With kind permission of the Archives of the Mathematisches Forschungsinstitut Oberwolfach.
Sergei Natanovich Bernstein was a twentieth-century Russian mathematician who worked mostly in probability theory.

Exercise 272
Let V and W be nontrivial vector spaces over a field F. Show that $W = \sum \{\text{im}(\alpha) \mid \alpha \in \text{Hom}(V, W)\}$.

Exercise 273
Let W be the subspace of $\mathbb{R}^{\mathbb{R}}$ consisting of all twice-differentiable functions and let $\alpha : W \to \mathbb{R}^{\mathbb{R}}$ be the linear transformation $\alpha : f \mapsto f''$. Find $\alpha^{-1}(f_0)$, where $f_0 \in \mathbb{R}^{\mathbb{R}}$ is defined by $f_0 : x \mapsto x + 1$.

Exercise 274
Let W be the subspace of $\mathbb{R}^{\mathbb{R}}$ consisting of all differentiable functions and let $\alpha : W \to \mathbb{R}^{\mathbb{R}}$ be the function defined by $\alpha(f) : x \mapsto f'(x) + \cos(x) f(x)$. Show that α is a linear transformation and find its kernel.

Exercise 275

Let n be a positive integer and let V be a vector space over $\mathbb{C}$. Does there exist a linear transformation $\alpha : V \to \mathbb{C}^n$ other than the 0-function satisfying the condition that $\mathrm{im}(\alpha) \subseteq \mathbb{R}^n$?

Exercise 276

Let V and W be vector spaces over a field F and let $\alpha : V \to W$ be a linear transformation other than the 0-function. Find a linear transformation $\beta : V \to W$ satisfying $\mathrm{im}(\alpha) = \mathrm{im}(\beta) \neq \mathrm{im}(\alpha + \beta)$.

Exercise 277

Let V be a finite-dimensional vector space over a field F and let $\alpha, \beta \in \mathrm{Hom}(V, V)$ be linear transformations satisfying $\mathrm{im}(\alpha) + \mathrm{im}(\beta) = V = \ker(\alpha) + \ker(\beta)$. Show that $\mathrm{im}(\alpha) \cap \mathrm{im}(\beta) = \{0_V\} = \ker(\alpha) \cap \ker(\beta)$.

Exercise 278

Let V, W, and Y be vector spaces over a field F and let $\alpha \in \mathrm{Hom}(V, W)$ and $\beta \in \mathrm{Hom}(W, Y)$ satisfy the condition that $\ker(\alpha)$ and $\ker(\beta)$ are both finitely generated. Is $\ker(\beta\alpha)$ necessarily finitely generated?

Exercise 279

Find a linear transformation $\alpha : \mathbb{Q}^3 \to \mathbb{Q}^4$ satisfying

$$\mathrm{im}(\alpha) = \mathbb{Q} \left\{ \begin{bmatrix} 0.5 \\ -1 \\ 3 \\ 0 \end{bmatrix}, \begin{bmatrix} 2 \\ 1 \\ 1 \\ -4 \end{bmatrix} \right\}.$$

Exercise 280

Let $F = \mathrm{GF}(2)$ and let $\alpha \in \mathrm{Hom}(F^7, F^3)$ be given by

$$\begin{bmatrix} a_1 \\ \vdots \\ a_7 \end{bmatrix} \mapsto \begin{bmatrix} a_4 + a_5 + a_6 + a_7 \\ a_2 + a_3 + a_6 + a_7 \\ a_1 + a_3 + a_5 + a_7 \end{bmatrix}.$$

If v is a nonzero element of $\ker(\alpha)$, show that at least three entries in v are equal to 1.

Exercise 281

Let V and W be vector spaces finitely-generated over a field F and let $\alpha \in \mathrm{Hom}(V, W)$. If Y is a subspace of W, is it true that $\dim(\alpha^{-1}(Y)) \geq \dim(V) - \dim(W) + \dim(Y)$?

Exercise 282

Let V be a vector space over a field F and let $Y = V^\infty$. Let W be the subspace of Y consisting of all those sequences $[v_1, v_2, \ldots]$ in which $v_i = 0$ for all odd i

and let W' be the subspace of Y consisting of all those sequences in which $v_i = 0$ for all even i. Find a linear transformation from Y to itself, the kernel of which equals W and the image of which equals W'.

Exercise 283

Let W be the subspace of $\mathbb{R}^6$ composed of all vectors $\begin{bmatrix} a_1 \\ \vdots \\ a_6 \end{bmatrix}$ satisfying $\sum_{i=1}^6 a_i = 0$. Does there exist a monomorphism from W to $\mathbb{R}^4$?

Exercise 284

Let n be a positive integer and let $\alpha : \mathbb{Q}^n \to \mathbb{Q}^n$ be a linear transformation which is not a monomorphism. Does there necessarily exist a nonzero element of $\ker(\alpha)$ all the entries of which are integers?

Exercise 285

Let n be a positive integer and let W be the subspace of $\mathbb{C}[X]$ consisting of all polynomials of degree less than n. Let $a_1, \dots, a_n$ be distinct complex numbers and let $\alpha : W \to \mathbb{C}^n$ be the function defined by $\alpha : p(X) \mapsto \begin{bmatrix} p(a_1) \\ \vdots \\ p(a_n) \end{bmatrix}$. Is α a monomorphism? Is it an isomorphism?

Exercise 286

Let V be a vector space over a field F and let $\alpha : V \to V$ be a linear transformation satisfying the condition that $\alpha^2 = a\alpha + b\sigma_1$, where a and b are nonzero scalars. Show that α is a monomorphism.

Exercise 287

Let p be a prime integer and let F be a field of characteristic p. Let $(K, \bullet)$ be an associative and commutative unital F-algebra and let $\alpha : K \to K$ be the function defined by $\alpha : v \mapsto v^p$. Show that α is an isomorphism of unital F-algebras.

Exercise 288

Let F be a field and let K and K' be fields containing F. Show that every homomorphism of F-algebras $K \to K'$ is a homomorphism of unital F-algebras.

Exercise 289

Let $F = GF(7)$. How many distinct monomorphisms can one define from F^2 to F^4?

Exercise 290

Let V and W be vector spaces over a field F and let $\alpha, \beta \in \mathrm{Hom}(V, W)$ be monomorphisms. Is $\alpha + \beta$ necessarily a monomorphism?

Exercise 291

Let F be a field and let F' be a field containing F. Let $(K, \bullet)$ be an F-algebra and let $\alpha : F' \to K$ be a nontrivial homomorphism of F-algebras. Show that α is monic.

Exercise 292

Let V and W be vector spaces over a field F and let $\alpha \in \operatorname{Hom}(V, W)$ be an epimorphism. Show that there exists a linear transformation $\beta \in \operatorname{Hom}(W, V)$ satisfying the condition that $\alpha\beta$ is the identity function on W.

Exercise 293

Let V, W, and Y be vector spaces over a field F and let $\alpha \in \operatorname{Hom}(V, W)$ be an epimorphism. Show that for each linear transformation $\beta \in \operatorname{Hom}(Y, W)$ there exists a linear transformation $\theta \in \operatorname{Hom}(Y, V)$ such that $\beta = \alpha\theta$.

Exercise 294

Let V, W, and Y be vector spaces over a field F and let $\alpha \in \operatorname{Hom}(V, W)$ be a monomorphism. Show that for each linear transformation $\beta \in \operatorname{Hom}(V, Y)$ there exists a linear transformation $\theta \in \operatorname{Hom}(W, Y)$ such that $\beta = \theta\alpha$.

Exercise 295

Let V be a vector space finitely-generated over a field F, the dimension of which is even. Show that there exists an isomorphism $\alpha : V \to V$ satisfying the condition that $\alpha^2(v) = -v$ for all $v \in V$.

Exercise 296

Let $\alpha : V \to W$ be a linear transformation between vector spaces over a field F and let D be a nonempty linearly-independent subset of $\operatorname{im}(\alpha)$. Show that there exists a basis B of V satisfying the condition that $\{\alpha(v) \mid v \in B\} = D$.

Exercise 297

Let V and W be vector spaces over a field F and let $\alpha \in \operatorname{Hom}(V, W)$ satisfy the condition that $\alpha\beta\alpha$ is not the 0-function for any linear transformation $\beta : W \to V$ which is not the 0-function. Show that α is an isomorphism.

Exercise 298

Let F be a field and let $\alpha : F^3 \to F[X]$ be the linear transformation defined by
$\begin{bmatrix} a \\ b \\ c \end{bmatrix} \mapsto (a + b)X + (a + c)X^5$. Find the nullity and rank of α.

Exercise 299

Let F be a field and let $p(X) = X^2 + bX + c \in F[X]$ be a polynomial having distinct nonzero roots d_1 and d_2 in F. Let $\alpha : F^3 \to F$ be the linear transformation defined by $\alpha : \begin{bmatrix} a_1 \\ a_2 \\ a_3 \end{bmatrix} \mapsto a_3 + ba_2 + ca_1$ and let $\beta : F^\infty \to F^\infty$ be the linear trans-

formation defined by $\beta : [a_1, a_2, a_3, \ldots] \mapsto \left[\alpha \left(\begin{bmatrix} a_1 \\ a_2 \\ a_3 \end{bmatrix} \right), \alpha \left(\begin{bmatrix} a_2 \\ a_3 \\ a_4 \end{bmatrix} \right), \ldots \right]$.

Show that the nullity of β is at least 2.

Exercise 300
Let Ω be a nonempty set and let V be the collection of all subsets of Ω, considered as a vector space over GF(2). Show that this vector space is isomorphic to $GF(2)^\Omega$.

Exercise 301
Let F be a field and let V be the subspace of F^∞ consisting of all sequences $[a_1, a_2, a_3, \ldots]$ in which $a_i = 0$ for all even i. Let W be the subspace of F^∞ consisting of all sequences $[a_1, a_2, a_3, \ldots]$ in which $a_i = 0$ for all odd i. Show that $V \cong F^\infty \cong W$.

Exercise 302
Let V be a vector space over a field F having subspaces W and W'. Let $Y = \left\{ \begin{bmatrix} w \\ w' \end{bmatrix} \middle| w \in W \text{ and } w' \in W' \right\}$, which is a subspace of V^2. Let $\alpha : Y \to V$ be the linear transformation defined by $\alpha : \begin{bmatrix} w \\ w' \end{bmatrix} \mapsto w + w'$. Find the kernel of α, and show that it is isomorphic to $W \cap W'$.

Exercise 303
Let V be a vector space over a field F. Let W be a subspace of V and let W' be a complement of W in V. Let $\alpha : W \to W'$ be a linear transformation. Show that W isomorphic to the subspace $Y = \{w + \alpha(w) \mid w \in W\}$ of V.

Exercise 304
Show that there is no vector space over any field F having precisely 15 elements.

Exercise 305
Let F be a field and let $V = F[X]$. Show that $V \cong V^2$.

Exercise 306
Let V, W, and Y be vector spaces over a field F. Let $\{\alpha_1, \ldots, \alpha_n\}$ be a finite subset of $\mathrm{Hom}(V, W)$ and let $\beta \in \mathrm{Hom}(V, Y)$ be a linear transformation satisfying $\bigcap_{i=1}^n \ker(\alpha_i) \subseteq \ker(\beta)$. Show that there exist linear transformations $\gamma_1, \ldots, \gamma_n$ in $\mathrm{Hom}(W, Y)$ satisfying $\beta = \sum_{i=1}^n \gamma_i \alpha_i$.

Exercise 307
Let V be a vector space over a field F and let W be a subspace of V. For each $v \in V$, let $v + W = \{v + w \mid w \in W\}$. Let V/W be the collection of all the sets of the form $v + W$ for $v \in V$ and define operations of addition and scalar multiplication on V/W by setting $(v + W) + (v' + W) = (v + v') + W$ and $c(v + W) = (cv) + W$ for all $v, v' \in V$ and $c \in F$. Show that:
(1) $v + W = v' + W$ if and only if $v - v' \in W$;
(2) V/W, with the given operations, is a vector space over F;
(3) The function $v \mapsto v + W$ is an epimorphism from V to W, the kernel of which equals W;
(4) Every complement of W in V is isomorphic to V/W;
(5) If $[v + W] \cap [v' + W] \neq \varnothing$ then $v + W = v' + W$.
The space V/W is called the *factor space* of V by W.

Exercise 308
Let F be a field and let $m > n$ be positive integers. Let A and B be fixed matrices in $\mathcal{M}_{n \times m}(F)$ and let $\theta : \mathcal{M}_{m \times n}(F) \to \mathcal{M}_{n \times m}(F)$ be the linear transformation defined by $\theta : C \mapsto ACB$. Show that θ is not an isomorphism.

Exercise 309
Let F be a field and let K and L be fields containing F as a subfield. Show that the set of homomorphisms of unital F-algebras from L to K is a linearly-independent subset of the vector space K^L over K.

Exercise 310
Let V and W be vector spaces over a field F, with V finitely generated, and let Y be a proper subspace of V. Let $\alpha \in \text{Hom}(V, W)$ and let β be the restriction of α to Y. Show that either $\ker(\beta) \subset \ker(\alpha)$ or $\text{im}(\beta) \subset \text{im}(\alpha)$.

Exercise 311
Let V be a vector space over a field F, and let $U \subseteq W$ be subspaces of V. Assume that there exist $x, y \in V$ satisfying the condition that the affine sets $x + U = \{x + u \mid u \in U\}$ and $y + W = \{y + w \mid w \in W\}$ have a vector in common. Show that $x + U \subseteq y + W$.

Exercise 312
Let V and W be vector spaces over a field F. A function $f : V \to W$ is *linearly independent* if and only if $\text{gr}(f)$ is a linearly-independent subset of $V \times W$.
(1) Show that if $f : V \to W$ is linearly independent and if $\alpha \in \text{Hom}(V, W)$ then $f + \alpha$ is linearly independent.
(2) Show that no linear transformation is linearly independent.

Exercise 313
Let V and W be a vector spaces over a field F. A linear transformation $\alpha : V \to W$ is said to have *algebraic degree* n if and only if the set $\{v, \alpha(v), \ldots, \alpha^n(v)\}$ is linearly dependent for any $v \in V$, but there exists an element v_0 of V such

that the set $\{v_0, \alpha(v_0), \dots, \alpha^{n-1}(v_0)\}$ is linearly independent. Find the algebraic

degree of $\alpha \in \mathrm{Hom}(\mathbb{R}^5, \mathbb{R}^5)$ defined by $\alpha : \begin{bmatrix} a \\ b \\ c \\ d \\ e \end{bmatrix} \mapsto \begin{bmatrix} a + 2b \\ b - c \\ a \\ c - a \\ c \end{bmatrix}$.

Exercise 314
Let n be a positive integer and let F be a field the characteristic of which does not divide n. Let W be the subspace of $\mathcal{M}_{n \times n}(F)$ generated by $\{AB - BA \mid A, B \in \mathcal{M}_{n \times n}(F)\}$. Show that $\dim(W) = n^2 - 1$.

Exercise 315
Let V be a vector space finitely generated over a field F and let $\alpha \in \mathrm{Hom}(V, V)$. Show that there exists a positive integer t satisfying $V = \mathrm{im}(\alpha^t) \oplus \ker(\alpha^t)$.

The Endomorphism Algebra of a Vector Space 7

Let V be a vector space over a field F. A linear transformation α from V to itself is called an *endomorphism* of V. We will denote the set of all endomorphisms of V by $\text{End}(V)$. This set is nonempty, since it includes the functions of the form $\sigma_c : v \mapsto cv$ for $c \in F$. In particular, it includes the 0-endomorphism $\sigma_0 : v \mapsto 0_V$ and the identity endomorphism $\sigma_1 : v \mapsto v$. If V is nontrivial, these functions are not the same. We see that we have two operations defined on $\text{End}(V)$: addition and multiplication (given by composition). Indeed, as a direct consequence of the definitions we conclude the following:

Proposition 7.1 *If V is a nontrivial vector space over a field F, then $\text{End}(V)$ is an associative unital F-algebra with σ_0 being the identity element for addition and σ_1 being the identity element for multiplication.*

If V is a nontrivial vector space over a field F then there exists a function $\sigma : F \to \text{End}(V)$ defined by $\sigma : c \mapsto \sigma_c$ for all $c \in F$. This function is monic, for if $\sigma_c = \sigma_d$ then for any $0_V \neq v \in V$ we have $cv = \sigma_c(v) = \sigma_d(v) = dv$ and hence $(c - d)v = 0_V$. Since $0_V \neq v$, this implies that $c - d = 0$ and so $c = d$. Moreover, if $c, d \in F$ then $\sigma_c + \sigma_d = \sigma_{c+d}$ and $\sigma_c \sigma_d = \sigma_{cd}$ so σ is a monic homomorphism of unital F-algebras. We can use this function to identify F with its image under σ and consider it a subalgebra of the F-algebra $\text{End}(V)$.

If $\alpha, \beta \in \text{End}(V)$ and if $c \in F$, then we have already seen that the functions $\alpha + \beta$, $\alpha\beta$, and $c\alpha$ all belong to $\text{End}(V)$. Therefore, we see that if $p(X) = \sum_{i=0}^{n} a_i X^i \in F[X]$ then $p(\alpha) = \sum_{i=0}^{n} a_i \alpha^i$ is an endomorphism of V, and, indeed, the set $F[\alpha]$ of all endomorphisms of V of this form is an F-subalgebra of $\text{End}(V)$. The function from $F[X]$ to $F[\alpha]$ given by $p(X) \mapsto p(\alpha)$ is immediately seen to be an epic homomorphism of unital F-algebras for any $\alpha \in \text{End}(V)$.

Example Let $F = \text{GF}(2)$ and let $p(X) = X^2 + X \in F[X]$. Then $p(a) = 0$ for every $a \in F$. However, $p(\alpha) \neq \sigma_0$, where $\alpha \in \text{End}(F^2)$ is defined by $\alpha : \begin{bmatrix} a \\ b \end{bmatrix} \mapsto \begin{bmatrix} b \\ a \end{bmatrix}$.

J.S. Golan, *The Linear Algebra a Beginning Graduate Student Ought to Know*,
DOI 10.1007/978-94-007-2636-9_7, © Springer Science+Business Media B.V. 2012

Example Structures of the form $F[\alpha]$ are important in many areas of mathematics. For example, let V be the collection of all infinitely-differentiable functions from $\mathbb{R}$ to $\mathbb{R}$ and let δ be the differentiation endomorphism on V. If $p(X) = \sum_{i=0}^{n} a_i X^i \in \mathbb{R}[X]$, then we have $p(\delta) : f \mapsto a_0 f + \sum_{i=1}^{n} a_i f^{[i]}$, where $f^{[i]}$ denotes the ith derivative of f. Such an endomorphism is called a *differential operator with constant coefficients* on V. If $c \in \mathbb{R}$ and if $f_c \in V$ is the function given by $f_c : x \mapsto e^{cx}$, then $\delta(f_c) = c f_c$ and so $p(\delta) : f_c \mapsto \sum_{i=0}^{n} a_i c^i e^{cx} = (\sum_{i=0}^{n} a_i c^i) f_c = p(c) f_c$. Thus, $p(\delta)$ is the 0-function whenever c is as root of $p(X)$. Hence $f_c \in \ker(p(\delta))$ for each root c of $p(X)$.

Example Let V be the convolution algebra on $\mathbb{R}$ and let $h \in V$ be the constant function $t \mapsto 1$. Then h defines an endomorphism of V given by $f \mapsto h * f$, called the *integration endomorphism* since $h * f : t \mapsto \int_0^t f(u)\,du$.

Example Let F be a field and let $(K, \bullet)$ be a nonassociative F-algebra. An endomorphism $\delta \in \mathrm{End}(K)$ is a *derivation* if and only if $\delta(v \bullet w) = [\delta(v)] \bullet w + v \bullet [\delta(w)]$. Thus, for example, if K is a Lie algebra then, as a consequence of the Jacobi identity, we see that every $y \in K$ defines a derivation δ_y of K given by $\delta_y : v \mapsto y \bullet v$. Also, if K is the $\mathbb{R}$-algebra consisting of all infinitely-differentiable functions in $\mathbb{R}^{\mathbb{R}}$, then the endomorphism of K which assigns to each function in K its derivative is a derivation. The set of all derivations defined on K is a subspace of $\mathrm{End}(K)$. If δ and δ' are derivations on K, then $\delta\delta'$ is not, in general, a derivation on K, but the Lie product $\delta\delta' - \delta'\delta$ is always a derivation on K, and so the set of all derivations on K is a Lie algebra over F.

Given a nontrivial vector space V over a field F, we note that the F-algebra $\mathrm{End}(V)$ is neither necessarily commutative nor necessarily entire, as the following examples show:

Example Let F be a field and let $V = F^3$. Let $\alpha, \beta \in \mathrm{End}(V)$ be the endomorphisms defined by $\alpha : \begin{bmatrix} a \\ b \\ c \end{bmatrix} \mapsto \begin{bmatrix} b \\ a \\ c \end{bmatrix}$ and $\beta : \begin{bmatrix} a \\ b \\ c \end{bmatrix} \mapsto \begin{bmatrix} a \\ 0 \\ 0 \end{bmatrix}$. Then $\beta\alpha : \begin{bmatrix} a \\ b \\ c \end{bmatrix} \mapsto \begin{bmatrix} b \\ 0 \\ 0 \end{bmatrix}$ and $\alpha\beta : \begin{bmatrix} a \\ b \\ c \end{bmatrix} \mapsto \begin{bmatrix} 0 \\ a \\ 0 \end{bmatrix}$, so $\beta\alpha \neq \alpha\beta$.

Example Let F be a field and let $V = F^3$. Let $\alpha, \beta \in \mathrm{End}(V)$ be the endomorphisms defined by $\alpha : \begin{bmatrix} a \\ b \\ c \end{bmatrix} \mapsto \begin{bmatrix} 0 \\ 0 \\ c \end{bmatrix}$ and $\beta : \begin{bmatrix} a \\ b \\ c \end{bmatrix} \mapsto \begin{bmatrix} a \\ 0 \\ 0 \end{bmatrix}$. Then $\beta\alpha = \sigma_0 = \alpha\beta$.

We do, however, have the following:

Proposition 7.2 *Let V be a vector space over a field F. Then for all $\alpha \in \text{End}(V)$ and all $c \in F$ we have $\alpha \sigma_c = \sigma_c \alpha$.*

Proof If $v \in V$ then $\alpha \sigma_c(v) = \alpha(cv) = c\alpha(v) = \sigma_c \alpha(v)$. $\square$

An endomorphism of a vector space V over a field F which is also an isomorphism (i.e., which is both monic and epic) is called an *automorphism* of V. Since $\alpha(0_V) = 0_V$ for any endomorphism α of V, we see that any automorphism of V induces a permutation of $V \smallsetminus \{0_V\}$. Similarly, a homomorphism of F-algebras which is also an isomorphism is an *automorphism of F-algebras*.

By what we have already seen, we know that $\alpha \in \text{End}(V)$ is an automorphism if and only if there exists an endomorphism $\alpha^{-1} \in \text{End}(V)$ satisfying $\alpha\alpha^{-1} = \sigma_1 = \alpha^{-1}\alpha$. We will denote the set of all automorphisms of V by $\text{Aut}(V)$. This set is nonempty, since $\sigma_1 \in \text{Aut}(V)$, where $\sigma_1^{-1} = \sigma_1$. Moreover, if $\alpha, \beta \in \text{Aut}(V)$ then $(\alpha\beta)(\beta^{-1}\alpha^{-1}) = \alpha(\beta\beta^{-1})\alpha^{-1} = \alpha\alpha^{-1} = \sigma_1$ and similarly $(\beta^{-1}\alpha^{-1})(\alpha\beta) = \sigma_1$. Thus $\alpha\beta \in \text{Aut}(V)$, with $(\alpha\beta)^{-1} = \beta^{-1}\alpha^{-1}$. It is also clear that if $\alpha \in \text{Aut}(V)$ then $\alpha^{-1} \in \text{Aut}(V)$. If $\alpha \in \text{Aut}(V)$ and $0 \neq c \in F$, then $c\alpha \in \text{Aut}(V)$ and $(c\alpha)^{-1} = c^{-1}\alpha^{-1}$.

Example Let V be a vector space over a field F and let $n > 1$ be an integer. Any permutation π of the set $\{1, \ldots, n\}$ defines an automorphism α_π of V^n given by

$$\alpha_\pi : \begin{bmatrix} v_1 \\ v_2 \\ \vdots \\ v_n \end{bmatrix} \mapsto \begin{bmatrix} v_{\pi(1)} \\ v_{\pi(2)} \\ \vdots \\ v_{\pi(n)} \end{bmatrix} \quad \text{which rearranges the entries of each vector according to}$$

the permutation π. More generally, if V is a vector space over a field F having a basis $B = \{v_i \mid i \in \Omega\}$ and if π is a permutation of Ω, then there is an automorphism of V defined by $\sum_{i \in \Lambda} a_i v_i \mapsto \sum_{i \in \Lambda} a_{\pi(i)} v_{\pi(i)}$ for each finite subset Λ of Ω.

Example Let F be a field and let n be a positive integer. We have already seen that the function $A \mapsto A^T$ is an automorphism of $\mathcal{M}_{n \times n}(F)$, considered as a vector space over F.

Example Let V be a vector space having finite dimension n over a field F and let v and y be nonzero elements of V. Then there exist bases $\{v_1, \ldots, v_n\}$ and $\{y_1, \ldots, y_n\}$ of V satisfying $v_1 = v$ and $y_1 = y$. The function $\alpha : V \to V$ defined by $\alpha : \sum_{i=1}^n a_i v_i \mapsto \sum_{i=1}^n a_i y_i$ is thus an automorphism of V satisfying $\alpha(v) = y$.

Let V be a vector space over a field F and let n be a positive integer. We will list several types automorphisms, called *elementary automorphisms*, of a vector space of the form V^n. These automorphisms will play an important part in our ensuing discussion.

(1) If $1 \leq h \neq k \leq n$, we define $\varepsilon_{hk} \in \mathrm{Aut}(V^n)$ by

$$\begin{bmatrix} v_1 \\ \vdots \\ v_n \end{bmatrix} \mapsto \begin{bmatrix} w_1 \\ \vdots \\ w_n \end{bmatrix}, \quad \text{where } w_i = \begin{cases} v_k & \text{if } i = h \\ v_h & \text{if } i = k, \\ v_i & \text{otherwise.} \end{cases}$$

This automorphism satisfies $\varepsilon_{hk}^{-1} = \varepsilon_{hk}$.

(2) If $1 \leq h \leq n$, and if $0 \neq c \in F$, we define $\varepsilon_{h;c} \in \mathrm{Aut}(V^n)$ by

$$\begin{bmatrix} v_1 \\ \vdots \\ v_n \end{bmatrix} \mapsto \begin{bmatrix} w_1 \\ \vdots \\ w_n \end{bmatrix}, \quad \text{where } w_i = \begin{cases} cv_i & \text{if } i = h, \\ v_i & \text{otherwise.} \end{cases}$$

This automorphism satisfies $\varepsilon_{h;c}^{-1} = \varepsilon_{h,c^{-1}}$.

(3) If $1 \leq h \neq k \leq n$ and if $c \in F$, we define $\varepsilon_{hk;c} \in \mathrm{Aut}(V^n)$ by

$$\begin{bmatrix} v_1 \\ \vdots \\ v_n \end{bmatrix} \mapsto \begin{bmatrix} w_1 \\ \vdots \\ w_n \end{bmatrix}, \quad \text{where } w_i = \begin{cases} v_i + cv_k & \text{if } i = h, \\ v_i & \text{otherwise.} \end{cases}$$

This automorphism satisfies $\varepsilon_{hk;c}^{-1} = \varepsilon_{hk;-c}$.

Identifying the automorphisms of a finite-dimensional vector space V over a field F is a problem which will be of major importance to us later, and so it is important to characterize these functions.

Proposition 7.3 *Let V be a vector space of finite dimension n over a field F. Then the following conditions on an endomorphism α of V are equivalent:*
(1) *α is an automorphism of V;*
(2) *α is monic;*
(3) *α is epic.*

Proof By definition, (1) implies (2). Now assume (2). By Proposition 6.10, we see that the rank of α equals n and so $\mathrm{im}(\alpha) = V$ by Proposition 5.11, proving (3). Now assume (3). By Proposition 6.10, we see that the nullity of α equals $n - n = 0$ and so $\ker(\alpha) = \{0_V\}$, proving that α is monic as well, and so is bijective. This proves (1). $\quad\square$

Proposition 7.4 *Let V be a finite-dimensional vector space over a field F and let $\alpha \in \mathrm{End}(V)$. If there exists a $\beta \in \mathrm{End}(V)$ satisfying $\alpha\beta = \sigma_1$ or $\beta\alpha = \sigma_1$, then $\alpha \in \mathrm{Aut}(V)$ and $\beta = \alpha^{-1}$.*

Proof If $\beta\alpha = \sigma_1$ then $\ker(\alpha) \subseteq \ker(\sigma_1) = \{0_V\}$ and so, by Proposition 7.3, $\alpha \in \mathrm{Aut}(V)$. Similarly, if $\alpha\beta = \sigma_1$ then $\mathrm{im}(\alpha) \supseteq \mathrm{im}(\sigma_1) = V$ and so, by Proposition 7.3, $\alpha \in \mathrm{Aut}(V)$. Moreover, if $\alpha\beta = \sigma_1$ we see that $\alpha^{-1} = \alpha^{-1}\sigma_1 = \alpha^{-1}(\alpha\beta) = \beta$ and similarly $\alpha^{-1} = \beta$ when $\beta\alpha = \sigma_1$. $\square$

Example Proposition 7.3 and Proposition 7.4 are no longer true if we remove the condition of finite dimensionality. For example, let F be a field and let $V = F[X]$. Define the endomorphisms α and β of V by setting $\alpha : \sum_{i=0}^{n} a_i X^i \mapsto \sum_{i=0}^{n} a_i X^{i+1}$ and $\beta : \sum_{i=0}^{n} a_i X^i \mapsto \sum_{i=1}^{n} a_i X^{i-1}$. Then $\alpha, \beta \notin \mathrm{Aut}(V)$, despite the fact that α is monic and β is epic. Moreover, $\beta\alpha = \sigma_1$ but $\alpha\beta \neq \sigma_1$.

Let V be a vector space over a field F and let $\alpha \in \mathrm{End}(V)$. A subspace W of V is *invariant* under α if and only if $\alpha(w) \in W$ for all $w \in W$ or, in other words, if and only if $\alpha(W) \subseteq W$. Thus, W is invariant under α if and only if the restriction of α to W is an endomorphism of W. It is clear that V and $\{0_V\}$ are both invariant under every endomorphism of V. If $\alpha \in \mathrm{End}(V)$ then $\mathrm{im}(\alpha)$ and $\ker(\alpha)$ are both invariant under α.

Example Let F be a field and, for each positive integer k, let W_k be the subspace of $F[X]$ composed of all polynomials of degree at most k. Let δ be the *formal differentiation* endomorphism of $F[X]$, namely the endomorphism defined by $\delta : \sum_{i=0}^{n} a_i X^i \mapsto \sum_{i=0}^{n} i a_i X^{i-1}$. Then each of the subspaces W_k is invariant under δ. Now assume that F is of characteristic 0. If $p(X) = \sum_{i=0}^{n} a_i X^i \in W_k$ and if $a \in F$ then it is easy to check that $p(X) = p(a) + \sum_{h=1}^{n} \frac{1}{h!}[\delta^h(p)(a)](X-a)^h$. The coefficients $\frac{1}{h!}[\delta^h(p)(a)]$ are known as the *Taylor coefficients* of $p(X)$ around a.

Example Let $V = \mathbb{R}^2$ and let α be the automorphism of V defined by $\alpha : \begin{bmatrix} a \\ b \end{bmatrix} \mapsto \begin{bmatrix} b \\ -a \end{bmatrix}$. Let W be a proper subspace of V which is invariant under α. Then $\dim(W) \leq 1$ and so there exists a vector $w = \begin{bmatrix} c \\ d \end{bmatrix}$ satisfying $W = \mathbb{R}w$. Since $\alpha(w) = \begin{bmatrix} d \\ -c \end{bmatrix}$, it follows that there exists a real number e such that $\alpha(w) = ew$. That is to say, $ec = d$ and $ed = -c$. From this we learn that $ce^2 = -c$, and so $c = d = 0$. This proves that $W = \left\{ \begin{bmatrix} 0 \\ 0 \end{bmatrix} \right\}$, and so we see that V has no proper nontrivial subspaces invariant under α.

Example Let F be a field and let n be a positive integer. Let α be the automorphism of F^n defined by $\alpha : \begin{bmatrix} a_1 \\ a_2 \\ \vdots \\ a_n \end{bmatrix} \mapsto \begin{bmatrix} a_n \\ a_1 \\ \vdots \\ a_{n-1} \end{bmatrix}$. A subspace W invariant under α is *cyclic*.

Cyclic subspaces of F^n, where F is a finite field, are important in defining certain families of error-correcting codes.

Let F be a field. An element a of an F-algebra $(K, \bullet)$ is *idempotent* if and only if $a^2 = a$. If V is a vector space over F, then an idempotent element of $\mathrm{End}(V)$ is called a *projection*. Note that if $\alpha \in \mathrm{End}(V)$ is a projection and if $w = \alpha(v) \in \mathrm{im}(\alpha)$ then $\alpha(w) = \alpha^2(v) = \alpha(v) = w$, so that the restriction of α to its image is just σ_1. The converse is also true. If $\alpha \in \mathrm{End}(V)$ satisfies the condition that the restriction of α to its image is just σ_1, then for each $v \in V$ we have $\alpha^2(v) = \alpha(\alpha(v)) = \sigma_1(\alpha(v)) = \alpha(v)$ and so α is a projection.

Example If F is a field then the endomorphism of F^3 defined by

$$\begin{bmatrix} a \\ b \\ c \end{bmatrix} \mapsto \begin{bmatrix} 3a - 2c \\ -a + b + c \\ 3a - 2c \end{bmatrix}$$

is a projection.

Example The sum of two projections need not be a projection. For example, if $V = \mathbb{R}^3$ then the endomorphisms α and β of V defined by

$$\alpha : \begin{bmatrix} a \\ b \\ c \end{bmatrix} \mapsto \begin{bmatrix} a \\ b \\ 0 \end{bmatrix} \quad \text{and} \quad \beta : \begin{bmatrix} a \\ b \\ c \end{bmatrix} \mapsto \begin{bmatrix} 0 \\ b \\ c \end{bmatrix}$$

are projections, but $\alpha + \beta$ is not a projection.

Example If W is a subspace of a vector space V over a field F having a complement Y in V, we know that every element $v \in V$ can be written in a unique way in the form $w + y$, where $w \in W$ and $y \in Y$. The endomorphism of V defined by $v \mapsto w$ is a projection the image of which is W. Statisticians often consider data in $V = \mathbb{R}^n$ and use a projection in $\mathrm{End}(V)$ to project it onto a subspace W of V that best preserves the variance in the data. This standard method in data analysis is called *principle component analysis* and there exist several efficient algorithms for performing it.

In fact, all projections of a vector space are of the form in the previous example, as the following example shows.

Proposition 7.5 *Let V be a vector space over a field F and let $\alpha \in \mathrm{End}(V)$ be a projection. Then $V = \mathrm{im}(\alpha) \oplus \ker(\alpha)$.*

Proof If $v \in \mathrm{im}(\alpha) \cap \ker(\alpha)$ then there exists an element y of V satisfying $v = \alpha(y)$ and so $v = \alpha(v) = 0_V$. Thus $\mathrm{im}(\alpha)$ and $\ker(\alpha)$ are disjoint. If v is an arbitrary vector in V then $v = [v - \alpha(v)] + \alpha(v) \in \ker(\alpha) + \text{im}(\alpha)$. Therefore, $V = \text{im}(\alpha) \oplus \ker(\alpha)$. $\qquad\square$

Proposition 7.6 *Let V be a vector space over a field F and let $\alpha \in \text{End}(V)$. A subspace W of V is invariant under α if and only if $\beta\alpha\beta = \alpha\beta$ for each projection β of V the image of which is W.*

Proof Assume that W is invariant under α and let β be a projection of V the image of which is W. By Proposition 7.5, we have $V = W \oplus \ker(\beta)$. If $v \in V$, we can therefore write $v = w + y$, where $w \in W$ and $y \in \ker(\beta)$. Hence $\alpha\beta(v) = \alpha\beta(w) + \alpha\beta(y) = \alpha(w) + 0_V = \alpha(w) = \beta\alpha(w) = \beta\alpha\beta(v)$, showing that $\beta\alpha\beta = \alpha\beta$. Conversely, if $\beta\alpha\beta = \alpha\beta$ for each projection β of V the image of which is W then, for each such β, we have $w = \beta(w)$ for all $w \in W$ and so $\alpha(w) = \alpha\beta(w) = \beta\alpha\beta(w) \in W$, showing that W is invariant under α. $\qquad\square$

Proposition 7.7 *Let V be a vector space over a field F and let $\{W_1, \ldots, W_n\}$ be a set of subspaces of V. Then the following conditions are equivalent:*
(1) $V = W_1 \oplus \cdots \oplus W_n$;
(2) There exist projections $\alpha_1, \ldots, \alpha_n$ in $\text{End}(V)$ with $W_i = \text{im}(\alpha_i)$ for all $1 \leq i \leq n$, which satisfy the conditions $\alpha_i \alpha_j = \sigma_0$ for $i \neq j$ and $\alpha_1 + \cdots + \alpha_n = \sigma_1$.

Proof (1) $\Rightarrow$ (2): From (1) it follows that every $v \in V$ can be written in a unique manner as $\sum_{i=1}^{n} w_i$, where $w_i \in W_i$ for all $1 \leq i \leq n$. Define α_i to be the projection $v \mapsto w_i$ for each i. It is easy to verify that these linear transformations do indeed satisfy the required conditions.

(2) $\Rightarrow$ (1): Since $\alpha_1 + \cdots + \alpha_n = \sigma_1$, we surely have $V = \sum_{i=1}^{n} \text{im}(\alpha_i) = \sum_{i=1}^{n} W_i$. If $0_V \neq v \in W_h \cap \sum_{j \neq h} W_j$ then there exists an $i \neq h$ such that $\alpha_i(v) \neq 0_V$. But $\alpha_h(v) = v$ so $\alpha_i \alpha_h \neq \sigma_0$, a contradiction. Therefore, $W_h \cap \sum_{j \neq h} W_j = \{0_V\}$ for each $1 \leq h \leq n$, proving (1). $\qquad\square$

Proposition 7.8 *Any two complements of a subspace W of a vector space V over a field F are isomorphic.*

Proof Let U and Y be complements of W in V. By Proposition 7.7, we know that there exists a projection $\beta \in \text{End}(V)$ the image of which is U and the kernel of which is W. Let α be the restriction of β to Y. The linear transformation α is a monomorphism since $\ker(\alpha) \subseteq \ker(\beta) \cap Y = W \cap Y = \{0_V\}$. Any vector $u \in U$ can be written as $w + y$, where $w \in W$ and $y \in Y$, and we have $\alpha(y) = \beta(y) = $

$\beta(w) + \beta(y) = \beta(w + y) = \beta(u) = u$. Thus we see that α is also epic and hence is the desired isomorphism. $\square$

We now introduce a notion which is basic in all branches of mathematics. A relation $\equiv$ defined on a given nonempty set U is called an *equivalence relation* if and only if the following conditions are satisfied:
(1) $u \equiv u$ for all $u \in U$;
(2) $u \equiv u'$ if and only if $u' \equiv u$;
(3) If $u \equiv u'$ and $u' \equiv u''$ then $u \equiv u''$.

Example Let B be a nonempty subset of a set A and define a relation $\equiv_B$ on A by setting $a \equiv_B a'$ if and only if $a = a'$ or both a and a' belong to B. Then $\equiv_B$ is an equivalence relation on A. In particular, if W is a subspace of a vector space V then the relation $\equiv_W$ defined on V by setting $v \equiv_W v'$ if and only if $v - v' \in W$ is an equivalence relation on V.

Example Let V and W be a vector spaces over a field F and let $\alpha \in \mathrm{Hom}_F(V, W)$. Define a relation $\equiv$ on V by setting $v \equiv v'$ if and only if $\alpha(v) = \alpha(v')$. This is easily seen to be an equivalence relation.

Let V be a vector space over a field F. A subset G of $\mathrm{Aut}(V)$ is a *group of automorphisms* if it is closed under taking products, contains σ_1, and satisfies the condition that $\alpha^{-1} \in G$ whenever $\alpha \in G$. Clearly, $\mathrm{Aut}(V)$ itself is such a group. The notion of a group of automorphisms is very important in linear algebra and its applications, but here we will only touch on it.

Example Let V be a vector space over a field F and let $\alpha \in \mathrm{Aut}(V)$. Then $\{\alpha^i \mid i \in \mathbb{Z}\}$ is surely a group of automorphisms.

Example Let V be a vector space over a field F and let Ω be a nonempty set. Every permutation π of Ω defines an automorphism α_π of the vector space V^Ω over F defined by $\alpha_\pi(f) : i \mapsto f(\pi(i))$ for all $i \in \Omega$ and all $f \in V^\Omega$. The collection G of all such automorphisms is a group of automorphisms in $\mathrm{Aut}(V^\Omega)$.

Proposition 7.9 *If V is a vector space over a field F and if G is a group of automorphisms of V then G defines an equivalence relation $\sim_G$ on V by setting $v \sim_G v'$ if and only if there exists an element α of G satisfying $\alpha(v) = v'$.*

Proof If $v \in V$ then $\sigma_1(v) = v$, and so $v \sim_G v$. If $v, v' \in V$ satisfy $v \sim_G v'$ then there exists an element α of G satisfying $\alpha(v) = v'$, and so $v' = \alpha^{-1}(v)$. Thus $v' \sim_G v$. Finally, if $v, v', v'' \in V$ satisfy $v \sim_G v'$ and $v' \sim_G v''$ then there exist elements α and β of G satisfying $\alpha(v) = v'$ and $\beta(v') = v''$, and so $\beta\alpha(v) = v''$. Thus $v \sim_G v''$. $\square$

Proposition 7.10 *If V is a vector space over a field F and if G is a group of automorphisms of V then all elements of G have the same rank.*

Proof If $\alpha \in G$ then, by Proposition 6.11, $\mathrm{rk}(\sigma_1) = \mathrm{rk}(\alpha\alpha^{-1}) \leq \mathrm{rk}(\alpha) = \mathrm{rk}(\alpha\sigma_1) \leq \mathrm{rk}(\sigma_1)$ and so $\mathrm{rk}(\alpha) = \mathrm{rk}(\sigma_1)$. $\qquad\square$

Exercises

Exercise 316
Let V be a vector space over GF(3). Find an endomorphism α of V satisfying $\alpha(v) + \alpha(v) = v$ for all $v \in V$.

Exercise 317
Let V be a vector space finitely generated over a field F and let $\alpha, \beta, \gamma \in \mathrm{End}(V)$. Find necessary and sufficient conditions for there to exist an endomorphism θ of V satisfying $\alpha\gamma\beta = \beta\theta\alpha$.

Exercise 318
Let $F = \mathrm{GF}(2)$ and let n be a positive integer. Let $\alpha : F^n \to F^n$ be the function defined by $\alpha : \begin{bmatrix} a_1 \\ \vdots \\ a_n \end{bmatrix} \mapsto \begin{bmatrix} a'_1 \\ \vdots \\ a'_n \end{bmatrix}$, where $0' = 1$ and $1' = 0$. Is α an endomorphism of F^n?

Exercise 319
Let $\alpha, \beta : \mathbb{Q}[X] \to \mathbb{Q}[X]$ be defined by $\alpha : p(X) \mapsto Xp(X)$ and $\beta : p(X) \mapsto X^2 p(X)$. Show that α, β, and $\alpha - \beta$ are all monic endomorphisms of $\mathbb{Q}[X]$.

Exercise 320
Let V be a finitely-generated vector space over a field F and let $\alpha \in \mathrm{End}(V)$. Show that α is not monic if and only if there exists an endomorphism $\beta \neq \sigma_0$ of V satisfying $\alpha\beta = \sigma_0$.

Exercise 321
Let V be a vector space over a field F and let $\alpha \in \mathrm{End}(V)$. Show that $\ker(\alpha) = \ker(\alpha^2)$ if and only if $\ker(\alpha)$ and $\mathrm{im}(\alpha)$ are disjoint.

Exercise 322
Let V be a vector space over a field F and let $\alpha \in \mathrm{End}(V)$. Show that $\mathrm{im}(\alpha) = \mathrm{im}(\alpha^2)$ if and only if $V = \ker(\alpha) + \mathrm{im}(\alpha)$.

Exercise 323
Let V be a vector space over a field F and let $K = F \times V \times \text{End}(V)$, which is again a vector space over F. Define an operation $\diamond$ on K by setting $(a, v, \alpha) \diamond (b, w, \beta) = (ab, aw + \beta(v), \beta\alpha)$. Is $(K, \diamond)$ an F-algebra? Is it associative? Is it unital?

Exercise 324
Let V be a vector space over a field F, and let $\text{Aff}(V, V)$ be the set of all affine transformations from V to itself. Is $\text{Aff}(V, V)$, on which we have defined the operations of addition and composition of functions, an associative unital F-algebra?

Exercise 325
Let $\alpha \in \text{Aut}(\mathbb{R}^2)$ be defined by $\alpha : \begin{bmatrix} a \\ b \end{bmatrix} \mapsto \begin{bmatrix} -b \\ a \end{bmatrix}$. Show that $\mathbb{R}\{\alpha, \sigma_1\}$ is a unital subalgebra of $\text{End}(\mathbb{R}^2)$. Show that it is proper by giving an example of an endomorphism of $\mathbb{R}^2$ not in this subalgebra.

Exercise 326
Let V be the space of all real-valued functions on the interval $[-1, 1]$ which are infinitely differentiable, and let δ be the endomorphism of V which assigns to each function f its derivative. Find the kernel and image of δ.

Exercise 327
Let $\alpha : \mathbb{C} \to \mathbb{C}$ be the function defined by $\alpha : a + bi \mapsto -b + ai$. Is α an endomorphism of $\mathbb{C}$ considered as a vector space over $\mathbb{R}$? Is it an endomorphism of $\mathbb{C}$ considered as a vector space over itself?

Exercise 328
Let V be a vector space of finite dimension n over a field F and let $\alpha \in \text{End}(V)$. Show that there exists an automorphism β of V satisfying $\alpha\beta\alpha = \alpha$.

Exercise 329
Let $V = \mathcal{M}_{2\times2}(\mathbb{R})$, which is a vector space over $\mathbb{R}$. Let $\alpha \in V^V$ be defined by $\alpha : \begin{bmatrix} a_{11} & a_{12} \\ a_{21} & a_{22} \end{bmatrix} \mapsto \begin{bmatrix} |a_{11}| & |a_{12}| \\ |a_{21}| & |a_{22}| \end{bmatrix}$. Is α an endomorphism of V?

Exercise 330
Consider $\mathbb{R}$ as a vector space over $\mathbb{Q}$ and let α be an endomorphism of this space satisfying the condition that there exists an $a_0 \in \mathbb{R}$ such that α is continuous at a_0. Show that α is continuous at every $a \in \mathbb{R}$.

Exercise 331
Let A be a nonempty set and let V be the collection of all subsets of A, considered as a vector space over $\text{GF}(2)$. For which subsets C of A is the function $B \mapsto B \cup C$ an endomorphism of V?

Exercise 332

Let V be a vector space of finite dimension n over a field F and let $\{\alpha_{ij} \mid 1 \le i, j \le n\}$ be a collection of endomorphisms of V, not all of which are equal to σ_0, satisfying the condition that

$$\alpha_{ij}\alpha_{kh} = \begin{cases} \alpha_{ih} & \text{if } j = k, \\ \sigma_0 & \text{otherwise.} \end{cases}$$

Show that there exists a basis $\{v_1, \ldots, v_n\}$ of V such that

$$\alpha_{jk}(v_i) = \begin{cases} v_j & \text{if } i = k, \\ 0_V & \text{otherwise.} \end{cases}$$

Exercise 333

Let V be a vector space of finite dimension n over a field F and choose an element $\alpha \in \text{End}(V)$. Let $\varphi : \text{End}(V) \to \text{End}(V)$ be the function defined by $\beta \mapsto \beta\alpha$. This is an endomorphism of $\text{End}(V)$, considered as a vector space over F. Show that a positive integer n satisfies $\alpha^n = \sigma_0$ if and only if φ^n is the 0-function.

Exercise 334

Let α be an endomorphism of $\mathbb{R}^3$ satisfying the condition that $\alpha^2 = \sigma_0$. Show that there exists a linear transformation $\beta : \mathbb{R}^3 \to \mathbb{R}$ and that there exists a vector $y \in \mathbb{R}^3$ satisfying $\alpha(v) = \beta(v)y$ for all $v \in \mathbb{R}^3$.

Exercise 335

For each $0 \ne a \in \mathbb{R}$, let $\beta_a : \mathbb{C} \to \mathbb{C}$ be the function defined by $\beta_a : z \mapsto z + a\bar{z}$. Show that β_a is an endomorphism of $\mathbb{C}$ considered as a vector space over $\mathbb{R}$, and describe its image and kernel.

Exercise 336

Let V be a vector space finitely generated over $\mathbb{Q}$ and let $\alpha, \beta \in \text{End}(V)$ satisfy $3\alpha^3 + 7\alpha^2 - 2\alpha\beta + 4\alpha - \sigma_1 = \sigma_0$. Show that $\alpha\beta = \beta\alpha$.

Exercise 337

Let F be a field of characteristic other than 2 and let V be a vector space of finite dimension n over F. Let α be an endomorphism of V satisfying the condition that $\alpha^2 = \sigma_1$. Show that $\text{rk}(\sigma_1 - \alpha) + \text{rk}(\sigma_1 + \alpha) = n$.

Exercise 338

Let V be a vector space over a field F which is not finitely generated, and let $\sigma_0 \ne \alpha \in \text{End}(V)$. Set $A = \{\beta \in \text{End}(V) \mid \alpha\beta = \sigma_1\}$. Show that if A has more than one element then it is infinite.

Exercise 339

Let V be a vector space over a field F having dimension greater than 1. Show that there exists a function $\alpha \in V^V$ which is not an endomorphism of V but which nonetheless satisfies the condition that $\alpha(av) = a\alpha(v)$ for all $a \in F$ and all $v \in V$.

Exercise 340
Let V be a vector space over a field F satisfying the condition that $\alpha\beta = \beta\alpha$ for all $\alpha, \beta \in \mathrm{End}(V)$. Show that $\dim(V) = 1$.

Exercise 341
Let $V = \mathcal{M}_{2\times 2}(\mathbb{R})$, considered as a vector space over $\mathbb{R}$. Let $\alpha : V \to V$ be the function defined by $\alpha : \begin{bmatrix} a & b \\ c & d \end{bmatrix} \mapsto \begin{bmatrix} a + 2b + c + 2d & 2a + 4b + 3c + 5d \\ 3a + 6b + 2c + 5d & a + 2b + c + 2d \end{bmatrix}$.
Is α an endomorphism of V? Is it an automorphism of V?

Exercise 342
Let V be the vector space of all continuous functions from $\mathbb{R}$ to itself and let $\alpha : V \to V$ be the function defined by $\alpha : f(x) \mapsto [x^2 + \sin(x) + 2] f(x)$. Show that α is an automorphism of V.

Exercise 343
Let F be a field and let $\alpha : F[X] \to F[X]$ be the function defined by $\alpha : p(X) \mapsto p(X + 1)$. Is α an endomorphism of $F[X]$? Is it an automorphism?

Exercise 344
Let F be a field and, for each $a \in F$, let θ_a be the endomorphism of $F[X]$ defined by $\theta_a : p(X) \mapsto p(X + a)$. Let $\alpha \in \mathrm{End}(F[X])$ satisfy $\alpha(X) \in F$ and $\alpha\theta_a = \theta_a\alpha$ for all $a \in F$. Can α be a monomorphism?

Exercise 345
Let $\alpha \in \mathrm{End}(\mathbb{R}^3)$ be given by $\alpha : \begin{bmatrix} a \\ b \\ c \end{bmatrix} \mapsto \begin{bmatrix} a - 2b \\ c \\ a - b \end{bmatrix}$. Is α an automorphism of $\mathbb{R}^3$?

Exercise 346
Let α be the endomorphism of $\mathbb{R}^{(\infty)}$ defined by

$$\alpha : [a_1, a_2, a_3, \ldots] \mapsto [b_1, b_2, b_3, \ldots],$$

where $b_h = \sum_{j \leq h} (-1)^{j-1} \binom{h-1}{j-1} a_j$ for each $h \geq 1$. Show that α is an automorphism satisfying $\alpha = \alpha^{-1}$.

Exercise 347
Let V be a vector space finitely generated over $\mathbb{R}$ and let α be an endomorphism of V satisfying $\alpha^3 + 4\alpha^2 + 2\alpha + \sigma_1 = \sigma_0$. Show that $\alpha \in \mathrm{Aut}(V)$.

Exercise 348

Let V be a vector space over a field F and let $\alpha, \beta \in \mathrm{End}(V)$ satisfy $\alpha\beta = \sigma_1$. Set $\varphi = \sigma_1 - \beta\alpha$. Show that for every integer $n \geq 1$ we have $\sigma_1 = \sum_{k=0}^{n-1} \beta^k \varphi \alpha^k + \beta^n \alpha^n$.

Exercise 349

Let V be the space of all polynomial functions from the interval $[0, 1]$ on the real line to $\mathbb{R}$. Let α and β be the endomorphisms of V defined by $\alpha(f) : x \mapsto \int_0^x f(t)\,dt$ and $\beta(f) : x \mapsto \int_x^1 f(t)\,dt$. Find $\mathrm{im}(\alpha + \beta)$. Is it true that $\alpha\beta = \beta\alpha$?

Exercise 350

Let F be a field and let $V = F^\infty$. Let $n > 1$ be an integer. Each vector

$$y = \begin{bmatrix} d_1 \\ \vdots \\ d_n \end{bmatrix} \in F^n$$ defines an endomorphism θ_y of V by $\theta_y : [a_1, a_2, \ldots] \mapsto$

$[b_1, b_2, \ldots]$, where $b_h = \sum_{i=1}^n a_{h-1+i} d_i$, for $h = 1, 2, \ldots$. Show that if θ_y is a monomorphism then the polynomial $p(X) = \sum_{i=1}^n d_i X^{i-1} \in F[X]$ has no roots in F.

Exercise 351

Let $F = \mathrm{GF}(5)$ and let $V = F^3$. How many endomorphisms α of V satisfy the

conditions $\alpha\left(\begin{bmatrix} 1 \\ 0 \\ 0 \end{bmatrix}\right) = \begin{bmatrix} 2 \\ 1 \\ 0 \end{bmatrix}$ and $\alpha\left(\begin{bmatrix} 0 \\ 3 \\ 0 \end{bmatrix}\right) = \begin{bmatrix} 1 \\ 1 \\ 1 \end{bmatrix}$?

Exercise 352

Let V be the set of all continuous functions from $\mathbb{R}$ to itself, which is a vector space over $\mathbb{R}$. Let $\alpha : V \to V$ be the function defined by $\alpha(f) : x \mapsto f(\frac{x}{2})$ for all $x \in \mathbb{R}$ and all $f \in V$. Is α an automorphism of V?

Exercise 353

Let $V = \mathbb{R}^\infty$ and let W be the subspace of V consisting of all convergent sequences. Let $\alpha \in \mathrm{End}(V)$ be defined by $\alpha : [a_1, a_2, \ldots] \mapsto [b_1, b_2, \ldots]$, where $b_h = \frac{1}{h}(\sum_{i=1}^h a_i)$ for all $h \geq 1$. If $v \in V$ satisfies $\alpha(v) \in W$, is v itself necessarily in W?

Exercise 354

Let V be a vector space over a field F and let $\alpha \in \mathrm{Aut}(V)$. Let $W_1, \ldots, W_k$ be subspaces of V satisfying $V = \bigoplus_{i=1}^k W_i$. For each $1 \leq i \leq k$, let $Y_i = \{\alpha(w) \mid w \in W_i\}$. Is $V = \bigoplus_{i=1}^k Y_i$?

Exercise 355

Consider $\mathbb{R}$ as a vector space over $\mathbb{Q}$. An endomorphism α of this space is *bounded* if and only if there exists a nonnegative real number $m(\alpha)$ satisfying the condition that $|\alpha(x)| \leq m(\alpha)|x|$ for all $x \in \mathbb{R}$. Does the set of all bounded endomorphisms of $\mathbb{R}$ form an $\mathbb{R}$-subalgebra of $\text{End}(\mathbb{R})$?

Exercise 356
Let F be a field, let n be a positive integer, and let $V = \mathcal{M}_{n \times n}(F)$. Given a matrix $B \in V$, is the function $\alpha_B : V \to V$ defined by $\alpha_B : A \mapsto AB + BA$ an endomorphism of V?

Exercise 357
Let F be a field and let $V = F[X]$. Let $\delta \in \text{End}(V)$ be the formal differentiation function and let $\alpha \in \text{End}(V)$ be defined by $\alpha : p(X) \mapsto Xp(X)$. Show that $\alpha\delta - \delta\alpha = \sigma_1$.

Exercise 358
Let V be a nontrivial vector space over a field F. Is the set of all automorphisms of V a subspace of the vector space $\text{End}(V)$ over F?

Exercise 359
Consider GF(3) as a vector space over itself. Does there exist an automorphism of this space other than σ_1?

Exercise 360
Let $V = F^\infty$ for some field F. Each $w = [c_1, c_2, \ldots] \in V$ defines a function $\beta_w :$ $V \to V$ by $\beta_w : [a_1, a_2, \ldots] \mapsto [a_1, a_1 c_1 + a_2, (a_1 c_1 + a_2)c_2 + a_3, \ldots]$. Show that β_w is an automorphism of V.

Exercise 361
Let V be a vector space over a field F; let $\alpha \in \text{End}(V)$ and let $\beta \in \text{Aut}(V)$. Define the function $\theta : V^2 \to V^2$ by setting $\theta : \begin{bmatrix} v \\ v' \end{bmatrix} \mapsto \begin{bmatrix} \beta(v) \\ \alpha(v) + v' \end{bmatrix}$. Is θ necessarily an automorphism of V^2?

Exercise 362
Let $F = $ GF(5) and let $\alpha \in \text{Aut}(F^2)$ be defined by $\alpha : \begin{bmatrix} a \\ b \end{bmatrix} \mapsto \begin{bmatrix} 2b \\ a + 2b \end{bmatrix}$. Show that there exists a positive integer h satisfying $\alpha^{h+1} = \alpha$ and find the smallest such integer h.

Exercise 363
Let F be a field of characteristic other than 2. Let V be a vector space over F and let $\alpha, \beta, \gamma, \delta$ be endomorphisms of V satisfying the condition that $\alpha - \beta$ and $\alpha + \beta$ are automorphisms of V. Show that there exist endomorphisms φ and ψ of V satisfying $\varphi\alpha + \psi\beta = \gamma$ and $\psi\alpha + \varphi\beta = \delta$.

Exercise 364

Let V be a vector space of finite dimension n over a field F. Let $\alpha \in \text{End}(V)$ and assume that there exists a vector in $y \in V$ satisfying the condition that $D = \{\alpha(y), \alpha^2(y), \ldots, \alpha^n(y)\}$ is a basis for V. Show that $D' = \{y, \alpha(y), \ldots, \alpha^{n-1}(y)\}$ is also a basis for V and that $\alpha \in \text{Aut}(V)$.

Exercise 365

Let F be a field and let $V = F^{(\mathbb{Z})}$. Let α be the endomorphism of V defined by $\alpha(f): i \mapsto f(i+1)$ for all $f \in V$. Show that $\alpha - c\sigma_1 \notin \text{Aut}(V)$ for all $0 \neq c \in F$.

Exercise 366

Let V be a vector space of finite dimension n over a field F, and let $0 < k < n$ be a positive integer. Let A_k be the set of all subspaces of V having dimension k. Let $\alpha \in \text{Aut}(V)$ and, for each $W \in A_k$, let $\theta_\alpha(W) = \{\alpha(w) \mid w \in W\}$. Show that the function θ_α is a permutation of A_k.

Exercise 367

Let r, s, and t be distinct real numbers and let α be the endomorphism of $\mathbb{R}^3$ defined by $\alpha : \begin{bmatrix} a \\ b \\ c \end{bmatrix} \mapsto \begin{bmatrix} a + br + cr^2 \\ a + bs + cs^2 \\ a + bt + ct^2 \end{bmatrix}$. Is α an automorphism of $\mathbb{R}^3$?

Exercise 368

Let V be a vector space over a field F and let $\alpha \in \text{End}(V)$. Show that $W = \bigcup_{i=1}^{\infty} \ker(\alpha^i)$ is a subspace of V which is invariant under α.

Exercise 369

Let α and β be the endomorphisms of $\mathbb{Q}^4$ defined by

$$\alpha : \begin{bmatrix} a \\ b \\ c \\ d \end{bmatrix} \mapsto \begin{bmatrix} 2a - 2b - 2c - 2d \\ 5b - c - d \\ -b + 5c - d \\ -b - c + 5d \end{bmatrix} \quad \text{and} \quad \beta : \begin{bmatrix} a \\ b \\ c \\ d \end{bmatrix} \mapsto \begin{bmatrix} 0 \\ -b + 2c + 3d \\ 2b - 3c + 6d \\ 3b + 6c + 2d \end{bmatrix}.$$

Find two nontrivial proper subspaces of $\mathbb{Q}^4$ which are invariant both under α and under β.

Exercise 370

Let F be a field and let $V = F^4$. Let α be the endomorphism of V defined by $\alpha : \begin{bmatrix} a \\ b \\ c \\ d \end{bmatrix} \mapsto \begin{bmatrix} a - b \\ a - b \\ a - b \\ c - b - d \end{bmatrix}$. Does there exist a two-dimensional subspace of V invariant under α?

Exercise 371

Let V be a vector space over a field F and let $\alpha \in \text{End}(V)$. If W and Y are subspaces of V which are invariant under α, show that both $W + Y$ and $W \cap Y$ are invariant under α.

Exercise 372

Let W be a subspace of a vector space V over a field F and let S be the set of all $\alpha \in \text{End}(V)$ such that W is invariant under α. Is S necessarily an F-subalgebra of $\text{End}(V)$?

Exercise 373

Let V be a vector space over a field F and let $\alpha \in \text{End}(V)$. If W is a subspace of V, show that the set of all subspaces of W which are invariant under α, partially ordered by inclusion, has a maximal element.

Exercise 374

Let $V = \mathbb{R}^\infty$ and let W be the subspace of V consisting of all sequences $[a_1, a_2, \ldots]$ for which the series $\sum_{i=1}^\infty a_i$ converges. Let σ be a permutation of the set of all positive integers and let $\alpha \in \text{End}(V)$ be defined by $\alpha : [a_1, a_2, \ldots] \to [a_{\sigma(1)}, a_{\sigma(2)}, \ldots]$. Is W invariant under α?

Exercise 375

Let V be a vector space over a field F. Let $0 \neq c \in F$ and let $\alpha \in \text{End}(V)$. Let $\{x_0, x_1, \ldots, x_n\}$ be a set of vectors in V satisfying $\alpha(x_0) = cx_0$ and $\alpha(x_i) - cx_i = x_{i-1}$ for all $1 \leq i \leq n$. Show that $F\{x_0, x_1, \ldots, x_n\}$ is a subspace of V which is invariant under α.

Exercise 376

Let F be a field which is not finite and let V be a vector space over F having dimension greater than 1. For each $0 \neq c \in F$, show that there exist infinitely-many distinct subspaces of V which are invariant under the endomorphism σ_c of V.

Exercise 377

Let V be a vector space of finite dimension n over a field F. Let $\alpha \in \text{End}(V)$ and let $\beta \in \text{End}(V)$ satisfy $\beta^2 = \alpha$. Find a positive integer k such that $\text{rk}(\beta) \leq \frac{1}{k}[\text{rk}(\alpha) + n]$.

Exercise 378

Let α and β be endomorphisms of a vector space V over a field F and let $\theta \in \text{Aut}(V)$ satisfy $\theta\alpha = \beta\theta$. Show that a subspace W of V is invariant under α if and only if $W' = \{\theta(w) \mid w \in W\}$ is invariant under β.

Exercise 379

Let α and β be endomorphisms of a vector space V over a field F satisfying $\alpha\beta = \beta\alpha$. Is $\ker(\alpha)$ invariant under β?

Exercise 380
Let V be a vector space over a field F and let $\alpha \in \mathrm{End}(V)$ be a projection. Show that $\sigma_1 - \alpha$ is also a projection.

Exercise 381
Let V be a vector space finitely generated over a field F and let $\alpha \in \mathrm{End}(V)$ satisfy the condition $\alpha^2(\sigma_1 - \alpha) = \sigma_0$. Is α necessarily a projection?

Exercise 382
Let V be the space of all continuous functions from $\mathbb{R}$ to itself and let $W = \mathbb{R}\{\sin(x), \cos(x)\} \subseteq V$. Let δ be the endomorphism of W which assigns to each function its derivative. Find a polynomial $p(X) \in \mathbb{R}[X]$ of degree 2 satisfying $p(\delta) = \sigma_0$.

Exercise 383
Let V be a vector space finitely generated over $\mathbb{Q}$ and assume that there exists an $\alpha \in \mathrm{Aut}(V)$ satisfying $\alpha^{-1} = \alpha^2 + \alpha$. Show that $\dim(V)$ is divisible by 3.

Exercise 384
Let n be a positive integer and let $G = \left\{ \begin{bmatrix} a_1 \\ \vdots \\ a_n \end{bmatrix} \in \mathbb{R}^n \;\middle|\; a_i \geq 0 \text{ for all } 1 \leq i \leq n \right\}$.
Let α be an endomorphism of $\mathbb{R}^n$ satisfying the condition that $\alpha(v) \in G$ implies that $v \in G$. Show that $\alpha \in \mathrm{Aut}(\mathbb{R}^n)$.

Exercise 385
Let V be a vector space over a field F and let W and Y be subspaces of V satisfying $W + Y = V$. Let Y' be a complement of Y in V and let Y'' be a complement of $W \cap Y$ in W. Show that $Y' \cong Y''$.

Exercise 386
Let F be a field of characteristic other than 2 and let V be a vector space over F. Let $\alpha, \beta \in \mathrm{End}(V)$ be projections satisfying the condition that $\alpha + \beta$ is also a projection. Show that $\alpha\beta = \beta\alpha = \sigma_0$.

Exercise 387
Let V be a vector space over F and let $\alpha, \beta \in \mathrm{End}(V)$. Show that α and β are projections satisfying $\ker(\alpha) = \ker(\beta)$ if and only if $\alpha\beta = \alpha$ and $\beta\alpha = \beta$.

Exercise 388
Let V be a vector space finitely generated over a field F and let $\alpha \neq \sigma_1$ be an endomorphism of V which is a product of projections. Show that $\alpha \notin \mathrm{Aut}(V)$.

Exercise 389
Let V be a vector space over $\mathbb{Q}$ and let $\alpha \in \mathrm{End}(V)$. Show that α is a projection if and only if $(2\alpha - \sigma_1)^2 = \sigma_1$.

Exercise 390

Let α and β be endomorphisms of a vector space V over a field F and let $f(X) \in F[X]$ satisfy $f(\alpha\beta) = \sigma_0$. Set $g(X) = Xf(X)$. Show that $g(\beta\alpha) = \sigma_0$.

Exercise 391

Let $V = \mathbb{R}^{\mathbb{R}}$ and let $g \in V$. Find necessary and sufficient conditions on g for the endomorphism $f \mapsto gf$ of V to be a projection.

Exercise 392

Let W be a subspace of a vector space V over a field F which is invariant under an endomorphism α of V. Let $\beta \in \text{End}(V)$ be a projection satisfying the condition that $\text{im}(\beta) = W$. Show that $\beta\alpha\beta = \alpha\beta$.

Exercise 393

Let V be a vector space of finite dimension n over a field F and let $\alpha \in \text{End}(V)$. Show that there exists an automorphism β of V and a projection θ of V satisfying $\alpha = \beta\theta$.

Exercise 394

Let F be a field of characteristic other than 2 and let V be a vector space over F. Let $\alpha \in \text{End}(V)$ be a projection satisfying the condition that $\alpha - \beta$ is a projection for all $\beta \in \text{End}(V)$. Show that $\alpha = \sigma_1$.

Exercise 395

Let V be a vector space over F and let $\alpha, \beta \in \text{End}(V)$ be projections satisfying the condition that $\text{im}(\alpha)$ and $\text{im}(\beta)$ are disjoint. Is it necessarily true that $\alpha\beta = \beta\alpha$?

Exercise 396

Let V be a vector space of finite dimension n over a field F and let $S = \text{End}(V) \smallsetminus \text{Aut}(V)$. For $\alpha, \beta \in S$, show that $\text{im}(\alpha) = \text{im}(\beta)$ if and only if $\{a\theta \mid \theta \in S\} = \{\beta\varphi \mid \varphi \in S\}$.

Exercise 397

Let F be a field. Does there exist an endomorphism α of F^3 which is not a projection satisfying the condition that α^2 is a projection equal neither to σ_0 nor to σ_1.

Exercise 398

Let V be a vector space finite dimensional over a field F and let α be an endomorphism of V. Show that there exist a positive integer k such that $\text{im}(\alpha^k)$ and $\ker(\alpha^k)$ are disjoint.

Exercise 399

Let F be a field of characteristic other than 2, and let V be a vector space over F. Let $\alpha \in \mathrm{End}(V)$ satisfy $\alpha^3 = \alpha$. Show that $V = W_1 \oplus W_2 \oplus W_3$, where $W_1 = \{v \in V \mid \alpha(v) = v\}$, $W_2 = \{v \in V \mid \alpha(v) = -v\}$, and $W_3 = \ker(\alpha)$.

Exercise 400

Let F be a field of characteristic other than 2 and let V be a finitely-generated vector space over F. Show that every endomorphism of V is the sum of two automorphisms of V.

Exercise 401

Let $n > 1$ be an integer and let $\theta : \mathbb{R}^n \to \mathbb{R}$ be the function defined by

$$\theta : \begin{bmatrix} a_1 \\ \vdots \\ a_n \end{bmatrix} \mapsto \sum_{i=1}^n a_i^2.$$ Assume that we can define an operation $\bullet$ on $\mathbb{R}^n$ satisfying the condition that $(\mathbb{R}^n, \bullet)$ is an associative unital $\mathbb{R}$-algebra with multiplicative identity e, and also satisfying the condition that $\theta(v \bullet w) = \theta(v)\theta(w)$ for all $v, w \in \mathbb{R}^n$. Show that $(\mathbb{R}^n, +, \bullet)$ is a division algebra over $\mathbb{R}$.

Exercise 402

Any sequence $v - [a_1, a_2, \ldots] \in \mathbb{R}^\infty$ defines an endomorphism α_v of $\mathbb{R}[X]$ which acts on elements of the canonical basis of $\mathbb{R}[X]$ according to the rule $\alpha_v : X^n \mapsto \sum_{k=0}^n \binom{n}{k}(k!)a_{k+1}X^{n-k}$ for each nonnegative integer n. Given $a \in \mathbb{R}$, find $v, w \in \mathbb{R}^\infty$ such that $\alpha_v : p(X) \mapsto p(X + a)$ and $\alpha_w : p(X) \mapsto p(X + a) - p(a)$.

Exercise 403

Let V be a vector space over a field F and let G be a group of automorphisms of V. For $v \in V$, define the *stabilizer* of v in G to be $G_v = \{\alpha \in G \mid \alpha(v) = v\}$. Is this necessarily a group of automorphisms of V?

Representation of Linear Transformations by Matrices

8

In this chapter, we show how we can study linear transformations between finitely-generated vector spaces by studying matrices. Let V and W be finitely-generated vector spaces over a field F, where $\dim(V) = n$ and $\dim(W) = k$. Fix bases $B = \{v_1, \ldots, v_n\}$ of V and $D = \{w_1, \ldots, w_k\}$ of W. From Proposition 5.4, we know that if we are given a linear transformation $\alpha \in \mathrm{Hom}(V, W)$ then for each $1 \le j \le n$ there exist scalars $a_{1j}, \ldots, a_{kj}$ satisfying the condition $\alpha(v_j) = \sum_{i=1}^{k} a_{ij} w_i$, and that these scalars are in fact uniquely determined by α. Thus α defines a matrix $[a_{ij}] \in \mathcal{M}_{k \times n}(F)$. Conversely, assume we have a matrix $A = [a_{ij}] \in \mathcal{M}_{k \times n}(F)$. Then we know that every vector v in V can be written in a unique way in the form $\sum_{j=1}^{n} b_j v_j$, and so A defines a linear transformation $\alpha \in \mathrm{Hom}(V, W)$ by setting $\alpha : v \mapsto \sum_{i=1}^{k} (\sum_{j=1}^{n} a_{ij} b_j) w_i$. Moreover, it is clear that different linear transformations in $\mathrm{Hom}(V, W)$ define different matrices in $\mathcal{M}_{k \times n}(F)$ and different matrices in $\mathcal{M}_{k \times n}(F)$ define different linear transformations in $\mathrm{Hom}(V, W)$. We summarize the above remarks in the following proposition.

With kind permission of the Special collections, Fine Arts Library, Harvard University.

The theory of matrices and their relation to linear transformations was developed in detail by the nineteenth-century British mathematician **Sir Arthur Cayley**, one of the most prolific researchers in history.

Proposition 8.1 *Let V be a vector space of finite dimension n over a field F and let W be a vector space of finite dimension k over F. For every basis B of V and every basis D of W there exists a bijective function $\Phi_{BD} : \mathrm{Hom}(V, W) \to \mathcal{M}_{k \times n}(F)$, which is an isomorphism of vector spaces over F.*

J.S. Golan, *The Linear Algebra a Beginning Graduate Student Ought to Know*, DOI 10.1007/978-94-007-2636-9_8, © Springer Science+Business Media B.V. 2012

Proof We have already seen that if $B = \{v_1, \ldots, v_n\}$ and $D = \{w_1, \ldots, w_k\}$, then the function Φ_{BD} is defined by $\Phi_{BD}(\alpha) = [a_{ij}]$, where $\alpha(v_j) = \sum_{i=1}^{k} a_{ij} w_i$ for all $1 \le j \le n$, and that this function is bijective. We are therefore left to show that this is a linear transformation. Indeed, if $\Phi_{BD}(\alpha) = [a_{ij}]$ and $\Phi_{BD}(\beta) = [b_{ij}]$ then

$$(\alpha + \beta)(v_j) = \sum_{i=1}^{k} (a_{ij} + b_{ij}) w_i = \sum_{i=1}^{k} a_{ij} w_i + \sum_{i=1}^{k} b_{ij} w_i = \alpha(v_j) + \beta(v_j)$$

for all $1 \le j \le n$, and so $\Phi_{BD}(\alpha + \beta) = \Phi_{BD}(\alpha) + \Phi_{BD}(\beta)$. Similarly, if $c \in F$ then $(c\alpha)(v_j) = \sum_{i=1}^{k} c a_{ij} w_i = c(\sum_{i=1}^{k} a_{ij} w_i) = c(\alpha(v_j))$ for all $1 \le j \le n$, and so $\Phi_{BD}(c\alpha) = c\Phi_{BD}(\alpha)$. Thus we see that Φ_{BD} is indeed a linear transformation and thus also an isomorphism. $\quad\square$

We have already seen that, in the above situation, $\dim(\mathcal{M}_{k \times n}(F)) = kn$ and so, by Proposition 6.9, we also see that $\dim(\mathrm{Hom}(V, W)) = kn$.

Example Let $V = \mathbb{R}^3$ and let B be the canonical basis on V. Each vector $v = \begin{bmatrix} a_1 \\ a_2 \\ a_3 \end{bmatrix} \in V$ defines a linear transformation $\alpha_v : V \to V$ given by $\alpha_v : w \mapsto v \times w$. Then $\Phi_{BB}(\alpha_v) = \begin{bmatrix} 0 & -a_3 & a_2 \\ a_3 & 0 & -a_1 \\ -a_2 & a_1 & 0 \end{bmatrix}$.

Example Let $V = \mathbb{R}^3$ and $W = \mathbb{R}^2$. Choose bases

$$B = \left\{ \begin{bmatrix} 0.5 \\ -0.5 \\ 0 \end{bmatrix}, \begin{bmatrix} 0.5 \\ 0 \\ -0.5 \end{bmatrix}, \begin{bmatrix} 0 \\ 0.5 \\ 0.5 \end{bmatrix} \right\}$$

of V and of $D = \left\{ \begin{bmatrix} 1 \\ 1 \end{bmatrix}, \begin{bmatrix} 1 \\ 0 \end{bmatrix} \right\}$ of W. If $\begin{bmatrix} r \\ s \\ t \end{bmatrix} \in \mathbb{R}^3$ then there exist $b_1, b_2, b_3 \in \mathbb{R}$ satisfying $\begin{bmatrix} r \\ s \\ t \end{bmatrix} = b_1 \begin{bmatrix} 0.5 \\ -0.5 \\ 0 \end{bmatrix} + b_2 \begin{bmatrix} 0.5 \\ 0 \\ -0.5 \end{bmatrix} + b_3 \begin{bmatrix} 0 \\ 0.5 \\ 0.5 \end{bmatrix} = \frac{1}{2} \begin{bmatrix} b_1 + b_2 \\ -b_1 + b_3 \\ -b_2 + b_3 \end{bmatrix}$, and so we have $2r = b_1 + b_2$, $2s = -b_1 + b_3$, and $2t = -b_2 + b_3$. From this we get $b_1 = r - s + t$, $b_2 = r + s - t$, and $b_3 = r + s + t$.

The matrix $A = \begin{bmatrix} 3 & 5 & 7 \\ 4 & 8 & 2 \end{bmatrix}$ defines a linear transformation $\alpha \in \mathrm{Hom}(V, W)$ given by

$$\alpha : b_1 \begin{bmatrix} 0.5 \\ -0.5 \\ 0 \end{bmatrix} + b_2 \begin{bmatrix} 0.5 \\ 0 \\ -0.5 \end{bmatrix} + b_3 \begin{bmatrix} 0 \\ 0.5 \\ 0.5 \end{bmatrix}$$

$$\mapsto (3b_1 + 5b_2 + 7b_3) \begin{bmatrix} 1 \\ 1 \end{bmatrix} + (4b_1 + 8b_2 + 2b_3) \begin{bmatrix} 1 \\ 0 \end{bmatrix},$$

so

$$\alpha\left(\begin{bmatrix} r \\ s \\ t \end{bmatrix}\right) = (15r + 9s + 5t) \begin{bmatrix} 1 \\ 1 \end{bmatrix} + (14r + 6s - 2t) \begin{bmatrix} 1 \\ 0 \end{bmatrix}$$

$$= \begin{bmatrix} 29r + 15s + 3t \\ 15r + 9s + 5t \end{bmatrix}.$$

It is very important to emphasize that the matrix representation of a linear transformation depends on the bases which we fixed at the beginning, and on the order in which the elements of the bases are written! If we choose different bases or write the elements of a chosen basis in a different order, we will get a different matrix. Shortly, we will consider the relation between the matrices which represent a given linear transformation with respect to different bases.

Let V be a vector space finitely generated over a field F, let α be an endomorphism of V, and let W be a subspace of V which is invariant under α. As we have already seen, the restriction β of α to W is an endomorphism of W. Now, let $B = \{v_1, \dots, v_k\}$ be a basis for W, which we can expand to a basis $D = \{v_1, \dots, v_n\}$ for all of V. If $\Phi_{DD}(\alpha) = [a_{ij}]$ then for all $1 \leq j \leq k$ we have $\alpha(v_j) = \sum_{i=1}^{k} a_{ij} v_i$, and so $a_{ij} = 0$ whenever $1 \leq j \leq k$ and $k < i \leq n$. Thus we see that the matrix $\Phi_{DD}(\alpha)$ is of the form $\begin{bmatrix} A_{11} & A_{21} \\ O & A_{22} \end{bmatrix}$, where $A_{11} = \Phi_{BB}(\beta)$. The subspace $Y = F\{v_{k+1}, \dots, v_n\}$ of V is a complement of W in V. If it too is invariant under α then we would also have $A_{21} = O$, and so α is represented by a matrix composed of two square matrices "strung out" along the diagonal. From a computational point of view, such a representation has distinct advantages.

Beside addition and scalar multiplication of matrices, we can also define the product of two matrices, provided that these matrices are of suitable sizes. Let $(K, \bullet)$ be an associative unital algebra over a field F. If $A = [v_{ij}] \in \mathcal{M}_{k \times n}(K)$ and $B = [w_{jh}] \in \mathcal{M}_{n \times t}(K)$ for some positive integers k, n, and t, we define the matrix AB to be the matrix $[y_{ih}] \in \mathcal{M}_{k \times t}(K)$ where, for each $1 \leq i \leq k$ and all $1 \leq h \leq t$, we set $y_{ih} = \sum_{j=1}^{n} v_{ij} \bullet w_{jh}$. For the most part, we will be interested in this construction for the case $K = F$, but sometimes we will have need of the more general construction. Note that a necessary condition for the product of two matrices to be defined is that the number of columns in the first matrix be equal to the number of rows in the second matrix.

Example If $A = \begin{bmatrix} 2 & 3 & 2 \\ -1 & 2 & 1 \end{bmatrix} \in \mathcal{M}_{2\times3}(\mathbb{Q})$ and $B = \begin{bmatrix} -1 & 0 & 2 & -1 \\ 1 & 0 & 1 & -2 \\ 2 & 1 & 0 & -3 \end{bmatrix} \in$

$\mathcal{M}_{3\times4}(\mathbb{Q})$ then $AB = \begin{bmatrix} 5 & 2 & 7 & -14 \\ 5 & 1 & 0 & -6 \end{bmatrix} \in \mathcal{M}_{2\times4}(\mathbb{Q})$ but BA is not defined.

Example If we consider the matrices

$$A = \begin{bmatrix} 2 & 3 & 2 \\ -1 & 2 & 1 \end{bmatrix} \in \mathcal{M}_{2\times3}(\mathbb{Q}) \quad \text{and} \quad B = \begin{bmatrix} -1 & 0 \\ 1 & 0 \\ 2 & 1 \end{bmatrix} \in \mathcal{M}_{3\times2}(\mathbb{Q})$$

then $AB = \begin{bmatrix} 5 & 2 \\ 5 & 1 \end{bmatrix} \in \mathcal{M}_{2\times2}(\mathbb{Q})$ and $BA = \begin{bmatrix} -2 & -3 & -2 \\ 2 & 3 & 2 \\ 3 & 8 & 5 \end{bmatrix} \in \mathcal{M}_{3\times3}(\mathbb{Q})$.

Suppose that $A = [a_{ij}] \in \mathcal{M}_{k\times n}(F)$ and $v = \begin{bmatrix} b_1 \\ \vdots \\ b_n \end{bmatrix} \in F^n$. Then $Av = \begin{bmatrix} c_1 \\ \vdots \\ c_k \end{bmatrix} \in F^k$

where, for each $1 \le i \le k$, we have $c_i = \sum_{j=1}^{n} a_{ij}b_j$. Denoting the columns of A by $u_1, \dots, u_n$, we see that $Av = \sum_{j=1}^{n} b_j u_j$. So we conclude that if there exists

a nonzero vector v such that $Av = \begin{bmatrix} 0 \\ \vdots \\ 0 \end{bmatrix}$, then the columns of A must be linearly

dependent. If every element of F^k is of the form Av for some $v \in F^n$, then the columns of A must form a generating set for F^k.

Let $(K, \bullet)$ be an associative unital algebra over a field F and let n be a pos-

itive integer. If $v = \begin{bmatrix} v_1 \\ \vdots \\ v_n \end{bmatrix}$ and $w = \begin{bmatrix} w_1 \\ \vdots \\ w_n \end{bmatrix}$ are elements of K^n then $v^T w =$

$[\sum_{i=1}^{n} v_i \bullet w_i] \in \mathcal{M}_{1\times1}(K)$. This is called the *interior product* of v and w. This 1×1 matrix is usually identified with the scalar $\sum_{i=1}^{n} v_i \bullet w_i \in K$, which we will denote by $v \odot w$, in a departure from usual notation.[1]

Dually, the *exterior product* of v and w is defined to be the matrix $vw^T = [y_{ij}] \in \mathcal{M}_{n\times n}(K)$, where $y_{ij} = v_i \bullet w_j$. We will denote the exterior product of v and w by $v \wedge w$. Notice that the exterior product is not commutative, but rather $v \wedge w = (w \wedge v)^T$. Exterior products of vectors are encountered far less often than interior products, but have important applications in many areas, among them physics (in the Dirac model of quantum physics, interior products are called *bra-ket products*, whereas exterior products are called *ket-bra products*).

---

[1] The usual notation is $v \cdot w$, but that can cause confusion with the dot product, which we will study later, in the case that $F = \mathbb{C}$. For that reason, also, we use the term "interior product" rather than the often-seen "inner product".

In particular, we note the following: let K be an algebra over a field F, let $A \in \mathcal{M}_{k \times n}(K)$, and let $B \in \mathcal{M}_{n \times t}(K)$. Let $v_1^T, \ldots, v_k^T$ be the rows of A and let $w_1, \ldots, w_t$ be the columns of B. Then $AB = [c_{ij}]$, where $c_{ij} = v_i \odot w_j$ for all $1 \leq i \leq k$ and all $1 \leq j \leq t$.

Let F be a field, let n be a positive integer, let $v = \begin{bmatrix} a_1 \\ \vdots \\ a_n \end{bmatrix}$ and $w = \begin{bmatrix} ba_1 \\ \vdots \\ b_n \end{bmatrix}$ belong to F^n, and let $C = [c_{ij}] \in \mathcal{M}_{n \times x}(F)$. Then the computation of $v \odot Cw = \sum_{i=1}^{n} a_i (\sum_{j=1}^{n} c_{ij} b_j)$ requires $n^2 + n$ multiplications and $n^2 - 1$ additions. However, if we can find vectors $u = \begin{bmatrix} u_1 \\ \vdots \\ u_n \end{bmatrix}$ and $y = \begin{bmatrix} y_1 \\ \vdots \\ n_n \end{bmatrix}$ in F^n such that $C = u \wedge y$, then, by the distributive law, $v \odot Cw = \sum_{i=1}^{n} a_i (\sum_{j=1}^{n} u_i y_j b_j) = (\sum_{i=1}^{n} a_i u_i)(\sum_{j=1}^{n} y_j b_j)$ and this requires only $2n + 1$ multiplications and $2n - 2$ additions. Similarly, if we can find vectors $u, u', y, y' \in F^n$ such that $C = u \wedge y + u' \wedge y'$, then the computation of $v \odot Cw$ requires $4n + 2$ multiplications and $4n - 4$ additions. For large values of n, this can result in considerable saving, especially if the computation is to be repeated frequently.

Example Combinatorial optimization is the area of mathematics dealing with the computational issues arising from finding optimal solutions to such problems as the traveling salesman problem, testing Hamiltonian graphs, sphere packing, etc. The general form of combinatorial optimization problems is the following: Let F be a subfield of $\mathbb{R}$ and let n be a positive integer. Assume that we have a nonempty finite (and in general very large) subset S of $\mathbb{N}^n \subseteq F^n$. Usually, the set S arises from the characteristic functions of certain subsets of $\{1, \ldots, n\}$ of interest in the problem. Then, given a vector $v = \begin{bmatrix} a_1 \\ \vdots \\ a_n \end{bmatrix} \in F^n$, we want to find $\min\{s \odot v \mid s \in S\}$. Note that if we consider F not as a subset of $\mathbb{R}$ but as a subset of the optimization algebra $\mathbb{R}_\infty$, then the problem becomes one of computing $p(a_1, \ldots, a_n)$, where

$$p(X_1, \ldots, X_n) = \sum \left\{ X_1^{i_1} \cdots X_n^{i_n} \;\middle|\; \begin{bmatrix} i_1 \\ \vdots \\ i_n \end{bmatrix} \in S \right\}$$

is a polynomial in several indeterminates over $\mathbb{R}_\infty$ (polynomials with coefficients in a semifield are defined in the same way as polynomials with coefficients in a field).

Observe that multiplying a $k \times n$ matrix by an $n \times t$ matrix requires $kt(n-1)$ arithmetic operations. If these numbers are all very large, as is often the case in real-life applications of matrix theory, the computational overhead—and risk of accumulated errors due to rounding and truncation—is substantial.[2] We will keep this in mind throughout our discussion, and try to consider strategies of minimizing this risk. In this connection, we should note that the product of two matrices has an important property: let $(K, \bullet)$ be an associative unital algebra over a field F assume that $A = [v_{ij}] \in M_{k \times n}(K)$ and $B = [w_{ij}] \in M_{n \times t}(K)$, where k, n, and t are positive integers. Furthermore, let us pick positive integers

$$1 = k(1) < k(2) < \cdots < k(p+1) = k,$$

$$1 = n(1) < n(2) < \cdots < n(q+1) = n,$$

$$1 = t(1) < t(2) < \cdots < t(r+1) = t.$$

For all $1 \le i \le p$ and all $1 \le j \le q$, let $A_{ij} = \begin{bmatrix} v_{k(i),n(j)} & \cdots & v_{k(i),n(j+1)} \\ \vdots & \ddots & \vdots \\ v_{k(i+1),n(j)} & \cdots & v_{k(i+1),n(j+1)} \end{bmatrix}$.

This allows us to write A in *block form* $\begin{bmatrix} A_{11} & \cdots & A_{1q} \\ \vdots & \ddots & \vdots \\ A_{p1} & \cdots & A_{pq} \end{bmatrix}$. Note that these blocks are not necessarily square matrices. In the same way, we can write B as a matrix $\begin{bmatrix} B_{11} & \cdots & B_{1t} \\ \vdots & \ddots & \vdots \\ B_{q1} & \cdots & B_{qt} \end{bmatrix}$. Then $AB = \begin{bmatrix} C_{11} & \cdots & C_{1t} \\ \vdots & \ddots & \vdots \\ C_{p1} & \cdots & C_{pt} \end{bmatrix}$ where, for each $1 \le i \le p$ and each $1 \le h \le t$, we have $C_{ih} = \sum_{j=1}^{q} A_{ij} B_{jh}$. A sophisticated use of this method can substantially decrease the number of operations needed to multiply two matrices, as we shall see. Moreover, skilled partitioning of matrices can allow us to make use efficiently the aspects of modern computer architecture such as cache memories to further increase the speed of computation.

Needless to say, this seemingly odd definition of the product of two matrices was not chosen at random. Indeed, it satisfies certain important properties. Thus, if $(K, \bullet)$ is an associative unital algebra over a field F, if k, n, t, and p are positive integers, and if we have matrices $A \in M_{k \times n}(K)$, $B, B_1, B_2 \in M_{n \times t}(K)$, and $C \in M_{t \times p}(K)$, then

(1) $A(BC) = (AB)C$;

(2) $A(B_1 + B_2) = AB_1 + AB_2$;

(3) $(B_1 + B_2)C = B_1C + B_2C$.

As a consequence, we see that if $B \in M_{n \times t}(K)$ is given, then the function from $M_{k \times n}(K)$ to $M_{k \times t}(K)$ defined by $A \mapsto AB$ is a linear transformation of vector spaces.

---

[2] We will often mention large matrices, without being too specific as to what that means. As a rule of thumb, a matrix is "large", and calls for special treatment as such, when it cannot be stored in the RAM memory of whatever computer we are using for our computations. Such matrices occur in sufficiently-many applications that considerable research is devoted to dealing with them.

We also note that if $A = [v_{ij}] \in \mathcal{M}_{k \times n}(K)$ and $B = [w_{jh}] \in \mathcal{M}_{n \times t}(K)$, then $A^T \in \mathcal{M}_{n \times k}(K)$ and $B^T \in \mathcal{M}_{t \times n}(K)$ so $B^T A^T \in \mathcal{M}_{t \times k}(K)$. Indeed, $B^T A^T = [y_{hi}]$, where $y_{hi} = \sum_{j=1}^{n} w_{jh} \bullet v_{ij}$. Hence, if K is also commutative (and in particular if $K = F$), we have $B^T A^T = (AB)^T$.

Matrix multiplication was first defined by the nineteenth-century French mathematician **Jacques Philippe Binet**. It took some getting used to; many decades later, the father of astrophysics, **Sir Arthur Eddington**, still wrote "I cannot believe that anything so ugly as multiplication of matrices is an essential part of the scheme of nature".

The definition of matrix multiplication is in fact a direct consequence of the relation between matrices and linear transformations, which we have already observed. This is best seen in the following result.

Proposition 8.2 *Let V be a vector space of finite dimension n over a field F for which we have chosen a basis $B = \{v_1, \ldots, v_n\}$, let W be a vector space of finite dimension k over F for which we have chosen a basis $D = \{w_1, \ldots, w_k\}$, and let Y be a vector space of finite dimension t over F, for which we have chosen a basis $E = \{y_1, \ldots, y_t\}$. If $\alpha \in \mathrm{Hom}(V, W)$ and $\beta \in \mathrm{Hom}(W, Y)$ then $\Phi_{BE}(\beta\alpha) = \Phi_{DE}(\beta)\Phi_{BD}(\alpha)$.*

Proof Assume that $\Phi_{BD}(\alpha) = [a_{ij}]$ and $\Phi_{DE}(\beta) = [b_{hi}]$. Then

$$\alpha : v \mapsto \sum_{i=1}^{k} \sum_{j=1}^{n} c_j a_{ij} w_i \quad \text{and} \quad \beta\alpha : v \mapsto \sum_{h=1}^{t} \sum_{i=1}^{k} \sum_{j=1}^{n} c_j b_{hi} a_{ij} y_h,$$

showing the desired equality. $\qquad\qquad\qquad\qquad\qquad\qquad\qquad\qquad\qquad\Box$

We can extend the definition of matrix multiplication as follows: let h, k, and n be positive integers, and let V be a vector space over a field F. If $A = [a_{ij}] \in \mathcal{M}_{h \times k}(F)$ and if $M = [v_{jt}] \in \mathcal{M}_{k \times n}(V)$, we can define $AM \in \mathcal{M}_{h \times n}(V)$ to be the matrix $[u_{it}]$, where $u_{it} = \sum_{j=1}^{k} a_{ij} v_{jt}$ for all $1 \leq i \leq h$ and $1 \leq t \leq n$. Notice that if $A, B \in \mathcal{M}_{k \times k}(F)$ and if $M, N \in \mathcal{M}_{k \times n}(V)$ then

(1) $A(BM) = (AB)M$;

(2) $A(M + N) = AM + AN$;

(3) $(A + B)M = AM + BM$.

In general, and especially when we are talking of actual computations, it is easier to work with matrices than with linear transformations, and indeed most of the modern computer software and hardware are designed to facilitate easy and speedy matrix computation. Therefore, given finitely-generated vector spaces V and W over a field F, it is usual to fix bases for them and then identify $\mathrm{Hom}(V, W)$ with the space of all matrices over F of the appropriate size. The choice of the correct bases then becomes critical, and we will focus on that throughout the following discussions. Such a choice usually depends on the problem at hand. In particular, the automatic choice of canonical bases, when they exist, may not be the best for a given problem, and can entail a considerable cost both in computational time and numerical accuracy.

Exercises

Exercise 404
Let V be the vector space over $\mathbb{R}$ composed of all polynomials in $\mathbb{R}[X]$ having degree less than 3 and let W be the vector space over $\mathbb{R}$ composed of all polynomials in $\mathbb{R}[X]$ having degree less than 4. Let $\alpha : V \to W$ be the linear transformation defined by

$$\alpha : a + bX + cX^2 \mapsto (a+b) + (b+c)X + (a+c)X^2 + (a+b+c)X^3.$$

Select bases $B = \{1, X + 1, X^2 + X + 1\}$ for V and

$$D = \left\{ X^3 - X^2, X^2 - X, X - 1, 1 \right\}$$

for W. Find the matrix $\Phi_{BD}(\alpha)$.

Exercise 405
Let $K = \mathcal{M}_{2 \times 2}(\mathbb{R})$ and let $A = \begin{bmatrix} a & b \\ c & d \end{bmatrix} \in K$. Let D be the canonical basis of K. If $\alpha, \beta \in \mathrm{End}(K)$ are defined by $\alpha : X \mapsto XA$ and $\beta : X \mapsto AX$, find $\Phi_{DD}(\alpha)$ and $\Phi_{DD}(\beta)$.

Exercise 406
Given the matrix $A = \begin{bmatrix} 0 & 1 & 2 & 3 \\ 1 & 3 & 4 & 0 \\ 3 & 2 & 0 & 1 \end{bmatrix} \in \mathcal{M}_{3 \times 4}(\mathbb{R})$, find the set of all matrices

$B \in \mathcal{M}_{4 \times 3}(\mathbb{R})$ satisfying $AB = \begin{bmatrix} 1 & 0 & 0 \\ 0 & 1 & 0 \\ 0 & 0 & 1 \end{bmatrix}$.

Exercise 407
Given the matrix $A = \begin{bmatrix} 1 & 8 \\ 3 & 5 \\ 2 & 2 \end{bmatrix} \in \mathcal{M}_{3 \times 2}(\mathbb{Q})$, find the set of all matrices

$B \in \mathcal{M}_{2 \times 3}(\mathbb{Q})$ satisfying $AB = \begin{bmatrix} 1 & 0 & 0 \\ 0 & 1 & 0 \\ 0 & 0 & 1 \end{bmatrix}$ and find the set of all matrices

$C \in \mathcal{M}_{2\times3}(\mathbb{Q})$ satisfying $CA = \begin{bmatrix} 1 & 0 \\ 0 & 1 \end{bmatrix}$.

Exercise 408

Given the matrix $A = \begin{bmatrix} 1 & -1 & 2 \\ 0 & 1 & 2 \\ -1 & 3 & 1 \\ 2 & 1 & -1 \end{bmatrix} \in \mathcal{M}_{4\times3}(\mathbb{R})$, find the set of all matrices

$B \in \mathcal{M}_{3\times4}(\mathbb{R})$ satisfying $BA = \begin{bmatrix} 1 & 0 & 0 \\ 0 & 1 & 0 \\ 0 & 0 & 1 \end{bmatrix}$.

Exercise 409

Find the matrix representing the linear transformation $\alpha : \begin{bmatrix} a \\ b \\ c \end{bmatrix} \mapsto \begin{bmatrix} a+b+c \\ b+c \end{bmatrix}$

from $\mathbb{R}^3$ to $\mathbb{R}^2$ with respect to the bases $\left\{ \begin{bmatrix} -1 \\ 0 \\ 2 \end{bmatrix}, \begin{bmatrix} 0 \\ 1 \\ 1 \end{bmatrix}, \begin{bmatrix} 3 \\ -1 \\ 0 \end{bmatrix} \right\}$ of $\mathbb{R}^3$ and

$\left\{ \begin{bmatrix} -1 \\ 1 \end{bmatrix}, \begin{bmatrix} 1 \\ 0 \end{bmatrix} \right\}$ of $\mathbb{R}^2$.

Exercise 410

Find the set of all matrices $A \in \mathcal{M}_{4\times3}(\mathbb{R})$ satisfying the condition

$$\begin{bmatrix} 0 & 1 & 0 & 0 \\ 0 & 0 & 1 & 0 \\ 0 & 0 & 0 & 1 \\ 0 & 0 & 0 & 0 \end{bmatrix} A = A \begin{bmatrix} 0 & 1 & 0 \\ 0 & 0 & 1 \\ 0 & 0 & 0 \end{bmatrix}.$$

Exercise 411

Let α be an endomorphism of $\mathbb{R}^3$ represented with respect to some basis by the

matrix $\begin{bmatrix} 0 & 2 & -1 \\ -2 & 5 & -2 \\ -4 & 8 & -3 \end{bmatrix}$. Is α a projection?

Exercise 412

Find the real numbers missing from the following equation:

$$\begin{bmatrix} 1 & -3 \\ 1 & * \\ 1 & * \\ * & 1 \end{bmatrix} \begin{bmatrix} -1 & * & 7 & * \\ * & 1 & * & 0 \end{bmatrix} = \begin{bmatrix} -25 & -1 & 1 & 3 \\ -1 & * & * & * \\ * & * & 5 & * \\ * & * & * & 0 \end{bmatrix}.$$

Exercise 413

Let $\alpha : \begin{bmatrix} a \\ b \\ c \end{bmatrix} \mapsto \begin{bmatrix} 3a + 2b \\ -a - c \\ a + 3b \end{bmatrix}$ be an endomorphism of $\mathbb{R}^3$. Find the matrix repre-

senting α with respect to the basis $B = \left\{ \begin{bmatrix} 1 \\ -1 \\ 0 \end{bmatrix}, \begin{bmatrix} 1 \\ 0 \\ -1 \end{bmatrix}, \begin{bmatrix} 0 \\ 1 \\ 0 \end{bmatrix} \right\}$ of $\mathbb{R}^3$.

Exercise 414

Let $D = \{v_1, v_2, v_3\}$ be a basis for $\mathbb{R}^3$ and let α be the endomorphism of $\mathbb{R}^3$

satisfying $\Phi_{DD}(\alpha) = \begin{bmatrix} -1 & -1 & -3 \\ -5 & -2 & -6 \\ 2 & 1 & 3 \end{bmatrix}$. Find $\ker(\alpha)$.

Exercise 415

Let V be the subspace of $\mathbb{R}[X]$ consisting of all polynomials of degree less than 3 and choose the basis $B = \{1, X, X^2\}$ for V. Let $\alpha \in \mathrm{End}(V)$ satisfy

$\Phi_{BB}(\alpha) = \begin{bmatrix} 1 & 1 & 1 \\ 0 & 2 & 2 \\ 0 & 0 & 3 \end{bmatrix}$. Let D be the basis $\{1, X + 1, 2X^2 + 4X + 3\}$ for V.

What is $\Phi_{DD}(\alpha)$?

Exercise 416

Let $\alpha \in \mathrm{End}(\mathbb{R}^3)$ be represented with respect to the canonical basis by the matrix $\begin{bmatrix} 2 & 2 & 0 \\ 1 & 1 & 2 \\ 1 & 1 & 2 \end{bmatrix}$. Find a real number a such that α is represented with respect to the

basis $\left\{ \begin{bmatrix} a \\ -1 \\ 0 \end{bmatrix}, \begin{bmatrix} -2 \\ a \\ 1 \end{bmatrix}, \begin{bmatrix} 1 \\ 1 \\ a \end{bmatrix} \right\}$ by the matrix $\begin{bmatrix} 0 & 0 & 0 \\ 0 & 1 & 0 \\ 0 & 0 & 4 \end{bmatrix}$.

Exercise 417

Let V be the subspace of $\mathbb{R}[X]$ consisting of all polynomials of degree less than 3 and let $\alpha \in \mathrm{End}(V)$ be defined by

$$\alpha : aX^2 + bX + c \mapsto (a + 2b + c)X^2 + (3a - b)X + (b + 2c).$$

Find $\Phi_{DD}(\alpha)$, where $D = \{X^2 + X + 1, X^2 + X, X^2\}$.

Exercise 418

Find all rational numbers a for which there exists a nonzero matrix $B \in$

$\mathcal{M}_{4\times3}(\mathbb{Q})$ satisfying $B \begin{bmatrix} a & 1 & 1 \\ 1 & 1 & a \\ 1 & a & 1 \end{bmatrix} = \begin{bmatrix} 0 & 0 & 0 \\ 0 & 0 & 0 \\ 0 & 0 & 0 \\ 0 & 0 & 0 \end{bmatrix}$.

Exercise 419
For which real numbers a does there exist a real number b (depending on a)
satisfying $\begin{bmatrix} a & 1 & 1 \\ 1 & 1 & a \end{bmatrix} \begin{bmatrix} b & -1 \\ 1 & -1 \\ 1 & b \end{bmatrix} = \begin{bmatrix} 1 & 0 \\ 0 & 1 \end{bmatrix}$?

Exercise 420
Let $V = \mathbb{R}^{\mathbb{R}}$ and let W be the subspace of V generated by the linearly-independent set $B = \{1, x, e^x, xe^x\}$. Let δ be the endomorphism of W which assigns to each function its derivative. Find $\Phi_{BB}(\delta)$.

Exercise 421
Let $B = \{1 + i, 2 + i\}$, which is a basis for $\mathbb{C}$ as a vector space over $\mathbb{R}$. Let α be the endomorphism of this space defined by $\alpha : z \mapsto \overline{z}$. Find $\Phi_{BB}(\alpha)$.

Exercise 422
Let $F = GF(3)$ and let $\alpha : F^3 \to F^2$ be the linear transformation defined by
$\alpha : \begin{bmatrix} a \\ b \\ c \end{bmatrix} \mapsto \begin{bmatrix} a - b \\ 2a - c \end{bmatrix}$. Let $\beta : F^2 \to F^4$ be the linear transformation defined by
$\beta : \begin{bmatrix} a \\ b \end{bmatrix} \mapsto \begin{bmatrix} b \\ a \\ 2b \\ 2a \end{bmatrix}$. Find the matrix representing $\beta\alpha$ with respect to the canonical bases.

Exercise 423
Let $\alpha \in \text{End}(\mathbb{R}^4)$ be represented with respect to the canonical basis by the matrix
$\begin{bmatrix} 3 & 1 & 0 & 0 \\ -1 & 2 & -1 & 0 \\ 0 & -1 & 2 & -1 \\ 0 & 0 & -1 & 1 \end{bmatrix}$. Given a vector $v \in \mathbb{R}^4$ satisfying the condition that all
entries of $\alpha(v)$ are nonnegative, show that all entries of v are nonnegative.

Exercise 424
Let V and W be vector spaces over a field F and choose bases $\{v_i \mid i \in \Omega\}$ and $\{w_j \mid j \in \Lambda\}$ for V and W, respectively. Let $p : \Omega \times \Lambda \to F$ be a function satisfying the condition that the set $\{j \in \Lambda \mid p(i, j) \neq 0\}$ is finite for each $i \in \Omega$. Let $\alpha_p : V \to W$ be the function defined as follows: if $v = \sum_{i \in \Gamma} a_i v_i$, where Γ is a finite subset of Ω and where the a_i are scalars in F, then $\alpha_p(v) = \sum_{i \in \Gamma} \sum_{j \in \Lambda} a_i p(i, j) w_j$. Show that α_p is a linear transformation and that every linear transformation from V to W is of this form.

Exercise 425
Let k and n be positive integers, let $v \in \mathbb{R}^n$, and let $A \in \mathcal{M}_{k \times n}(\mathbb{R})$. Show that $Av = O$ if and only if $A^T Av = O$.

Exercise 426

Let $A \in \mathcal{M}_{3\times 2}(\mathbb{R})$ and $B \in \mathcal{M}_{2\times 3}(\mathbb{R})$ be matrices satisfying

$$AB = \begin{bmatrix} 8 & 2 & -2 \\ 2 & 5 & 4 \\ -2 & 4 & 5 \end{bmatrix}.$$

Calculate BA.

Exercise 427

Find matrices $A \in \mathcal{M}_{3\times 2}(\mathbb{R})$ and $B \in \mathcal{M}_{2\times 3}(\mathbb{R})$ satisfying

$$AB = \begin{bmatrix} 1 & 1 & 1 \\ -2 & 0 & -6 \\ 0 & 1 & -2 \end{bmatrix}.$$

Exercise 428

Let F be a field and let $k \neq n$ be positive integers. Let $A, B \in \mathcal{M}_{k\times n}(F)$ and let $\alpha : \mathcal{M}_{n\times k}(F) \to \mathcal{M}_{k\times n}(F)$ be the linear transformation defined by $\alpha : C \mapsto ACB$. Under which conditions is α an isomorphism?

Exercise 429

Let F be a field and let n be a positive integer. Let W be a nontrivial subspace of the vector space $V = \mathcal{M}_{n\times n}(F)$ satisfying the condition that if $A \in W$ and $B \in V$ then AB and BA both belong to W. Show that $W = V$.

Exercise 430

Let $a, b, c, a', b', c' \in \mathbb{C}$ satisfy the condition that $aa' + bb' + cc' = 2$, and let $A = I - \begin{bmatrix} a \\ b \\ c \end{bmatrix} \begin{bmatrix} a' & b' & c' \end{bmatrix}$. Calculate A^2.

Exercise 431

Find a nonzero matrix A in $\mathcal{M}_{2\times 2}(\mathbb{R})$ satisfying $v \odot Av = 0$ for all $v \in \mathbb{R}^2$.

Exercise 432

Let α be the endomorphism of $\mathbb{Q}^4$ represented with respect to the canonical basis by the matrix $\begin{bmatrix} 1 & 0 & 1 & -1 \\ 2 & 1 & 2 & 1 \\ 0 & 1 & 6 & 1 \\ 3 & 1 & 3 & 4 \end{bmatrix}$. Find a two-dimensional subspace of $\mathbb{Q}^4$ which is invariant under α.

Exercise 433

Let n be a positive integer and let F be a field. For each $v, w \in F^n$, consider the function $\tau_{v,w} : F^n \to Fv$ defined by $\tau_{v,w} : y \mapsto (w \odot y)v$ (this function is called the *dyadic product* function). Show that $\tau_{v,w}$ a linear transformation. Is the function $F^n \to \mathrm{Hom}(F^n, Fv)$ defined $w \mapsto \tau_{v,w}$ a linear transformation?

Exercise 434

Let $k < n$ be positive integers and let F be a field. Given a matrix $A \in \mathcal{M}_{k \times n}(F)$, do there necessarily exist matrices $B, C \in \mathcal{M}_{n \times k}(F)$ satisfying the condition that $AB = O \in \mathcal{M}_{k \times k}(F)$ and $CA = O \in \mathcal{M}_{n \times n}(F)$?

Exercise 435

Let $A \in \mathcal{M}_{n \times n}(\mathbb{Q})$ be a matrix satisfying the condition that if $v \in \mathbb{Q}^n$ is a vector all of the components of which are nonnegative, then all of the components of Av are nonnegative. Are all of the entries in A necessarily nonnegative?

Exercise 436

Find the set of all matrices A in $\mathcal{M}_{3 \times 3}(\mathbb{R})$ satisfying $A^2 = \begin{bmatrix} 0 & 1 & 0 \\ 0 & 0 & 0 \\ 0 & 0 & 0 \end{bmatrix}$.

Exercise 437

Find the set of all real numbers a such that the endomorphism of $\mathbb{R}^3$ represented by the matrix $\begin{bmatrix} 1 & a & a \\ 2 & 2a & 4 \\ 3 & a & 6 \end{bmatrix}$ with respect to the canonical basis is an automorphism.

Exercise 438

Find the set of all real numbers a and b such that the endomorphism of $\mathbb{R}^3$ represented by the matrix $\begin{bmatrix} 1 & a & b \\ 0 & a & 1 \\ 0 & a & 1 \end{bmatrix}$ with respect to the canonical basis is a projection.

Exercise 439

Let $A = [a_{ij}] \in \mathcal{M}_{n \times n}(\mathbb{R})$ satisfy the condition that for each $v \in \mathbb{R}^n$ there exists a vector $y \in \mathbb{R}^n$ all entries in which are nonnegative satisfying $Av = v + y$. Show that $A = I$.

Exercise 440

Let $F = GF(2)$ and let K be the subset of $\mathcal{M}_{3 \times 3}(F)$ consisting of O, I, and the following matrices:

$$\begin{bmatrix} 0 & 0 & 1 \\ 1 & 0 & 0 \\ 0 & 1 & 1 \end{bmatrix}, \quad \begin{bmatrix} 1 & 0 & 1 \\ 1 & 1 & 0 \\ 0 & 1 & 0 \end{bmatrix}, \quad \begin{bmatrix} 0 & 1 & 1 \\ 0 & 0 & 1 \\ 1 & 1 & 1 \end{bmatrix}, \quad \begin{bmatrix} 1 & 1 & 1 \\ 0 & 1 & 1 \\ 1 & 1 & 0 \end{bmatrix},$$

$$\begin{bmatrix} 0 & 1 & 0 \\ 1 & 0 & 1 \\ 1 & 0 & 0 \end{bmatrix}, \quad \text{and} \quad \begin{bmatrix} 1 & 1 & 0 \\ 1 & 1 & 1 \\ 1 & 0 & 1 \end{bmatrix}.$$

Show that K, together with matrix addition and multiplication, is a field. What is its characteristic? Does there exist an element A of K such that every nonzero element of K is a power of A?

The Algebra of Square Matrices

9

We are now going to concentrate on the algebraic structure of sets of the form $\mathcal{M}_{n\times n}(K)$, where n is a positive integer and $(K, \bullet)$ is an associative unital algebra over a field F. From what we have already seen, this is again an associative unital F-algebra, which will not be commutative if $n > 1$. The additive identity of this algebra is the matrix all of the entries of which equal 0_K. The additive inverse of a matrix $A = [d_{ij}] \in \mathcal{M}_{n\times n}(K)$ is the matrix $[-a_{ij}]$. The multiplicative identity of $\mathcal{M}_{n\times n}(K)$ is the matrix $E = [d_{ij}]$ given by

$$d_{ij} = \begin{cases} e & \text{if } i = j, \\ 0 & \text{otherwise,} \end{cases}$$

where e is the multiplicative identity of $(K, \bullet)$.

The most important case is, of course, that of $K = F$. In this case, the additive identity is O and the multiplicative identity is the matrix $I = [a_{ij}]$ defined by

$$a_{ij} = \begin{cases} 1 & \text{if } i = j, \\ 0 & \text{otherwise.} \end{cases}$$

If K is a vector space of dimension n over F and if B is a basis of K, then it is straightforward to verify that the function $\Phi_{BB} : \text{End}(K) \to \mathcal{M}_{n\times n}(F)$ is an isomorphism of unital F-algebras.

If F is a field and if n is a positive integer then, corresponding to the associative F-algebra $\mathcal{M}_{n\times n}(F)$, we have the Lie algebra $\mathcal{M}_{n\times n}(F)^-$. This Lie algebra is called the *general Lie algebra* defined by F^n.

Example Let F be a field and let $A = [a_{ij}] \in \mathcal{M}_{4\times 4}(F)$. Then A can also be written in block form as $\begin{bmatrix} \begin{bmatrix} a_{11} & a_{12} \\ a_{21} & a_{22} \end{bmatrix} & \begin{bmatrix} a_{13} & a_{14} \\ a_{23} & a_{24} \end{bmatrix} \\ \begin{bmatrix} a_{31} & a_{32} \\ a_{41} & a_{42} \end{bmatrix} & \begin{bmatrix} a_{33} & a_{34} \\ a_{43} & a_{44} \end{bmatrix} \end{bmatrix} \in \mathcal{M}_{2\times 2}(K)$, where

J.S. Golan, *The Linear Algebra a Beginning Graduate Student Ought to Know*,
DOI 10.1007/978-94-007-2636-9_9, © Springer Science+Business Media B.V. 2012

$K = \mathcal{M}_{2\times2}(F)$. Addition and multiplication of matrices are so defined (and not accidentally!) that they give the same results whether performed in $\mathcal{M}_{4\times4}(F)$ or in $\mathcal{M}_{2\times2}(K)$.

Example The set K of all analytic functions from $\mathbb{C}$ to itself is clearly an algebra over $\mathbb{C}$. At the beginning of the twentieth century, G. D. Birkhoff made use of matrices in $\mathcal{M}_{n\times n}(K)$ to study the properties of analytic functions.

With kind permission of the Archives of the Mathematisches Forschungsinstitut Oberwolfach.

George D. Birkhoff was one of the leading American mathematicians at the beginning of the twentieth century, who worked in many areas of analysis.

We begin by identifying some particularly-important square matrices over a unital associative F-algebra K, and with them some significant subalgebras of $\mathcal{M}_{n\times n}(K)$.

Let $(K, \bullet)$ is an associative unital F-algebra and let n be a positive integer. A matrix $A = [d_{ij}] \in \mathcal{M}_{n\times n}(K)$ is a *diagonal matrix* if and only if there exist elements $c_1, \dots, c_n$ of K such that

$$d_{ij} = \begin{cases} c_i & \text{if } i = j, \\ 0_K & \text{otherwise.} \end{cases}$$

The matrices O and E are diagonal. Moreover, the sum and product of diagonal matrices are diagonal matrices, and so the set of all diagonal matrices is an F-subalgebra of $\mathcal{M}_{n\times n}(K)$. If K is commutative (and, in particular, if $K = F$) then this algebra is also commutative. The units of the subalgebra are all diagonal matrices in which each c_i is a unit of K (and hence surely nonzero). In this case,

$$\begin{bmatrix} c_1 & \cdots & 0_K \\ \vdots & \ddots & \vdots \\ 0_K & \cdots & c_n \end{bmatrix}^{-1} = \begin{bmatrix} c_1^{-1} & \cdots & 0_K \\ \vdots & \ddots & \vdots \\ 0_K & \cdots & c_n^{-1} \end{bmatrix}.$$

Example Let F be a field, let $(K, \bullet)$ is an associative unital F-algebra, and let n be a positive integer. A matrix $A = [a_{ij}] \in \mathcal{M}_{n\times n}(K)$ is a *scalar matrix* if and only if there exists a scalar $c \in K$ such that $a_{ij} = c$ when $i = j$ and $a_{ij} = 0_K$ otherwise. We denote this matrix by cE (and, in particular, cI when $K = F$). Scalar matrices are surely diagonal matrices, and both O and E are scalar matrices. Moreover, the sum and product of scalar matrices are scalar matrices. If $c, d \in K$ then $(cE)(dE) = (dE)(cE)$ and if $0_K \neq c \in K$ is a unit, then $(cE)(c^{-1}E) = E$. Hence the set of all scalar matrices over F forms an F-subalgebra of $\mathcal{M}_{n \times n}(F)$, which is in fact a field. The function $F \rightarrow \mathcal{M}_{n \times n}(F)$ defined by $c \mapsto cI$ is a monic homomorphism of F-algebras, and so we can identify F with the subfield of all scalar matrices of $\mathcal{M}_{n \times n}(F)$. Moreover, it is also easy to see that $(cI)A = A(cI) = cA$ for any $A \in \mathcal{M}_{n \times n}(F)$.

Let $(K, \bullet)$ is an associative unital F-algebra, let n be a positive integer, and let d be a positive integer less than n. A matrix $A = [v_{ij}] \in \mathcal{M}_{n \times n}(K)$ is a *band matrix of width* $2d - 1$ if and only if $v_{ij} = 0_K$ whenever $|i - j| > d - 1$. Thus, the band matrices of width 1 are the diagonal matrices. The matrix

$$\begin{bmatrix} 1 & 2 & 0 & 0 & 0 \\ 2 & 3 & 1 & 0 & 0 \\ 0 & 0 & 0 & 0 & 0 \\ 0 & 0 & 0 & 0 & 1 \\ 0 & 0 & 0 & 4 & 4 \end{bmatrix} \in \mathcal{M}_{5 \times 5}(\mathbb{R})$$

is an example of a band matrix of width 3. The set of band matrices of fixed width is closed under addition and contains O and I, but is not necessarily closed under multiplication, and so is not a subalgebra of $\mathcal{M}_{n \times n}(K)$. However, it is closed under scalar multiplication and so is a subspace of the vector space $\mathcal{M}_{n \times n}(K)$ over F.

Band matrices over a field are very important for numerical computations, especially when d is small relative to n. Of particular importance are band matrices of width 3, which are also known as *tridiagonal matrices*, and have important use in the computation of quadratic splines and in the computation of extremal eigenvalues of matrices; they also appear very often in methods of solution of differential equations. Tridiagonal matrices have the added advantage of being easily stored in a computer, since all we need to do is keep the three diagonals in which nonzero entries can occur. For example, a tridiagonal matrix in $\mathcal{M}_{1000 \times 1000}(\mathbb{R})$ has $1,000,000$ entries, of which at most 2998 are nonzero.

A special type of tridiagonal matrix in $\mathcal{M}_{2n \times 2n}(F)$, which we will see again later, is one of the form $\begin{bmatrix} A_{11} & O & \cdots & O \\ O & A_{22} & \cdots & O \\ \vdots & \vdots & \ddots & \vdots \\ O & O & \cdots & A_{nn} \end{bmatrix}$, where the A_{ii} are 2×2 blocks. Note that this matrix can also be thought of as a diagonal matrix in $\mathcal{M}_{n \times n}(K)$, where $K = \mathcal{M}_{2 \times 2}(F)$. More generally, if d and n are positive integers, then any diagonal matrix in $\mathcal{M}_{n \times n}(L)$, where $L = \mathcal{M}_{d \times d}(K)$, is a band matrix of width $2d - 1$ in $\mathcal{M}_{dn \times dn}(K)$.

Let $(K, \bullet)$ is an associative unital F-algebra and let n be a positive integer. A matrix $A = [c_{ij}] \in \mathcal{M}_{n \times n}(K)$ is an *upper-triangular matrix* if and only if $c_{ij} = 0_K$

whenever $i > j$. Thus, the matrix $\begin{bmatrix} 1 & 2 & 6 & 3 & 7 \\ 0 & 3 & 1 & 0 & 0 \\ 0 & 0 & 0 & 0 & 0 \\ 0 & 0 & 0 & 0 & 1 \\ 0 & 0 & 0 & 0 & 4 \end{bmatrix} \in \mathcal{M}_{5 \times 5}(\mathbb{R})$ is upper trian-

gular. The set of all upper-triangular matrices includes the set of diagonal matrices, is closed under addition, and contains O and E. Moreover, it is closed under multiplication, and so is an F-subalgebra of $\mathcal{M}_{n \times n}(K)$. In the case that $K = F$, we see that the dimension of $\mathcal{M}_{n \times n}(F)$ as a vector space over F equals $\frac{n}{2}(n + 1)$. Upper-triangular matrices arise naturally in many applications, as we will see below. In a similar manner, we say that a matrix $A = [c_{ij}] \in \mathcal{M}_{n \times n}(K)$ is a *lower-triangular matrix* if and only if $c_{ij} = 0_K$ whenever $i < j$. Again, the set of all lower-triangular matrices is a subspace of the vector space $\mathcal{M}_{n \times n}(K)$ over F and, indeed, an F-subalgebra. Note that a matrix A is upper triangular if and only if A^T is lower triangular.

A matrix $A = [c_{ij}] \in \mathcal{M}_{n \times n}(K)$ is *symmetric* if and only if $A = A^T$. That is, A is symmetric if and only if $c_{ij} = c_{ji}$ for all $1 \le i, j \le n$. If B is any matrix in $\mathcal{M}_{n \times n}(K)$ then $B + B^T$ is symmetric. If K is commutative and if $C \in \mathcal{M}_{k \times n}(K)$ for any positive integers k and n, then $CC^T \in \mathcal{M}_{k \times k}(K)$ and $C^T C \in \mathcal{M}_{n \times n}(K)$ are symmetric. If n is a positive integer and F is a field, then $v \wedge v$ is a symmetric matrix in $\mathcal{M}_{n \times n}(F)$ for all $v \in F^n$. Diagonal matrices are clearly symmetric and the set of symmetric matrices in $\mathcal{M}_{n \times n}(K)$ is closed under taking sums and scalar multiples, and so it is a subspace of the vector space $\mathcal{M}_{n \times n}(K)$ over F. In the case $K = F$, the dimension of $\mathcal{M}_{n \times n}(F)$ equals $\frac{n}{2}(n + 1)$. However, the set of symmetric matrices is not closed under products. For example, the matrices $A = \begin{bmatrix} 2 & 5 & 1 \\ 5 & 2 & 0 \\ 1 & 0 & 1 \end{bmatrix}$

and $B = \begin{bmatrix} 1 & 2 & 1 \\ 2 & 0 & 0 \\ 1 & 0 & 3 \end{bmatrix}$ in $\mathcal{M}_{3 \times 3}(\mathbb{R})$ are symmetric, but $AB = \begin{bmatrix} 13 & 4 & 5 \\ 9 & 10 & 5 \\ 2 & 2 & 4 \end{bmatrix}$ is

not. In fact, in Chap. 13 we will show that if $n > 1$ then every matrix in $\mathcal{M}_{n \times n}(\mathbb{C})$ is a product of two symmetric matrices.

We note, however, that if A and B are a commuting pair of symmetric matrices then $(AB)^T = (BA)^T = A^T B^T = AB$, so AB is again symmetric.

A matrix $A = [c_{ij}] \in \mathcal{M}_{n \times n}(K)$ is *skew symmetric* if and only if $A = -A^T$. The set of all skew-symmetric matrices in $\mathcal{M}_{n \times n}(K)$ is again a subspace of $\mathcal{M}_{n \times n}(K)$. Note that if F is a field having characteristic other than 2, then any matrix $A \in \mathcal{M}_{n \times n}(K)$ can be written as the sum of a symmetric matrix and a skew-symmetric matrix, since $A = \frac{1}{2}(A + A^T) + \frac{1}{2}(A - A^T)$. In one of the examples after Proposition 5.14, we saw that this representation is in fact unique. The Lie product of two skew-symmetric matrices is again skew-symmetric.

Example Let n be a positive integer. A matrix $A = [a_{ij}] \in \mathcal{M}_{n \times n}(\mathbb{R})$ is a *Markov matrix* if and only if $a_{ij} \ge 0$ for all $1 \le i, j \le n$ and $\sum_{j=1}^{n} a_{hj} = 1$ for each $1 \le h \le n$; it is a *stochastic matrix* if and only if both A and A^T are Markov matrices. It is easy to show that the product of two Markov matrices is again a Markov matrix and the product of two stochastic matrices in $\mathcal{M}_{n \times n}(\mathbb{R})$ is again a stochastic matrix.

Markov matrices arise naturally in probability theory. In particular, if we have a system which, at each tick of a (discrete) clock, is in one of the distinct states $s_1, \ldots, s_n$ and if, for each $1 \leq i, j \leq n$, we denote by p_{ij} the probability that if the situation is in state i at a given time t then it will be in state j at time $t + 1$, the matrix $[p_{ij}]$ is a Markov matrix.

Russian mathematician **Andrei Andreyevich Markov** made major contributions to probability theory at the beginning of the twentieth century.

As we have already pointed out, a matrix $O \neq A \in \mathcal{M}_{n \times n}(K)$ is not necessarily a unit. The units of $\mathcal{M}_{n \times n}(K)$ are known as *nonsingular* matrices; the other matrices are *singular* matrices. By what we have already noted, the product of nonsingular matrices is again nonsingular and if A is nonsingular then surely so is A^{-1}. A matrix A satisfying $A^2 = I$ is certainly nonsingular. Such matrices are called *involutory matrices*.

With kind permission of the Harvard University Archives, HUP.

These terms were first used by American mathematician **Maxime Bôcher** in 1907. He was also the first to popularize the terms "linearly dependent" and "linearly independent".

Example If $a, b \in \mathbb{R}$ with $b \neq 0$ then $\begin{bmatrix} a & b \\ b^{-1}(1-a) & -a \end{bmatrix}$ is involutory.

Example We have already noted that if n is a positive integer then a diagonal matrix in $A = [a_{ij}] \in \mathcal{M}_{n \times n}(\mathbb{C})$ is nonsingular when all of the diagonal entries a_{ii} are nonzero. It therefore seems reasonable to conjecture that a matrix will be nonsingular if the diagonal entries are all "much greater" than the other entries. Indeed, this is true in the following sense: A sufficient condition for a matrix $A = [a_{ij}] \in \mathcal{M}_{n \times n}(\mathbb{C})$ to be nonsingular is that for each $1 \leq i \leq n$ we have $|a_{ii}| > \sum_{j \neq i} |a_{ij}|$. This result is known as the Diagonal Dominance Theorem. A proof of this theorem will be given in Chap. 15.

Let V be a vector space over F of dimension n and let B be a basis of V. Then there exists an endomorphism α of V such that $A = \Phi_{BB}(\alpha)$. If A is nonsingular then there also exists an endomorphism β of V satisfying $A^{-1} = \Phi_{BB}(\beta)$. This means that $I = AA^{-1} = \Phi_{BB}(\alpha)\Phi_{BB}(\beta) = \Phi_{BB}(\alpha\beta)$ and so $\alpha\beta = \sigma_1$, and similarly $\beta\alpha = \sigma_1$. Therefore, $\alpha \in \mathrm{Aut}(V)$ and $\beta = \alpha^{-1}$.

Example Let F be a field and let n be a positive integer. If $c \in F$ and $v, w \in F^n$, then the matrix $A = I + c(v \wedge w)$ is nonsingular if and only if the scalar $1 + c(v \odot w)$ is nonzero. Indeed, direct computation shows that if $1 + c(v \odot w) \neq 0$, then $A^{-1} = I + d(v \wedge w)$, where $d = -c[1 + c(v \odot w)]^{-1}$ and if $1 + c(v \odot w) = 0$ then $Av = v + c(v \odot w)v$ is the 0-vector, and so A must be singular.

Example The multiplicative inverse of a "nice" nonsingular matrix may not be "nice". Thus, if $A \in M_{n \times n}(\mathbb{R})$ is a nonsingular matrix all of the entries of which are nonnegative, it does not follow that all of the entries of A^{-1} are nonnegative. For example, if we choose $A = \begin{bmatrix} 1 & 1 & 1 \\ 1 & 2 & 1 \\ 1 & 1 & 2 \end{bmatrix}$ then direct computation shows us that $A^{-1} = \begin{bmatrix} 3 & -1 & -1 \\ -1 & 1 & 0 \\ -1 & 0 & 1 \end{bmatrix}$. If $A = [a_{ij}]$ is the $n \times n$ tridiagonal matrix with $a_{ii} = 2$ for all $1 \leq i \leq n$ and $a_{ij} = -1$ whenever $|i - j| = 1$, then not only is A^{-1} not tridiagonal, but in fact no entries in A^{-1} equal 0, for any $n > 1$.

Example If a matrix $A \in M_{n \times n}(F)$ can be written in block form $[A_{ij}]$, where A_{ii} is a nonsingular square matrix and $A_{ij} = O$ for $i \neq j$, then A is nonsingular, and $A^{-1} = [B_{ij}]$, where $B_{ii} = A_{ii}^{-1}$ for each i and $B_{ij} = O$ for each $i \neq j$. In particular, if each A_{ii} is involutory, then so is A.

Example Let n be a prime positive integer. The complex number $c_n = \cos(\frac{2\pi}{n}) + i \sin(\frac{2\pi}{n})$ is called a *primitive root of unity* of degree n, since it easy to check that $c_n^n = 1$ but $c_n^h \neq 1$ for all $0 < h < n$. Therefore, $c_n^{-1} = c_n^{n-1}$ for all n. For each $z \in \mathbb{C}$, let $F(z) \in M_{n \times n}(\mathbb{C})$ be the matrix $[a_{ij}]$ defined by $a_{ij} = z^{(i-1)(j-1)}$ for all $1 \leq i, j \leq n$. It is straightforward to show that the matrix $F(c_n)$ is nonsingular and, indeed, $F(c_n)^{-1} = \frac{1}{n}F(c_n^{-1})$. The endomorphism φ_n of $\mathbb{C}^n$ which is represented with respect to the canonical basis by the matrix $F(c_n)$ is called the *discrete Fourier transform* of $\mathbb{C}^n$. This endomorphism is of great importance in applied mathematics. An algorithm, known as the *fast Fourier transform* (FFT), introduced by J.W. Cooley and John Tukey in 1965, allows one to calculate $\varphi_n(v)$ in an order of $n \log(n)$ arithmetic operations, rather than n^2, as one would anticipate. This facilitates the use of Fourier transforms in applications. A similar construction is also possible over finite fields, and especially over fields of the form $GF(p)$. We will look at this example again in Chap. 15.

A closely-related endomorphism, the *discrete cosine transform*, is used in defining the JPEG algorithm for image compression.

Joseph Fourier was a close friend of Napoleon and served for many years as permanent secretary of the Parisian Academy of Sciences. He worked primarily in applied mathematics, and developed many important tools in this area. John Tukey was a twentieth-century American statistician who developed many advanced mathematical tools in statistics.

Let $(K, \bullet)$ be an algebra over a field F. A *matrix representation* of K by matrices over F is a homomorphism of F-algebras from K to $\mathcal{M}_{n \times x}(F)$ for some positive integer n. Matrix representations are a very important tool in studying the structure of algebras over fields. More generally, a *representation* of K over F is a homomorphism of F-algebras from K to $\text{End}(V)$ for some vector space V, not necessarily finitely generated, over F.

Example Recall that $\mathbb{C}$ is an algebra of dimension 2 over $\mathbb{R}$. The function $\gamma : \mathbb{C} \to \mathcal{M}_{2 \times 2}(\mathbb{R})$ defined by $\gamma : a + bi \mapsto \begin{bmatrix} a & b \\ -b & a \end{bmatrix}$ is a matrix representation of $\mathbb{C}$ by matrices over $\mathbb{R}$. In fact, this representation is clearly monic and its image is the subalgebra T of $\mathcal{M}_{2 \times 2}(\mathbb{R})$ consisting of all matrices of the form $\begin{bmatrix} a & b \\ -b & a \end{bmatrix}$, so γ is an $\mathbb{R}$-algebra isomorphism from $\mathbb{C}$ to T.

Let $(K, \bullet)$ be an associative unital algebra over a field F having multiplicative identity e, and let n be a positive integer. Let E be the multiplicative identity of $\mathcal{M}_{n \times n}(K)$ A matrix $A = [c_{ij}] \in \mathcal{M}_{n \times n}(K)$ is an *elementary matrix* if and only if it is of one of the following forms:
(1) E_{hk}, the matrix formed from E by interchanging the hth and kth columns, where $h \neq k$;
(2) $E_{h;c}$, the matrix formed from E by multiplying the hth column by $0_K \neq c \in K$;
(3) $E_{hk;c}$, the matrix formed from E by adding c times the kth column to the hth column, where $h \neq k$, where $c \in K$.

It is easy to verify that matrices of the form E_{hk} and $E_{hk;c}$ are always nonsingular, with $E_{hk}^{-1} = E_{hk}$ and $E_{kh;c}^{-1} = E_{hk;-c}$. If c is a unit in K, then matrices of the form $E_{h;c}$ are nonsingular, with $E_{h;c}^{-1} = E_{h,c^{-1}}$. Thus, if K is a field (and in particular, if $K = F$), every elementary matrix of the form $E_{h;c}$ is nonsingular. We note that the transpose of an elementary matrix is again an elementary matrix. Indeed, $E_{hk}^T = E_{hk}$ and $E_{h;c}^T = E_{h;c}$ for all $1 \leq h, k \leq n$ and $0_K \neq c \in K$, while $E_{hk;c}^T = E_{kh;c}$ for all $1 \leq h \neq k \leq n$ and all $c \in K$.

As the name clearly implies, there is a connection between the elementary automorphisms which we defined previously and the elementary matrices. Indeed, if $K = F$ and if B is the canonical basis of F^n, then $E_{hk} = \Phi(\varepsilon_{hk})$, $E_{h;c} = \Phi(\varepsilon_{h;c})$, and $E_{hk;c} = \Phi(\varepsilon_{hk;c})$.

Let us see what happens when one multiplies an arbitrary matrix in $\mathcal{M}_{n \times n}(K)$ on the left by an elementary matrix:

(1) If $B \in \mathcal{M}_{n \times n}(K)$ then $E_{hk}B$ is the matrix obtained from B by interchanging the hth and kth rows of B. Thus, for example, in $\mathcal{M}_{4 \times 4}(\mathbb{Q})$ we look at the effect of E_{24}:

$$\begin{bmatrix} 1 & 0 & 0 & 0 \\ 0 & 0 & 0 & 1 \\ 0 & 0 & 1 & 0 \\ 0 & 1 & 0 & 0 \end{bmatrix} \begin{bmatrix} 5 & 6 & 4 & 1 \\ 3 & 2 & 2 & 2 \\ 0 & 4 & 2 & 7 \\ 3 & 3 & 2 & 2 \end{bmatrix} = \begin{bmatrix} 5 & 6 & 4 & 1 \\ 3 & 3 & 2 & 2 \\ 0 & 4 & 2 & 7 \\ 3 & 2 & 2 & 2 \end{bmatrix}.$$

(2) If $B \in \mathcal{M}_{n \times n}(K)$ then $E_{h;c}B$ is the matrix obtained from B by multiplying the hth row of B by c. Thus, for example, in $\mathcal{M}_{4 \times 4}(\mathbb{Q})$ we look at the effect of $E_{2;5}$:

$$\begin{bmatrix} 1 & 0 & 0 & 0 \\ 0 & 5 & 0 & 0 \\ 0 & 0 & 1 & 0 \\ 0 & 0 & 0 & 1 \end{bmatrix} \begin{bmatrix} 5 & 6 & 4 & 1 \\ 3 & 2 & 2 & 2 \\ 0 & 4 & 2 & 7 \\ 3 & 3 & 2 & 2 \end{bmatrix} = \begin{bmatrix} 5 & 6 & 4 & 1 \\ 15 & 10 & 10 & 10 \\ 0 & 4 & 2 & 7 \\ 3 & 3 & 2 & 2 \end{bmatrix}.$$

(3) If $B \in \mathcal{M}_{n \times n}(K)$ then $E_{hk;c}B$ is the matrix obtained from B by adding c times the hth row to the kth row. Thus, for example, in $\mathcal{M}_{4 \times 4}(\mathbb{Q})$ we look at the effect of $E_{13;2}$:

$$\begin{bmatrix} 1 & 0 & 0 & 0 \\ 0 & 1 & 0 & 0 \\ 2 & 0 & 1 & 0 \\ 0 & 0 & 0 & 1 \end{bmatrix} \begin{bmatrix} 5 & 6 & 4 & 1 \\ 3 & 2 & 2 & 2 \\ 0 & 4 & 2 & 7 \\ 3 & 3 & 2 & 2 \end{bmatrix} = \begin{bmatrix} 5 & 6 & 4 & 1 \\ 3 & 2 & 2 & 2 \\ 10 & 16 & 10 & 9 \\ 3 & 3 & 2 & 2 \end{bmatrix}.$$

Proposition 9.1 *If F is a field, if n is a positive integer, and if $A, B, C, D \in M_{n \times n}(F)$ then:*

(1) *When A and B are nonsingular, so is AB, with $(AB)^{-1} = B^{-1}A^{-1}$;*

(2) *When AB is nonsingular, both A and B are nonsingular;*

(3) *When A and B are nonsingular, $A^{-1} + B^{-1} = A^{-1}(B + A)B^{-1}$;*

(4) *When $I + AB$ is nonsingular, so is $I + BA$, and $(I + BA)^{-1} = I - B(I + AB)^{-1}A$;*

(5) *(Guttman's Theorem) If A is nonsingular and if $v, w \in F^n$ satisfy the condition that $1 + w \odot A^{-1}v \neq 0$, then the matrix $A + v \wedge w \in M_{n \times n}(F)$ is nonsingular and satisfies $(A + v \wedge w)^{-1} = A^{-1} - (1 + w \odot A^{-1}v)^{-1}(A^{-1}[v \wedge w]A^{-1})$.*

(6) *(Sherman–Morrison–Woodbury Theorem) When the matrices C, D, $D^{-1} + AC^{-1}B$, and $C + BDA$ are nonsingular, then $(C + BDA)^{-1} = C^{-1} - C^{-1}B(D^{-1} + AC^{-1}B)^{-1}AC^{-1}$.*

Proof (1) This is a special case of a general remark about units in associative F-algebras, which we have already noted.

(2) Let V a vector space of dimension n over F, and let D be a basis of V. Then there exist endomorphisms α and β of V satisfying $A = \Phi_{DD}(\alpha)$ and $B = \Phi_{DD}(\beta)$, and so $AB = \Phi_{DD}(\alpha\beta)$. Since AB is nonsingular, we know that $\alpha\beta \in \text{Aut}(V)$. Then there exists an automorphism γ of V satisfying $\gamma(\alpha\beta) = \sigma_1 = (\alpha\beta)\gamma$. Then $(\gamma\alpha)\beta = \sigma_1 = \alpha(\beta\gamma)$ and so, by Proposition 7.4, we know that both α and β are automorphisms of V, and hence both A and B are nonsingular.

(3) This is an immediate consequence of the fact that $A(A^{-1} + B^{-1})B = B + A$.

(4) We note that

$$(I + BA)\big[I - B(I + AB)^{-1}A\big] = I + BA - (B + BAB)(I + AB)^{-1}A$$

$$= I + BA - B(I + AB)(I + AB)^{-1}A$$

$$= I + BA - BA = I.$$

(5) A simple calculation shows us that if $x, y \in F^n$ satisfy the condition that $c = 1 + y \odot x$ is nonzero, then

$$\big(I - c^{-1}[x \wedge y]\big)\big(I + [x \wedge y]\big) = I + x \wedge y - c^{-1}[x \wedge y] - c^{-1}[x \wedge y]^2$$

$$= I + x \wedge y - c^{-1}c[x \wedge y] = I,$$

and so $(I + [x \wedge y])^{-1} = I - c^{-1}[x \wedge y]$. Therefore, if we set $d = 1 + w \odot A^{-1}v$, then

$$(A + v \wedge w)^{-1} = \big[A(I + A^{-1}[v \wedge w])\big]^{-1} = (I + A^{-1}[v \wedge w])^{-1}A^{-1}$$

$$= \big[I - d^{-1}(A^{-1}[v \wedge w])\big]A^{-1}$$

$$= A^{-1} - d^{-1}(A^{-1}[v \wedge w]A^{-1}),$$

as required.

(6) First, note that $I + C^{-1}BDA = C^{-1}(C + BDA)$ and so, by (1), this matrix is nonsingular as well. By (4), $(I + C^{-1}BDA)^{-1} = I - C^{-1}B(I + (DAC^{-1}B)DA$, and so

$$(C + BDA)^{-1} = \big[C(I + C^{-1}BDA)\big]^{-1}$$

$$= \big[I - C^{-1}B(I + DAC^{-1}B)^{-1}DA\big]C^{-1}$$

$$= C^{-1} - C^{-1}B\big[D^{-1}(I + DAC^{-1}B)\big]^{-1}AC^{-1}$$

$$= C^{-1} - C^{-1}B(D^{-1} + AC^{-1}B)^{-1}AC^{-1},$$

as required.   □

With kind permission of Nurit Guttman.

Louis Guttman was a twentieth-century American/Israeli statistician and sociologist who developed many advanced mathematical tools for use in statistics. The Sherman–Morrison–Woodbury Theorem was in fact first published by British aeronautics professor W.J. Duncan, but is named after the twentieth-century American statisticians Jack Sherman, Winifred J. Morrison, and Max Woodbury who used it extensively.

Guttman's Theorem is important in the following context: assume we have calculated A^{-1} for some square matrix A and now we have to calculate B^{-1}, where B differs from A in only one entry. With the help of this result, we can make use of our knowledge of A^{-1} to calculate B^{-1} with relative ease and speed. The Sherman–Morrison–Woodbury Theorem has similar uses.

In particular, we note from Proposition 9.1 that if $A, B \in M_{n \times n}(F)$ then AB is nonsingular if and only if BA is nonsingular. We should also note that if $A, B \in M_{n \times n}(F)$ then $B^T A^T = (AB)^T$, and so if A is a nonsingular matrix and $B = A^{-1}$ then $AB = I$, and so $B^T A^T = I^T = I$. Thus A^T is also nonsingular. Moreover, this also shows that $(A^T)^{-1} = (A^{-1})^T$ for every nonsingular matrix $A \in M_{n \times n}(F)$.

Proposition 9.2 *Let F be a field, let n be a positive integer, and let $A, B \in M_{n \times n}(F)$, where A is nonsingular. Then there exist unique matrices C and D in $M_{n \times n}(F)$ satisfying $CA = B = AD$.*

Proof Define $C = BA^{-1}$ and $D = A^{-1}B$. Then surely $CA = B = AD$. If C' and D' are matrices satisfying $C'A = B = AD'$ then $C' = (C'A)A^{-1} = BA^{-1} = C$ and $D' = A^{-1}(AD') = A^{-1}B = D$, and so we have uniqueness. $\square$

Example The matrices C and D in Proposition 9.2 need not be the same. For example, if $A, B \in M_{2 \times 2}(\mathbb{R})$ are defined by $A = \begin{bmatrix} 1 & 3 \\ 0 & 1 \end{bmatrix}$ and $B = \begin{bmatrix} 1 & 0 \\ 3 & 1 \end{bmatrix}$ then $A^{-1} = \begin{bmatrix} 1 & -3 \\ 0 & 1 \end{bmatrix}$, $C = \begin{bmatrix} 1 & -3 \\ 3 & -8 \end{bmatrix}$, and $D = \begin{bmatrix} -8 & -3 \\ 3 & 1 \end{bmatrix}$.

Proposition 9.3 *Let F be a field, let n be a positive integer, and let $A = [a_{ij}] \in M_{n \times n}(F)$. Then the following conditions are equivalent:*
(1) *A is nonsingular;*
(2) *The columns of A are distinct and the set of these columns is a linearly-independent subset of F^n;*
(3) *The rows of A are distinct and the set of these rows is a linearly-independent subset of $M_{1 \times n}(F)$.*

Proof (1) $\Leftrightarrow$ (2): Denote the columns of A by $y_1, \ldots, y_n$. Let $V = F^n$ and let $B = \{v_1, \ldots, v_n\}$ be the canonical basis of V. If two columns of A are equal or if the set of columns is linearly dependent, there exist scalars $c_1, \ldots, c_n$, not all equal to 0, such that $A \begin{bmatrix} c_1 \\ \vdots \\ c_n \end{bmatrix} = \sum_{i=1}^n c_i y_i = \begin{bmatrix} 0 \\ \vdots \\ 0 \end{bmatrix}$. But if (1) holds, then

$$\begin{bmatrix} c_1 \\ \vdots \\ c_n \end{bmatrix} = A^{-1} \begin{bmatrix} 0 \\ \vdots \\ 0 \end{bmatrix} = \begin{bmatrix} 0 \\ \vdots \\ 0 \end{bmatrix},$$ which is a contradiction. Therefore, (2) holds. Con-

versely, assume (2) holds. Then the endomorphism α of V given by $v \mapsto Av$ is a monic and so an automorphism of V. But $A = \Phi_{BB}(\alpha)$, and so, as we have seen, A is nonsingular.

(1) $\Leftrightarrow$ (3): This follows directly from the equivalence of (1) and (2), given the fact that a matrix A is nonsingular if and only if A^T is nonsingular. $\qquad\square$

Example Let F be a field and let $n > 1$ be an integer. If $v, w \in F^n$, then the columns of $v \wedge w \in \mathcal{M}_{n \times n}(F)$ are linearly dependent and so $v \wedge w$ is always singular.

Example If F is a field and if $U = [u_{ij}] \in \mathcal{M}_{n \times n}(F)$ is an upper-triangular matrix satisfying the condition that $u_{ii} \neq 0$ for all $1 \le i \le n$ then, by Proposition 9.3, it is clear that U is nonsingular. We claim that, moreover, U^{-1} is again upper triangular. Let us prove this contention by induction on n. It is clearly true for $n = 1$. Assume therefore that $n > 1$ and that we have already shown that the inverse of any upper-triangular matrix in $\mathcal{M}_{(n-1) \times (n-1)}(F)$ is upper-triangular. Write

$$U = \begin{bmatrix} A & y \\ z & u_{nn} \end{bmatrix},$$ where $A \in \mathcal{M}_{(n-1) \times (n-1)}(F)$, $y \in F^{n-1}$, and $z = \begin{bmatrix} 0 \\ \vdots \\ 0 \end{bmatrix}^T$. As-

sume that $U^{-1} = \begin{bmatrix} B & x \\ w^T & b \end{bmatrix}$, where $B \in \mathcal{M}_{(n-1) \times (n-1)}(F)$ and $w, x \in F^{n-1}$. Then $AB + y \wedge w = I$, $Ax + by = z^T$, $u_{nn} w^T = z$, and $u_{nn} b = 1$, so we must have $b = u_{nn}^{-1} \neq 0$ and $w^T = z$. Therefore, $y \wedge w = O$ and so $B = A^{-1}$. By hypothesis, B is upper triangular and so U^{-1} is again upper triangular. A similar argument holds for lower-triangular matrices.

Proposition 9.4 *Let F be a field and let n be a positive integer. A matrix in $\mathcal{M}_{n \times n}(F)$ is nonsingular if and only if it is a product of elementary matrices.*

Proof Since each of the elementary matrices is nonsingular, we know that any product of elementary matrices is also nonsingular. Conversely, let $A = [a_{ij}]$ be a nonsingular matrix in $\mathcal{M}_{n \times n}(F)$ and let $B = [b_{ij}]$ be A^{-1}. Then B is also nonsingular and so, by Proposition 9.3, the columns of B are distinct and the set of columns is linearly independent in F^n. In particular, there exists a nonzero entry b_{h1} in the first column of B. Multiply B on the left by E_{h1} to get a new matrix in which the $(1, 1)$-entry nonzero. Now multiply it on the left by $E_{1;c}$, where $c = b_{h1}^{-1}$, in order to get

a matrix of the form $\begin{bmatrix} 1 & * & \cdots & * \\ * & * & \cdots & * \\ \vdots & \vdots & \ddots & \vdots \\ * & * & \cdots & * \end{bmatrix}$. Now let $1 < t \le n$, and let $d(t)$ be the additive inverse of the $(t, 1)$-entry of the matrix. Multiplying the matrix on the left by $E_{t1;d(t)}$, we will get a matrix with 0 in the $(t, 1)$-entry and so, after this for each such t, we see that a matrix $B' = C_1 B$, where C_1 is a product of elementary matrices, and

which is of the form $\begin{bmatrix} 1 & * & \cdots & * \\ 0 & * & \cdots & * \\ \vdots & \vdots & \ddots & \vdots \\ 0 & * & \cdots & * \end{bmatrix}$. This matrix is still nonsingular, since it is

a product of two nonsingular matrices, and so its columns are distinct and form a linearly-independent subset of F^n. Therefore, there exists a nonzero entry b'_{h2} in the second column, with $h > 1$. Repeating the above procedure, we can find a matrix C_2 which is a product of elementary matrices and such that $C_2 C_1 B$ is of the form

$\begin{bmatrix} 1 & 0 & * & \cdots & * \\ 0 & 1 & * & \cdots & * \\ 0 & 0 & * & \cdots & * \\ \vdots & \vdots & \vdots & \ddots & \vdots \\ 0 & 0 & * & \cdots & * \end{bmatrix}$. Continuing in this manner, we obtain matrices $C_1, \ldots, C_n$,

each of them a product of elementary matrices, such that $C_n \cdots C_1 B = I$. Therefore, $C_n \cdots C_1 = B^{-1} = A$, as we wanted to show.   □

Example Let F be a field and let n be a positive integer. Every permutation π of the set $\{1, \ldots, n\}$ defines a matrix $A_\pi = [a_{ij}] \in \mathcal{M}_{n \times n}(F)$ by setting $a_{ij} = 1$ if $j = \pi(i)$ and $a_{ij} = 0$ otherwise, called the *permutation matrix* defined by π. This matrix is clearly a result of multiplying I by a number of elementary matrices of the form E_{hk}, and so is nonsingular.

The order of multiplication given in Proposition 9.4 is not unique. Indeed, we claim that it is possible to write any nonsingular matrix $A \in \mathcal{M}_{n \times n}(F)$ in the form PC, where P is a permutation matrix and C is a product of elementary matrices of the form $E_{i;c}$ and $E_{ij;c}$. To see how this is done, we note that if $1 \leq i, h, k \leq m$ and if $c \in F$ then $E_{i;c} E_{hk} = E_{hk} E_{i;c}$ if $i \notin \{h, k\}$ and $E_{h;c} E_{hk} = E_{hk} E_{k;c}$ and a similar result holds for elementary matrices of the form $E_{ij;c}$ and E_{hk}. Thus, one by one, the elementary matrices the form E_{hk} can be "moved to the left" until we obtain the desired decomposition.

Proposition 9.4 allows us to construct an algorithm for computing A^{-1} when A is a nonsingular matrix in $\mathcal{M}_{n \times n}(F)$. First of all, we construct the matrix $[I \ A] \in \mathcal{M}_{n \times 2n}(F)$ and on this matrix we perform a series of *elementary operations*, namely operations which are the result of multiplying it on the left by elementary matrices, which bring the right-hand block into the form I. Then the left-hand block is A^{-1}. To calculate A^{-1} by this method, we use $n^3 - 2n^2 + n$ additions and n^3 multiplications.

Example Consider the matrix $A = \begin{bmatrix} 1 & 2 & 3 \\ 2 & 3 & 0 \\ 0 & 1 & 2 \end{bmatrix} \in \mathcal{M}_{3\times3}(\mathbb{Q})$. Therefore, we begin

with the matrix $\begin{bmatrix} 1 & 0 & 0 & 1 & 2 & 3 \\ 0 & 1 & 0 & 2 & 3 & 0 \\ 0 & 0 & 1 & 0 & 1 & 2 \end{bmatrix} \in \mathcal{M}_{3\times6}(\mathbb{Q})$. Then we

(1) Get $\begin{bmatrix} 1 & 0 & 0 & 1 & 2 & 3 \\ -2 & 1 & 0 & 0 & -1 & -6 \\ 0 & 0 & 1 & 0 & 1 & 2 \end{bmatrix}$ after multiplying the first row by -2 and

adding it to the second row;

(2) Get $\begin{bmatrix} 1 & 0 & 0 & 1 & 2 & 3 \\ -2 & 1 & 0 & 0 & -1 & -6 \\ -2 & 1 & 1 & 0 & 0 & -4 \end{bmatrix}$ after adding the second row to the third row;

(3) Get $\begin{bmatrix} 1 & 0 & 0 & 1 & 2 & 3 \\ 2 & -1 & 0 & 0 & 1 & 6 \\ 0.5 & -0.25 & -0.25 & 0 & 0 & 1 \end{bmatrix}$ after multiplying the second row by

-1 and then multiplying the third row by -0.25;

(4) Get $\begin{bmatrix} 1 & 0 & 0 & 1 & 2 & 3 \\ -1 & 0.5 & 1.5 & 0 & 1 & 0 \\ 0.5 & -0.25 & -0.25 & 0 & 0 & 1 \end{bmatrix}$ after multiply the third row by -6 and

adding it to the second row;

(5) Get $\begin{bmatrix} -0.5 & 0.75 & 0.75 & 1 & 2 & 0 \\ -1 & 0.5 & 1.5 & 0 & 1 & 0 \\ 0.5 & -0.25 & -0.25 & 0 & 0 & 1 \end{bmatrix}$ after multiplying the third row by

-3 and adding it to the first row;

(6) Finally, get $\begin{bmatrix} 1.5 & -0.25 & -2.25 & 1 & 0 & 0 \\ -1 & 0.5 & 1.5 & 0 & 1 & 0 \\ 0.5 & -0.25 & -0.25 & 0 & 0 & 1 \end{bmatrix}$ after multiplying the second

row by -2 and adding it to the first row.

Therefore, we see that $A^{-1} = \frac{1}{4} \begin{bmatrix} 6 & -1 & -9 \\ -4 & 2 & 6 \\ 2 & -1 & -1 \end{bmatrix}$.

Example When one uses computer to compute matrix inverses, one must always be aware of hardware limitations. For example, one can show that *Nievergelt's matrix* $A = \begin{bmatrix} 888445 & 887112 \\ 887112 & 885871 \end{bmatrix} \in \mathcal{M}_{2\times2}(\mathbb{Q})$ is nonsingular, while the matrix $B = A - \begin{bmatrix} c & c \\ c & c \end{bmatrix}$, where $c = \frac{1}{3548450}$ (which is approximately 2.818×10^{-7}) is not. Nonetheless, a computer or calculator capable of only 12-digit accuracy cannot differentiate between the two.

Example For each positive integer n, let $H_n \in \mathcal{M}_{n\times n}(\mathbb{Q})$ be the matrix $[a_{ij}]$ in which $a_{ij} = \frac{1}{i+j-1}$. This matrix is called the $n \times n$ *Hilbert matrix*. Hilbert matrices are all nonsingular but, while their entries all lie between 0 and 1, the entries in their inverses are very large. For example, H_6^{-1} equals

$$\begin{bmatrix}
36 & -630 & 3360 & -7560 & 7560 & -2772 \\
-630 & 14700 & -88200 & 211680 & -220500 & 83160 \\
3360 & -88200 & 564480 & -1411200 & 1512000 & -582120 \\
-7560 & 211680 & -1411200 & 3628800 & -3969000 & 1552320 \\
7560 & -220500 & 1512000 & -3969000 & 4410000 & -1746360 \\
-2772 & 83160 & -582120 & 1552320 & -1746360 & 698544
\end{bmatrix}.$$

Therefore, these matrices are often used as benchmarks to judge the efficiency and accuracy of computer programs to calculate matrix inverses. In particular if the computer we are using has only 7-digit accuracy, it is reasonable to assume that we will have a 100% error in computing H_6^{-1}.

With kind permission of the Archives of the Mathematisches Forschungsinstitut Oberwolfach.

German **David Hilbert** was one of the foremost mathematicians in the world at the beginning of the twentieth century. He and his students were among the first to study infinite-dimensional vector spaces.

It is sometimes possible to use a representation of a nonsingular matrix A in block form in order to calculate A^{-1}. Indeed, suppose that $A \in \mathcal{M}_{n \times n}(F)$ is a matrix which can be written in block form $\begin{bmatrix} A_{11} & A_{12} \\ A_{21} & A_{22} \end{bmatrix}$, where $A_{11} \in \mathcal{M}_{k \times k}(F)$. If A_{11} and $C = A_{22} - A_{21}A_{11}^{-1}A_{12}$ are both nonsingular, then A is also nonsingular, with

$$A^{-1} = \begin{bmatrix} I & -A_{11}^{-1}A_{12} \\ O & I \end{bmatrix} \begin{bmatrix} A_{11}^{-1} & O \\ O & C^{-1} \end{bmatrix} \begin{bmatrix} I & O \\ -A_{21}A_{11}^{-1} & O \end{bmatrix}.$$

Similarly, if A_{22} and $D = A_{11} - A_{12}A_{22}^{-1}A_{21}$ are both nonsingular, then A is also nonsingular, with

$$A^{-1} = \begin{bmatrix} I & O \\ -A_{22}^{-1}A_{21} & O \end{bmatrix} \begin{bmatrix} D^{-1} & O \\ O & A_{22}^{-1} \end{bmatrix} \begin{bmatrix} I & A_{12}A_{22}^{-1} \\ O & I \end{bmatrix}.$$

The matrices C and D are, respectively, the **Schur complements** of A_{11} and A_{22} in A. These conditions, however, are sufficient but not necessary for A to be nonsingular, as the following example shows.

Issai Schur was a twentieth-century German mathematician who is known primarily for his work in group theory.

Example The matrix $A = \begin{bmatrix} \begin{bmatrix} 1 & 0 \\ 0 & 0 \end{bmatrix} & \begin{bmatrix} 0 & 0 \\ 1 & 0 \end{bmatrix} \\ \begin{bmatrix} 0 & 1 \\ 0 & 0 \end{bmatrix} & \begin{bmatrix} 0 & 0 \\ 0 & 1 \end{bmatrix} \end{bmatrix} \in \mathcal{M}_{4\times 4}(\mathbb{Q})$ is nonsingular, despite the fact that all of the given 2×2 blocks are singular.

It is important to make clear, however, that it is hardly ever necessary, in applications, to actually compute the inverse of a nonsingular matrix. One is more likely to have to compute a product of the form $A^{-1}B$, which can usually be done without explicitly computing A^{-1} first.

Let F be a field and let k and n be positive integers. Two matrices $B, C \in \mathcal{M}_{k\times n}(F)$ are *equivalent* if and only if there exist nonsingular matrices $P \in \mathcal{M}_{k\times k}(F)$ and $Q \in \mathcal{M}_{n\times n}(F)$ such that $PBQ = C$. This is, indeed, an equivalence relation on $\mathcal{M}_{k\times n}(F)$ since:

(1) $IBI = B$ for each such matrix B, showing that B is equivalent to itself;

(2) If $PBQ = C$ then $P^{-1}CQ^{-1} = B$;

(3) If $PBQ = C$ and $P'CQ' = D'$ then $(P'P)B(QQ') = D$, where we note that both $P'P$ and QQ' are again nonsingular.

Similarly, we say that B and C are *row equivalent* if and only if there exists a nonsingular matrix $P \in \mathcal{M}_{k\times k}(F)$ satisfying $PB = C$, and we say that B and C are *column equivalent* if and only if there exists a nonsingular matrix $Q \in \mathcal{M}_{n\times n}(F)$ satisfying $BQ = C$. Both of these relations are also equivalence relations on $\mathcal{M}_{k\times n}(F)$, and it is clear that if B and C are row equivalent then they are equivalent (take $Q = I$) and if they are column equivalent then they are equivalent (take $P = I$).

Equivalence of matrices is a very strong concept. Indeed, it is easy to show that any matrix $B \in \mathcal{M}_{k\times n}(F)$ is equivalent to one which is in block form $\begin{bmatrix} I & O \\ O & O \end{bmatrix}$. Therefore, it is more useful to consider row equivalence of matrices as our basic tool.

Now let V be a vector space of dimension n over a field F and choose bases $B = \{v_1, \ldots, v_n\}$ and $D = \{w_1, \ldots, w_n\}$ of V. For each $1 \leq j \leq n$ there exist elements $q_{1j}, \ldots, q_{nj}$ of F satisfying $w_j = \sum_{i=1}^{n} q_{ij} v_i$. By Proposition 9.3, we know that the matrix $Q = [q_{ij}]$ is nonsingular. If $v = \sum_{i=1}^{n} a_i v_i = \sum_{j=1}^{n} b_j w_j$ is an element of V, then we see that $v = \sum_{j=1}^{n} b_j w_j = \sum_{j=1}^{n} b_j (\sum_{i=1}^{n} q_{ij} v_i) = \sum_{i=1}^{n} (\sum_{j=1}^{n} q_{ij} b_j) v_i$

and so we must have $a_i = \sum_{j=1}^n q_{ij} b_j$ for all $1 \le i \le n$. Thus we see that

$$\begin{bmatrix} a_1 \\ \vdots \\ a_n \end{bmatrix} = Q \begin{bmatrix} b_1 \\ \vdots \\ b_n \end{bmatrix}.$$

The matrix Q is called the *change-of-basis matrix* from D to B.

Example Let F be a field, let n be a positive integer, and let V be the subspace of the vector space $F[X]$ made up of all polynomials of degree at most $n - 1$. Then $\dim(V) = n$, and it has a canonical basis $B = \{1, X, \dots, X^{n-1}\}$. Let $c_1, \dots, c_n$ be distinct scalars, and for each $1 \le i \le n$, consider the polynomial

$$p_i(X) = \prod_{j \ne i} \frac{1}{c_i - c_j}(X - c_j) \in V.$$

This polynomial is called the ith *Lagrange interpolation polynomial*, and we will return to these polynomials below in another context. It is clear that

$$p_i(c_j) = \begin{cases} 1 & \text{if } i = j, \\ 0 & \text{otherwise.} \end{cases}$$

Thus, for example, if $n = 4$ and if we choose $c_1 = 1$, $c_2 = 3$, $c_3 = 5$, and $c_4 = 7$, we obtain

$$p_1(X) = -\frac{1}{48}X^3 + \frac{5}{16}X^2 - \frac{71}{48}X + \frac{35}{16},$$

$$p_2(X) = \frac{1}{16}X^3 - \frac{13}{16}X^2 + \frac{47}{16}X - \frac{35}{16},$$

$$p_3(X) = -\frac{1}{16}X^3 + \frac{11}{16}X^2 - \frac{31}{16}X + \frac{21}{16},$$

$$p_4(X) = \frac{1}{48}X^3 - \frac{3}{16}X^2 + \frac{23}{48}X - \frac{5}{16}.$$

Returning to the general case, we see that the set $D = \{p_1(X), \dots, p_n(X)\}$ of Lagrange interpolation polynomials is linearly independent since, if we have $\sum_{i=1}^n a_i p_i(X) = 0$, then for each $1 \le h \le n$ we have $a_h = \sum_{i=1}^n a_i p_i(c_h) = 0$. Therefore, D is also a basis of V. If $q(X)$ is an arbitrary polynomial in V then there exist scalars $a_1, \dots, a_n$ satisfying $q(X) = \sum_{i=1}^n a_i p_i(X)$. Again, this implies that $a_i = q(c_i)$ for all i. In particular, if $q(X) = X^k$ we see that $X^k = \sum_{i=1}^n c_i^k p_i(X)$.

Therefore, the change of basis matrix from D to B is $\begin{bmatrix} 1 & c_1 & \cdots & c_1^n \\ 1 & c_2 & \cdots & c_2^n \\ \vdots & \vdots & \ddots & \vdots \\ 1 & c_n & \cdots & c_n^n \end{bmatrix}$. A matrix of this form is called a *Vandermonde matrix*, and such matrices are always nonsingular.

With kind permission of ETH-Bibliothek Zurich, Image Archive (Lagrange).

Joseph-Louis Lagrange was one of the applied mathematicians who surrounded Napoleon, and his book on analytical mechanics is considered a mathematical classic. **Alexandre-Théophile Vandermonde** was an eighteenth century French chemist and mathematician who studied determinants of matrices. Vandermonde matrices do not appear in his work, and it is not clear why they are named after him.

Lagrange interpolation allows us to represent a polynomial $p(X)$ of degree less than n in a computer not by its list of coefficients but rather by a list of its values $p(a_1), \ldots, p(a_n)$ at n preselected elements of F. Such representations can be used to obtain algorithms for rapid multiplication of polynomials, especially in the case the field F is finite (having n elements, of course). Indeed, if $p(X)$ and $q(X)$ are polynomials in $F[X]$ of positive degree satisfying $\deg(p) + \deg(q) = h < n$, then $p(X)q(X)$ is the unique polynomial $t(X)$ of degree h satisfying $t(a_i) = p(a_i)q(a_i)$ for all $1 \leq i \leq h+1$.

Let us now return to the matter of change of basis, and now let us assume that we have a linear transformation $\alpha : V \to Y$, where V is a vector space of dimension n over a field F and Y is a vector space of dimension k over F. We have bases $B = \{v_1, \ldots, v_n\}$ and $D = \{w_1, \ldots, w_n\}$ of V. Choose a basis $E = \{y_1, \ldots, y_k\}$ of Y. Then $\Phi_{BE}(\alpha)$ is a matrix $C = [c_{ij}]$. If $Q = [q_{ij}]$ is the change of basis matrix from D to B then for each $1 \leq j \leq n$ we have $\alpha(w_j) = \alpha(\sum_{h=1}^{n} q_{hj} v_h) = \sum_{h=1}^{n} q_{hj} \alpha(v_h) = \sum_{h=1}^{n} q_{hj}(\sum_{i=1}^{k} c_{ih} y_i) = \sum_{i=1}^{k}(\sum_{h=1}^{n} c_{ih} q_{hj}) y_i$, and so $\Phi_{DE}(\alpha) = CQ$, showing that $\Phi_{DE}(\alpha)$ and C are column equivalent. In the same manner, if we have another basis $G = \{z_1, \ldots, z_k\}$ of Y and if $P = [p_{ij}]$ is the change of basis matrix from E to G, then $z_j = \sum_{i=1}^{k} p_{ij} y_i$ for all $1 \leq j \leq k$. If $\Phi_{BG}(\alpha)$ is the matrix $C' = [c'_{ij}]$, then for all $1 \leq j \leq n$ we have $\alpha(v_j) = \sum_{h=1}^{k} e_{hj} z_h = \sum_{h=1}^{k} e_{hj}(\sum_{i=1}^{k} p_{ih} y_i) = \sum_{i=1}^{k}(\sum_{h=1}^{k} p_{ih} e_{hj}) y_i$ and this equals $\sum_{i=1}^{k} c_{ij} y_i$, implying $C = PC'$, and so $C' = P^{-1}C$. Thus $\Phi_{BG}(\alpha)$ and C are row equivalent. If we put both of these results together, we see that $\Phi_{DG}(\alpha) = P^{-1}\Phi_{BE}(\alpha)Q$, and so $\Phi_{DG}(\alpha)$ and $\Phi_{BE}(\alpha)$ are equivalent.

Example Let $\alpha : \mathbb{R}^3 \to \mathbb{R}^2$ be the linear transformation given by

$$\alpha : \begin{bmatrix} a \\ b \\ c \end{bmatrix} \mapsto \begin{bmatrix} a+b \\ b+c \end{bmatrix}.$$

Choose bases $B = \left\{ \begin{bmatrix} 1 \\ 0 \\ 0 \end{bmatrix}, \begin{bmatrix} 0 \\ 1 \\ 1 \end{bmatrix}, \begin{bmatrix} 0 \\ -1 \\ 0 \end{bmatrix} \right\}$ and $D = \left\{ \begin{bmatrix} 1 \\ 0 \\ 0 \end{bmatrix}, \begin{bmatrix} 1 \\ 0 \\ 1 \end{bmatrix}, \begin{bmatrix} 0 \\ 1 \\ -1 \end{bmatrix} \right\}$ of

$\mathbb{R}^3$ and bases $E = \left\{ \begin{bmatrix} 1 \\ 0 \end{bmatrix}, \begin{bmatrix} 0 \\ -1 \end{bmatrix} \right\}$ and $G = \left\{ \begin{bmatrix} 1 \\ 1 \end{bmatrix}, \begin{bmatrix} 1 \\ -1 \end{bmatrix} \right\}$ of $\mathbb{R}^2$. Then $\Phi_{BE}(\alpha) =$

$\begin{bmatrix} 1 & 1 & -1 \\ 0 & -2 & 1 \end{bmatrix}$ since

$$\alpha\left(\begin{bmatrix} 1 \\ 0 \\ 0 \end{bmatrix}\right) = \begin{bmatrix} 1 \\ 0 \end{bmatrix} = 1\begin{bmatrix} 1 \\ 0 \end{bmatrix} + 0\begin{bmatrix} 0 \\ -1 \end{bmatrix},$$

$$\alpha\left(\begin{bmatrix} 0 \\ 1 \\ 1 \end{bmatrix}\right) = \begin{bmatrix} 1 \\ 2 \end{bmatrix} = 1\begin{bmatrix} 1 \\ 0 \end{bmatrix} - 2\begin{bmatrix} 0 \\ -1 \end{bmatrix},$$

$$\alpha\left(\begin{bmatrix} 0 \\ -1 \\ 0 \end{bmatrix}\right) = \begin{bmatrix} -1 \\ -1 \end{bmatrix} = (-1)\begin{bmatrix} 1 \\ 0 \end{bmatrix} + 1\begin{bmatrix} 0 \\ -1 \end{bmatrix},$$

and similarly, $\Phi_{DG}(\alpha) = \frac{1}{2}\begin{bmatrix} 1 & 2 & 1 \\ 1 & 0 & 1 \end{bmatrix}$ since

$$\alpha\left(\begin{bmatrix} 1 \\ 0 \\ 0 \end{bmatrix}\right) = \begin{bmatrix} 1 \\ 0 \end{bmatrix} = \frac{1}{2}\begin{bmatrix} 1 \\ 1 \end{bmatrix} + \frac{1}{2}\begin{bmatrix} 1 \\ -1 \end{bmatrix},$$

$$\alpha\left(\begin{bmatrix} 1 \\ 0 \\ 1 \end{bmatrix}\right) = \begin{bmatrix} 1 \\ 1 \end{bmatrix} = 1\begin{bmatrix} 1 \\ 1 \end{bmatrix} + 0\begin{bmatrix} 1 \\ -1 \end{bmatrix},$$

$$\alpha\left(\begin{bmatrix} 0 \\ 1 \\ -1 \end{bmatrix}\right) = \begin{bmatrix} 1 \\ 0 \end{bmatrix} = \frac{1}{2}\begin{bmatrix} 1 \\ 1 \end{bmatrix} + \frac{1}{2}\begin{bmatrix} 1 \\ -1 \end{bmatrix}.$$

Further, we also see that

$$\begin{bmatrix} 1 \\ 0 \\ 0 \end{bmatrix} = 1\begin{bmatrix} 1 \\ 0 \\ 0 \end{bmatrix} + 0\begin{bmatrix} 0 \\ 1 \\ 1 \end{bmatrix} + 0\begin{bmatrix} 0 \\ -1 \\ 0 \end{bmatrix},$$

$$\begin{bmatrix} 1 \\ 0 \\ 1 \end{bmatrix} = 1\begin{bmatrix} 1 \\ 0 \\ 0 \end{bmatrix} + 1\begin{bmatrix} 0 \\ 1 \\ 1 \end{bmatrix} + 1\begin{bmatrix} 0 \\ -1 \\ 0 \end{bmatrix},$$

and $\begin{bmatrix} 0 \\ 1 \\ -1 \end{bmatrix} = 0\begin{bmatrix} 1 \\ 0 \\ 0 \end{bmatrix} - 1\begin{bmatrix} 0 \\ 1 \\ 1 \end{bmatrix} - 2\begin{bmatrix} 0 \\ -1 \\ 0 \end{bmatrix}$, so $Q = \begin{bmatrix} 1 & 1 & 0 \\ 0 & 1 & -1 \\ 0 & 1 & -2 \end{bmatrix}$.

Moreover, $\begin{bmatrix} 1 \\ 1 \end{bmatrix} = 1 \begin{bmatrix} 1 \\ 0 \end{bmatrix} - 1 \begin{bmatrix} 0 \\ -1 \end{bmatrix}$ and $\begin{bmatrix} 1 \\ -1 \end{bmatrix} = 1 \begin{bmatrix} 1 \\ 0 \end{bmatrix} + 1 \begin{bmatrix} 0 \\ -1 \end{bmatrix}$ so $P =$ $\begin{bmatrix} 1 & 1 \\ -1 & 1 \end{bmatrix}$ and $P^{-1} = \frac{1}{2} \begin{bmatrix} 1 & -1 \\ 1 & 1 \end{bmatrix}$. Note that $P^{-1} \Phi_{BE}(\alpha) Q = \Phi_{DG}(\alpha)$.

Example We will now see an application of linear algebra to calculus. Let V be the vector space over $\mathbb{R}$ consisting of all infinitely-differentiable functions $f \in \mathbb{R}^{\mathbb{R}}$, and let $\delta \in \mathrm{End}(V)$ be the differentiation endomorphism.

(1) If a and b are given real numbers, not both equal to 0, then the functions $f_0 : x \mapsto e^{ax} \sin(bx)$ and $f_1 : x \mapsto e^{ax} \cos(bx)$ belong to V and the subspace $W = \mathbb{R}\{f_0, f_1\}$ of V is invariant under δ. The restriction of δ to W can be represented with respect to the basis $\{f_0, f_1\}$ of W by the nonsingular matrix $A = \begin{bmatrix} a & -b \\ b & a \end{bmatrix}$. It is easy to check that $A^{-1} = (a^2 + b^2)^{-1} \begin{bmatrix} a & b \\ -b & a \end{bmatrix}$. Therefore,

$$\int f_0(t)\, dt = \delta^{-1}(f_0) = \left(\frac{1}{a^2 + b^2} \right) [af_0 - bf_1] \quad \text{and}$$

$$\int f_1(t)\, dt = \delta^{-1}(f_1) = \left(\frac{1}{a^2 + b^2} \right) [bf_0 + af_1].$$

(2) The functions $g_0 : x \mapsto x^2 e^x$, $g_1 : x \mapsto xe^x$, and $g_2 : x \mapsto e^x$ all belong to V and the subspace $Y = \mathbb{R}\{g_0, g_1, g_2\}$ of V is invariant under δ. The restriction of δ to Y can be represented with respect to the basis $\{g_0, g_1, g_2\}$ of Y by the nonsingular matrix $B = \begin{bmatrix} 1 & 0 & 0 \\ 2 & 1 & 0 \\ 0 & 1 & 1 \end{bmatrix}$. Since $B^{-1} = \begin{bmatrix} 1 & 0 & 0 \\ -2 & 1 & 0 \\ 2 & -1 & 1 \end{bmatrix}$, we see that

$$\int g_0(t)\, dt = \delta^{-1}(g_0) = g_0 - 2g_1 + 2g_2,$$

$$\int g_1(t)\, dt = \delta^{-1}(g_1) = g_1 - g_2, \quad \text{and}$$

$$\int g_2(t)\, dt = \delta^{-1}(g_2) = g_2.$$

Let us turn to problems connected with the implementation of this theory. Let F be a field and let n be a positive integer. Let $A = [a_{ij}]$ and $B = [b_{ij}]$ belong to $\mathcal{M}_{n \times n}(F)$ and let $C = AB$. In order to calculate each one of the n^2 entries in C, we need n multiplications and $n - 1$ additions/subtractions, and so to calculate C we need n^3 multiplications and $n^3 - 2n^2 + n$ additions/subtractions. Putting this in another way, the total number of operations needed to calculate AB from the definition is on the order of n^c, where $c = 3$. If n is very large, this can entail considerable computational overhead and leaves room for the introduction of significant error due to roundoff and truncation in the course of the calculation. It is therefore very important to find a more sophisticated method of matrix multiplication, if possible. One such method is the *Strassen–Winograd algorithm*.

With kind permission of Volker Strassen (Strassen); With kind permission of the Department of Computer Science, City University of Hong Kong (Winograd).

Variants of this algorithm were discovered by the contemporary German mathematician **Volker Strassen** and the contemporary Israeli mathematician **Shmuel Winograd** who later served as director of mathematical research at IBM.

To illustrate the Strassen–Winograd algorithm, let us first begin with the special case $n = 2$. First, calculate

$$p_0 = (a_{11} + a_{12})(b_{11} + b_{12}), \quad p_1 = (a_{11} + a_{22})b_{11}, \quad p_2 = a_{11}(b_{12} - b_{22}),$$
$$p_3 = (a_{21} - a_{11})(b_{11} + b_{12}), \quad p_4 = (a_{11} + a_{12})b_{22}, \quad p_5 = a_{22}(b_{21} - b_{11}),$$
$$p_6 = (a_{12} - a_{22})(b_{21} + b_{22}),$$

and then note that $C = \begin{bmatrix} p_0 + p_5 - p_4 + p_6 & p_2 + p_4 \\ p_1 + p_5 & p_0 - p_1 + p_2 + p_3 \end{bmatrix}$. In this calculation, we used 7 multiplications and 18 additions/subtractions (Winograd's variant of this algorithm uses only 15 additions/subtractions, but these are more interdependent, and so the algorithm is less amenable to implementation on parallel computers) instead of 8 multiplications and 4 additions/subtractions. In the early days of computers, when multiplication was several orders of magnitude slower than addition, this in itself was a great accomplishment. If $n = 4$, we write our matrices in block form: $A = \begin{bmatrix} A_{11} & A_{12} \\ A_{21} & A_{22} \end{bmatrix}$ and $B = \begin{bmatrix} B_{11} & B_{12} \\ B_{21} & B_{22} \end{bmatrix}$, where each block is a 2×2 matrix. We now calculate 2×2 matrices $P_0, \ldots, P_6$ and then construct $C = AB$ as above. To do this, we need 49 multiplications and 198 additions/subtractions, as opposed to 64 multiplications and 46 additions/subtractions if one goes according to the definition. We continue recursively. If $n = 2^h$, then the number of multiplications needed is $M(h) = 7^h$ and the number of additions/subtractions needed is $A(h) = 6(7^h - 4^h)$ and so $M(h) + A(h) < 7^{h+1}$. (If n is not a power of 2, we can add rows and columns of 0's in order to enlarge it to the desired size.) Thus, we see that the number of arithmetic operations needed to calculate AB is on the order of n^c, where $c \le \log_2 7 = 2.807\ldots$ and so, for large n, we have a definite advantage over multiplication following from the definition. Using even more sophisticated techniques, it is possible to reduce the number of arithmetic operations to the order of n^c, where $c \le 2.376\ldots$, as was done by Winograd and Coppersmith in 1986. Recent results by American mathematicians Chris Ulmas and Henry Cohn, using sophisticated group-theoretic techniques, suggest that c can be reduced still further, but their methods are not, as yet, practical for all but matrices of immense size.

For *sparse* matrices—namely matrices in which a very large majority of the entries are 0—these algorithms can be combined with other sophisticated techniques to produce even faster multiplication. If the matrices are in $\mathcal{M}_{n \times n}(F)$ but have no more than n nonzero entries, then one can multiply them in an order of $n^{2+k(n)}$ operations, where $k(n) \to 0$ as $n \to \infty$.

The size of matrices for which the Strassen–Winograd algorithm is significantly faster than the regular method depends, of course, on the particular hardware on which it is being used. The Strassen–Winograd algorithm can also be modified to multiplication of matrices which are not necessarily square.

Unfortunately, the Strassen–Winograd algorithm is no less susceptible to round-off and truncation errors than the regular algorithm. On a computer with seven-digit accuracy, the product

$$\begin{bmatrix} 211 & 2 & 3 & 4 \\ 1 & 2 & 3 & 4 \\ 0.001 & 0.032 & 0.043 & 0.044 \\ 311 & 0.0032 & 1233 & 0.0324 \end{bmatrix} \begin{bmatrix} 50 & 0.32 & 0.0023 & 421 \\ 60 & 0.023 & 0.033 & 982 \\ 23 & 0.032 & 0.03 & 623 \\ 33 & 0.043 & 0.022 & 44 \end{bmatrix}$$

equals
$$\begin{bmatrix} 10871 & 67.834 & 0.7293 & 92840 \\ 371 & 0.634 & 0.2463 & 4430 \\ 4.411 & 0.0043 & 0.0033 & 60.57 \\ 43910.3 & 138.977 & 37.7061 & 899094.0 \end{bmatrix}$$
using the ordinary method of matrix multiplication, whereas, using the Strassen–Winograd algorithm, we obtain
$$\begin{bmatrix} 10871 & 68.54 & 0.6294 & 92840 \\ 370.9 & 1.0 & 0.2463 & 4430.18 \\ 4.411 & 0.0043 & 0 & 62.0 \\ 43910.3 & 139.047 & 37.7 & 899095.0 \end{bmatrix}$$
. This problem can be overcome to some extent by stopping the recursion in the Strassen–Winograd algorithm early, and doing the bottom-level matrix multiplication using the ordinary method. Another disadvantage of this algorithm is that it requires a much larger amount of scratch memory space to perform its calculations.

There are other tricks that can be used to reduce the computations necessarily in matrix multiplication. For example, if n is a positive integer and if $A, B, C, D \in \mathcal{M}_{n \times n}(\mathbb{R})$, then the matrix product $(A + iB)(C + iD)$ in $\mathcal{M}_{n \times n}(\mathbb{C})$ can be calculated using only three matrix multiplications in $\mathcal{M}_{n \times n}(\mathbb{R})$, rather than the expected four, by noting that

$$(A + iB)(C + iD) = AC - BD + i\big[(A + B)(C + D) - AC - BD\big].$$

If we have a parallel-processing computational system at our disposal, matrix multiplication can be done much more rapidly. There exist parallel algorithms to multiply two $n \times n$ matrices in an order of $\log(n)$ time, on the condition that we have n^3 processors working in parallel. Given the availability of such parallel computational power, one can also invert a nonsingular $n \times n$ matrix in an order of $\log^2(n)$ time. The first such algorithm was developed by Laszlo Csanky in 1977, though this algorithm has the disadvantage of being wildly unstable.

Again, we keep in mind that real and complex numbers are represented in a computer by approximations having a limited degree of accuracy. The longer calculations become, the error due to roundoff and truncation increases and limits the correctness of the calculations. It is possible to reduce the effect of roundoff and truncation errors as much as possible. Let us recall how our algorithm for inverting a matrix A worked:

(1) We formed the matrix $[I \; A] = [b_{ij}]$;
(2) We interchanged the first row which one of the rows below it, if necessary, such that $b_{1,n+1} \neq 0$; we then multiplied this row by $b_{1,n+1}^{-1}$ so that this element is now equal to 1, and we subtracted multiples of this row from the rows below it, in order to make $b_{i,n+1}$ equal to 0 for all $1 < i \leq n$.
(3) We now go iterate this process for the elements $b_{h,n+h}$, where $h = 2, 3, \ldots$ and so forth. If we cannot do it, i.e., if there exists an h such that $b_{i,n+h}$ for all $h \leq i \leq n$, the matrix A is nonsingular. Otherwise, at the end of the process, we have brought the matrix to the form $[A^{-1} \; I]$.

The elements $b_{h,n+h}$ are called *pivots* of the algorithm. If we are working over $\mathbb{R}$ or $\mathbb{C}$, we can minimize roundoff and truncation errors, to some extent, by making sure that each time we interchange rows we choose to bring into the pivot position a nonzero number having maximal absolute value. This strategy is known as *partial pivoting*. We could do better by also interchanging columns in order to bring into the pivot position $b_{h,n+h}$ the element b_{ij} $(h \leq i, j \leq n)$ having maximal absolute value. This strategy is known as *full pivoting*; it requires a certain amount of computational overhead on the side so that the columns can be returned to their proper positions at the end of the algorithm. Although there are matrices so pathological that full pivoting rather than partial pivoting is needed in order to invert them, most experts believe that it is not worth the effort and the computational overhead and that for such matrices one should use other methods altogether. Partial pivoting also does not work well on parallel or systolic-array computers, since it requires many nonlocal data movements. Several variants of pivoting strategies for matrices having specific structures have, however, been developed and are in wide use.

Indeed, let us now consider another method. It is clearly easier to invert a nonsingular upper-triangular or lower-triangular matrix—namely a matrix in one of these forms all of the diagonal elements of which are nonzero. Therefore, our job would be much easier if we could write A in the form LU, where L is lower triangular nonsingular and U is upper triangular nonsingular, for then $A^{-1} = U^{-1}L^{-1}$. This is not always possible. For example, one can see that there is no way of writing the matrix $\begin{bmatrix} 0 & 1 \\ 1 & 0 \end{bmatrix} \in \mathcal{M}_{2 \times 2}(\mathbb{R})$ in this form. However, it is always possible to write A in the form LU when A equals a product of elementary matrices of the form $E_{i;c}$ and $E_{ij;c}$ only.

How can this be done? Assume that $A = [a_{ij}]$, $U = [u_{ij}]$, and $L = [v_{ij}]$ and that $A = LU$, where U is upper-triangular nonsingular and L is lower-triangular nonsingular. Then for each $1 \leq i, j \leq n$ we have $a_{ij} = \sum_{h=1}^{n} v_{ih}u_{hj}$. In each of L and U there are only $\frac{1}{2}(n^2 + n)$ entries which can be nonzero and so our problem is one of solving n^2 nonlinear equations in $n^2 + n$ unknowns. This means that we can allow ourself to choose the value of n of these variables arbitrarily, and we will do so by insisting that $v_{ii} = 1$ for all $1 \leq i \leq n$. Now we have a system of $n^2 + n$ nonlinear equations in $n^2 + n$ unknowns, which can be solved by a method known as *Crout's algorithm*:

(1) First set $v_{ii} = 1$ for all $1 \leq i \leq n$;
(2) For all $2 \leq j \leq n$ and all $1 \leq i \leq j$, first calculate $u_{ij} = a_{ij} - \sum_{h=1}^{i-1} v_{ih}u_{hj}$ and then $v_{ij} = \frac{1}{u_{jj}}\left(a_{ij} - \sum_{h=1}^{j-1} v_{ih}u_{hj}\right)$ for all $j < i \leq n$.

With kind permission of the National Portrait Gallery (Turing); With kind permission of Sir Peter Swinnerton-Dyer (Swinnerton-Dyer).

The LU method was devised by the British mathematician **Alan Turing** who is better known as the founder of automata theory and one of the fathers of the electronic computer. It appears implicitly in the work of Jacobi on bilinear forms. The first computer algorithm to compute LU factorizations using partial pivoting was described by the contemporary British mathematicians D.W. Barron and **Sir Peter Swinnerton-Dyer**. Prescott Crout was a twentieth-century American mathematician.

We note that if A is a nonsingular matrix which can be written in the form LU, where $L = [v_{ij}]$ is a lower-triangular nonsingular matrix satisfying $v_{ii} = 1$ for all $1 \leq i \leq n$ and $U = [u_{ij}]$ is upper-triangular and nonsingular, then this factorization must be unique. Indeed, assume that $L_1U_1 = L_2U_2$ where the L_h are lower triangular matrices with 1's on the diagonal, and the U_h are nonsingular upper-triangular matrices. Then $L_2^{-1}L_1 = U_2U_1^{-1}$. Since the product of lower-triangular matrices is lower triangular and the product of upper-triangular matrices is upper triangular, this matrix must be a diagonal matrix. But then $L_2^{-1}L_1 = I$ and so $L_1 = L_2$ and that implies that $U_1 = U_2$, proving uniqueness.

Example Some singular matrices may also be written in the form LU, but for them the above uniqueness result is no longer necessarily true. For example,

$$\begin{bmatrix} 1 & -1 & 2 \\ -1 & 1 & -1 \\ 2 & -2 & 4 \end{bmatrix} = \begin{bmatrix} 1 & 0 & 0 \\ -1 & 1 & 0 \\ 2 & b & 1 \end{bmatrix}\begin{bmatrix} 1 & -1 & 2 \\ 0 & 0 & 1 \\ 0 & 0 & -b \end{bmatrix}$$

for any scalar $b \in \mathbb{R}$.

As was previously remarked, not all nonsingular matrices can be written in the form LU. However, we have already noted that any nonsingular matrix A can be written in the form PC, where P is a permutation matrix and C is a product of elementary matrices of the form $E_{i;c}$ and $E_{ij;c}$ and C can be written in the desired LU form.

Example It is easy to verify that $\begin{bmatrix} 0 & 1 & 1 & -3 \\ -2 & 4 & 1 & 4 \\ 0 & 0 & 0 & 1 \\ 3 & 1 & 1 & 0 \end{bmatrix} = PLU$, where $P =$

$\begin{bmatrix} 0 & 1 & 0 & 0 \\ 1 & 0 & 0 & 0 \\ 0 & 0 & 0 & 1 \\ 0 & 0 & 1 & 0 \end{bmatrix}$ is a permutation matrix, $L = \begin{bmatrix} 1 & 0 & 0 & 0 \\ 0 & 1 & 0 & 0 \\ -\frac{3}{2} & 7 & 1 & 0 \\ 0 & 0 & 0 & 1 \end{bmatrix}$ is lower trian-

gular, and $U = \begin{bmatrix} -2 & 4 & 1 & 4 \\ 0 & 1 & 1 & -3 \\ 0 & 0 & -\frac{9}{2} & 27 \\ 0 & 0 & 0 & 1 \end{bmatrix}$ is upper triangular.

In general, the problem of factorization of a square matrix into a product of matrices of a more desirable form is one which arises often in computational matrix theory, and many techniques have been developed to facilitate such computations. One method, for example, is to associate with any matrix $A = [a_{ij}] \in \mathcal{M}_{n \times n}(F)$ an undirected graph Γ_A the vertices of which are $\{1, \ldots, n\}$ and in which there exists an edge connecting i and j if and only if $a_{ij} \neq 0$ or $a_{ji} \neq 0$. If this graph has nice structure—if it is a tree, for example—then this structure can be exploited to produce efficient factorization algorithms for A, as has recently been shown by Israeli computer scientist Sivan Toledo.

Exercises

Exercise 441
Let $F = \text{GF}(5)$. Calculate $\begin{bmatrix} 1 & 3 & 1 \\ 2 & 1 & 1 \\ 1 & 2 & 3 \end{bmatrix} \begin{bmatrix} 1 & 2 & 2 \\ 4 & 3 & 2 \\ 1 & 4 & 2 \end{bmatrix}$ in $\mathcal{M}_{3 \times 3}(F)$.

Exercise 442
Does there exist a real number b such that the matrices

$$A = \begin{bmatrix} 1 & 0 & -1 & 0 \\ 0 & 1 & 0 & -1 \\ 1 & 0 & -1 & 0 \\ 0 & 1 & 0 & -1 \end{bmatrix} \quad \text{and} \quad B = \begin{bmatrix} b & -1 & -1 & 0 \\ -1 & b & 0 & -1 \\ 1 & 0 & \frac{b}{2} & -1 \\ 0 & 1 & -1 & \frac{b}{2} \end{bmatrix}$$

are a commuting pair in $\mathcal{M}_{4 \times 4}(\mathbb{R})$?

Exercise 443

Let $F = GF(7)$ and let K be the subalgebra of $\mathcal{M}_{2\times2}(F)$ consisting of all matrices of the form $\begin{bmatrix} a & b \\ -b & a \end{bmatrix}$, for $a, b \in F$. Show that K is a field. Is it a field if $F = GF(5)$?

Exercise 444

Let $A = \begin{bmatrix} 1 & 1 \\ 0 & 1 \end{bmatrix} \in \mathcal{M}_{2\times2}(\mathbb{R})$. Find the set of all matrices $B \in \mathcal{M}_{n\times n}(\mathbb{R})$ satisfying $BA = AB$.

Exercise 445

Let $A = \begin{bmatrix} 1 & i \\ -i & 1 \end{bmatrix} \in \mathcal{M}_{2\times2}(\mathbb{C})$. Find a complex number c satisfying $(cA)^2 = A$.

Exercise 446

Let $A = \begin{bmatrix} 0 & 1 & 0 \\ 0 & 0 & 1 \\ 1 & 0 & 0 \end{bmatrix} \in \mathcal{M}_{3\times3}(\mathbb{R})$. Find a positive integer k satisfying $A^k - A^{-1}$.

Exercise 447

Let F be a field. Find all matrices $A \in \mathcal{M}_{3\times3}(F)$ satisfying $A^2 = \begin{bmatrix} 0 & 0 & 1 \\ 0 & 0 & 0 \\ 0 & 0 & 0 \end{bmatrix}$.

Exercise 448

Let n be a positive integer and let F be a field of characteristic 0. Show that $AB - BA \neq I$ for all $A, B \in \mathcal{M}_{n\times n}(F)$ (in other words, that I is not the product of any two elements of the Lie algebra $\mathcal{M}_{n\times n}(F)^-$).

Exercise 449

Show that there are infinitely-many pairs (a, b) of real numbers satisfying the condition $\begin{bmatrix} a & 0 & 0 \\ 0 & 1 & 0 \\ 0 & 0 & b \end{bmatrix} \begin{bmatrix} 1 & 0 & 1 \\ 0 & 1 & 0 \\ 1 & 0 & 1 \end{bmatrix} = \begin{bmatrix} 1 & 0 & 1 \\ 0 & 1 & 0 \\ 1 & 0 & 1 \end{bmatrix} \begin{bmatrix} a & 0 & 0 \\ 0 & 1 & 0 \\ 0 & 0 & b \end{bmatrix}$.

Exercise 450

Does there exist a positive integer k satisfying

$$\begin{bmatrix} 0 & 1 & 0 \\ 0 & 0 & 1 \\ 1 & 0 & 0 \end{bmatrix} \begin{bmatrix} 0 & 0 & 1 \\ 0 & 1 & 0 \\ 1 & 0 & 0 \end{bmatrix}^k = \begin{bmatrix} 0 & 0 & 1 \\ 1 & 0 & 0 \\ 0 & 1 & 0 \end{bmatrix}?$$

Exercise 451

Let $F = GF(3)$. Show that there exist at least 27 distinct matrices A in $\mathcal{M}_{3\times 3}(F)$ satisfying $A^3 = I$.

Exercise 452

If $F = GF(2)$, find the set of all pairs (A, B) of matrices in $\mathcal{M}_{2\times 2}(F)$ satisfying $AB - BA = I$.

Exercise 453

For a field F, find $\{A \in \mathcal{M}_{2\times 2}(F) \mid A^2 = O\}$.

Exercise 454

Find a matrix $A \in \mathcal{M}_{3\times 3}(\mathbb{R})$ satisfying $A \begin{bmatrix} 1 & 1 & -1 \\ 2 & 1 & 0 \\ 1 & -1 & 1 \end{bmatrix} = \begin{bmatrix} 1 & -1 & 3 \\ 4 & 3 & 2 \\ 1 & -2 & 5 \end{bmatrix}$.

Exercise 455

Show that if $A = \begin{bmatrix} a & 1 & 0 \\ 0 & a & 1 \\ 0 & 0 & a \end{bmatrix} \in \mathcal{M}_{3\times 3}(\mathbb{R})$ then for each $n > 1$ we have $A^n = \begin{bmatrix} a^n & na^{n-1} & \frac{n(n-1)}{2}a^{n-2} \\ 0 & a^n & na^{n-1} \\ 0 & 0 & a^n \end{bmatrix}$.

Exercise 456

Let $(K, \bullet)$ be an associative unital algebra over a field F and let S be the subset of $\mathcal{M}_{3\times 3}(K)$ consisting of all matrices of the form $\begin{bmatrix} v_{11} & 0_K & v_{13} \\ 0_K & v_{22} & 0_K \\ v_{31} & 0_K & v_{33} \end{bmatrix}$. Is S an F-subalgebra of $\mathcal{M}_{3\times 3}(K)$?

Exercise 457

Let n be a positive integer and let F be a field. A matrix in $\mathcal{M}_{n\times n}(F)$ of the form $\begin{bmatrix} a_1 & a_2 & \cdots & a_n \\ a_n & a_1 & \cdots & a_{n-1} \\ \vdots & \vdots & \ddots & \vdots \\ a_2 & a_3 & \cdots & a_1 \end{bmatrix}$ is called a *circulant matrix*. Determine if the set of all circulant matrices in $\mathcal{M}_{n\times n}(F)$ is an F-subalgebra of $\mathcal{M}_{n\times n}(F)$.

Circulant matrices, which have many important applications, were first studied by the nineteenth-century French mathematician **Eugène Catalan**.

Exercise 458
Let n be a positive integer and let F be a field. If $A \in \mathcal{M}_{n \times n}(F)$ is a nonsingular circulant matrix, is A^{-1} necessarily a circulant matrix?

Exercise 459
Let K be the subset of $\mathcal{M}_{2 \times 2}(\mathbb{Q})$ consisting of all matrices of the form $\begin{bmatrix} a & b \\ 2b & a \end{bmatrix}$, where $a, b \in \mathbb{Q}$. Show that K is a $\mathbb{Q}$-subalgebra of $\mathcal{M}_{2 \times 2}(\mathbb{Q})$ which is, in fact, a field.

Exercise 460
Find a matrix $A \in \mathcal{M}_{2 \times 2}(\mathbb{R})$ satisfying $A^2 = \begin{bmatrix} 1 & 3 \\ 0 & 1 \end{bmatrix}$.

Exercise 461
Find all matrices $A \in \mathcal{M}_{3 \times 3}(\mathbb{R})$ satisfying $A \begin{bmatrix} 1 & 1 & 1 \\ 2 & 2 & 2 \\ 0 & 1 & 1 \end{bmatrix} = O$.

Exercise 462
Let $A = \begin{bmatrix} a & b \\ c & d \end{bmatrix} \in \mathcal{M}_{2 \times 2}(\mathbb{R})$ be an idempotent matrix. Show that $a + d \in \{0, 1, 2\}$.

Exercise 463
Show that $\begin{bmatrix} 1 & 1 \\ 1 & 1 \end{bmatrix}^n = \begin{bmatrix} 2^{n-1} & 2^{n-1} \\ 2^{n-1} & 2^{n-1} \end{bmatrix}$ for all $n \geq 1$.

Exercise 464
Let F be a field and let $A = \begin{bmatrix} 0 & b \\ c & 0 \end{bmatrix} \in \mathcal{M}_{2 \times 2}(F)$. Find A^n for all $n \geq 1$.

Exercise 465
Find matrices $A, B \in \mathcal{M}_{2 \times 2}(\mathbb{Q})$ for which

$$(A - B)(A + B) \neq A^2 - B^2.$$

Exercise 466

Let F be a field and let $A = \begin{bmatrix} 1 & 0 & 0 \\ 1 & 0 & 1 \\ 0 & 1 & 0 \end{bmatrix} \in \mathcal{M}_{3\times3}(F)$. Show that $A^{k+2} = A^k +$ $A^2 - I$ for all positive integers k.

Exercise 467

Let n be a positive integer and let F be a field. Let $A, B \in \mathcal{M}_{n\times n}(F)$ satisfy $A + B = I$. Show that $AB = O$ if and only if A and B are idempotent.

Exercise 468

Let n be a positive integer and let $(K, \bullet)$ be an associative unital algebra over a field F. Define a new operation $\boxdot$ on $\mathcal{M}_{n\times n}(K)$, called the *Schur product* (sometimes also called the *Hadamard product*, especially in the context of statistics), by setting $[v_{ij}] \boxdot [w_{ij}] = [v_{ij} \bullet w_{ij}]$, for all $1 \leq i, j \leq n$. Is $(\mathcal{M}_{n\times n}(K), +, \boxdot)$ an F-algebra? Is it associative? Is it unital? When is it commutative?

Exercise 469

Let n be a positive integer and for each $A = [a_{ij}] \in \mathcal{M}_{n\times n}(\mathbb{R})$, let $\mu(A) = \max_{1 \leq i,j \leq n} |a_{ij}|$. Show that $\mu(A^2) \leq n\mu(A)^2$ for all $A \in \mathcal{M}_{n\times n}(\mathbb{R})$.

Exercise 470

Let F be a field. Find a matrix $A \in \mathcal{M}_{3\times3}(F)$ satisfying $A^2 = \begin{bmatrix} 0 & 1 & 0 \\ 0 & 0 & 0 \\ 0 & 0 & 0 \end{bmatrix}$ or

show that no such matrix exists.

Exercise 471

Find a matrix $A \in \mathcal{M}_{2\times2}(\mathbb{Q})$ satisfying $A \begin{bmatrix} 0 & c \\ c & 0 \end{bmatrix} A^T = \begin{bmatrix} 2c & 0 \\ 0 & -\frac{c}{2} \end{bmatrix}$ for all $c \in \mathbb{Q}$.

Exercise 472

Let F be a field and let n be a positive integer. Show that $H_{11}AH_{11}BH_{11} = H_{11}BH_{11}AH_{11}$ for all $A, B \in \mathcal{M}_{n\times n}(F)$.

Exercise 473

Let F be a field and let n be a positive integer. Show that $(\sum_{i=1}^{n} \sum_{j=1}^{n} H_{ij}AH_{ji})B = B(\sum_{i=1}^{n} \sum_{j=1}^{n} H_{ij}AH_{ji})$ for all $A, B \in \mathcal{M}_{n\times n}(F)$.

Exercise 474

Is the set $\left\{ \begin{bmatrix} 1 & 1 & 0 \\ 0 & 0 & 0 \\ 0 & 0 & 0 \end{bmatrix}, \begin{bmatrix} 1 & 1 & 1 \\ 0 & 0 & 0 \\ 0 & 0 & 0 \end{bmatrix}, \begin{bmatrix} 0 & 0 & 0 \\ 1 & 1 & 0 \\ 0 & 0 & 0 \end{bmatrix}, \begin{bmatrix} 0 & 0 & 0 \\ 1 & 1 & 1 \\ 0 & 0 & 0 \end{bmatrix} \right\}$ of matrices in $\mathcal{M}_{3\times3}(\mathbb{Q})$ closed under taking products?

Exercise 475
Find infinitely-many triples (A, B, C) of nonzero matrices in $M_{3\times 3}(\mathbb{Q})$, the entries of which are nonnegative integers, satisfying the condition $A^3 + B^3 = C^3$.

Exercise 476
Let F be a field. Find a matrix $A \in M_{4\times 4}(F)$ satisfying $A^4 = I \neq A^3$.

Exercise 477
Let n be a positive integer and let $F = GF(p)$ for some prime integer p. Show that for any $A \in M_{n\times n}(F)$ there exist positive integers $k > h$ satisfying $A^k = A^h$. Would this also be true if we chose $F = \mathbb{Q}$?

Exercise 478
Let $A = [a_{ij}] \in M_{2\times 2}(\mathbb{C})$ be a matrix satisfying the condition that $\frac{1}{2}[a_{11} + a_{22}] \neq \sqrt{a_{11}a_{22} - a_{12}a_{21}}$. Show that there exist four distinct matrices $B \in M_{2\times 2}(\mathbb{C})$ satisfying $B^2 = A$.

Exercise 479
Let c be a given complex number. Find the set of all matrices $A \in M_{2\times 2}(\mathbb{C})$ satisfying $(A - cI)^2 = O$.

Exercise 480
Show that $\begin{bmatrix} 3-4c & 2-4c & 2-4c \\ -1+2c & 2c & -1+2c \\ -3+2c & -3+2c & -2+2c \end{bmatrix}$ is involutory for all complex numbers c.

Exercise 481
Let n be a positive integer and let F be a field. How many matrices $A = [a_{ij}] \in M_{n\times n}(F)$ having entries in $\{0, 1\}$ satisfy the condition that each row and each column contain exactly one 1.

Exercise 482
Show that for an integer $n \geq 4$ and for a field F there exist matrices A and B in $M_{n\times n}(F)$ satisfying $A^2 = B^2 = O$ but $AB = BA \neq O$.

Exercise 483
Let $F = GF(2)$ and let F' be a field of characteristic other than 2. Define a function $\varphi : M_{2\times 2}(F') \to M_{2\times 2}(F)$ as follows: If $A = [a_{ij}] \in M_{2\times 2}(F')$ then set $\varphi(A) = [b_{ij}]$, where

$$b_{ij} = \begin{cases} 1 & \text{if } a_{ij} \neq 0, \\ 0 & \text{otherwise.} \end{cases}$$

Is $\varphi(A + A') = \varphi(A) + \varphi(A')$ for all $A, A' \in M_{2\times 2}(F')$? Is $\varphi(AA') = \varphi(A)\varphi(A')$ for all $A, A' \in M_{2\times 2}(F')$?

Exercise 484

Find a matrix $I \neq A \in \mathcal{M}_{3\times3}(\mathbb{Q})$ satisfying

$$A\begin{bmatrix} 1 & 0 & 0 \\ 1 & 1 & 0 \\ 0 & 0 & 1 \end{bmatrix} = \begin{bmatrix} 1 & 0 & 0 \\ 1 & 1 & 0 \\ 0 & 0 & 1 \end{bmatrix}A \quad \text{and} \quad A\begin{bmatrix} 1 & 0 & 0 \\ 0 & 1 & 0 \\ 0 & 1 & 1 \end{bmatrix} = \begin{bmatrix} 1 & 0 & 0 \\ 0 & 1 & 0 \\ 0 & 1 & 1 \end{bmatrix}A.$$

Exercise 485

For each real number a, find a matrix $B(a) \in \mathcal{M}_{2\times2}(\mathbb{R})$ satisfying

$$\begin{bmatrix} \cos(a) & -\sin(a) \\ \sin(a) & \cos(a) \end{bmatrix} = B(a)\begin{bmatrix} 1 & 0 \\ \sin(a) & 1 \end{bmatrix}$$

or show that such matrices need not exist.

Exercise 486

Let $A = \begin{bmatrix} 5 & 7 \\ -3 & -4 \end{bmatrix} \in \mathcal{M}_{2\times2}(\mathbb{R})$. What is A^{1024}?

Exercise 487

Find all pairs (a, b) of rational numbers such that the matrix $A = \begin{bmatrix} 2a & -a \\ 2b & -b \end{bmatrix} \in$ $\mathcal{M}_{2\times2}(\mathbb{Q})$ is idempotent.

Exercise 488

Let F be a field and let n be a positive integer. Show that there do not exist nonsingular matrices $P, Q \in \mathcal{M}_{n\times n}(F)$ satisfying $PAQ = A^T$ for all $A \in \mathcal{M}_{n\times n}(F)$.

Exercise 489

Let F be a field and let $A, B \in \mathcal{M}_{n\times n}(F)$ be a commuting pair of matrices, where B is nonsingular. Is (A, B^{-1}) necessarily a commuting pair?

Exercise 490

Let F be a field. Is $S = \left\{ \begin{bmatrix} a & b \\ c & d \end{bmatrix} \,\middle|\, a+c = b+d \right\}$ an F-subalgebra of $\mathcal{M}_{2\times2}(F)$?

Exercise 491

Let F be a field of characteristic other than 2, let n be a positive integer, and let $A \in \mathcal{M}_{n\times n}(F)$ be an involutory matrix. For each $c \in F$, let $B_c = c(A + I)$. For which values of c do we have $B_c^2 = B_c$?

Exercise 492

Find all rational numbers a, b, and d satisfying the condition that $\begin{bmatrix} a & b \\ 1 & d \end{bmatrix} \in$ $\mathcal{M}_{2\times2}(\mathbb{Q})$ is involutory.

Exercise 493
Let $F = \text{GF}(3)$ and let $A = \begin{bmatrix} 0 & 2 \\ 1 & 0 \end{bmatrix} \in \mathcal{M}_{3\times3}(F)$. Show that the subset $\{O, I, 2I, A, I+A, 2I+A, 2A, I+2A, 2I+2A\}$ of $\mathcal{M}_{3\times3}(F)$ is a field under addition and multiplication of matrices.

Exercise 494
Let $F = \text{GF}(p)$, where p is a prime integer, and let K be the subset of $\mathcal{M}_{2\times2}(F)$ consisting of all matrices of the form $\begin{bmatrix} a & b \\ -b & a \end{bmatrix}$, where $a, b \in F$. Show that K, together with the operations of matrix addition and multiplication, is a field when $p = 3$ and is not a field when $p = 5$. What happens when $p = 7$?

Exercise 495
Let n be a positive integer, let F be a field, and let $O \neq A, B \in \mathcal{M}_{n\times n}(F)$. Show that there exists a matrix $C \in \mathcal{M}_{n\times n}(F)$ satisfying $ACB \neq O$.

Exercise 496
Find all matrices $A, B \in \mathcal{M}_{2\times2}(\mathbb{R})$, the entries of which are nonnegative integers, which satisfy $AB = \begin{bmatrix} 1 & 1 \\ 0 & 1 \end{bmatrix}$.

Exercise 497
Let $V = \mathcal{M}_{3\times3}(\mathbb{Q})$. For each rational number t, let $\alpha_t : V \to V$ be the linear transformation $A \mapsto A \begin{bmatrix} 0 & 1 & 3 \\ t & 0 & 0 \\ 0 & -1 & 4 \end{bmatrix}$. Is the function $t \mapsto \alpha_t$ a linear transformation from $\mathbb{Q}$ to $\text{End}(V)$, both considered as vector spaces over $\mathbb{Q}$?

Exercise 498
Let n be a positive integer, let F be a field, and for some fixed $c \in F$, let $A = [a_{ij}]$ be the matrix in $\mathcal{M}_{n\times n}(F)$ defined by

$$a_{ij} = \begin{cases} c & \text{when } i+j \text{ is even,} \\ 0 & \text{otherwise.} \end{cases}$$

Show that the subset $\{A, A^2, A^3\}$ of $\mathcal{M}_{n\times n}(F)$ is linearly dependent.

Exercise 499
Let $F = \text{GF}(2)$ and let $A = \begin{bmatrix} 0 & 0 & 0 & 1 \\ 1 & 0 & 0 & 1 \\ 0 & 1 & 0 & 0 \\ 0 & 0 & 1 & 0 \end{bmatrix} \in \mathcal{M}_{4\times4}(F)$. Let $L = \{O\} \cup \{A^i \mid i \geq 0\} \subseteq \mathcal{M}_{4\times4}(F)$. Show that L is closed under addition. Is L, under the usual definitions of addition and multiplication of matrices, a field?

Exercise 500

Let K be the set of all matrices in $\mathcal{M}_{2\times 2}(\mathbb{Q})$ of the form $\begin{bmatrix} a & -3b \\ b & a \end{bmatrix}$. Show that K is a subalgebra of $\mathcal{M}_{2\times 2}(\mathbb{Q})$ which is in fact a field.

Exercise 501

Find the set of all matrices $A \in \mathcal{M}_{2\times 2}(\mathbb{Q})$ which satisfy $A^2 + A = \begin{bmatrix} 1 & 1 \\ 1 & 1 \end{bmatrix}$.

Exercise 502

Let $A = \begin{bmatrix} 0 & 1 \\ -1 & 0 \end{bmatrix} \in \mathcal{M}_{2\times 2}(\mathbb{Q})$ and let B and C be matrices in $\mathcal{M}_{2\times 2}(\mathbb{Q})$ satisfying $AB = BA$ and $AC = CA$. Show that $BC = CB$.

Exercise 503

Find infinitely-many matrices $A \in \mathcal{M}_{3\times 3}(\mathbb{Q})$ satisfying

$$A \begin{bmatrix} 1 & -1 & 2 \\ 2 & 0 & 1 \\ 3 & -1 & 3 \end{bmatrix} = \frac{1}{2} \begin{bmatrix} 2 & 0 & 1 \\ 0 & 2 & -3 \\ 0 & 0 & 0 \end{bmatrix}.$$

Exercise 504

Let $A = \begin{bmatrix} 1 & -1 & -1 \\ -1 & 1 & -1 \\ -1 & -1 & 1 \end{bmatrix} \in \mathcal{M}_{3\times 3}(\mathbb{Q})$. Find functions f and g from the set of

all positive integers to $\mathbb{Q}$ satisfying the condition that $A^n = \begin{bmatrix} f(n) & g(n) & g(n) \\ g(n) & f(n) & g(n) \\ g(n) & g(n) & f(n) \end{bmatrix}$

for all $n \geq 1$.

Exercise 505

Let $F = \mathrm{GF}(2)$. Do there exist matrices $A = [a_{ij}]$ and $B = [b_{ij}]$ in $\mathcal{M}_{2\times 2}(F)$ satisfying $a_{11} + a_{22} = 1$, $b_{11} + b_{22} = 0$, and $AB = I$?

Exercise 506

Let F be a field and let G be the set of all matrices in $\mathcal{M}_{3\times 3}(F)$ of the form $\begin{bmatrix} 1 & 0 & 0 \\ a & 0 & 0 \\ 0 & 0 & b \end{bmatrix}$, where $a, b \in F$. Is G closed under matrix multiplication? Does there exist a matrix J in G satisfying the condition that $AJ = A$ for all $A \in G$? If such a matrix J exists, is it necessarily true that $JA = A$ for all $A \in G$?

Exercise 507

Let n be a positive integer and let F be a field. Let A and B be matrices in $\mathcal{M}_{n\times n}(F)$ of the form $\begin{bmatrix} I & A' \\ O & I \end{bmatrix}$ and $\begin{bmatrix} I & B' \\ O & I \end{bmatrix}$, respectively, where A' and B'

are (not-necessarily square) matrices of the same size. Find necessary conditions for A and B to satisfy $AB = BA$.

Exercise 508
Let F be a field and let $A, B \in \mathcal{M}_{2 \times 2}(F)$. Show that $(AB - BA)^2$ is a diagonal matrix.

Exercise 509
Let n be a positive integer and let F be a field. Let $A \in \mathcal{M}_{n \times n}(F)$ be a diagonal matrix having distinct entries on the diagonal. Let $B \in \mathcal{M}_{n \times n}(F)$ be a matrix satisfying $AB = BA$. Show that B is also a diagonal matrix.

Exercise 510
Let n be a positive integer and let F be a field. For each integer $-n < t < n$, let $D_t(F)$ be the set of all matrices $A = [a_{ij}] \in \mathcal{M}_{n \times n}(F)$ satisfying the condition that $a_{ij} = 0$ when $j \neq i + t$. Thus, for example, $D_0(F)$ is the set of all diagonal matrices in $\mathcal{M}_{n \times n}(F)$. If $A \in D_t(F)$ and $B \in D_s(F)$, does there necessarily exist an integer $-n < u < t$ such that $AB \in D_u(F)$?

Exercise 511
Let $A = \begin{bmatrix} 1 & 2 & 3 \\ -1 & -2 & -3 \\ 2 & 4 & 6 \end{bmatrix}$ and $B = \begin{bmatrix} 1 & 2 & 3 \\ 0 & 0 & 0 \\ 0 & 0 & 0 \end{bmatrix}$ be matrices in $\mathcal{M}_{3 \times 3}(\mathbb{R})$.
Find infinitely-many lower-triangular matrices C satisfying $A = CB$.

Exercise 512
Let n be a positive integer and let F be a field. Let $A_1, \ldots, A_n$ be upper-triangular matrices in $\mathcal{M}_{n \times n}(F)$ satisfying the condition that the (i, i)-entry in A_i is equal to 0 for $1 \leq i \leq n$. Show that $A_1 \cdots A_n = O$.

Exercise 513
Let F be a field in which we have elements $a \neq 0$ and b. Show that there exists an upper-triangular matrix $C \in \mathcal{M}_{2 \times 2}(F)$ satisfying $\begin{bmatrix} 0 & a \\ 0 & 0 \end{bmatrix} C = \begin{bmatrix} 0 & b \\ 0 & 0 \end{bmatrix}$. Is C necessarily unique?

Exercise 514
Let F be a field. Find an element A of $\mathcal{M}_{2 \times 2}(F)$ satisfying $AA^T \neq A^T A$.

Exercise 515
Let F be a field and let $n > 1$. If a matrix $A \in \mathcal{M}_{n \times n}(F)$ satisfies $AA^T = O$, does it necessarily follow that $A^T A = O$?

Exercise 516
Let n be a positive integer, let F be a field, and let $A \in \mathcal{M}_{n \times n}(F)$ satisfy the condition $A = AA^T$. Show that $A^2 = A$.

Exercise 517

Let n be a positive integer, let F be a field, and let $A, B \in M_{n \times n}(F)$ be symmetric matrices. Is ABA necessarily symmetric?

Exercise 518

Let n be a positive integer and let F be a field. If $A \in M_{n \times n}(F)$ is symmetric, is A^h symmetric for all $h > 1$?

Exercise 519

Show that $\left\{ \begin{bmatrix} 1 & -2 \\ -2 & 1 \end{bmatrix}, \begin{bmatrix} 1 & 3 \\ 3 & 6 \end{bmatrix}, \begin{bmatrix} -1 & 1 \\ 1 & -3 \end{bmatrix} \right\}$ forms a basis for the subspace of $M_{2 \times 2}(\mathbb{Q})$ consisting of all symmetric matrices.

Exercise 520

Does there exist a matrix $A \in M_{2 \times 2}(\mathbb{R})$ satisfying $AA^T = \begin{bmatrix} 1 & 9 \\ 9 & 1 \end{bmatrix}$?

Exercise 521

Given real numbers a, b, and c, find all real numbers d such that

$$\begin{bmatrix} 0 & 0 & 0 & -1 \\ 0 & 0 & -1 & a \\ 0 & -1 & a & b \\ -1 & a & b & c \end{bmatrix} \begin{bmatrix} a & b & c & 1 \\ 1 & 0 & 0 & 0 \\ 0 & d & 0 & 0 \\ 0 & 0 & 1 & 0 \end{bmatrix}$$

is symmetric.

Exercise 522

Find a matrix $B \in M_{2 \times 2}(\mathbb{Q})$ such that the Nievergelt's matrix equals $B^T B$.

Exercise 523

Calculate $\begin{bmatrix} 1 & 2 & -3 \\ 0 & 1 & 2 \\ 0 & 0 & 1 \end{bmatrix}^{-1}$ in $M_{3 \times 3}(\mathbb{R})$.

Exercise 524

Let $a \in \mathbb{R} \setminus \{1, -2\}$. Calculate $\begin{bmatrix} a & 1 & 1 \\ 1 & a & 1 \\ 1 & 1 & a \end{bmatrix}^{-1} \in M_{3 \times 3}(\mathbb{R})$.

Exercise 525

Does there exist an $a \in \mathbb{R}$ such that $\begin{bmatrix} -3 & 4 & 0 \\ 8 & 5 & -2 \\ a & -7 & 6 \end{bmatrix} \in M_{3 \times 3}(\mathbb{R})$ is singular?

Exercise 526

Let n be a positive integer. Each complex number c defines a matrix $A(c) = [a_{ij}] \in M_{n \times n}(\mathbb{C})$ given by $a_{ij} = c^{(i-1)(j-1)}$ for all $1 \leq i, j \leq n$. If $w = e^{2\pi i/n} \in \mathbb{C}$, show that $A(w)$ is nonsingular and satisfies $A(w)^{-1} = \frac{1}{n}A(w^{-1})$.

Exercise 527

Let n be a positive integer and let F be a field. Given a matrix $B \in M_{n \times n}(F)$, do there exist vectors $u, v \in F^n$ such that the matrix $\begin{bmatrix} B & -Bv \\ -u^T B & u^T Bv \end{bmatrix}$ is nonsingular?

Exercise 528

Is the matrix $\begin{bmatrix} 1+X & -X \\ X & 1-X \end{bmatrix} \in M_{2 \times 2}(\mathbb{Q}[X])$ nonsingular?

Exercise 529

Is the matrix $\begin{bmatrix} 1-a^2 & 1-a & 0 \\ 0 & 1-a^2 & 1-a \\ 1-a & 0 & 1-a^2 \end{bmatrix} \in M_{3 \times 3}(\mathbb{C})$ nonsingular, where $a = -\frac{1}{2} + \frac{1}{2}\sqrt{-3} \in \mathbb{C}$.

Exercise 530

Let n be a positive integer and let F be a field. If $A \in M_{n \times n}(F)$ nonsingular, is the same necessarily true for $A + A^T$?

Exercise 531

Let n be a positive integer and let F be a field. Let $A = [a_{ij}] \in M_{n \times n}(F)$ satisfy the condition that $\sum_{i=1}^{n} a_{ij} = 1$ for all $1 \leq j \leq n$. Show that the matrix $I - A$ is singular.

Exercise 532

Let n be a positive integer and let F be a field. If $A \in M_{n \times n}(F)$ is a Markov matrix, is A^{-1} necessarily a Markov matrix?

Exercise 533

Let n be a positive integer and let F be a field. For $A \in M_{n \times n}(F)$, show that A^2 is nonsingular if and only if A^3 is nonsingular.

Exercise 534

Let $F = \text{GF}(p)$, where p is a prime integer, and let n be a positive integer. What is the probability that a matrix in $M_{n \times n}(F)$, chosen at random, is nonsingular?

Exercise 535

Let $P = \begin{bmatrix} 0 & -1 \\ 1 & -1 \end{bmatrix} \in \mathcal{M}_{2 \times 2}(\mathbb{R})$ and let A and Q be nonsingular matrices in $\mathcal{M}_{2 \times 2}(\mathbb{R})$. Set $B = AQ^{-1}PQ$. Show that B is nonsingular and $A^{-1} + B^{-1} = (A + B)^{-1}$.

Exercise 536

Show that there are infinitely-many involutory matrices in $\mathcal{M}_{2 \times 2}(\mathbb{Q})$.

Exercise 537

Let $F = \mathrm{GF}(2)$. Is the sum of all nonsingular matrices in $\mathcal{M}_{2 \times 2}(F)$ nonsingular?

Exercise 538

Let F be a field and let U be the set of all nonsingular matrices in $\mathcal{M}_{2 \times 2}(F)$. Is the function $\theta : U \to U$ defined by $\theta : A \mapsto A^2$ a permutation of U?

Exercise 539

Let n be a positive integer, let F be a field, and let $A \in \mathcal{M}_{2n \times 2n}(F)$ be a matrix which can be written in the form $\begin{bmatrix} A_{11} & A_{12} \\ A_{21} & A_{22} \end{bmatrix}$, where each $A_{ij} \in \mathcal{M}_{n \times n}(F)$ is nonsingular. Is A necessarily nonsingular?

Exercise 540

Let n be a positive integer and let F be a field. Do there exist matrices $A, B \in \mathcal{M}_{n \times n}(F)$ such that the matrix $\begin{bmatrix} A^2 & AB \\ BA & B^2 \end{bmatrix} \in \mathcal{M}_{2n \times 2n}(F)$ is nonsingular?

Exercise 541

Let n be a positive integer and let F be a field. For $A, B \in \mathcal{M}_{n \times n}(F)$ with A nonsingular, show that $(A + B)A^{-1}(A - B) = (A - B)A^{-1}(A + B)$.

Exercise 542

Let n and p be positive integers and let F be a field. Let $A \in \mathcal{M}_{n \times n}(F)$ and let $B, C \in \mathcal{M}_{n \times p}(F)$ be matrices satisfying the condition that A and $(I + C^T A^{-1}B)$ are nonsingular. Show that $A + BC^T$ is nonsingular, and that

$$\left(A + BC^T\right)^{-1} = A^{-1} - A^{-1}B\left(I + C^T A^{-1}B\right)^{-1}C^T A^{-1}.$$

Exercise 543

Let n be a positive integer and let F be a field. If $\begin{bmatrix} 0 \\ \vdots \\ 0 \end{bmatrix} \neq v \in F^n$, show that there exists a nonsingular matrix in $\mathcal{M}_{n \times n}(F)$ the rightmost column of which is v.

Exercise 544

Let F be a field. Show that every nonsingular matrix in $\mathcal{M}_{2\times2}(F)$ can be written as a product of matrices of the form $\begin{bmatrix} 0 & 1 \\ 1 & 0 \end{bmatrix}, \begin{bmatrix} 1 & 1 \\ 0 & 1 \end{bmatrix}$, or $\begin{bmatrix} a & 0 \\ 0 & 1 \end{bmatrix}$ for $a \in F$.

Exercise 545

For each real number t, let $A(t) = \begin{bmatrix} 1 & 0 & t \\ -t & 1 & -\frac{1}{2}t^2 \\ 0 & 0 & 1 \end{bmatrix} \in \mathcal{M}_{3\times3}(\mathbb{R})$. Show that each such matrix is nonsingular and that the set of all such matrices is closed under taking products.

Exercise 546

Let n be a positive integer and let F be a field. Let $A \in \mathcal{M}_{n\times n}(F)$ be a matrix for which there exists a positive integer k satisfying $A^k = O$. Show that the matrix $I - A$ is nonsingular and find $(I - A)^{-1}$.

Exercise 547

Let n be a positive integer and let F be a field. Let $A \in \mathcal{M}_{n\times n}(F)$ be a matrix for which there exists a matrix $B \in \mathcal{M}_{n\times n}(F)$ satisfying $I + A + AB = O$. Show that A is nonsingular.

Exercise 548

Let n be a positive integer and let F be a field. Let $A, B \in \mathcal{M}_{n\times n}(F)$ satisfy the condition that A and $A + B$ are nonsingular. Show that $I + A^{-1}B$ is nonsingular and that $(I + A^{-1}B)^{-1} = (A + B)^{-1}A$.

Exercise 549

Find matrices A and B in $\mathcal{M}_{2\times2}(\mathbb{R})$ satisfying $A^2 = B^2 = O$ such that $A + iB$ is a nonsingular matrix in $\mathcal{M}_{2\times2}(\mathbb{C})$.

Exercise 550

Let F be a field and let $A = \begin{bmatrix} 1 & 0 & b \\ 0 & 1 & 0 \\ a & 0 & 1 \end{bmatrix} \in \mathcal{M}_{3\times3}(F)$, where $ab \neq 1$. Show that A is nonsingular and calculate A^{-1}.

Exercise 551

Let $c \neq 0$ be an element of a field F and let $A = \begin{bmatrix} c & 1 & 0 & 0 \\ 0 & c & 1 & 0 \\ 0 & 0 & c & 1 \\ 0 & 0 & 0 & c \end{bmatrix} \in \mathcal{M}_{4\times4}(F)$. Is A is nonsingular? If so, find A^{-1}.

Exercise 552

Let $n > 1$ and let $B \in \mathcal{M}_{n \times n}(\mathbb{Q})$ be the matrix all of the entries of which are equal to 1. Show that there exists a matrix $A \in \mathcal{M}_{n \times n}(\mathbb{Q})$ satisfying the condition that $A + cB$ is nonsingular for all rational numbers c.

Exercise 553

Let $n > 1$ and let $B \in \mathcal{M}_{n \times n}(\mathbb{Q})$ be the matrix all of the entries of which are equal to 1. Find a rational number t such that $(I - B)^{-1} = I - tB$.

Exercise 554

Let n be a positive integer and let $A = [a_{ij}] \in \mathcal{M}_{n \times n}(\mathbb{R})$ be the matrix defined by $a_{ij} = \min\{i, j\}$ for all $1 \leq i, j \leq n$. Show that A is nonsingular.

Exercise 555

Let $A = [a_{ij}] \in \mathcal{M}_{4 \times 4}(\mathbb{R})$ be the matrix defined by

$$a_{ij} = \begin{cases} 2 & \text{if } i = j - 1, \\ 1 & \text{otherwise.} \end{cases}$$

Show that A is nonsingular and calculate A^{-1}.

Exercise 556

For each real number a, let $G(a) = \begin{bmatrix} \cos(a) & \sin(a) \\ -\sin(a) & \cos(a) \end{bmatrix} \in \mathcal{M}_{2 \times 2}(\mathbb{R})$. Given real numbers a, b, and c, show that $G(a, b, c) = \begin{bmatrix} G(a) & G(b) \\ O & G(c) \end{bmatrix} \in \mathcal{M}_{4 \times 4}(\mathbb{R})$ is nonsingular, and find $G(a, b, c)^{-1}$.

Exercise 557

Find a singular matrix in $\mathcal{M}_{3 \times 3}(\mathbb{Q})$ the entries of which (in some order) are the integers $1, 2, \ldots, 9$.

Exercise 558

Let n be a positive integer and let F be a field. Given elements $b, c \in F$, let $A = [a_{ij}] \in \mathcal{M}_{n \times n}(F)$ be the matrix defined by

$$a_{ij} = \begin{cases} b & \text{if } i = j, \\ c & \text{otherwise.} \end{cases}$$

Find necessary and sufficient conditions for A to be nonsingular.

Exercise 559

Give an example of a singular matrix in $\mathcal{M}_{3 \times 3}(\mathbb{Q})$ the entries of which are distinct prime positive integers, or show that no such matrix can exist.

Exercise 560

Let F be a field and let $D = \begin{bmatrix} 0 & 1 \\ -1 & 0 \end{bmatrix} \in \mathcal{M}_{2\times 2}(F)$. Let $A \in \mathcal{M}_{2\times 2}(F)$ satisfy the condition that $A^T D A = D$. Show that A is nonsingular.

Exercise 561

Let n be a positive integer and let F be a field. Is the set of all singular matrices in $\mathcal{M}_{n\times n}(F)$ closed under taking products?

Exercise 562

Let n be a positive integer, let F be a field, and let $A, B \in \mathcal{M}_{n\times n}(F)$. Show that A and B are both nonsingular if and only if the matrix $\begin{bmatrix} A & O \\ O & B \end{bmatrix} \in \mathcal{M}_{2n\times 2n}(F)$ is nonsingular.

Exercise 563

Write the matrix $\begin{bmatrix} 1 & -2 \\ 2 & 2 \end{bmatrix} \in \mathcal{M}_{2\times 2}(\mathbb{R})$ as a product of elementary matrices.

Exercise 564

Find the change of basis matrix from the canonical basis B of $\mathbb{R}^3$ to the basis $D = \left\{ \begin{bmatrix} 1 \\ 1 \\ 1 \end{bmatrix}, \begin{bmatrix} 1 \\ 1 \\ 0 \end{bmatrix}, \begin{bmatrix} 1 \\ 0 \\ 0 \end{bmatrix} \right\}$ and the change of basis matrix from D to B.

Exercise 565

Let $G = \left\{ \begin{bmatrix} a & a \\ a & a \end{bmatrix} \,\middle|\, 0 \neq a \in \mathbb{R} \right\}$. Show that there exists a matrix $E \in G$ satisfying the condition that $EA = A = AE$ for all $A \in G$. For each $A \in G$, show that there exists a matrix $A^{\ddagger} \in G$ satisfying $AA^{\ddagger} = E = A^{\ddagger}A$.

Exercise 566

Let F be a field. Given matrices $A, B \in \mathcal{M}_{2\times 2}(F)$, find the set of all matrices $C \in \mathcal{M}_{2\times 2}(F)$ satisfying $(AB - BA)C = C(AB - BA)$.

Exercise 567
Let F be a field and let G be the set of all automorphisms of F^2 which are represented with respect to the canonical basis by a matrix of the form $\begin{bmatrix} a & b \\ 0 & a^{-1} \end{bmatrix}$.
Is G a group of automorphisms of F^2?

Exercise 568
Let G be the set of all automorphisms of $\mathbb{Q}^2$ which are represented with respect to the canonical basis by a matrix of the form $\begin{bmatrix} a & b \\ 0 & d \end{bmatrix}$, where $a, d > 0$. Is G a group of automorphisms of $\mathbb{Q}^2$?

Exercise 569
Let $W_1 \subseteq W_2 \subseteq \cdots \subseteq W_n$ be a fixed sequence of subspaces of a vector space V finitely generated over a field F. If $\alpha \in \text{Aut}(V)$, we say that given sequence is an α-*fan* if and only if each of the W_i is invariant under α. Show that $G = \{\alpha \in \text{Aut}(V) \mid \text{the given sequence is an } \alpha\text{-fan}\}$ is a group of automorphisms of V.

Exercise 570
For any real number t and any positive integer n, we can define the matrix $P(n, t) \in \mathcal{M}_{n \times n}(\mathbb{R})$ to equal the identity matrix I in the case $t = 0$ and otherwise to equal the matrix $[p_{ij}]$ defined by

$$p_{ij} = \begin{cases} 0 & \text{if } i < j, \\ \binom{i-1}{j-1} t^{i-j} & \text{otherwise.} \end{cases}$$

Show that $P(n, s)P(n, t) = P(n, s + t)$ for all $s, t \in \mathbb{R}$. In particular, show that each matrix $P(n, t)$ is nonsingular.

Exercise 571
Let F be a field and let X be an indeterminate over F. Find matrices P and Q in $\mathcal{M}_{2 \times 2}(F[X])$ such that the matrix $P \begin{bmatrix} 1 + X^2 & X \\ X & 1 + X \end{bmatrix} Q$ is a diagonal matrix.

Exercise 572
Let n be a positive integer and let $\alpha : \mathcal{M}_{2 \times 2}(\mathbb{C}) \to \mathcal{M}_{4 \times 4}(\mathbb{R})$ be the function defined by $\alpha : \begin{bmatrix} a + bi & c + di \\ e + fi & g + hi \end{bmatrix} \mapsto \begin{bmatrix} a & b & c & d \\ -b & a & -d & c \\ e & f & g & h \\ -f & e & -h & g \end{bmatrix}$. Show that α is a linear transformation of vector spaces over $\mathbb{R}$. Is it a homomorphism of unital $\mathbb{R}$-algebras?

Exercise 573

Let F be a field and let $A \in \mathcal{M}_{2 \times 2}(F)$. Explicitly find a nonsingular matrix $P \in \mathcal{M}_{2 \times 2}(F)$ satisfying $PAP^{-1} = A^T$.

Exercise 574

Let Y be the subspace of $\mathcal{M}_{3 \times 3}(\mathbb{R})$ consisting of all skew-symmetric matrices. Show that Y is isomorphic to $\mathbb{R}^3$ and find an isomorphism $\alpha : \mathbb{R}^3 \rightarrow Y$ satisfying the condition that $\alpha(v)w = v \times w$ for all $v, w \in \mathbb{R}^3$.

Exercise 575

Let $A = \begin{bmatrix} 0 & 1 \\ 0 & 0 \end{bmatrix} \in \mathcal{M}_{2 \times 2}(\mathbb{R})$. Does there exist a matrix $B \in \mathcal{M}_{2 \times 2}(\mathbb{R})$ satisfying $B^2 = A$? Does there exist a matrix $C \in \mathcal{M}_{4 \times 4}(\mathbb{R})$ satisfying $C^2 = \begin{bmatrix} A & O \\ O & A \end{bmatrix}$?

Exercise 576

Let $A = \begin{bmatrix} 1 & 1 & 0 & 0 \\ 0 & 1 & 0 & 0 \\ 0 & 0 & 1 & 1 \\ 0 & 0 & 0 & 1 \end{bmatrix} \in \mathcal{M}_{4 \times 4}(\mathbb{Q})$. Find the set of all monic polynomials $p(X) \in \mathbb{Q}[X]$ of degree 2 satisfying the condition that $p(A)^2 = A$. (Caution: this set may be empty.)

Exercise 577

(Simpson's rule) Let $a < b$ be real numbers and let $c = \frac{1}{2}(a + b)$ be the midpoint of the interval $[a, b]$. Given a continuous function $f \in \mathbb{R}^{[a,b]}$, use Lagrange interpolation to show that $\int_a^b f(t)\,dt$ is approximately equal to $\frac{b-a}{6}[f(a) + 4f(c) + f(b)]$.

The eighteenth-century British mathematician **Thomas Simpson** was noted for his work on numerical approximations in calculus.

Exercise 578

Let F be a field and let $k < n$ be positive integers. Let $A \in \mathcal{M}_{n \times n}(F)$ be written in block form as $\begin{bmatrix} A_{11} & A_{12} \\ A_{21} & A_{22} \end{bmatrix}$, where A_{11} is nonsingular. Let $v, w \in F^k$

and $v', w' \in F^{n-k}$ and let $B = \begin{bmatrix} A_{11}^{-1} & -A_{11}^{-1}A_{12} \\ A_{21}A_{11}^{-1} & A_{22} - A_{21}A_{11}^{-1}A_{12} \end{bmatrix}$. Show that

$A \begin{bmatrix} v \\ v' \end{bmatrix} = \begin{bmatrix} w \\ w' \end{bmatrix}$ if and only if $B \begin{bmatrix} w \\ v' \end{bmatrix} = \begin{bmatrix} v \\ w' \end{bmatrix}$.

Systems of Linear Equations

Let k and n be positive integers. The classical problem of linear algebra is to find all solutions (if any exist) to a *system of k linear equations in n unknowns* of the form

$$a_{11}X_1 + \cdots + a_{1n}X_n = b_1,$$
$$a_{21}X_1 + \cdots + a_{2n}X_n = b_2,$$
$$\vdots$$
$$a_{k1}X_1 + \cdots + a_{kn}X_n = b_k,$$

where the a_{ij} and the b_i are scalars belonging to some field F and the X_j are variables which take values in the field.

What about infinite systems of equations? The study of infinite systems of linear equations over $\mathbb{R}$ was indeed initiated by Hill and formalized by Poincaré but has since been subsumed into functional analysis and will not be considered here. It is known that every finite subsystem of an infinite system of linear equations over an arbitrary field F has a solution over F if and only if the infinite system has a solution over F.

With kind permission of the American Mathematical Society (Hill); With kind permission of the AIP Emilio Segre Visual Archives, Physics Today Collection and Tenn Collection (Poincaré).

George William Hill was a nineteenth-century American mathematical astronomer. French mathematician **Jules Henri Poincaré** was one of the foremost mathematical geniuses of the late nineteenth century.

Example Let $a < b$ be real numbers and let $V = C(a, b)$. If W is a subspace of V of dimension n then the *interpolation problem* of V is the following: given a function $f \in V$ and given real numbers $a \le t_1 < \cdots < t_n \le b$, find a function $g \in W$ satisfying $f(t_j) = g(t_j)$ for $1 \le j \le n$. If we are given a basis $\{g_1, \ldots, g_n\}$ of W

J.S. Golan, *The Linear Algebra a Beginning Graduate Student Ought to Know*,
DOI 10.1007/978-94-007-2636-9_10, © Springer Science+Business Media B.V. 2012

then we want to find real numbers $c_1, \ldots, c_n$ satisfying $\sum_{i=1}^{n} c_i g_i(t_j) = f(t_j)$ for all $1 \leq j \leq n$. In other words, we want to solve a system of linear equations of the above form, where $k = n$, $a_{ij} = g_j(t_i)$ and $b_i = f(t_i)$ for all $1 \leq i, j \leq n$.

Example In Proposition 4.2, we noted that if F is a field and if $f(X)$ and $g(X) \neq 0$ are elements of $F[X]$, then there exist unique polynomials $u(X)$ and $v(X)$ in $F[X]$ satisfying $f(X) = g(X)u(X) + v(X)$ and $\deg(v) < \deg(g)$. If we set

$$g(X) = \sum_{i=0}^{k} a_i X^i \quad \text{and} \quad f(x) = \sum_{i=1}^{n} b_i X^i,$$

then the coefficients of $u(X) = \sum_{i=0}^{n-k} c_i X^i$ are found by solving the system of linear equations

$$a_k Y_0 + a_{k-1} Y_1 + \cdots + a_0 Y_k = b_k,$$
$$a_k Y_1 + a_{k-1} Y_2 + \cdots + a_0 Y_{k+1} = b_{k-1},$$
$$\vdots$$
$$a_k Y_{n-k-1} + a_{k-1} Y_{n-k} = b_{n-1},$$
$$a_k Y_{n-k} = b_n$$

by any of the methods we will discuss.

Example Sometimes we can transform systems of nonlinear equations into systems of linear equations. For example, suppose that we want to find positive real numbers r_1, r_2, and r_3 satisfying the following nonlinear system of equations:

$$r_1 r_2 r_3 = 1,$$
$$r_1^3 r_2^2 r_3^2 = 27,$$
$$r_3 / r_1 r_2 = 81.$$

Since each of the integers on the right is a power of 3, we can take the logarithm to the base 3 of both sides of each equation. Setting $X_i = \log_3(r_i)$ for $1 \leq i \leq 3$, the system now becomes linear

$$X_1 + X_2 + X_3 = 0,$$
$$3X_1 + 2X_2 + 2X_3 = 3,$$
$$-X_1 - X_2 + X_3 = 4,$$

and this has a unique solution (which we can find by methods to be discussed in this chapter) $X_1 = 3$, $X_2 = -5$, and $X_3 = 2$, showing that the original system has a solution $r_1 = 27$, $r_2 = 1/243$, and $r_3 = 9$.

A system of linear equations of the above form is *homogeneous* if and only if $b_i = 0$ for all $1 \leq i \leq k$; otherwise it is *nonhomogeneous*. At this stage, we do not yet know answers to the following questions:

(1) Does a given system of linear equations have a solution?
(2) If it has a solution, is that solution unique?
(3) If the solution is not unique, can we characterize the set of all solutions?
(4) If there are solutions, how do we compute them efficiently?

In order to answer these questions, we have to move to the language of matrices. The use of matrices for this purpose was developed in Europe in the nineteenth century by Cayley, Sylvester, and Laguerre. However, the real pioneers were the Chinese and Japanese mathematicians. During the time of the Han dynasty in China, around 2000 years ago, the *Nine Chapters on the Mathematical Art* (Jiuzhang Suanshu) presented a method for solving systems of linear equations using matrices. A major commentary on this was subsequently written by Liu Hui. This, in turn, formed the basis for the later work of Seki.

Edmond Laguerre, a nineteenth-century French mathematician, wrote an important book on systems of linear equations in 1867. **Liu Hui** lived in the third century in the Kingdom of Wei in north-central China. He added proofs and computational algorithms using counting rods. **Takakazu Seki Kowa** was a seventeenth-century Japanese mathematician, the son of a samurai warrior family, who developed matrix-based methods based on Chinese texts.

To see how this is done, let us write the above system in the form

$$\begin{bmatrix} a_{11} & \cdots & a_{1n} \\ \vdots & \ddots & \vdots \\ a_{k1} & \cdots & a_{kn} \end{bmatrix} \begin{bmatrix} X_1 \\ \vdots \\ X_n \end{bmatrix} = \begin{bmatrix} b_1 \\ \vdots \\ b_k \end{bmatrix}.$$

The matrix $A = [a_{ij}] \in \mathcal{M}_{k \times n}(F)$ is the *coefficient matrix* of the system. If we set

$w = \begin{bmatrix} b_1 \\ \vdots \\ b_k \end{bmatrix} \in F^k$, then the matrix $[A \ w] \in \mathcal{M}_{k \times (n+1)}(F)$ is called the *extended*

coefficient matrix of the system. The set of all vectors $v = \begin{bmatrix} d_1 \\ \vdots \\ d_n \end{bmatrix} \in F^n$ satisfying

$Av = w$ is the *solution set* of the system. This is clearly equal to $\alpha^{-1}(w)$, where $\alpha : F^n \to F^k$ is the linear transformation satisfying $\Phi_{BD}(\alpha) = A$, where B and D are the canonical bases of F^n and F^k, respectively. In particular, if the system is homogeneous then its solution set is just the kernel of α, and is called the *solution space* of the system.

We note the following simple but important point: if F is a subfield of a field K and if k and n are positive integers, then any matrix A in $\mathcal{M}_{k \times n}(F)$ also belongs to $\mathcal{M}_{k \times n}(K)$ and any vector $v \in F^n$ also belongs to K^n. Therefore, if $w \in F^k$, any element of the solution set of $Av = w$, considered as a system of linear equations over F, remains a solution when we consider this as a system of linear equations over K.

Proposition 10.1 *The solution set of a homogeneous system of linear equations in n unknowns is a subspace of F^n.*

Proof This is a direct consequence of Proposition 6.4. $\square$

For nonhomogeneous systems, the situation is a bit more complicated.

Proposition 10.2 *Let $AX = w$ be a nonhomogeneous system of linear equations in n unknowns over a field F and let $v_0 \in F^n$ be a solution to this systems. Then the solution set of the system is the set of all vectors in F^n of the*

form $v_0 + v$, where v is a solution to the homogeneous system $AX = \begin{bmatrix} 0 \\ \vdots \\ 0 \end{bmatrix}$.

Proof This is an immediate consequence of Proposition 6.6. $\square$

We should emphasize that the solution set of a nonhomogeneous system of linear equations is not a subspace of F^n but rather an affine subset of that space.

Example If we identify $\mathbb{R}^2$ with the Euclidean plane by associating each vector $\begin{bmatrix} a \\ b \end{bmatrix}$ with the point with coordinates (a, b), then we see its subspaces of dimension 1 are precisely the straight lines going through the origin. The solutions of linear equations of the form $a_1 X_1 + a_2 X_2 = b$, where $b \neq 0$, and at least one of the a_i is also nonzero, are the straight lines in the plane which do not go through the origin.

We are still left with the question of how to actually find a solution to a system of linear equations. Here we can distinguish between two approaches:

Direct Methods These methods involve the manipulation of the matrix A, either replacing it with another matrix which is easier to work with or factoring it into a product of matrices which are easier to work with, and thus reducing the difficulty of the problem.

Iterative Methods These methods involve selecting a likely solution for the system and then repeatedly modifying it to obtain a sequence of vectors which (hopefully) will converge to an actual solution to the system. Such methods work, of course, only if our vector space is one in which the notion of convergence is meaningfully defined. As we shall see, this is possible when the field of scalars is $\mathbb{R}$ or $\mathbb{C}$.

We begin by looking at direct methods. Let P be a nonsingular matrix in $\mathcal{M}_{k \times k}(F)$. A vector $v \in F^n$ is a solution to the system $AX = w$ over F if and only if it is a solution to the system $(PA)X = Pw$. In particular, this is true for elementary matrices. Thus, given a system of linear equations, we can change the order of the equations, multiply one of the equations by a nonzero scalar, or add a scalar multiple of one equation to another, without changing the solution set of the system, so long as we do the same thing on both sides of the equal sign. In order to do this efficiently, it is best to work with the extended coefficient matrix $[A \ w]$ and perform elementary operations on it to reduce it to a convenient form.

Let F be a field, let k and n be positive integers, and let $B = [b_{ij}] \in \mathcal{M}_{k \times n}(F)$. The matrix B is in *row echelon form* if and only if for each $1 \le i \le k$ there exists an integer $1 \le s(i) \le n+1$ such that
(1) $b_{ij} = 0$ for all $1 \le j < s(i)$ but $b_{i,s(i)} \ne 0$ if $s(i) \le n$; and
(2) $s(1) < s(2) < \cdots < s(k)$.

Example The matrices
$$\begin{bmatrix} 1 & 6 & 7 & 7 & 1 \\ 0 & 9 & 2 & 1 & 1 \\ 0 & 0 & 0 & 2 & 2 \\ 0 & 0 & 0 & 0 & 1 \end{bmatrix} \text{ and } \begin{bmatrix} 8 & 0 & 0 & 0 & 0 \\ 0 & 0 & 0 & 2 & 6 \\ 0 & 0 & 0 & 0 & 0 \\ 0 & 0 & 0 & 0 & 0 \end{bmatrix} \text{ are in row}$$
echelon form. The matrix
$$\begin{bmatrix} 1 & 5 & 2 & 9 & 0 \\ 0 & 0 & 1 & 5 & 4 \\ 0 & 0 & 1 & 0 & 1 \\ 0 & 0 & 0 & 0 & 7 \end{bmatrix} \text{ is not in row echelon form.}$$

Example If n is a positive integer and if $B \in \mathcal{M}_{n \times n}(F)$ is in row echelon form, then B is surely upper triangular. However,
$$\begin{bmatrix} 1 & 0 & 2 & 7 \\ 0 & 0 & 3 & 8 \\ 0 & 0 & 9 & 0 \\ 0 & 0 & 0 & 5 \end{bmatrix} \text{ is an upper-triangular}$$
matrix which is not in row echelon form.

We claim that for any matrix $A = [a_{ij}] \in \mathcal{M}_{k \times n}(F)$ is row equivalent to a matrix in row echelon form. By Proposition 9.4, this is equivalent to saying that A can be transformed into a matrix in row echelon form by a series of elementary operations, as follows:
(1) Find the leftmost column of A which has a nonzero entry and interchange rows if necessary, so that this entry is in the first row. Thus we now have a matrix A in which $a_{1h} \ne 0$ and $a_{ij} = 0$ for all $1 \le i \le k$ and all $1 \le j < h$.
(2) For each $1 < i \le k$, if $a_{ih} \ne 0$ then we multiply the first row by $-a_{ij}a_{1h}^{-1}$ and add it to the ith row, which creates a new row in which the (i, h)-entry is equal to 0. Thus, we now have a matrix in which $a_{ih} = 0$ for all $1 < i \le k$.

(3) Now consider the submatrix of A from which we deleted the first row and the first h columns, and repeat the above procedure.

Example Let us begin with the matrix $A = \begin{bmatrix} 1 & 2 & 3 & 1 \\ 2 & 1 & 4 & 2 \\ 1 & -1 & 1 & 1 \end{bmatrix} \in \mathcal{M}_{3 \times 4}(\mathbb{R})$. We already have $a_{11} \neq 0$. Multiplying the first row by -2 and adding it to the second row, we obtain $\begin{bmatrix} 1 & 2 & 3 & 1 \\ 0 & -3 & -2 & 0 \\ 1 & -1 & 1 & 1 \end{bmatrix}$ and then multiplying the first row by -1 and adding it to the third row, we obtain $\begin{bmatrix} 1 & 2 & 3 & 1 \\ 0 & -3 & -2 & 0 \\ 0 & -3 & -2 & 0 \end{bmatrix}$. We also already have $a_{22} \neq 0$. Multiplying the second row by -1 and adding it to the third row, we obtain $\begin{bmatrix} 1 & 2 & 3 & 1 \\ 0 & -3 & -2 & 0 \\ 0 & 0 & 0 & 0 \end{bmatrix}$, and this is in row echelon form.

If $A = [a_{ij}] \in \mathcal{M}_{k \times n}(F)$ is a matrix in row echelon form, and if the hth row of A contains nonzero entries, then the leftmost nonzero entry of the row is the *leading entry*. The matrix A is in *reduced row echelon form* if it is in row echelon form and, in addition, satisfies the following additional conditions:
(1) The leading entry in each nonzero row is equal to 1;
(2) If a_{hj} is a leading entry, then $a_{ij} = 0$ for all $i \neq h$.
Any matrix in row echelon is row-equivalent to one in reduced row echelon form; that is to say, such a matrix can be converted to one in reduced row echelon form by performing additional elementary operations: first, we multiply each nonzero row by the multiplicative inverse of its leading entry, to obtain a matrix in which the leading entry of each nonzero row equals 1. Then, if a_{hj} is a leading entry and if $i < h$, we multiply the hth row by $-a_{ij}$ and add it to the ith row, which will give us a matrix with the (i, j)-entry equal to 0. The reduced row echelon form of any given matrix is clearly unique.

Example Let us go back and look at the matrix $\begin{bmatrix} 1 & 2 & 3 & 1 \\ 0 & -3 & -2 & 0 \\ 0 & 0 & 0 & 0 \end{bmatrix}$ in row echelon form. The leading entry of the first row is already equal to 1. Multiplying the second row by $-\frac{1}{3}$ to obtain, $\begin{bmatrix} 1 & 2 & 3 & 1 \\ 0 & 1 & \frac{2}{3} & 0 \\ 0 & 0 & 0 & 0 \end{bmatrix}$, a matrix in which the leading entry of the second row is equal to 1 as well. Now multiply the second row by -2 and add it to the first row, to obtain $\begin{bmatrix} 1 & 0 & \frac{8}{3} & 1 \\ 0 & 1 & \frac{2}{3} & 0 \\ 0 & 0 & 0 & 0 \end{bmatrix}$, which is in reduced row echelon form.

Example Even this very simple algorithm can lead to computational problems. Let n be a positive integer and let $A = [a_{ij}] \in \mathcal{M}_{n \times n}(\mathbb{R})$ be the matrix defined as follows:

$$a_{ij} = \begin{cases} 1 & \text{if } i = j \text{ or } j = n, \\ -1 & \text{if } i > j, \\ 0 & \text{otherwise.} \end{cases}$$

If we use the above method to reduce A to reduced row echelon form we obtain a matrix $B = [b_{ij}]$ where

$$b_{ij} = \begin{cases} 1 & \text{if } i = j < n, \\ 2^{i-1} & \text{for } j = n, \\ 0 & \text{otherwise.} \end{cases}$$

If n is sufficiently large, the element b_{nn} may be considerably corrupted due to roundoff and truncation error.

Reduction of a matrix in $\mathcal{M}_{k \times n}(F)$ to reduced row-echelon form depends strongly on the fact that every nonzero element in a field has a multiplicative inverse. If we are considering matrices in $\mathcal{M}_{k \times n}(K)$, where K is the unital commutative associative algebra of polynomials in one or several variables over F, this now longer holds. In such situations, however, it is possible to reduce a matrix to a form known as Howell Canonical Form, which is equivalent to row-echelon form with leading entries equal to 1 in the case we are working over a field. This is important for computations since, as we will see, algebras of the form $\mathcal{M}_{k \times n}(F[X])$ have an important part to play in the theory we are developing.

Now let us return to the system of linear equations $AX = w$ in n unknowns and consider methods of solution. The most well-known is *Gaussian elimination* or the *Gauss–Jordan method*. In this method, we first perform elementary operations on the extended coefficient matrix $[A \ w]$ to bring it to reduced row echelon form. Having done this, we now have a new system of linear equations $A'X = w'$, the solution set of which is the same as that of the original system. Let t be the greatest integer i such that the ith row has nonzero entries. There are several possibilities:

(1) $b_t \neq 0$ but $a'_{tj} = 0$ for all $1 \leq j \leq n$. Then the system has no solutions, and we are done.

(2) There is precisely one index j such that $a'_{tj} \neq 0$. Then this must in fact be the leading entry of the tth row and so $a'_{tj} = 1$. This means that in any element of the solution set of the system we must have the jth entry equal to b_j. We can therefore substitute b_j for X_j in each of the other equations, and reduce the system to one of equations of $n - 1$ unknowns.

(3) There are several indices j such that $a'_{tj} \neq 0$, say those in columns $h_1 < h_2 < \cdots < h_m$. Then a'_{th_1} is the leading entry of the tth row and so equals 1. Moreover, for any values $z_1, \ldots, z_m$ we substitute for $X_{h_2}, \ldots, X_{h_m}$, we will get a solution to the system with these values and with $b_t - \sum_{s=2}^{m} z_s$ substituted

for X_{h_1}. Thus we can consider the z_i as parameters of a general solution and again reduce the system to one in a smaller number of unknowns.

(4) Having reduced the system, we now recursively apply the previous steps until the system is solved.

With kind permission of the Archives of the Mathematisches Forschungsinstitut Oberwolfach © Universität Göttingen, Sammlung Sternwarte

Carl Friedrich Gauss, who lived in Germany at the beginning of the nineteenth century, is considered to be the leading mathematician of all times, as well as a physicist and astronomer of the first rank. He developed this method in connection with his work in astronomy in 1809. Gaussian elimination first appeared in print in a handbook by German geodesist **Wilhelm Jordan**, who applied the method to problems in surveying. The first computer program to solve a system of linear equations by Gaussian elimination was written by Lady Augusta Ada Lovelace, a student of De Morgan and daughter of the poet Lord Byron, who developed software for Charles Babbage's (never completed) mechanical computer in the nineteenth century. Her program was capable of solving systems of 10 linear equations in 10 unknowns.

Strassen's insight that Gaussian elimination may not be the optimal method of solving systems of linear equations, as had been previously thought, led to the development of his method of matrix multiplication.

Example Let us consider the system of linear equations

$$3X_1 + 2X_2 + X_3 = 0,$$

$$-2X_1 + X_2 - X_3 = 2,$$

$$2X_1 - X_2 + 2X_3 = -1$$

over the field $\mathbb{R}$. The extended coefficient matrix of this system is

$$\begin{bmatrix} 3 & 2 & 1 & 0 \\ -2 & 1 & -1 & 2 \\ 2 & -1 & 2 & -1 \end{bmatrix}$$

and this is row equivalent to the matrix $\begin{bmatrix} 1 & \frac{2}{3} & \frac{1}{3} & 0 \\ 0 & 7 & -1 & 6 \\ 0 & 0 & 1 & 1 \end{bmatrix}$ in row echelon form,

which is in turn row equivalent to the matrix $\begin{bmatrix} 1 & 0 & 0 & -1 \\ 0 & 1 & 0 & 1 \\ 0 & 0 & 1 & 1 \end{bmatrix}$ in reduced row

echelon form. Thus we see that the solution set of the system is $\left\{ \begin{bmatrix} -1 \\ 1 \\ 1 \end{bmatrix} \right\}$.

Example Let us consider the system of linear equations

$$X_1 + X_2 = 1,$$
$$X_1 - X_2 = 3,$$
$$-X_1 + 2X_2 = -2$$

over the field $\mathbb{R}$. The extended coefficient matrix of this system equals

$$\begin{bmatrix} 1 & 1 & 1 \\ 1 & -1 & 3 \\ -1 & 2 & -2 \end{bmatrix},$$

and this is row equivalent to the matrix $\begin{bmatrix} 1 & 1 & 1 \\ 0 & -2 & 2 \\ 0 & 0 & 2 \end{bmatrix}$ in row echelon form, which is

row equivalent to the matrix $\begin{bmatrix} 1 & 0 & 0 \\ 0 & 1 & 0 \\ 0 & 0 & 1 \end{bmatrix}$ in reduced row echelon form. Therefore,

this system has no solutions at all.

Example Let us consider the system of linear equations

$$X_1 + 2X_2 + X_3 = -1,$$
$$2X_1 + 4X_2 + 3X_3 = 3,$$
$$3X_1 + 6X_2 + 4X_3 = 2$$

over $\mathbb{R}$. The extended coefficient matrix of this system is $\begin{bmatrix} 1 & 2 & 1 & -1 \\ 2 & 4 & 3 & 3 \\ 3 & 6 & 4 & 2 \end{bmatrix}$ and

this is row equivalent to $\begin{bmatrix} 1 & 2 & 1 & -1 \\ 0 & 0 & 1 & 5 \\ 0 & 0 & 0 & 0 \end{bmatrix}$ in row echelon form, which is in turn

row equivalent to $\begin{bmatrix} 1 & 2 & 0 & -6 \\ 0 & 0 & 1 & 5 \\ 0 & 0 & 0 & 0 \end{bmatrix}$ in reduced row echelon form. From the second

row, we see that we must have $X_3 = 5$. From the first row, we have $X_1 + 2X_2 = -6$
and so, for each value $X_2 = z$, we have a solution with $X_1 = -6 - 2z$. Therefore,

the solution set to our system is $\left\{ \begin{bmatrix} -6 - 2z \\ z \\ 5 \end{bmatrix} \,\middle|\, z \in \mathbb{R} \right\}$.

Gaussian elimination can also be used to check if a set of vectors in F^k is linearly independent. Let $\{v_1, \ldots, v_n\}$ be a set of vectors in F^k, where $v_j = \begin{bmatrix} a_{1j} \\ \vdots \\ a_{kj} \end{bmatrix}$ for all j.

We want to know if there are scalars $b_1, \ldots, b_n$ in F, not all equal to 0, satisfying $\sum_{j=1}^n b_j v_j = \begin{bmatrix} 0 \\ \vdots \\ 0 \end{bmatrix}$. That is, we want to know if the homogeneous systems of linear equations $AX = \begin{bmatrix} 0 \\ \vdots \\ 0 \end{bmatrix}$ has a nonzero solution, where $A = [a_{ij}] \in \mathcal{M}_{k \times n}(F)$.

Example Let us check if the subset $\left\{ \begin{bmatrix} 1 \\ -1 \\ 3 \\ 4 \end{bmatrix}, \begin{bmatrix} 3 \\ -3 \\ 6 \\ 4 \end{bmatrix}, \begin{bmatrix} -1 \\ 1 \\ 0 \\ 4 \end{bmatrix} \right\}$ of $\mathbb{Q}^4$ is linearly

dependent, and to do so we need to consider the matrix $A = \begin{bmatrix} 1 & 3 & -1 & 0 \\ -1 & -3 & 1 & 0 \\ 3 & 6 & 0 & 0 \\ 4 & 4 & 4 & 0 \end{bmatrix}$.

This matrix is row equivalent to the matrix $\begin{bmatrix} 1 & 0 & 2 & 0 \\ 0 & 1 & -1 & 0 \\ 0 & 0 & 0 & 0 \\ 0 & 0 & 0 & 0 \end{bmatrix}$ in reduced row ech-

elon form. Therefore, the set of solutions to the homogeneous system $AX = \begin{bmatrix} 0 \\ \vdots \\ 0 \end{bmatrix}$

is $\left\{ \begin{bmatrix} -2z \\ z \\ z \end{bmatrix} \mid z \in \mathbb{Q} \right\}$ so that if we pick one such nonzero element, say $\begin{bmatrix} -2 \\ 1 \\ 1 \end{bmatrix}$, we

see that $(-2) \begin{bmatrix} 1 \\ -1 \\ 3 \\ 4 \end{bmatrix} + \begin{bmatrix} 3 \\ -3 \\ 6 \\ 4 \end{bmatrix} + \begin{bmatrix} -1 \\ 1 \\ 0 \\ 4 \end{bmatrix} = \begin{bmatrix} 0 \\ 0 \\ 0 \\ 0 \end{bmatrix}$, showing that the set is indeed linearly dependent.

We note that if $A \in \mathcal{M}_{k \times n}(F)$ then the number of arithmetic operations needed so solve a system of linear equations of the form $AX = w$ using Gaussian elimination, is no more than $\frac{1}{6} k(k - 1)(3n - k - 2)$ if $k < n$ and no more than $\frac{1}{6} n[3kn + 3(k - n) - n^2 - 2]$ otherwise. Of course, if the matrix A is of a special form, this procedure can be much faster. For example, if $A \in \mathcal{M}_{n \times n}(F)$ is a tridiagonal matrix, then a system of equations of the form $AX = w$ can be solved using $3n$ additions/subtractions and $5n$ multiplications.

If $A \in M_{k \times n}(F)$ is a nonsingular matrix which can be written in the form LU, where L is lower triangular and U is upper triangular, then a system of linear equations of the form $UX = w$ is easy to solve using Gaussian elimination, since U is already in row-echelon form. Moreover, since U must also be nonsingular, this system has a unique solution $y = U^{-1}w$. Then the system $AX = w$ has a unique solution, which is also the solution to the system $LX = y$ and that system too is easy to solve. We therefore see the importance of the LU-decomposition of matrices, assuming that one exists.

Given a matrix $A \in M_{k \times n}(F)$, we define the *column space* of A to be the subspace of F^k generated by the set of all columns of A. The dimension of the column space of A is called the *rank* of A. Moreover, there exists a linear transformation $\alpha : F^n \to F^k$ satisfying the condition that $\Phi_{BD}(\alpha) = A$, where B and D be the canonical bases of F^n and F^k, respectively, and it is clear that the column space of A is just $\operatorname{im}(\alpha)$. Similarly, we define the *row space* of A to be the subspace of $M_{1 \times n}(F)$ generated by the rows of A. We will show that the dimension of this space is also equal to the rank of A.

Proposition 10.3 *Let F be a field, let k and n be positive integers, and let $A \in M_{k \times n}(F)$ and let $w = \begin{bmatrix} b_1 \\ \vdots \\ b_k \end{bmatrix} \in F^k$. Then the system of linear equations $AX = w$ has a solution if and only if w belongs to the column space of A.*

Proof If $v = \begin{bmatrix} d_1 \\ \vdots \\ d_n \end{bmatrix}$ is a solution of the system $AX = w$ then

$$w = \begin{bmatrix} \sum_{j=1}^{n} a_{1j}d_j \\ \vdots \\ \sum_{j=1}^{n} a_{kj}d_j \end{bmatrix} = \sum_{j=1}^{n} d_j \begin{bmatrix} a_{1j} \\ \vdots \\ a_{kj} \end{bmatrix}$$

and so w is a linear combination of the columns of A. Conversely, if we assume that there exist scalars $d_1, \dots, d_n$ in F such that $w = \sum_{j=1}^{n} d_j \begin{bmatrix} a_{1j} \\ \vdots \\ a_{kj} \end{bmatrix}$, then $v = \begin{bmatrix} d_1 \\ \vdots \\ d_n \end{bmatrix}$ is a solution of the given system. $\square$

In particular, we get the following consequence of this result.

Proposition 10.4 *Let F be a field, let k and n be positive integers, and let $A \in M_{k \times n}(F)$ and let $w = \begin{bmatrix} b_1 \\ \vdots \\ b_k \end{bmatrix} \in F^k$. Then the system of linear equations $AX = w$ has a solution if and only if the rank of the coefficient matrix A is equal to the rank of the extended coefficient matrix.*

Now let us return to the problem of identifying the solution sets of homogeneous systems of linear equations.

Proposition 10.5 *Let F be a field, let k and n be positive integers, and let $A \in M_{k \times n}(F)$ be a matrix the columns of which are vectors $y_1, \ldots, y_n$ in F^k. Assume these columns are arranged such that $\{y_1, \ldots, y_r\}$ is a basis for the column space of A, for some $r \le n$. Moreover, for all $r < h \le n$, let us select scalars $b_{h1}, \ldots, b_{hn}$ such that:*
(1) $y_h = b_{h1} y_1 + \cdots + b_{hr} y_r$;
(2) $b_{hh} = -1$;
(3) $b_{hj} = 0$ otherwise.

For each $r < h \le n$, let $v_h = \begin{bmatrix} b_{h1} \\ \vdots \\ b_{hn} \end{bmatrix} \in F^n$. Then $\{v_{r+1}, \ldots, v_n\}$ is a basis for the solution space of the homogeneous system of linear equations $AX = \begin{bmatrix} 0 \\ \vdots \\ 0 \end{bmatrix}$.

(*Comment before the proof*: Since $\{y_1, \ldots, y_n\}$ is a set of generators for the column space of A, it contains a subset that is a basis. The assumption that this is $\{y_1, \ldots, y_r\}$ is for notational convenience only.)

Proof If $r = n$ then the solution space of the system of linear equations is $\left\{ \begin{bmatrix} 0 \\ \vdots \\ 0 \end{bmatrix} \right\}$, and so the result is immediate. Hence let us assume that $r < n$. If $r < h \le n$, then $Av_h = \sum_{j=1}^{r} b_{hj} y_j - y_h = \begin{bmatrix} 0 \\ \vdots \\ 0 \end{bmatrix}$, and so each v_h belongs to the solution space of $AX = \begin{bmatrix} 0 \\ \vdots \\ 0 \end{bmatrix}$. Moreover, the set $\{v_{r+1}, \ldots, v_n\}$ is linearly independent, since if

$\sum_{j=r+1}^{n} c_j v_j = \begin{bmatrix} 0 \\ \vdots \\ 0 \end{bmatrix}$ then for each $r < h \leq n$ we note that the hth entry on the

left-hand side is $-c_h$ whereas the corresponding entry on the right-hand side is 0, proving that $c_h = 0$ for all $r < h \leq n$.

We are therefore left to show that $\{v_{r+1}, \ldots, v_n\}$ is a generating set for the solution space of the given homogeneous system. And, indeed, let $w = \begin{bmatrix} d_1 \\ \vdots \\ d_n \end{bmatrix}$ be

a vector in this solution space. Then $w + \sum_{h=r+1}^{n} d_h v_h = \begin{bmatrix} e_1 \\ \vdots \\ e_n \end{bmatrix}$, where $e_{r+1} =$

$\cdots = e_n = 0$. Therefore, this vector belongs to solution space of the system, and

so $\sum_{h=q}^{r} e_h y_h = \begin{bmatrix} 0 \\ \vdots \\ 0 \end{bmatrix}$. However, since the set $\{y_1, \ldots, y_r\}$ is linearly independent,

this implies that $e_1 = \cdots = e_r = 0$ as well. Therefore, $w = -\sum_{h=r+1}^{n} d_h v_h$, showing that $\{v_{r+1}, \ldots, v_n\}$ is a generating set for the solution space, as required. $\square$

As an immediate consequence of Proposition 10.5, we obtain the following result.

Proposition 10.6 *Let F be a field, let k and n be positive integers, and let $A \in M_{k \times n}(F)$. Then the dimension of the solution space of the homogeneous*

system of linear equations $AX = \begin{bmatrix} 0 \\ \vdots \\ 0 \end{bmatrix}$ is $n - r$, where r is the rank of the

coefficient matrix A.

We are now ready to prove the characterization of rank which we mentioned before.

Proposition 10.7 *Let F be a field, let k and n be positive integers, and let $A \in M_{k \times n}(F)$. Then the rank of A equals the dimension of the row space of A.*

Proof Let $v_1, \ldots, v_k$ be the rows of A, which generate a subspace of $M_{1 \times n}(F)$. We can reorder these rows in such a way that $\{v_1, \ldots, v_t\}$ is a basis for the row space, for some $1 \leq t \leq k$. This, as we know, does not change the solution space of the homogeneous system of linear equations $AX = \begin{bmatrix} 0 \\ \vdots \\ 0 \end{bmatrix}$ and hence does not

change the rank r_A of A. Let $B \in \mathcal{M}_{t \times n}(F)$ be the matrix obtained from A by deleting rows $t+1, \ldots, k$. The columns of B belong to F^t and so the rank r_B of B satisfies $r_B \le t$, which implies that $n - t \le n - r_B$. But we have already seen that

the homogeneous systems of linear equations $AX = \begin{bmatrix} 0 \\ \vdots \\ 0 \end{bmatrix}$ and $BX = \begin{bmatrix} 0 \\ \vdots \\ 0 \end{bmatrix}$ have

the same solution space and so, by Proposition 10.6, $n - t \le n - r_A$. From this we conclude that $r_A \le t$. We have thus shown that the rank of any matrix is less than or equal to the dimension of its row space. In particular, this is also true for A^T. But the rank of A^T is t, while the dimension of its row space is r_A, and so we have $t \le r_A$ as well, proving equality. $\qquad\qquad\qquad\qquad\qquad\qquad\qquad\qquad\qquad\qquad \square$

Example Let us find a basis for the solution space of the system of linear equations

$\begin{bmatrix} 1 & 2 & -3 & 1 \\ 1 & 1 & 1 & 1 \end{bmatrix} \begin{bmatrix} X_1 \\ X_2 \\ X_3 \\ X_4 \end{bmatrix} = \begin{bmatrix} 0 \\ 0 \end{bmatrix}$ over $\mathbb{R}$. We know that the coefficient matrix is

row-equivalent to the matrix $\begin{bmatrix} 1 & 0 & 5 & 1 \\ 0 & 1 & -4 & 0 \end{bmatrix}$ in reduced row echelon form, and this matrix has rank 2. Therefore, the solution space of the system has dimension

$4 - 2 = 2$. Indeed, it is easy to check that $\left\{ \begin{bmatrix} -5 \\ 4 \\ 1 \\ 0 \end{bmatrix}, \begin{bmatrix} -1 \\ 0 \\ 0 \\ 1 \end{bmatrix} \right\}$ is a basis for this

solution space.

Gaussian elimination requires an order of magnitude of n^3 arithmetic operations to solve a system of n linear equations in n unknowns. This computational overhead is quite significant if n is large (say, over 10,000), even with the use of supercomputers. As a result, there is considerable continuing research into finding faster methods of computation, especially in those cases in which we have additional information on the structure of the matrix of coefficients, originating in knowledge of the particular problem from which the system arose. Often this structural information is immediately noticeable, but sometimes it appears only after a sophisticated consideration of the problem.

Example It is often possible to show that the matrix we are interested in, while not itself having a special structure, is equal to the product of two matrices having a special structure, a situation which arises in many mathematical models. Let us consider one such case. An $n \times n$ *symmetric Toeplitz matrix* is a matrix $B = [b_{ij}] \in \mathcal{M}_{n \times n}(\mathbb{R})$ satisfying the condition that there exist real numbers $c_0, \ldots, c_{n-1}$ such that $b_{ij} = c_h$

whenever $|i - j| = h$. Thus, for example, the matrix $\begin{bmatrix} 1 & 2 & 0 & 7 \\ 2 & 1 & 2 & 0 \\ 0 & 2 & 1 & 2 \\ 7 & 0 & 2 & 1 \end{bmatrix}$ is a symmetric

Toeplitz matrix. Clearly, the set of all symmetric Toeplitz matrices is a subspace of $\mathcal{M}_{n \times n}(\mathbb{R})$. However, it is not a subalgebra, since the product of two such matrices need not be a symmetric Toeplitz matrix. They are also convenient to store in a computer, since we need to keep in memory only the n scalars $c_0, \ldots, c_{n-1}$. Note that symmetric Toeplitz matrices are symmetric with respect to both main diagonals.

Many mathematical models in economics are built around solving systems of linear equations of the form $AX = w$, where A is a product of two symmetric Toeplitz matrices—a fact which emerges from a knowledge of economic theory.

With kind permission of the Archives of the Mathematisches Forschungsinstitut Oberwolfach.

Otto Toeplitz was a twentieth-century German mathematician who studied endomorphisms of infinite-dimensional vector spaces.

The proper use of mathematical techniques, and especially computational techniques, also depends very much on a deep understanding of the particular problem one is dealing with. Also, it is crucial to emphasize once again that any method we use to solve a system of linear equations on a computer will induce errors as a result of roundoff and truncation in our computations. With some methods—such as Gaussian elimination—these errors tend to accumulate, whereas with others they often cancel each other out, within certain limits. It is therefore necessary, especially when we are dealing with large matrices, to have on hand several methods of handling such systems of equations and to be able to keep track of the way in which errors can propagate in each of the different methods at one's disposal. The matter of the numerical stability of solutions to such systems was investigated by Wilkinson, among many others.

© Sergei Vostok (Faddeev); © Dr. Vera Simonova (Faddeeva).

The problem computing solutions of systems of linear equations was the subject of considerable research in the early days of computers. Among the contributors were the Russian husband-and-wife team of **Dimitri Konstantinovich Faddeev** and **Vera Nikolaevna Faddeeva**.

The following is a useful trick which we will need later.

Proposition 10.8 *Let F be a subfield of a field K. Let k and n be positive integers and let $A \in \mathcal{M}_{k \times n}(F)$. Suppose that there exists a nonzero vector*

$$x = \begin{bmatrix} x_1 \\ \vdots \\ x_n \end{bmatrix} \in K^n \text{ satisfying } Ax = \begin{bmatrix} 0 \\ \vdots \\ 0 \end{bmatrix}. \text{ Then there exists a nonzero vector}$$

$$y \in F^n \text{ satisfying } Ay = \begin{bmatrix} 0 \\ \vdots \\ 0 \end{bmatrix}.$$

Proof Let $V = F\{x_1, \ldots, x_n\}$, which is a subspace of K, considered as a vector space over F. Let $E = \{v_1, \ldots, v_p\}$ be a basis for V over F and set

$$v = \begin{bmatrix} v_1 \\ \vdots \\ v_p \end{bmatrix} \in K^p. \text{ Then there exists a nonzero matrix } B \in \mathcal{M}_{n \times p}(F) \text{ satisfying}$$

$Bv = x$ and so $ABv = Ax = \begin{bmatrix} 0 \\ \vdots \\ 0 \end{bmatrix}$. But E is linearly independent and so we must

have $AB = O$. Now take y to be any nonzero column of B. $\square$

We now turn to iterative methods of solution of systems of linear equations. For simplicity, we will assume that our field of scalars is always $\mathbb{R}$. The basic idea is, as we have already noted, to guess a possible solution and then use this initial guess to compute a sequence of further approximations to the solution which, hopefully, will converge (in some topology) with relative rapidity. Usually, the initial guess is based on knowledge of the real-life problem which gave rise to the system of equations, something that can often be done with good accuracy. In very large and computationally-difficult situations (for example, weather prediction, chip design, large-scale economic models, computational acoustics, or the modeling the chemistry of polymer chains), one can even use Monte Carlo methods, based on statistical sampling and estimation techniques, to come up with an initial guess or even an approximate solution.

To illustrate this approach, let us consider the problem of solving a system of linear equations of the form $AX = w$, where $A = [a_{ij}] \in \mathcal{M}_{n \times n}(\mathbb{R})$ is a nonsingular matrix and $w = \begin{bmatrix} b_1 \\ \vdots \\ b_n \end{bmatrix} \in \mathbb{R}^n$. We know that this system has a unique solution, namely $A^{-1}w$, but inverting the matrix A may be computationally time-consuming and prone to error, so we are looking for another method. Suppose that we can write $A = E - D$, where E is some matrix which is easy to invert. Then if $v \in \mathbb{R}^n$ satisfies $Av = w$, we know that $Ev = Dv + w$ and so $v = E^{-1}(Dv + w)$.

We now guess a value for v, call it $v^{(0)}$. Then, using this formula, we can define new vectors $v^{(1)}, v^{(2)}, \ldots$ iteratively by setting $v^{(h)} = E^{-1}(Dv^{(h-1)} + w)$ for each $h > 0$. This can be done relatively quickly since, by assumption, E^{-1} was relatively easy to compute and, having computed it once for the first step of the iteration, we don't need to recompute it for subsequent steps. Our hope is that the sequence $v^{(0)}, v^{(1)}, v^{(2)}, \ldots$ will in fact converge. Indeed, if this sequence does converge to some vector v then it is easy to verify that v must be the unique solution of $AX = w$.

For example, let us assume that the diagonal entries a_{ii} of A are all nonzero, and let us choose E to be the diagonal matrix having these entries on the diagonal. Then E^{-1} is also a diagonal matrix having the entries a_{ii}^{-1} on the diagonal. If our initial guess is $v^{(0)} = \begin{bmatrix} c_1^{(0)} \\ \vdots \\ c_n^{(0)} \end{bmatrix}$, then it is easy to see that for $h > 0$ we have $v^{(h)} = \begin{bmatrix} c_1^{(h)} \\ \vdots \\ c_n^{(h)} \end{bmatrix}$, where $c_i^{(h+1)} = a_{ii}^{-1}\left[b_i - \sum_{j \neq i} a_{ij} c_j^{(h)}\right]$ for all $1 \leq i \leq n$.

This method is known as the *Jacobi iteration method*. Another possibility, again under the assumption that the diagonal entries a_{ii} of A are all nonzero, is to choose E to be the upper-triangular matrix $[e_{ij}]$ defined by setting $e_{ij} = a_{ij}$ if $i \leq j$. Given an initial guess $v^{(0)} = \begin{bmatrix} c_1^{(0)} \\ \vdots \\ c_n^{(0)} \end{bmatrix}$, we see that $v^{(h)} = \begin{bmatrix} c_1^{(h)} \\ \vdots \\ c_n^{(h)} \end{bmatrix}$ for $h > 0$, where

$$c_i^{(h+1)} = a_{ii}^{-1}\left[b_i - \sum_{j=1}^{i-1} a_{ij} c_j^{(h+1)} - \sum_{j=i+1}^{n} a_{ij} c_j^{(h)}\right]$$ for all $1 \leq i \leq n$. This method is known as the *Gauss–Seidel iteration method*, since it was discovered independently by Gauss and by Jacobi's student Philipp Ludwig von Seidel.

With kind permission of the Archives of the Mathematisches Forschungsinstitut Oberwolfach.

Carl Gustav Jacob Jacobi was a nineteenth-century German mathematician, who worked mostly in analysis and applied mathematics. His work in astronomy led him to solve large systems of linear equations, and his papers on determinants helped make them well-known.

In both of the above methods, and in other iteration methods (and there are many of these), there is no guarantee that the sequence of approximations will always converge or that, even if it does converge, it will do so rapidly. Understanding the conditions for convergence and analyzing the speed of convergence requires sophisticated techniques in numerical analysis, and indeed there are many examples of matrices for which one iteration scheme converges whereas another doesn't, as well as various necessary and sufficient conditions for a given iteration method to converge. For example, a sufficient condition for the Jacobi iteration method to converge for a matrix $A = [a_{ij}]$ is that, $\sum_{j \neq i} |a_{ij}| < |a_{ii}|$ for all $1 \leq i \leq n$. It is also known that if the matrix A is tridiagonal, then the Jacobi method converges if and only if the Gauss–Seidel method converges, but the latter always converges faster.

The convergence and accuracy of the Gauss–Seidel iteration method was studied in detail by the Russian mathematician and engineer **Alexander Ivanovich Nekrasov** at the beginning of the twentieth century, long before the use of electronic computers.

Example Let $A = \begin{bmatrix} 4 & 2 & 1 \\ -1 & 1 & 2 \\ 0 & 1 & 3 \end{bmatrix} \in \mathcal{M}_{3\times 3}(\mathbb{R})$ and let $w = \begin{bmatrix} 7 \\ 2 \\ 4 \end{bmatrix}$. The system of

linear equations $Ax = w$ has a unique solution $\begin{bmatrix} 1 \\ 1 \\ 1 \end{bmatrix}$. If we use the Jacobi iteration

method beginning with the initial guess $v^{(0)} = \begin{bmatrix} 0 \\ 0 \\ 0 \end{bmatrix}$, we get the sequence of vectors

(written to six-digit accuracy):

$$\begin{bmatrix} 0 \\ 0 \\ 0 \end{bmatrix}, \begin{bmatrix} 1.75000 \\ 2.00000 \\ 1.33333 \end{bmatrix}, \begin{bmatrix} 0.41667 \\ 1.08333 \\ 0.66667 \end{bmatrix}, \begin{bmatrix} 1.04167 \\ 1.08333 \\ 0.97222 \end{bmatrix}, \begin{bmatrix} 0.96528 \\ 1.09722 \\ 0.97222 \end{bmatrix},$$

$$\begin{bmatrix} 0.95833 \\ 1.02083 \\ 0.96759 \end{bmatrix}, \begin{bmatrix} 0.99768 \\ 1.02314 \\ 1.01157 \end{bmatrix}, \begin{bmatrix} 0.99016 \\ 1.01157 \\ 0.99228 \end{bmatrix}, \begin{bmatrix} 0.99614 \\ 1.00559 \\ 0.99614 \end{bmatrix},$$

$$\begin{bmatrix} 0.99816 \\ 1.00386 \\ 0.99814 \end{bmatrix}, \begin{bmatrix} 0.99853 \\ 1.00190 \\ 0.99871 \end{bmatrix}, \ldots$$

and if we use the Gauss–Seidel iteration method with the same initial guess, we get the sequence of vectors (written to six-digit accuracy):

$$\begin{bmatrix} 0 \\ 0 \\ 0 \end{bmatrix}, \begin{bmatrix} 1.75000 \\ -0.66667 \\ 1.33333 \end{bmatrix}, \begin{bmatrix} 1.04167 \\ 0.63889 \\ 1.55556 \end{bmatrix}, \begin{bmatrix} 1.06944 \\ 0.80093 \\ 1.12037 \end{bmatrix}, \begin{bmatrix} 1.01504 \\ 0.93672 \\ 1.06636 \end{bmatrix},$$

$$\begin{bmatrix} 1.00829 \\ 0.97287 \\ 1.02109 \end{bmatrix}, \begin{bmatrix} 1.00264 \\ 0.99020 \\ 1.00904 \end{bmatrix}, \begin{bmatrix} 1.00113 \\ 0.99611 \\ 1.00326 \end{bmatrix}, \begin{bmatrix} 1.00040 \\ 0.99853 \\ 1.00129 \end{bmatrix},$$

$$\begin{bmatrix} 1.00016 \\ 0.99943 \\ 1.00049 \end{bmatrix}, \quad \begin{bmatrix} 1.00006 \\ 0.99978 \\ 1.00019 \end{bmatrix}, \quad \dots$$

so we see that both methods converge, albeit quite differently.

Example Let $A = \begin{bmatrix} 1 & 2 & 0 \\ 2 & 1 & 2 \\ 0 & 2 & 1 \end{bmatrix} \in \mathcal{M}_{3\times 3}(\mathbb{R})$ and let $w = \begin{bmatrix} 0 \\ -1 \\ 3 \end{bmatrix}$. The system of

linear equations $Ax = w$ has a unique solution $\begin{bmatrix} -2 \\ 1 \\ 1 \end{bmatrix}$. If we try to solve this system

using the Gauss–Seidel method with the initial guess $\begin{bmatrix} 0 \\ 0 \\ 0 \end{bmatrix}$, we get the sequence of

vectors $\begin{bmatrix} 0 \\ -1 \\ 5 \end{bmatrix}, \begin{bmatrix} 2 \\ -15 \\ 33 \end{bmatrix}, \begin{bmatrix} 30 \\ -127 \\ 257 \end{bmatrix}, \begin{bmatrix} 254 \\ -1023 \\ 2049 \end{bmatrix}, \dots$ which clearly diverges.

A more sophisticated iteration technique is, at each stage, not to replace $v^{(i)}$ by the computed $v^{(i+1)}$ but rather by a linear combination of the form $rv^{(i+1)} + (1-r)v^{(i)}$, where $r \in \mathbb{R}$ is a *relaxation parameter*. Doing this with Jacobi iteration gives us the *Jacobi overrelaxation (JOR) method*, and doing it with the Gauss–Seidel method gives us the *successive overrelaxation (SOR) method*. The relaxation parameter r is chosen on the basis of certain properties of the matrix A. By choosing this parameter wisely, one can often achieve a considerable improvement in convergence. For the JOR method, one normally chooses $0 < r < 1$. In 1958, Kahan showed that the SOR method does not converge for r outside the open interval $(0, 2)$.

© Neville Miles, Imperial College London (Southwell); With kind permission of the Archives of the Mathematisches Forschungsinstitut Oberwolfach (Kahan, Young).

Relaxation methods were first developed by the twentieth-century British mathematician **Richard V. Southwell**. Contemporary Canadian mathematician **William Kahan** has made major contributions to numerical analysis and matrix computation. The optimal relaxation parameters for the SOR method were calculated by the twentieth-century American mathematician **David M. Young, Jr.**

As a rule of thumb, iteration methods work best for large sparse matrices, such as those arising from the solution of systems of partial differential equations. As previously remarked, in iteration methods truncation and roundoff errors tend to cancel each other out, rather than accumulate. While sparse matrices arise in many applications—as circuit simulation, analyses of chemical processes, and magnetic-field computation—there are also important situations, such as the matrices arising in radial-basis function interpolation, a technique of great important in computer graphics, which lead to very large matrices almost all entries of which are nonzero.

The Jacobi, Gauss–Seidel, JOR, and SOR methods are examples of iteration methods of the form $v^{(h+1)} = \alpha(v^{(h)})$, where α is an affine transformation of $\mathbb{R}^n$ that does not depend on h. Such methods are known as *stationary iteration methods*. In a later chapter, we shall also mention some iteration methods which are not stationary.

Example In the beginning of this chapter, we saw an example of how a nonlinear system of equations can be turned into a linear system. This can often be done in more general cases, producing large systems of linear equations of the form $AX = w$, where the matrix A is usually sparse and for which iteration methods are therefore appropriate. Consider, for example, the problem of finding real numbers a, b, and c such that the following conditions hold:

$$a^2 - b^2 + c^2 = 6,$$

$$ab + ac + 4bc = 29,$$

$$a^2 + 2ab - 2bc = -7,$$

$$2a^2 - 3ab + c^2 = 5,$$

$$b^2 - c^2 + 5ab = 5,$$

$$2ac - 3b^2 = -6.$$

To linearize this, we begin by assigning variables to all of the terms appearing in the equations: $X_1 = a^2$, $X_2 = b^2$, $X_3 = c^2$, $X_4 = ab$, $X_5 = ac$, and $X_6 = bc$. This then yields the system of linear equations

$$\begin{bmatrix} 1 & -1 & 1 & 0 & 0 & 0 \\ 0 & 0 & 0 & 1 & 1 & 4 \\ 1 & 0 & 0 & 2 & 0 & -2 \\ 2 & 0 & 1 & -3 & 0 & 0 \\ 0 & 1 & -1 & 5 & 0 & 0 \\ 0 & -3 & 0 & 0 & 2 & 0 \end{bmatrix} X = \begin{bmatrix} 6 \\ 29 \\ -7 \\ 5 \\ 5 \\ -6 \end{bmatrix}$$

which has a unique solution $X_1 = 1$, $X_2 = 4$, $X_3 = 9$, $X_4 = 2$, $X_5 = 3$, and $X_6 = 6$, from which we deduce that $a = 1$, $b = 2$, and $c = 3$.

The iterative methods we have discussed so far are all linear, in the sense that they involve only methods of linear algebra. There are, however, also families of nonlinear iterative methods, involving the calculus of functions of several variables, of which one should be aware. These include gradient (steepest-descent) methods and conjugate-direction methods. A discussion of these methods is beyond the scope of this book.

Finally, another important warning. When we attempt to solve systems of linear equations on a computer, it is important to remember that the system may be very sensitive, and small changes in the entries of the coefficient matrix may lead to large changes in the solution. Such systems are said to be *ill-conditioned*. Applied mathematicians and others who design mathematical models often take considerable pains to avoid creating ill-conditioned systems.

Example Let $A = \begin{bmatrix} 7 & 7 & 8 & 10 \\ 5 & 5 & 6 & 7 \\ 6 & 9 & 10 & 8 \\ 5 & 10 & 9 & 7 \end{bmatrix} \in \mathcal{M}_{4 \times 4}(\mathbb{R})$. This matrix is nonsingular,

with inverse equal to $\begin{bmatrix} 41 & 68 & -17 & 10 \\ -6 & 10 & -3 & 2 \\ 10 & -17 & 5 & -3 \\ 25 & -41 & 10 & -6 \end{bmatrix}$. Let $w = \begin{bmatrix} 32 \\ 23 \\ 33 \\ 31 \end{bmatrix}$. Then the sys-

tem of equations $AX = w$ has a unique solution $\begin{bmatrix} 1 \\ 1 \\ 1 \\ 1 \end{bmatrix}$. However, we also note that

$$A \begin{bmatrix} -7.2 \\ -0.1 \\ 2.9 \\ 6.0 \end{bmatrix} = \begin{bmatrix} 32.10 \\ 22.90 \\ 32.90 \\ 31.10 \end{bmatrix}.$$

Example Consider the system of linear equations

$$\begin{bmatrix} 1 & -10 & 0 & 0 & 0 & 0 \\ 0 & 1 & -10 & 0 & 0 & 0 \\ 0 & 0 & 1 & -10 & 0 & 0 \\ 0 & 0 & 0 & 1 & -10 & 0 \\ 0 & 0 & 0 & 0 & 1 & -10 \\ 0 & 0 & 0 & 0 & 0 & 1 \end{bmatrix} X = \begin{bmatrix} -9 \\ -9 \\ -9 \\ -9 \\ -9 \\ 1 \end{bmatrix}$$

over $\mathbb{R}$. This system has a unique solution, namely $\begin{bmatrix} 1 \\ 1 \\ 1 \\ 1 \\ 1 \\ 1 \end{bmatrix}$. However, if we alter the coefficient matrix by changing the $(6, 6)$-entry to $\frac{1}{1.001}$ (which is roughly equal to

$0.9990009)$, we will obtain a completely different solution, namely $\begin{bmatrix} 101 \\ 11 \\ 2 \\ 1.1 \\ 1.01 \\ 1.001 \end{bmatrix}$.

Since real-life computations are based, as a rule, on numbers gathered through some sort of a measurement process, which is, as a matter of fact, not completely accurate and certainly beyond our control, it is extremely important to know how sensitive the system is to possible small variations in the values of the entries. The numerical analysis of matrices deals extensively with this issue, and here we can only present a simplistic measure of this sensitivity for nonsingular square matrices over $\mathbb{R}$. To any matrix $A = [a_{ij}] \in \mathcal{M}_{n \times n}(\mathbb{R})$, we will assign the number $\theta(A)$ defined by $\theta(A) = \max_{1 \le j \le n}\{\sum_{i=1}^{n} |a_{ij}|\}$. The number $\theta(A)\theta(A^{-1})$ is the *condition number* of the matrix A. Note that A has the same condition number as A^{-1} and as cA, for any $0 \ne c \in \mathbb{R}$.

With kind permission of the American Mathematical Society.

Condition numbers were introduced by **John von Neumann**, one of the great mathematical geniuses of the twentieth century, who contributed to practically all branches of mathematics—pure and applied. Von Neumann was a major force in the introduction of digital computers after World War II and the development of numerical methods for them.

The condition number can be written in the form $g \times 10^t$. where $0.1 \le g < 1$. If $t > 0$ then, as a rule of thumb, one can expect that the solution of a system of linear equations $AX = w$ will have t significant digits *fewer* than that of the entries of A. Thus, if A is the matrix in the previous example, then $\theta(A) = 11$. Moreover,

$$A^{-1} = \begin{bmatrix} 1 & 10 & 100 & 1000 & 10000 & 100000 \\ 0 & 1 & 10 & 100 & 1000 & 10000 \\ 0 & 0 & 1 & 10 & 100 & 1000 \\ 0 & 0 & 0 & 1 & 10 & 100 \\ 0 & 0 & 0 & 0 & 1 & 10 \\ 0 & 0 & 0 & 0 & 0 & 1 \end{bmatrix}$$

and so $\theta(A^{-1}) = 111,111$. Therefore $\theta(A)\theta(A^{-1})$ is roughly 12×10^7, and so we cannot, as we have seen, expect any accuracy in our solution, if we assume our data is only good to 6-digit accuracy.

Similarly, Nievergelt's matrix $\begin{bmatrix} 888445 & 887112 \\ 887112 & 885871 \end{bmatrix}$, which we have already encountered, has condition number roughly equal to 0.39×10^5.

Of course, computing the condition number of a given matrix may also be a problem, since it involves calculating A^{-1}. Fortunately, there are many fairly-efficient condition number estimators, algorithms that give a good estimate of the condition number of a matrix with relatively low computational overhead.

Various techniques, going under the collective name of preconditioning techniques, are also often used to increase the speed of convergence and accuracy of various iterative methods. A discussion of these can be found in any advanced book on numerical matrix computation.

Exercises

Exercise 579

Are the matrices $\begin{bmatrix} -3 & 4 & 1 \\ -2 & -4 & -6 \\ 5 & 2 & 7 \end{bmatrix}$ and $\begin{bmatrix} 1 & 0 & 1 \\ 0 & 1 & 1 \\ 0 & 0 & 0 \end{bmatrix}$ in $\mathcal{M}_{3\times3}(\mathbb{R})$ row equivalent?

Exercise 580

Bring the matrix $\begin{bmatrix} 1 & 2 & 3 & 4 \\ 1 & 2 & 4 & 3 \\ 2 & 3 & 1 & 4 \end{bmatrix} \in \mathcal{M}_{3\times4}(\mathbb{R})$ to reduced row echelon form.

Exercise 581

Let $F = \mathrm{GF}(5)$. Bring the matrix $\begin{bmatrix} 1 & 2 & 1 & 0 \\ 2 & 3 & 1 & 1 \\ 1 & 2 & 4 & 0 \end{bmatrix} \in \mathcal{M}_{3\times4}(F)$ to reduced row echelon form.

Exercise 582
Solve the system of linear equations

$$(3-i)X_1 + (2-i)X_2 + (4+2i)X_3 = 2+6i,$$
$$(4+3i)X_1 - (5+i)X_2 + (1+i)X_3 = 2+2i,$$
$$(2-3i)X_1 + (1-i)X_2 + (2+4i)X_3 = 5i$$

over $\mathbb{C}$.

Exercise 583
Solve the system of linear equations

$$X_1 + 2X_2 + 4X_3 = 31,$$
$$5X_1 + X_2 + 2X_3 = 29,$$
$$3X_1 - X_2 + X_3 = 10$$

over $\mathbb{R}$.

Exercise 584

Solve the system of linear equations

$$3X_1 + 4X_2 + 10X_3 = 1,$$
$$2X_1 + 2X_2 + 2X_3 = 0,$$
$$X_1 + X_2 + 5X_3 = 1$$

over $GF(11)$.

Exercise 585

Find all solutions to the system

$$\begin{bmatrix} 1 & 2 & 3 & 4 \\ 2 & 1 & 2 & 3 \\ 3 & 2 & 1 & 2 \\ 4 & 3 & 2 & 1 \end{bmatrix} \begin{bmatrix} X_1 \\ X_2 \\ X_3 \\ X_4 \end{bmatrix} = \begin{bmatrix} 5 \\ 1 \\ 1 \\ -5 \end{bmatrix}$$

over $\mathbb{R}$.

Exercise 586

Find all solutions to the system

$$\begin{bmatrix} 1 & 2 & 3 & 4 & 5 \\ 2 & 1 & 2 & 3 & 4 \\ 2 & 2 & 1 & 2 & 3 \\ 2 & 2 & 2 & 1 & 2 \\ 2 & 2 & 2 & 2 & 1 \end{bmatrix} \begin{bmatrix} X_1 \\ X_2 \\ X_3 \\ X_4 \\ X_5 \end{bmatrix} = \begin{bmatrix} 13 \\ 10 \\ 11 \\ 6 \\ 3 \end{bmatrix}$$

over $\mathbb{R}$.

Exercise 587

Find all solutions to the system

$$\begin{bmatrix} 1 & 1 & 1 & 1 \\ 1 & 1 & 1 & 0 \\ 0 & 0 & 1 & 1 \end{bmatrix} \begin{bmatrix} X_1 \\ X_2 \\ X_3 \\ X_4 \end{bmatrix} = \begin{bmatrix} 1 \\ 0 \\ 1 \end{bmatrix}$$

over $GF(2)$.

Exercise 588

Find all solutions to the system

$$\begin{bmatrix} 1 & 3 & 2 \\ 2 & -1 & 3 \\ 3 & -5 & 4 \\ 1 & 17 & 4 \end{bmatrix} \begin{bmatrix} X_1 \\ X_2 \\ X_3 \end{bmatrix} = \begin{bmatrix} 0 \\ 0 \\ 0 \\ 0 \end{bmatrix}$$

over $\mathbb{R}$.

Exercise 589

Find a real number a so that the system of linear equations

$$2X_1 - X_2 + X_3 + X_4 = 1,$$
$$X_1 + 2X_2 - X_3 + 4X_4 = 2,$$
$$X_1 + 7X_2 - 4X_3 + 11X_4 = a$$

has a solution over $\mathbb{R}$.

Exercise 590

Find all real numbers c such that the system of equations

$$X_1 + X_2 - X_3 = 1,$$
$$X_1 + cX_2 + 3X_3 = 2,$$
$$2X_1 + 3X_2 + cX_3 = 3$$

has a unique solution over $\mathbb{R}$; find those real numbers c for which it has infinitely-many solutions over $\mathbb{R}$; find those real numbers c for which it has no solution over $\mathbb{R}$.

Exercise 591

Solve the system of linear equations

$$X_1 + 2X_2 + X_3 = 1,$$
$$X_1 + X_2 + X_3 = 0$$

over $GF(3)$.

Exercise 592

Solve the system of linear equations

$$X_1 + \left(\sqrt{2}\right)X_2 + \left(\sqrt{2}\right)X_3 = 3,$$
$$X_1 + \left(1 + \sqrt{2}\right)X_2 + X_3 = 3 + \sqrt{2},$$
$$X_1 + X_2 - \left(\sqrt{2}\right)X_3 = 4 + \sqrt{2}$$

over $\mathbb{Q}(\sqrt{2})$.

Exercise 593

Solve the system of linear equations

$$4X_1 - 3X_2 = 3,$$
$$2X_1 - X_2 + 2X_3 = 1,$$

$$3X_1 + 2X_3 = 4$$

over GF(5).

Exercise 594
Solve the system of linear equations

$$4X_1 + 6X_2 + 2X_3 = 8,$$
$$X_1 - aX_2 - 2X_3 = -5,$$
$$7X_1 + 3X_2 + (a - 5)X_3 = 7$$

over $\mathbb{R}$, for various values of the real number a.

Exercise 595
For a given real number a, solve the system

$$aX_1 + X_2 + X_3 = 1,$$
$$X_1 + aX_2 + X_3 = 1,$$
$$X_1 + X_2 + aX_3 = 1$$

over $\mathbb{R}$.

Exercise 596
For a given $a \in \mathbb{R}$, does the system of linear equations

$$aX_1 + X_2 + 2X_3 = 0,$$
$$X_1 - X_2 + aX_3 = 1,$$
$$X_1 + X_2 + X_3 = 1$$

have a unique solution in $\mathbb{R}$?

Exercise 597
Let a be an element of a field F. Find the set of all solutions to the system of linear equations

$$X_1 + X_2 + aX_3 = a,$$
$$X_1 + aX_2 - X_3 = 1,$$
$$X_1 + X_2 - X_3 = 1$$

over F.

Exercise 598
For which $a \in \mathbb{Q}$ does the system of linear equations

$$X_1 + 3X_2 - 2X_3 = 2,$$
$$3X_1 + 9X_2 - 2X_3 = 2,$$
$$2X_1 + 6X_2 + X_3 = a$$

have a unique solution in $\mathbb{Q}$?

Exercise 599
Find real numbers a, b, c, and d such that the points $(1, 2)$, $(-1, 6)$, $(-2, 38)$, and $(2, 6)$ all lie on the curve $y = ax^4 + bx^3 + cx^2 + d$ in the Euclidean plane.

Exercise 600
Find a polynomial $p(X) = a_2 X^2 + a_1 X + a_0 \in \mathbb{R}[X]$ satisfying $p(1) = -1$, $p(-1) = 9$, and $p(2) = -3$.

Exercise 601
Find a polynomial $p(X) = a_3 X^3 + a_2 X^2 + a_1 X + a_0 \in \mathbb{R}[X]$ satisfying $p(0) = 2$, $p(2) = 6$, $p(4) = 3$, and $p(6) = -5$.

Exercise 602
Let $F = GF(13)$. Find a homogeneous system of linear equations over F satisfying the condition that its solution space equals

$$F \left\{ \begin{bmatrix} 2 \\ 1 \\ 9 \\ 7 \\ 4 \end{bmatrix}, \begin{bmatrix} 8 \\ 3 \\ 10 \\ 5 \\ 12 \end{bmatrix}, \begin{bmatrix} 7 \\ 6 \\ 2 \\ 11 \\ 7 \end{bmatrix} \right\}.$$

Exercise 603
Let F be a field. Let $b \in F$ and let $A = \begin{bmatrix} a_{11} & a_{12} & a_{13} \\ a_{21} & a_{22} & a_{23} \end{bmatrix} \in \mathcal{M}_{2\times3}(F)$ be a matrix satisfying the condition that the sum of the entries in each row and each column of A equals b. Show that $b = 0$.

Exercise 604
Let $p(X) = X^5 - 7X^3 + 12 \in \mathbb{Q}[X]$. Find a polynomial $q(X) \in \mathbb{Q}[X]$ of degree at most 3 satisfying $p(a) = q(a)$ for all $a \in \{0, 1, 2, 3\}$.

Exercise 605
Find the rank of the matrix

$$\begin{bmatrix} 1 & -1 & 2 & 3 & 4 \\ 2 & 1 & -1 & 2 & 0 \\ -1 & 2 & 1 & 1 & 3 \\ 1 & 5 & -8 & -5 & -12 \\ 3 & -7 & 8 & 9 & 13 \end{bmatrix} \in \mathcal{M}_{5\times 5}(\mathbb{R}).$$

Exercise 606
Find the rank of the matrix

$$\begin{bmatrix} 1 & 0 & 1 & 0 & 0 \\ 1 & 1 & 0 & 0 & 0 \\ 0 & 1 & 1 & 0 & 0 \\ 0 & 0 & 1 & -1 & 0 \\ 0 & 1 & 0 & 1 & 1 \end{bmatrix} \in \mathcal{M}_{5\times 5}(\mathbb{R}).$$

Exercise 607
Let $F = \mathrm{GF}(2)$. Find the rank of the matrix $\begin{bmatrix} 1 & 1 & 0 \\ 0 & 1 & 1 \\ 1 & 0 & 1 \end{bmatrix} \in \mathcal{M}_{3\times 3}(F)$.

Exercise 608
Let $F = \mathrm{GF}(5)$. Find the rank of the matrix

$$\begin{bmatrix} 1 & 2 & 3 & 4 & a \\ 4 & 3 & a & 1 & 2 \\ a & 1 & 2 & 3 & 4 \\ 2 & 3a & 2 & 4a & 1 \end{bmatrix} \in \mathcal{M}_{4\times 5}(F)$$

for various values of $a \in F$.

Exercise 609
Do there exist a lower-triangular matrix L and an upper-triangular matrix U in $\mathcal{M}_{3\times 3}(\mathbb{Q})$ satisfying the condition $LU = \begin{bmatrix} 1 & -1 & 2 \\ 2 & -1 & 3 \\ 0 & 1 & 8 \end{bmatrix}$?

Exercise 610
Let $F = \mathrm{GF}(5)$. For which values of $a \in F$ do there exist a lower-triangular matrix L and an upper-triangular matrix U in $\mathcal{M}_{3\times 3}(F)$ satisfying the condition that $LU = \begin{bmatrix} 1 & 1 & a \\ 4 & 1 & 0 \\ a & 1 & 4 \end{bmatrix}$?

Exercise 611

Find the LU-decomposition of $\begin{bmatrix} 4 & 2 & 3 & 4 \\ 2 & 0 & 2 & 2 \\ 3 & 4 & -4 & 5 \\ -1 & 0 & 2 & 3 \end{bmatrix} \in \mathcal{M}_{4\times4}(\mathbb{R})$.

Exercise 612

Let $A \in \mathcal{M}_{n\times n}(\mathbb{R})$ be a tridiagonal matrix all diagonal entries of which are nonzero. Can we write $A = LU$, where L is a lower-triangular matrix and U is an upper-triangular matrix, both of which are also tridiagonal?

Exercise 613

Let F be a field and let $a, b, c \in F$. Find the rank of the matrix

$$\begin{bmatrix} 1 & 1 & 1 \\ b+c & c+a & a+b \\ bc & ca & ab \end{bmatrix} \in \mathcal{M}_{3\times3}(F).$$

Exercise 614

Let F be a field and let $a, b, c, d \in F$. Find the rank of the matrix

$$\begin{bmatrix} a & c & c \\ d & a+b & c \\ d & d & b \end{bmatrix} \in \mathcal{M}_{3\times3}(F).$$

Exercise 615

Find the rank of the matrix $\begin{bmatrix} 3 & 1 & 1 & 4 \\ a & 4 & 10 & 1 \\ 1 & 7 & 17 & 3 \\ 2 & 2 & 4 & 3 \end{bmatrix} \in \mathcal{M}_{4\times4}(\mathbb{R})$ for various values of

the real number a.

Exercise 616

Find the rank of $\begin{bmatrix} a & -1 & 2 & 1 \\ -1 & a & 5 & 2 \\ 10 & -6 & 1 & 1 \end{bmatrix} \in \mathcal{M}_{3\times4}(\mathbb{Q})$ for various values of the ra-

tional number a.

Exercise 617

Find the set of all real numbers a such that the rank of the matrix

$$\begin{bmatrix} a & 1 & 1 \\ -1 & -1 & -1 \\ 1 & 1 & a \end{bmatrix} \in \mathcal{M}_{3\times3}(\mathbb{R})$$

equals 2.

Exercise 618

Let n be a positive integer and, for each $A \in \mathcal{M}_{n \times n}(\mathbb{R})$, let $r(A)$ be the rank of A. Define a relation $\preceq$ on $\mathcal{M}_{n \times n}(\mathbb{R})$ by setting $B \preceq A$ if and only if $r(A - B) = r(A) - r(B)$. Is this a partial order relation?

Exercise 619

Let F be a subfield of a field K. Let k and n be positive integers and let $A \in \mathcal{M}_{k \times n}(F)$ be a matrix having rank r. If we now think of A as an element of $\mathcal{M}_{k \times n}(K)$, is its rank necessarily still equal to r?

Exercise 620

Find $k \in \mathbb{Z}$ such that the rank of $\begin{bmatrix} 1 & 7 & 17 & 3 \\ 4 & 4 & 8 & 6 \\ 3 & 1 & 1 & 4 \\ 2k & 8 & 20 & 2 \end{bmatrix} \in \mathcal{M}_{4 \times 4}(\mathbb{Q})$ is minimal.

Exercise 621

Let F be a field and let k and n be positive integers. For a matrix $A \in \mathcal{M}_{k \times n}(F)$ having rank h, show that there exist matrices $B \in \mathcal{M}_{k \times h}(F)$ and $C \in \mathcal{M}_{h \times n}(F)$ such that $A = BC$.

Exercise 622

Let k and n be positive integers and let F be a field. For matrices $A, B \in \mathcal{M}_{k \times n}(F)$, show that the rank of $A + B$ is no more than the sum of the ranks of A and of B.

Exercise 623

Let k and n be positive integers and let F be a field. Let $A, B \in \mathcal{M}_{k \times n}(F)$ be matrices satisfying the condition that he row space of A and the row space of B are disjoint. Does it follow from this that the rank of $A + B$ equals the sum of the rank of A and the rank of B?

Exercise 624

Find bases for the row space and column space of the matrix

$$\begin{bmatrix} 1 & 2 & -3 & -7 & -2 \\ -1 & -2 & 1 & 1 & 0 \\ 1 & 2 & 0 & 2 & 1 \end{bmatrix} \in \mathcal{M}_{3 \times 5}(\mathbb{R}).$$

Exercise 625

Find matrices $P, Q \in \mathcal{M}_{3 \times 3}(\mathbb{R})$ satisfying

$$P \begin{bmatrix} 1 & 2 & 3 \\ 2 & -2 & 1 \\ 3 & 0 & 4 \end{bmatrix} Q = \begin{bmatrix} 1 & 0 & 0 \\ 0 & 1 & 0 \\ 0 & 0 & 0 \end{bmatrix}.$$

Exercise 626

Write the rows of the matrix $A = \begin{bmatrix} 1 & 2 & 0 \\ i-1 & 2 & i \\ 0 & 2 & -i \end{bmatrix} \in \mathcal{M}_{3\times3}(\mathbb{C})$ as linear combinations of the rows of A^T.

Exercise 627

Calculate $\begin{bmatrix} 1 & 2 & 3 & 4 & 5 \\ 0 & 1 & 2 & 3 & 4 \\ 0 & 0 & 1 & 2 & 3 \\ 0 & 0 & 0 & 1 & 2 \\ 0 & 0 & 0 & 0 & 1 \end{bmatrix}^{-1} \in \mathcal{M}_{5\times5}(\mathbb{R})$.

Exercise 628

Let k and n be positive integers and let F be a field. Let $A = \begin{bmatrix} B & C \\ D & E \end{bmatrix}$ be a matrix in $\mathcal{M}_{k\times n}(F)$, where B is a nonsingular matrix in $\mathcal{M}_{r\times r}(F)$ for some $1 \le r < \min\{k, n\}$. Show that the rank of A equals r if and only of $DB^{-1}C = E$.

Exercise 629

Let F be a field and let a, b, c be distinct elements of F. Furthermore, let d, e, f be distinct elements of F. What is the rank of the matrix

$$\begin{bmatrix} 1 & a & d & ad \\ 1 & b & e & be \\ 1 & c & f & cf \end{bmatrix} \in \mathcal{M}_{3\times4}(F)?$$

Exercise 630

Let k and n be positive integers and let F be a field. Let $A \in \mathcal{M}_{k\times n}(F)$ and let $w \in F^k$ be such that the system of linear equations $AX = w$ has a nonempty set of solutions and that all of these solutions satisfy the condition that the hth entry in them is some fixed scalar c. What can we deduce about the columns of the matrix A?

Exercise 631

Let n be a positive integer and let F be a field. Let $O \ne A \in \mathcal{M}_{n\times n}(F)$. Show that there exists a nonnegative integer k such that the rank of A^h equals the rank of A^k for all $h > k$.

Exercise 632

Let $A = \begin{bmatrix} 1 & -1 & 1 \\ -1 & -1 & 1 \\ 1 & 1 & 1 \end{bmatrix} \in \mathcal{M}_{3\times3}(\mathbb{R})$. Find the condition number of A.

Exercise 633

Let a be a positive real number. It is necessarily true that the condition number

of $A = \begin{bmatrix} a & 1 & a \\ 0 & 0 & -1 \\ a & -1 & a \end{bmatrix} \in \mathcal{M}_{3\times3}(\mathbb{R})$ is greater than $2a + 1$?

Exercise 634

Find a positive real number a for which the condition number of

$$A = \begin{bmatrix} 1 & 1 & 0 \\ 0 & a & a \\ 1 & 1 & 1 \end{bmatrix} \in \mathcal{M}_{3\times3}(\mathbb{R})$$

is maximal.

Exercise 635

Does there exist a system $AX = w$ of linear equations in n unknowns (for some positive integer n) over $\mathbb{R}$ having precisely 35 distinct solutions?

Exercise 636

Can one find an integer h such that the condition number of the matrix

$$\begin{bmatrix} 1 & -1000 & 1 \\ 1 & -100 & 0 \\ 1 & h & 1 \end{bmatrix}$$

is greater than 10^6?

Determinants

Let F be a field and let n be a positive integer. We would like to find a function from $\mathcal{M}_{n \times n}(F)$ to F which will serve as an oracle of singularity, namely a function that will assign a value of 0 to singular matrices and a value other than 0 to nonsingular matrices. Indeed, let F be a field and let n be a positive integer. A function $\delta_n : \mathcal{M}_{n \times n}(F) \to F$ is a *determinant function* if and only if it satisfies the following conditions:

(1) $\delta_n(I) = 1$;
(2) $\delta_n(A) = 0$ if A is a matrix having a row all of the entries of which are 0;
(3) $\delta_n(E_{ij}A) = -\delta_n(A)$ for all $1 \leq i \neq j \leq n$;
(4) $\delta_n(E_{ij;c}A) = \delta_n(A)$ for all $1 \leq i \neq j \leq n$ and all $c \in F$;
(5) $\delta_n(E_{i;c}A) = c\delta_n(A)$ for all $1 \leq i \leq n$ and all $0 \neq c \in F$.

In particular, we note that for each $1 \leq i \neq j \leq n$ and all $c \in F$ we have $\delta_n(E_{ij}) = -1 = \delta_n(E_{ij}^T)$, $\delta_n(E_{ij;c}) = 1 = \delta_n(E_{ij;c}^T)$, and $\delta_n(E_{i;c}) = c = \delta_n(E_{i;c}^T)$.

We have yet to show that such functions exist for all values of n, but certainly they exist for a few small ones.

Example For $n = 1$, the function $\delta_1 : [a] \mapsto a$ is a determinant function. For $n = 2$, the function $\delta_2 : \begin{bmatrix} a_{11} & a_{12} \\ a_{21} & a_{22} \end{bmatrix} \mapsto a_{11}a_{22} - a_{12}a_{21}$ is a determinant function.

As an immediate consequence of parts (1) and (5) of the definition, we see that if $A = [a_{ij}] \in \mathcal{M}_{n \times n}(F)$ is a diagonal matrix and if $\delta_n : \mathcal{M}_{n \times n}(F) \to F$ is a determinant function, then $\delta_n(A) = \prod_{i=1}^n a_{ii} \delta_n(I) = \prod_{i=1}^n a_{ii}$.

We now want to show that for each positive integer n there exists a determinant function $\delta_n : \mathcal{M}_{n \times n}(F) \to F$, and indeed that this function is unique. We will first establish the uniqueness of these functions and check some of their properties, holding off on existence until later in this chapter.

> **Proposition 11.1** *Let F be a field. For each positive integer n there exists at most one determinant function $\delta_n : \mathcal{M}_{n \times n}(F) \to F$.*

J.S. Golan, *The Linear Algebra a Beginning Graduate Student Ought to Know*,
DOI 10.1007/978-94-007-2636-9_11, © Springer Science+Business Media B.V. 2012

Proof Let us assume that $\delta_n : \mathcal{M}_{n \times n}(F) \to F$ and $\eta_n : \mathcal{M}_{n \times n}(F) \to F$ are determinant functions and let $\beta = \eta_n - \delta_n$. Then the function β satisfies the following conditions:

(1) $\beta(I) = 0$;
(2) $\beta(A) = 0$ if A is a matrix having a row all of the entries of which are 0;
(3) $\beta(E_{ij}A) = -\beta(A)$ for all $1 \le i \ne j \le n$;
(4) $\beta(E_{ij;c}A) = \beta(A)$ for all $1 \le i \ne j \le n$ and all $c \in F$;
(5) $\beta(E_{i;c}A) = c\beta(A)$ for all $1 \le i \le n$ and all $0 \ne c \in F$.

In particular, if $A \in \mathcal{M}_{n \times n}(F)$ and E is an elementary matrix, then $\beta(A)$ and $\beta(EA)$ are either both equal to 0 or both of them are different from 0. But for any matrix A we know that there exist elementary matrices $E_1, \ldots, E_t$ in $\mathcal{M}_{n \times n}(F)$ such that either $E_1 \cdots E_t A = I$ or $E_1 \cdots E_t A$ is a matrix having at least one row all of the entries of which equal 0. Therefore, $\beta(A) = 0$ for every $A \in \mathcal{M}_{n \times n}(F)$. Thus β is the zero-function, and so $\delta_n = \eta_n$. $\qquad\square$

Proposition 11.2 *Let F be a field and let $\delta_n : M_{n \times n}(F) \to F$ be a determinant function. Then $\delta_n(A) \ne 0$ if and only if A is nonsingular.*

Proof If A is nonsingular, there exist elementary matrices $E_1, \ldots, E_t$ in $\mathcal{M}_{n \times n}(F)$ such that $E_1 \cdots E_t A = I$, and so, by the definition of the determinant function, $\delta_n(A) = c\delta_n(I) = c$, where $0 \ne c \in F$, and so $\delta_n(A) \ne 0$. Now assume that $\delta_n(A) \ne 0$ and that A is singular. Then there exist elementary matrices $E_1, \ldots, E_t$ in $\mathcal{M}_{n \times n}(F)$ such that $E_1 \cdots E_t A$ is a matrix having at least one row all of the entries of which equal 0. But then, for some $0 \ne c \in F$, we have $0 \ne \delta_n(A) = c\delta_n(E_1 \cdots E_t A) = c0 = 0$, which is a contradiction, proving that A must be nonsingular. $\qquad\square$

Thus we see that the determinant function, to the extent it exists, is the oracle we are seeking.

Example The subset $\left\{ \begin{bmatrix} a + bi \\ c + di \end{bmatrix}, \begin{bmatrix} -c + di \\ a - bi \end{bmatrix} \right\}$ of $\mathbb{C}^2$ is linearly dependent if and only if $A = \begin{bmatrix} a + bi & -c + di \\ c + di & a - bi \end{bmatrix} \in \mathcal{M}_{2 \times 2}(\mathbb{C})$ is singular. We have already noted that $\delta_2 : \begin{bmatrix} a_{11} & a_{12} \\ a_{21} & a_{22} \end{bmatrix} \mapsto a_{11}a_{22} - a_{12}a_{21}$ is a determinant function, and so this happens if and only if $\delta_2(A) = a^2 + b^2 + c^2 + d^2 = 0$, i.e., if and only if $a = b = c = d = 0$.

Proposition 11.3 *Let F be a field and let $\delta_n : M_{n \times n}(F) \to F$ be a determinant function. If A is a matrix in $M_{n \times n}(F)$ having two identical rows then $\delta_n(A) = 0$.*

Proof Suppose that rows h and k of A are identical. First, assume that the characteristic of F is other than 2. Then $A = E_{hk}(A)$ and so $\delta_n(A) = \delta_n(E_{hk}A) = -\delta_n(A)$, which implies that $\delta_n(A) = 0$. If the characteristic of F equals 2 then $\delta_n(A) = \delta_n(E_{hk;1}A)$, and $E_{hk;1}A$ is a matrix having a row in which the entries of one row are all 0. Therefore, by Proposition 11.2, $\delta_n(A) = 0$. $\square$

Proposition 11.4 *Let F be a field and let $\delta_n : M_{n \times n}(F) \to F$ be a determinant function. If $A, B \in M_{n \times n}(F)$ then*
(1) $\delta_n(AB) = \delta_n(A)\delta_n(B)$;
(2) $\delta_n(AB) = \delta_n(BA)$.

Proof (1) By Proposition 9.1, we know that AB is nonsingular if and only if both A and B are nonsingular. Therefore, $\delta_n(A) = 0$ or $\delta_n(B) = 0$ if and only if $\delta_n(AB) = 0$. If $\delta_n(A) \neq 0 \neq \delta_n(B)$ then there exist elementary matrices $E_1, \ldots, E_t, G_1, \ldots, G_s$ in $M_{n \times n}(F)$ such that $B = E_1 \cdots E_t I$ and $A = G_1 \cdots G_s I$ and so $AB = G_1 \cdots G_s E_1 \cdots E_t I$, which implies that $\delta_n(AB) = \delta_n(A)\delta_n(B)$ from the definition of a determinant function.

(2) This is an immediate consequence of (1), since $\delta_n(A)\delta_n(B) = \delta_n(B)\delta_n(A)$ in F. $\square$

Proposition 11.5 *Let F be a field and let $\delta_n : M_{n \times n}(F) \to F$ be a determinant function. If $A \in M_{n \times n}(F)$ is nonsingular then $\delta_n(A^{-1}) = \delta_n(A)^{-1}$.*

Proof By Proposition 11.4, we see that $\delta_n(A^{-1})\delta_n(A) = \delta_n(A^{-1}A) = \delta_n(I) = 1$ and from this the result follows immediately. $\square$

Proposition 11.6 *Let F be a field and let $\delta_n : M_{n \times n}(F) \to F$ be a determinant function. If $A \in M_{n \times n}(F)$ then:*
(1) $\delta_n(AE_{ij}) = -\delta_n(A)$ *for all* $1 \leq i \neq j \leq n$;
(2) $\delta_n(AE_{ij;c}) = \delta_n(A)$ *for all* $1 \leq i \neq j \leq n$ *and all* $c \in F$;
(3) $\delta_n(AE_{i;c}) = c\delta_n(A)$ *for all* $1 \leq i \leq n$ *and all* $0 \neq c \in F$.

Proof This is a direct consequence of the definition of the determinant function and Proposition 11.4(2). $\square$

Proposition 11.7 *Let F be a field and let $\delta_n : M_{n \times n}(F) \to F$ be a determinant function. If $A \in M_{n \times n}(F)$ then $\delta_n(A) = \delta_n(A^T)$.*

Proof If A is singular then so is A^T, and so $\delta_n(A) = 0 = \delta_n(A^T)$. If A is non-singular then there exist elementary matrices $E_1, \ldots, E_t$ in $\mathcal{M}_{n \times n}(F)$ such that $E_1 \cdots E_t A = I = I^T = A^T E_t^T \cdots E_1^T$. By our remarks in Chap. 9 concerning the transposes of elementary matrices, and by the remarks at the beginning of this chapter, we see that $\delta_n(A) = \delta_n(E_1 \cdots E_t A) = \delta_n(A^T E_t^T \cdots E_1^T) = \delta_n(A^T)$ and so $\delta_n(A) = \delta_n(A^T)$. $\square$

Of course, at this stage we do not know that determinant functions $\delta_n : \mathcal{M}_{n \times n}(F) \to F$ even exist for the case $n > 2$ and so we now have to construct them. Let us denote the set of all permutations of the set $\{1, \ldots, n\}$ by S_n. We note that any $\pi \in S_n$ is a bijective function from $\{1, \ldots, n\}$ to itself and so there exists a function $\pi^{-1} \in S_n$ satisfying the condition that $\pi \pi^{-1}$ and $\pi^{-1}\pi$ are equal to the identity function $i \mapsto i$. We also note that if $\pi, \pi' \in S_n$ then $\pi \pi' \in S_n$.

Proposition 11.8 *If n is a positive integer then the number of elements of S_n equals $n!$.*

Proof Suppose we wanted to construct an arbitrary element π of S_n. There are n possibilities for selecting $\pi(1)$. Once we have done that, there are $n-1$ ways of selecting $\pi(2)$, then $n-2$ ways of selecting $\pi(3)$, etc. Thus, the total number of ways in which we can define π is $n(n-1) \cdots 1 = n!$. $\square$

Now let $\pi \in S_n$ and let $1 \leq i < j \leq n$. The pair (i, j) is called an *inversion* with respect to π if and only if $\pi(i) > \pi(j)$. That is to say, (i, j) is an inversion with respect to π if and only if

$$\frac{i-j}{\pi(i) - \pi(j)} < 0.$$

We will denote the number of distinct inversions with respect to π by $h(\pi)$, and define the *signum* of π to be $\text{sgn}(\pi) = (-1)^{h(\pi)}$. Thus

$$\text{sgn}(\pi) = \begin{cases} 1 & \text{if there are an even number of inversions with respect to } \pi, \\ -1 & \text{if there are an odd number of inversions with respect to } \pi. \end{cases}$$

It is easy to check that $\text{sgn}(\pi) = \text{sgn}(\pi^{-1})$ for all $\pi \in S_n$. If $\text{sgn}(\pi) = 1$, the permutation π is *even*; if $\text{sgn}(\pi) = -1$, the permutation π is *odd*.

Example Let $\pi \in S_4$ be defined by $1 \mapsto 3$, $2 \mapsto 4$, $3 \mapsto 2$, and $4 \mapsto 1$. Then if we consider all possible pairs (i, j) with $1 \leq i < j \leq 4$ we get

(i, j)	$(\pi(i), \pi(j))$	inversion?
(1,2)	(3,4)	no
(1,3)	(3,2)	yes
(1,4)	(3,1)	yes
(2,3)	(4,2)	yes
(2,4)	(4,1)	yes
(3,4)	(2,1)	yes

and so we see that $\mathrm{sgn}(\pi) = -1$.

Now let n be a positive integer and let $(K, \bullet)$ be an associative and commutative unital F-algebra. Let $A = [a_{ij}] \in \mathcal{M}_{n \times n}(K)$. We then define the function $A \mapsto |A|$ from $\mathcal{M}_{n \times n}(K)$ to K by setting

$$|A| = \sum_{\pi \in S_n} \mathrm{sgn}(\pi) a_{\pi(1),1} \bullet a_{\pi(2),2} \bullet \cdots \bullet a_{\pi(n),n}.$$

Note that, by the commutativity of K, if $\tau = \pi^{-1}$ then

$$a_{\pi(1),1} \bullet a_{\pi(2),2} \bullet \cdots \bullet a_{\pi(n),n} = a_{1,\tau(1)} \bullet a_{2,\tau(2)} \bullet \cdots \bullet a_{n,\tau(n)}$$

and so $|A| = \sum_{\tau \in S_n} \mathrm{sgn}(\tau) a_{1,\tau(1)} \bullet a_{2,\tau(2)} \bullet \cdots \bullet a_{n,\tau(n)}$. Thus we see immediately that $|A| = |A^T|$ for every $A \in \mathcal{M}_{n \times n}(K)$. If $K = \mathbb{C}$ then, since $\overline{c + d} = \overline{c} + \overline{d}$ and $\overline{cd} = \overline{c}\overline{d}$, we also see that for $\overline{A} = [\overline{a}_{ij}]$ we have $|\overline{A}| = \overline{|A|}$. Defining this function for an arbitrary commutative and associative unital F-algebra is important for us, as we will need it in the case that $K = F[X]$, where F is a field.

Example If $A = [a_{ij}] \in \mathcal{M}_{3 \times 3}(K)$, for an associative and commutative unital F-algebra $(K, \bullet)$, then

$$|A| = a_{11} \bullet a_{22} \bullet a_{33} + a_{12} \bullet a_{23} \bullet a_{31} + a_{13} \bullet a_{21} \bullet a_{32}$$
$$- a_{11} \bullet a_{23} \bullet a_{32} - a_{13} \bullet a_{22} \bullet a_{31} - a_{12} \bullet a_{21} \bullet a_{33}.$$

Proposition 11.9 *Let F be a field, let $(K, \bullet)$ be an associative and commutative unital F-algebra, and let $A = [a_{ij}] \in M_{n \times n}(K)$. Pick $1 \leq h \leq n$ and write $a_{hj} = b_{hj} + c_{hj}$ in K for all $1 \leq j \leq n$. For all $1 \leq i \leq n$ satisfying $i \neq h$, set $b_{ij} = c_{ij} = a_{ij}$. Set $B = [b_{ij}]$ and $C = [c_{ij}]$, matrices in $M_{n \times n}(K)$. Then $|A| = |B| + |C|$.*

Proof From the definition of $|A|$, we have

$$|A| = \sum_{\pi \in S_n} \text{sgn}(\pi) a_{1,\pi(1)} \bullet \cdots \bullet a_{h,\pi(h)} \bullet \cdots \bullet a_{n,\pi(n)}$$

$$= \sum_{\pi \in S_n} \text{sgn}(\pi) a_{1,\pi(1)} \bullet \cdots \bullet [b_{h,\pi(h)} + c_{h,\pi(m)}] \bullet \cdots \bullet a_{n,\pi(n)}$$

$$= \sum_{\pi \in S_n} \text{sgn}(\pi) a_{1,\pi(1)} \bullet \cdots \bullet b_{h,\pi(h)} \bullet \cdots \bullet a_{n,\pi(n)}$$

$$+ \sum_{\pi \in S_n} \text{sgn}(\pi) a_{1,\pi(1)} \bullet \cdots \bullet c_{h,\pi(h)} \bullet \cdots \bullet a_{n,\pi(n)}$$

$$= |B| + |C|,$$

as required.   □

We are now ready to prove that determinant functions, in fact, always exist.

Proposition 11.10 *For an integer $n > 1$ and a field F, the function $M_{n \times n}(F) \to F$ defined by $A \mapsto |A|$ is a determinant function.*

Proof In order to simplify our notation, we will make the following temporary convention: if $\pi \in S_n$ and if $A = [a_{ij}] \in M_{n \times n}(F)$, we will write $u(\pi, A) = \text{sgn}(\pi) a_{1,\pi(1)} \cdots a_{n,\pi(n)}$. Now let us check the five conditions of a determinant function.

(1) Clearly, $u(\pi, I)$ equals 1 if π is the identity permutation and 0 otherwise, and so $|I| = 1$.

(2) Let A be a matrix one of the rows of which has all of its entries equal to 0. Since a factor from each row appears in every term $u(\pi, A)$, we conclude that all of these are equal to 0 and hence $|A| = 0$.

(3) Let A be a matrix and let $B = E_{ij}A$. Let $\rho \in S_n$ be the permutation which interchanges i and j and leaves all of the other numbers between 1 and n fixed. Then $\text{sgn}(\pi\rho) \neq \text{sgn}(\pi)$ for all $\pi \in S_n$ and so for each $\pi \in S_n$ we have $-u(\pi, A) = u(\pi\rho, A) = u(\pi, B)$. This implies that $|B| = -|A|$.

(4) Let A be a matrix and let $B = E_{ij;c}A$. Then $B = [b_{ht}]$, where $b_{ht} = a_{ht}$ when $h \neq j$ and $1 \leq t \leq n$, and where $b_{jt} = a_{jt} + ca_{it}$ for all $1 \leq t \leq n$. By Proposition 11.9, we have $|B| = |A| + |C|$, where C is the matrix all of the rows of which except the jth are identical with those of A, and where in the jth row we have $c_{jt} = ca_{it}$ for all $1 \leq t \leq n$. Then $|C| = c|D|$ where D is a matrix in which two rows, the ith and the jth, are equal. If the characteristic of F is other than 2, then $D = E_{ij}D$ and so, by (3), we get $|D| = -|D|$, and so we get $|C| = c|D| = 0$ and we have $|A| = |B|$, which is what we want. Therefore, let us assume that the characteristic of F equals 2. Let $\rho \in S_n$ be the permutation which interchanges i and

j and leaves all other numbers between 1 and n fixed. Let H be the set of all even permutations in S_n and let K be the set of all odd permutations. The function from H to K defined by $\pi \mapsto \rho\pi$ is bijective since $\rho\pi_1 = \rho\pi_2$ implies that $\pi_1 = \rho^{-1}\rho\pi_1 = \rho^{-1}\rho\pi_2 = \pi_2$. Moreover, since the characteristic of F is 2 and since $u(\pi, D) = u(\rho\pi, D)$ for all $\pi \in H$, we see that $u(\pi, D) + u(\rho\pi, D) = 0$ for all $\pi \in H$. Therefore, $|D| = \sum_{\pi \in H}[u(\pi, D) + u(\rho\pi, D)] = 0$ and this implies, again, that $|C| = 0$ and so $|A| = |B|$.

(5) It is clear from the definition of $|A|$ and if $B = E_{i;c}A$ then $|B| = c|A|$. □

Thus, in summary, we see that if F is a field and if n is a positive integer, then there exists a unique determinant function $\mathcal{M}_{n \times n}(F) \to F$, namely $A \mapsto |A|$. We call the scalar $|A|$ the *determinant* of the matrix A.

With kind permission of the Archives of the Mathematisches Forschungsinstitut Oberwolfach (Scherk).

Determinants were first used in the work of the seventeenth-century German mathematician, philosopher, and diplomat Gottfried von Leibnitz, who developed calculus along with Sir Isaac Newton. The common properties of determinants were first studied by the nineteenth-century German mathematician **Heinrich Scherk**, and the first systematic analysis of the theory of determinants was done by the nineteenth-century French mathematician **Augustin-Louis Cauchy**, relying on the work of many mathematicians who preceded him. His work was continued by Cayley and Sylvester. The term "determinant" was first used by Gauss in 1801, and was popularized by Jacobi.

Example Let $n > 1$ be an integer. If $c_1, \ldots, c_n$ are distinct elements of a field F and if $A = [a_{ij}] \in \mathcal{M}_{n \times n}(F)$ is the Vandermonde matrix defined by $a_{ij} = c_i^{j-1}$ for all $1 \leq i, j \leq n$, then it is easy to verify that $|A| = \prod_{i<j}(c_j - c_i) \neq 0$. This result can, in fact, be generalized. Suppose that, for $1 \leq h \leq n$, we have a polynomial $p_h(X) = \sum_{i=0}^{h} b_{hi}X^i \in F[X]$ with $b_{hh} \neq 0$. Let $c_1, \ldots, c_n$ be distinct elements of a field F and let $A = [a_{ij}] \in \mathcal{M}_{n \times n}(F)$ be defined by $a_{ij} = p_j(c_i)$ for all $1 \leq i, j \leq n$. Then $|A| = b_{11} \cdots b_{nn} \prod_{i<j}(c_j - c_i) \neq 0$.

Example As a consequence of Proposition 11.7, we note that if $n > 0$ is odd and if $A \in \mathcal{M}_{n \times n}(F)$ is a skew-symmetric matrix then $|A| = |A^T| = |-A| = -|A|$ and so $|A| = 0$. Therefore, by Proposition 11.2, A is singular. If n is even, then one can use the definition of $|A|$ to show that $|A| = b^2$ for some b which is a sum of products of the a_{ij}. Thus, for example,

$$\begin{vmatrix} 0 & a_{12} & a_{13} & a_{14} \\ -a_{12} & 0 & a_{23} & a_{24} \\ -a_{13} & -a_{23} & 0 & a_{34} \\ -a_{14} & -a_{24} & -a_{34} & 0 \end{vmatrix} = [a_{12}a_{34} - a_{13}a_{24} + a_{14}a_{23}]^2.$$

This number b is called the *Pfaffian* of the matrix A. Pfaffians arise naturally in combinatorics, differential geometry, and other areas of mathematics.

Pfaffians were first defined by Cayley, and named in honor of **Johann Pfaff**, an eighteenth-century German mathematician whose most famous doctoral student was Gauss.

We now give two examples of why it was worthwhile to define $|A|$ for matrices A with entries in an associative and commutative unital F-algebra, and not just a field.

Example Let V be a vector space of finite dimension n over a field F and let $B = \{v_1, \ldots, v_k\}$ be a linearly-independent subset of V. Let $y_1, \ldots, y_k$ be a list of vectors in V. We claim that there are at most finitely-many elements a of F satisfying the condition that the list $v_1 + ay_1, \ldots, v_k + ay_k$ is linearly dependent. To establish this claim, we will consider determinants of matrices over $F[X]$. Indeed, extend B to a basis $D = \{v_1, \ldots, v_n\}$ of V. Then, for each $1 \leq i \leq k$, we can write $y_i = \sum_{j=1}^{n} c_{ij} w_j$. For each $1 \leq i, j \leq k$, define the polynomial $p_{ij}(X) \in F[X]$ by setting

$$p_{ij}(X) = \begin{cases} c_{ii} X + 1 & \text{if } i = j, \\ c_{ij} X & \text{otherwise,} \end{cases}$$

and consider the matrix $B = [p_{ij}(X)] \in \mathcal{M}_{k \times k}(F[X])$. Then $|B|$ is a polynomial $q(X)$ in $F[X]$, which is not the 0-polynomial since $q(0) = 1$. Moreover, for any $a \in F$, we see that $q(a) = 0$ whenever the list $v_1 + ay_1, \ldots, v_k + ay_k$ is linearly dependent. Since a polynomial can have only finitely-many distinct roots, this can happen only for finitely-many values of a.

Example Let $n > 1$ be an integer and let U be an open interval of real numbers. Let K be the set of all functions in $\mathbb{R}^U$ which are differentiable at least $n - 1$ times. Then K is an associative and commutative unital $\mathbb{R}$-algebra which is not entire, let alone a field. We will denote the derivative of a function $f \in K$ by Df and, if $h > 1$, we will denote the hth derivative of f by $D^h f$. Given $f_1, \ldots, f_n \in K$, the function

$$W(f_1, \ldots, f_n) : t \mapsto \begin{vmatrix} f_1(t) & f_2(t) & \cdots & f_n(t) \\ (Df_1)(t) & (Df_2)(t) & \cdots & (Df_n)(t) \\ \vdots & \vdots & \ddots & \vdots \\ (D^{n-1}f_1)(t) & (D^{n-1}f_2)(t) & \cdots & (D^{n-1}f_n)(t) \end{vmatrix}$$

is called the *Wronskian* of $f_1, \ldots, f_n$. One can show that if we have $W(f_1, \ldots, f_n)(t) \neq 0$ for some $t \in U$ then the subset $\{f_1, \ldots, f_n\}$ of K is linearly independent over $\mathbb{R}$. The converse is false. To see this, let U be an open interval containing the origin, let $f_1 : t \mapsto t^3$, and let $f_2 : t \mapsto |t^3|$. Then $\{f_1, f_2\}$ is linearly independent over $\mathbb{R}$, but $W(f_1, f_2)(t) = 0$ for any $t \in U$.

The insight of **Josef Wronski**, a nineteenth-century Polish mathematician living in France, was obscured by his decidedly eccentric philosophical ideas and style of writing, and was recognized only after his death. The notion of a determinant of functions was first used by Jacobi.

Example Let n be a positive integer equal to 2 or divisible by 4. A matrix $A = [a_{ij}] \in \mathcal{M}_{n \times n}(\mathbb{C})$ with $|a_{ij}| \leq 1$ for all $1 \leq i, j \leq n$ having maximal possible determinant (in absolute value) is known as an *Hadamard matrix* (though, in fact, such matrices were studied by Sylvester, a generation before Hadamard considered them). For such a matrix, we have $|A| = n^{n/2}$, and the entries of A are all ± 1. Indeed, a matrix A is an Hadamard matrix precisely when all of its entries are ± 1 and $AA^T = nI$. Thus,

$$\begin{bmatrix} -1 & 1 & 1 & 1 \\ 1 & -1 & 1 & 1 \\ 1 & 1 & -1 & 1 \\ 1 & 1 & 1 & -1 \end{bmatrix} \text{ and } \begin{bmatrix} 1 & 1 & 1 & 1 \\ 1 & -1 & 1 & -1 \\ 1 & 1 & -1 & -1 \\ 1 & -1 & -1 & 1 \end{bmatrix}$$

are Hadamard matrices. Moreover, for each $t \geq 1$, there exists an Hadamard matrix H_t of size $2^t \times 2^t$, defined recursively by setting $H_1 = \begin{bmatrix} 1 & 1 \\ 1 & -1 \end{bmatrix}$ and $H_t = \begin{bmatrix} H_{t-1} & H_{t-1} \\ H_{t-1} & -H_{t-1} \end{bmatrix}$ for each $t > 1$.

We also note immediately that if A is an Hadamard matrix so are A^T and $-A$. Hadamard matrices have important applications in algebraic coding theory, especially in defining the error-correcting Reed–Muller codes. Needless to say, the determinants of Hadamard matrices get very big very quickly. If A is a 16×16 Hadamard matrix, then $|A| = 4{,}294{,}967{,}296$ and If B is a 32×32 Hadamard matrix, then $|B| = 1{,}208{,}925{,}819{,}614{,}629{,}174{,}706{,}176$.

We still are faced with the problem of actually computing the determinant of an $n \times n$ matrix A, especially when n is large. If we work using the definition, we see that we must add $n!$ summands, each of which requires $n - 1$ multiplications. The total number of arithmetic operations need is therefore $(n - 1)n! + (n! - 1) = n(n!) - 1$, which is a huge number even if n is relatively small. For example, if we are using a computer capable of performing a billion arithmetic operations per second, it would take us $12{,}200{,}000{,}000$ years of nonstop computation to compute the determinant of a 25×25 matrix, based on the definition. Thus we must find better methods of computing determinants, a task which became a high priority for many nineteenth-century mathematicians.

Example Let $A = [a_{ij}] \in \mathcal{M}_{n \times n}(F)$ be a matrix in which $a_{11} \neq 0$. Then Chiò, Dodgson, and others showed that $|A| = a_{11}^{2-n}|B|$, where $B \in \mathcal{M}_{(n-1) \times (n-1)}(F)$ is the matrix obtained from A by erasing the first row and first column and replacing each other a_{ij} by $\begin{vmatrix} a_{11} & a_{1j} \\ a_{i1} & a_{ij} \end{vmatrix}$. Thus, for example,

$$
\begin{vmatrix} 1 & 2 & 3 & 4 \\ 8 & 7 & 6 & 5 \\ 1 & 8 & 2 & 7 \\ 3 & 6 & 4 & 5 \end{vmatrix} =
\begin{vmatrix}
\begin{vmatrix} 1 & 2 \\ 8 & 7 \end{vmatrix} & \begin{vmatrix} 1 & 3 \\ 8 & 6 \end{vmatrix} & \begin{vmatrix} 1 & 4 \\ 8 & 5 \end{vmatrix} \\
\begin{vmatrix} 1 & 2 \\ 1 & 8 \end{vmatrix} & \begin{vmatrix} 1 & 3 \\ 1 & 2 \end{vmatrix} & \begin{vmatrix} 1 & 4 \\ 1 & 7 \end{vmatrix} \\
\begin{vmatrix} 1 & 2 \\ 3 & 6 \end{vmatrix} & \begin{vmatrix} 1 & 3 \\ 3 & 4 \end{vmatrix} & \begin{vmatrix} 1 & 4 \\ 3 & 5 \end{vmatrix}
\end{vmatrix}
$$

$$
= \begin{vmatrix} -9 & -18 & -27 \\ 6 & -1 & 3 \\ 0 & -5 & -7 \end{vmatrix}
$$

$$
= -144.
$$

This method can, of course, be iterated. The method of evaluating determinants in this way is known as the *method of condensation*.

© George E. Andrews (Andrews).

During the nineteenth century, matrix theory and the theory of determinants attracted many gifted mathematicians and mathematical amateurs. Felice Chiò was a nineteenth-century Italian mathematician and physicist. On the other hand, **Rev. Charles Lutwidge Dodgson** was an amateur who is better known by his pen name Lewis Carroll, the author of *Alice in Wonderland*. Dodgson published several works on mathematics and mathematical logic. In the twentieth century, ingenious ways for computing determinants of matrices arising from various combinatorial problems have been devised by American mathematician **George Andrews**.

Let $A = [a_{ij}] \in \mathcal{M}_{n \times n}(K)$, where K is an associative and commutative unital F-algebra. For each $1 \le i, j \le n$, we define the *minor* of the entry a_{ij} of A to be $|A_{ij}|$, where $A_{ij} \in \mathcal{M}_{(n-1) \times (n-1)}(K)$ is the matrix obtained from A by erasing the ith row and the jth column.

Example If $A = \begin{bmatrix} 4 & 3 & 1 \\ 2 & 8 & 9 \\ 7 & 3 & 4 \end{bmatrix}$, then $A_{13} = \begin{bmatrix} 2 & 8 \\ 7 & 3 \end{bmatrix}$ and $A_{22} = \begin{bmatrix} 4 & 1 \\ 7 & 4 \end{bmatrix}$.

Proposition 11.11 *Let F be a field, let $(K, \bullet)$ be an associative and commutative unital F-algebra. If n is a positive integer, and $A = [a_{ij}] \in M_{n \times n}(K)$, then $|A| = \sum_{j=1}^{n}(-1)^{t+j} a_{tj} \bullet |A_{tj}|$ for each $1 \leq t \leq n$.*

Proof In order to simplify our notation, let $\det(y_1, \ldots, y_n)$ denote the determinant of the matrix the rows of which are $y_1, \ldots, y_n$. We will first prove the theorem for the case $t = 1$. That is to say, we must show that $|A|$ equals $\sum_{j=1}^{n}(-1)^{1+j} a_{1j} \bullet |A_{1j}|$. For each $1 \leq h \leq n$, let $v_h \in M_{1 \times n}(K)$ be the matrix $[d_1 \ldots d_n]$ defined by

$$d_i = \begin{cases} 1 & \text{if } i = h, \\ 0 & \text{otherwise.} \end{cases}$$

Then the ith row of A can be written as $w_i = \sum_{j=1}^{n} a_{ij} v_j$ and so

$$|A| = \det(w_1, \ldots, w_n) = \det\left(\sum_{j=1}^{n} a_{1j} v_j, w_2, \ldots, w_n\right)$$

$$= \sum_{j-1}^{n} a_{1j} \bullet \det(v_j, w_2, \ldots, w_n).$$

Thus we will prove the desired result if we can show that

$$\det(v_j, w_2, \ldots, w_n) = (-1)^{1+j}|A_{1j}|$$

for each $1 \leq j \leq n$. Denote the matrix the rows of which are $v_j, w_2, \ldots, w_n$ by $B = [b_{ih}]$, where

$$b_{ih} = \begin{cases} 1 & \text{if } i = 1 \text{ and } h = j, \\ 0 & \text{if } i = 1 \text{ and } h \neq j, \\ a_{ih} & \text{if } i > 1. \end{cases}$$

For $1 \leq j \leq n$, set $G_{1j} = \{\pi \in S_n \mid \pi(1) = j\}$.

Suppose that $j = 1$. Then, in particular, there is a bijective correspondence between G_{11} and the set of all permutations of $\{2, \ldots, n\}$ which does not affect the signum of the permutation since if $\pi \in G_{11}$ then 1 does not appear in any inversion of π. Since $b_{11} = 1$ and $b_{1h} = 0$ if $h > 1$, we thus have

$$|B| = \sum_{\pi \in S_n} \text{sgn}(\pi) b_{1,\pi(1)} \bullet \cdots \bullet b_{n,\pi(n)}$$

$$= \sum_{\pi \in G_{11}} \text{sgn}(\pi) b_{1,\pi(1)} \bullet \cdots \bullet b_{n,\pi(n)}$$

$$= \sum_{\pi \in G_{11}} \text{sgn}(\pi) b_{2,\pi(2)} \bullet \cdots \bullet b_{n,\pi(n)} = |A_{11}|,$$

and so we have shown, as desired, that $|B| = (-1)^{1+1}|A_{11}|$. If $j > 1$ put column j of B in the position of the first column and shift columns 1 to $j - 1$ of B to the right by one column position. This involves $j - 1$ column interchanges, and we have

$$\det(v_j, w_2, \ldots, w_n) = (-1)^{j-1} \begin{vmatrix} 1 & 0 & \cdots & 0 \\ a_{2j} & a_{21} & \cdots & a_{2n} \\ \vdots & \vdots & \ddots & \vdots \\ a_{nj} & a_{n1} & \cdots & a_{nn} \end{vmatrix} = (-1)^{j+1}|A_{1j}|.$$

Now assume that $t > 1$. Again, we can interchange the tth row with the first row by $t - 1$ exchanges with the row above, and we get $|A| = (-1)^{t-1}|C|$, where C is a matrix satisfying $|C_{1j}| = |A_{tj}|$ for each $1 \leq j \leq n$. Therefore,

$$|A| = (-1)^{t-1}|C| = (-1)^{t-1} \sum_{j=1}^{n} (-1)^{j+1} c_{1j} \bullet |C_{1j}| = \sum_{j=1}^{n} (-1)^{j+t} a_{tj} \bullet |A_{tj}|$$

as desired.   □

Example For $A = \begin{bmatrix} 1 & 7 & 3 & 0 \\ 4 & 0 & 1 & 3 \\ 0 & 2 & 4 & 0 \\ 3 & 1 & 5 & 1 \end{bmatrix}$ we see that

$$|A| = 1 \begin{vmatrix} 0 & 1 & 3 \\ 2 & 4 & 0 \\ 1 & 5 & 1 \end{vmatrix} - 7 \begin{vmatrix} 4 & 1 & 3 \\ 0 & 4 & 0 \\ 3 & 5 & 1 \end{vmatrix} + 3 \begin{vmatrix} 4 & 0 & 3 \\ 0 & 2 & 0 \\ 3 & 1 & 1 \end{vmatrix} - 0 \begin{vmatrix} 4 & 0 & 1 \\ 0 & 2 & 4 \\ 3 & 1 & 5 \end{vmatrix}$$

$$= 16 + 140 - 30 + 0 = 126$$

and

$$|A| = 0 \begin{vmatrix} 7 & 3 & 0 \\ 0 & 1 & 3 \\ 1 & 5 & 1 \end{vmatrix} - 2 \begin{vmatrix} 1 & 3 & 0 \\ 4 & 1 & 3 \\ 3 & 5 & 1 \end{vmatrix} + 4 \begin{vmatrix} 1 & 7 & 0 \\ 4 & 0 & 3 \\ 3 & 1 & 1 \end{vmatrix} - 0 \begin{vmatrix} 1 & 7 & 3 \\ 4 & 0 & 1 \\ 3 & 1 & 5 \end{vmatrix}$$

$$= 0 - 2 + 128 - 0 = 126.$$

Even this method of computing determinants is not easy, however, unless there is a row (or column) of the matrix a significant number of the entries in which are equal to 0. To see the computational overhead of computing the determinant of a general $n \times n$ matrix using minors, let us denote the number of arithmetic operations needed to do so by p_n. Clearly $p_1 = 1$ and $p_2 = 3$. Suppose that we have already found p_{n-1}. Then, by Proposition 11.11, we see that in order to compute the determinant of an $n \times n$ matrix we have to compute the determinants of n matrices of size $(n - 1) \times (n - 1)$ and then perform n multiplications and $n - 1$ additions/subtractions. That is to say, we obtain the recursive formula

$$p_n = np_{n-1} + n + (n - 1) = np_{n-1} + 2n - 1,$$

when $n > 2$. Setting $t_n = \frac{1}{n!} p_n$, we see that

$$t_n - t_{n-1} = \frac{2}{(n-1)!} - \frac{1}{n!}$$

and so

$$t_n = [t_n - t_{n-1}] + [t_{n-1} - t_{n-2}] + \cdots + [t_3 - t_2] + t_2$$

$$= \left[\frac{2}{(n-1)!} - \frac{1}{n!} \right] + \cdots + \left[\frac{2}{2!} - \frac{1}{3!} \right] + 1$$

$$= 2 \left[\frac{1}{(n-1)!} + \cdots + \frac{1}{1!} \right] - \left[\frac{1}{n!} + \cdots + \frac{1}{1!} \right] + 1$$

$$= \frac{1}{(n-1)!} + \cdots + \frac{1}{1!} + 1 - \frac{1}{n!}$$

$$= \left[\frac{1}{n!} + \frac{1}{(n-1)!} + \cdots + \frac{1}{1!} + 1 \right] - \frac{2}{n!}$$

and thus we see that $p_n = n! [\frac{1}{n!} + \frac{1}{(n-1)!} + \cdots + \frac{1}{1!} + 1] - 2$. But from calculus we know that e, the base of the natural logarithms, has an expansion of the form

$$e = 1 + \frac{1}{1!} + \cdots + \frac{1}{n!} + \frac{e^c}{(n+1)!},$$

where $0 < c < 1$, and so $p_n = n! [e - \frac{e^c}{(n+1)!}] - 2$. If $n > 2$, we see that

$$0 < \frac{e^c}{n+1} < \frac{e}{n+1} \le \frac{e}{3} < 1$$

and so we conclude that $en! - 3 < p_n < en! - 2$. Since p_n is a positive integer, we see that $p_n = \lfloor n! \rfloor - 2$, where $\lfloor r \rfloor$ denotes the largest whole number less than or equal to r, for any real number r. In particular, we see that p_n grows even faster than exponentially, as a function of n, which is very rapid growth indeed. For example, $p_{10} = 9{,}864{,}094$ and $p_{15} = 3{,}554{,}625{,}081{,}047$.

Recently, sophisticated numerical techniques have been developed to compute the determinants of matrices with entries from a finite field.

In special cases, it is also possible to find bounds on the value of the determinant of a matrix, without actually computing it. For example, we will see below that if $A \in \mathcal{M}_{n \times n}(\mathbb{R})$ and if g is a positive real number greater than or equal to the absolute value of each of the entries of A, then the absolute value of $|A|$ is at most $g^n \sqrt{n^n}$. In 1980, American mathematicians Charles R. Johnson and Morris Newman proved a surprising bound. Let $A = [a_{ij}] \in \mathcal{M}_{n \times n}(\mathbb{R})$. For each $1 \le i \le n$, let b_i be the sum of all positive entries in the ith row of A and let c_i be the sum of all negative entries in the ith row of A (the sum of an empty list is taken to be 0). The absolute value of $|A|$ is then at most $\prod_{i=1}^{n} \max\{b_i, -c_i\} - \prod_{i=1}^{n} \min\{b_i, -c_i\}$.

Proposition 11.12 *Let n be a positive integer, let F be a field, and let $(K, \bullet)$ be an associative and commutative unital F-algebra. Let $A = [a_{ij}] \in \mathcal{M}_{n \times n}(K)$ be a matrix which can be represented in block form as*

$$
\begin{bmatrix}
B_{11} & O & \cdots & O \\
B_{21} & B_{22} & \cdots & O \\
\vdots & \vdots & \ddots & \vdots \\
B_{m1} & B_{m2} & \cdots & B_{mm}
\end{bmatrix},
$$

where $m > 1$ and each of the B_{hh} is square. Then $|A| = \prod_{h=1}^{m} |B_{hh}|$.

Proof Let us first consider the case $m = 2$, and assume that $B_{11} \in \mathcal{M}_{t \times t}(K)$ for some $t < n$. We will proceed by induction on t. If $t = 1$, then, by Proposition 11.11, $|A| = a_{11}|B_{22}| = |B_{11}| \bullet |B_{22}|$, and we are done. Now assume that $t > 1$ and that the result has been established for all matrices of the form $\begin{bmatrix} B_{11} & O \\ B_{21} & B_{22} \end{bmatrix}$, where $B_{11} \in \mathcal{M}_{(t-1) \times (t-1)}(K)$. Let C_j be the matrix obtained from B_{12} by deleting the jth column. Then, by Proposition 11.11 and the induction hypothesis,

$$
|A| = \sum_{j=1}^{t} (-1)^{j+1} a_{1j} \bullet \begin{vmatrix} (B_{11})_{1j} & O \\ C_j & B_{22} \end{vmatrix}
$$

$$
= \sum_{j=1}^{t} (-1)^{j+1} a_{1j} \bullet \left(\left| (B_{11})_{1j} \right| \bullet |B_{22}| \right)
$$

$$
= \left(\sum_{j=1}^{t} (-1)^{j+1} a_{1j} \bullet \left| (B_{11})_{1j} \right| \right) \bullet |B_{22}| = |B_{11}| \bullet |B_{22}|,
$$

which establishes this case.

Now assume, inductively, that the result has been established for m and consider a matrix $A \in \mathcal{M}_{n \times n}(K)$ which can be written in block form as

$$
\begin{bmatrix}
B_{11} & O & \cdots & O \\
B_{21} & B_{22} & \cdots & O \\
\vdots & \vdots & \ddots & \vdots \\
B_{m+1,1} & B_{m+1,2} & \cdots & B_{m+1,m+1}
\end{bmatrix}.
$$

If we set $C = \begin{bmatrix} B_{11} & O & \cdots & O \\ B_{21} & B_{22} & \cdots & O \\ \vdots & \vdots & \ddots & \vdots \\ B_{m1} & B_{m2} & \cdots & B_{mm} \end{bmatrix}$ then, by the case $m = 2$ and the induction

hypothesis, $|A| = |C| \bullet |B_{m+1,m+1}| = \prod_{h=1}^{m+1} |B_{hh}|$. $\square$

Note that, as an immediate consequence of Proposition 11.4, we see that if all of the matrices B_{hh} are of the same size, then $|A| = |B_{11} \cdots B_{mm}|$.

Let $A = [a_{ij}] \in \mathcal{M}_{n \times n}(K)$ for some associative and commutative unital F-algebra $(K, \bullet)$. We define the *adjoint* of A to be the matrix $\mathrm{adj}(A) = [b_{ij}] \in \mathcal{M}_{n \times n}(K)$, where $b_{ij} = (-1)^{i+j}|A_{ji}|$ for all $1 \leq i, j \leq n$.

Example If $A = \begin{bmatrix} 1 & 0 & 3 & 5 \\ -3 & 1 & 3 & 1 \\ 4 & 2 & 1 & 2 \\ 1 & 1 & 2 & 5 \end{bmatrix} \in \mathcal{M}_{4 \times 4}(\mathbb{R})$ then

$$\mathrm{adj}(A) = \begin{bmatrix} -20 & 9 & -17 & 25 \\ 50 & -18 & -16 & -40 \\ -40 & -18 & -16 & 50 \\ 10 & 9 & 13 & -35 \end{bmatrix}.$$

Proposition 11.13 *Let F be a field and let n be a positive integer. If $A = [a_{ij}] \in M_{n \times n}(F)$ then $A[\mathrm{adj}(A)] = |A|I$. In particular, if the matrix A is nonsingular then $A^{-1} = |A|^{-1} \mathrm{adj}(A)$.*

Proof Suppose that $\mathrm{adj}(A) = [b_{ij}]$. Then $A[\mathrm{adj}(A)] = [c_{ij}]$, where $c_{ij} = \sum_{k=1}^{n} a_{ik} b_{kj} = \sum_{k=1}^{n} (-1)^{j+k} a_{ik}|A_{jk}|$. If $i = j$, then, by Proposition 11.11, this is just $|A|$. If $i \neq j$, this is just $|A'|$, where A' is a matrix identical to A in all of its rows except the ith row, and that is equal to the jth row of A. Thus the matrix A' has two identical rows, and so by Proposition 11.3, $|A'|$ is equal to 0. Hence $A[\mathrm{adj}(A)] = |A|I$, from which we also immediately deduce the second statement since if A is nonsingular then $|A| \neq 0$. $\square$

In particular, we note that if A is nonsingular then so is $\mathrm{adj}(A)$.

Proposition 11.14 *Let F be a field, let $(K, \bullet)$ be an associative and commutative unital F-algebra, and let n be a positive integer. If $A = [a_{ij}] \in M_{n \times n}(K)$ is an upper-triangular matrix then $|A| = \prod_{i=1}^{n} a_{ii}$.*

Proof We can prove this by induction on n. For the case $n = 1$, it is immediate. Assume therefore that we have already established it for all matrices in $\mathcal{M}_{n \times n}(K)$. Then, by Proposition 11.11, $|A| = |A^T| = \sum_{j=1}^{n} (-1)^{1+j} a_{j1} \bullet |A_{j1}| = a_{11} \bullet |A_{11}|$. But, by the induction hypothesis, $|A_{11}| = \prod_{i=2}^{n} a_{ii}$, and we are done. $\square$

By Proposition 11.14, we see that in general, from a computational point of view, it is much faster to first perform elementary operations on a matrix to reduce it to upper-triangular form, and then calculate the determinant (making use of the fact that, from the definition of a determinant function and from Proposition 11.4, we easily know the determinants of the elementary matrices), than to calculate the determinant directly. When working in associative and commutative unital algebras over a field, or when working with matrices of integers, this presents somewhat of a problem since it is not always possible to divide by nonzero scalars in such contexts. However, various variants on Gaussian elimination which do not involve division have been developed to overcome this.

© The Daily Northwestern.

One of the major researchers instrumental in the development of such methods was the twentieth-century Swiss/American computer scientist **Erwin Bareiss**.

Combining Propositions 11.4 and 11.14, we see that if $A \in M_{n \times n}(F)$ can be written in the form LU, where L is a lower-triangular matrix and U is an upper-triangular matrix, then $|A|$ is the product of the diagonal elements of L and the diagonal elements of U.

Proposition 11.15 (Cramer's Theorem) *Let F be a field and let n be a positive integer. If $A = [a_{ij}] \in M_{n \times n}(F)$ is a nonsingular matrix and if $w = \begin{bmatrix} b_1 \\ \vdots \\ b_n \end{bmatrix} \in F^n$, then the system of linear equations $AX = w$ has the unique solution $v = \begin{bmatrix} d_1 \\ \vdots \\ d_n \end{bmatrix}$ in which, for each $1 \le i \le n$, we have $d_i = |A|^{-1}|A_{(i)}|$, where $A_{(i)}$ is the matrix formed from A by replacing the ith column of A by w.*

Proof If $Av = w$ then $|A|v = (|A|A^{-1})Av = \mathrm{adj}(A)Av = \mathrm{adj}(A)w$ and so for each $1 \le i \le n$, we have $|A|d_i = \sum_{j=1}^{n}(-1)^{i+j}b_j|A_{ji}|$. But the expression on the right-hand side of this equation is just, by Proposition 11.11, $|A_{(i)}|$, developed by minors on the ith column. $\qquad \square$

Gabriel Cramer was an eighteenth-century Swiss mathematician and friend of Johann Bernoulli (one of the formulators of calculus) who was among the first to study determinants and their use in solving systems of linear equations. Cramer's rule was also described independently by the eighteenth-century Scottish mathematician **Colin Maclaurin**.

Cramer's theorem, published in 1750, was the first systematic method for solving a system of linear equations, though special cases of it were known to Leibnitz 75 years earlier. While it is elegant mathematically, it is clearly not computationally feasible, even when n is only moderately large, as was immediately realized by mathematicians of the time. Indeed, solving a system of linear equations $AX = w$ by Cramer's method, where A is a nonsingular $n \times n$ matrix over a field F, requires $\frac{1}{3}n^4 - \frac{1}{6}n^3 - \frac{1}{3}n^2 + \frac{1}{6}n$ additions and $\frac{1}{3}n^4 + \frac{1}{3}n^3 + \frac{2}{3}n^2 + \frac{2}{3}n - 1$ multiplications, which is considerably worse than the methods we have previously studied, for which the number of arithmetic operations necessary grows as n^3, rather than as n^4.

Example Consider the system of linear equations $AX = \begin{bmatrix} 2 \\ 1 \\ 4 \end{bmatrix}$, where $A = \begin{bmatrix} 1 & -1 & 1 \\ 1 & 2 & 0 \\ 1 & 0 & -1 \end{bmatrix}$. Then $|A| = -5$ and

$$|A_{(1)}| = \begin{vmatrix} 2 & -1 & 1 \\ 1 & 2 & 0 \\ 4 & 0 & -1 \end{vmatrix} = -13, \qquad |A_{(2)}| = \begin{vmatrix} 1 & 2 & 1 \\ 1 & 1 & 0 \\ 1 & 4 & -1 \end{vmatrix} = 4, \quad \text{and}$$

$$|A_{(3)}| = \begin{vmatrix} 1 & -1 & 2 \\ 1 & 2 & 1 \\ 1 & 0 & 4 \end{vmatrix} = 7.$$

As a consequence, we see that the unique solution to the equation is $\frac{1}{5} \begin{bmatrix} 13 \\ -4 \\ -7 \end{bmatrix}$.

We note that if $A = [a_{ij}] \in \mathcal{M}_{n \times n}(F)$ then the polynomial

$$\sum_{\pi \in S_n} \text{sgn}(\pi) X_{\pi(1)} X_{\pi(2)} \cdots X_{\pi(n)} \in F[X_1, \ldots, X_n]$$

is flat and of degree n. This allows us to make an interesting use of Proposition 4.5.

Proposition 11.16 *Let F be a field of characteristic other than 2, let $A = [a_{ij}] \in \mathcal{M}_{n \times n}(F)$ be an arbitrary matrix, and let $C = [c_{ij}] \in \mathcal{M}_{n \times n}(F)$ be a diagonal matrix with nonzero entries on the diagonal. Then there exists a diagonal matrix $E = [e_{ij}] \in \mathcal{M}_{n \times n}(F)$ with diagonal entries ± 1 such that $EC + A$ is nonsingular.*

Proof Let $X_1, \dots, X_n$ be indeterminates over F and let $D = [d_{ij}] \in \mathcal{M}_{n \times n}(F[X_1, \dots, X_n])$ be the diagonal matrix with $d_{ii} = X_i$ for $1 \leq i \leq n$. Then $|DC + A|$ is a flat polynomial in $F[X_1, \dots, X_n]$ of degree n, and so the result follows immediately from Proposition 4.5.   □

Example If $A = [a_{ij}] \in \mathcal{M}_{n \times n}(\mathbb{R})$ and if $e > 0$ then, by Proposition 11.16, it is possible to "tweak" the diagonal of A to obtain a nonsingular matrix $[a'_{ij}]$, where

$$a'_{ij} = \begin{cases} a_{ij} \pm e & \text{if } i = j, \\ a_{ij} & \text{otherwise.} \end{cases}$$

The sum which appears in the definition of the determinant shows up in other contexts related to matrix algebras. An associative algebra $(K, \bullet)$ over a field F satisfies the *standard identity of degree n* if and only if $\sum_{\pi \in S_n} \text{sgn}(\pi) a_{\pi(1)} \bullet a_{\pi(2)} \bullet \cdots \bullet a_{\pi(n)} = 0$ for any list $a_1, \dots, a_n$ of elements of K. Thus, for example, the standard identity of degree 2 is $a_1 \bullet a_2 - a_2 \bullet a_1 = 0$. The algebra K satisfies this identity precisely when it is commutative. The *Amitsur–Levitzki Theorem* states that for any field F and any positive integer n, the F-algebra $\mathcal{M}_{n \times n}(F)$ satisfies the standard identity of degree k for each $k \geq 2n$. There are several proofs of this result, all beyond the scope of this book. Some of these are based on a generalization of the Cayley–Hamilton Theorem, which we shall see in the following chapter.

© Alexander Levitzki (Levitzki).

Yaakov Levitzki and his student **Shimshon Amitsur** were twentieth-century Israeli algebraists.

We end this chapter by showing how an important construction in analysis can be considered in terms of determinants of matrices over the $\mathbb{R}$-algebra $\mathbb{R}[X]$. Let $c_0, c_1, \dots$ be real numbers and let us consider the analytic function $f : x \mapsto \sum_{i=0}^{\infty} c_i x^i$, which converges for all x in some subset U of $\mathbb{R}$. We know that $U \neq \varnothing$ since surely $0 \in U$. Given positive integers k and n, we want to find polynomials

$p(X), q(X) \in \mathbb{R}[X]$ of degrees at most k and n, respectively, such that the function $x \mapsto p(x)q(x)^{-1} - f(x)$ also converges for all $x \in U$ and is representable there by a power series of the form $x \mapsto \sum_{i=1}^{\infty} d_i x^{k+n+i}$. If we find such p and q, then the function $x \mapsto p(x)q(x)^{-1}$ is called the *Padé approximant* to f of type k/n. Padé approximants are very important tools in differential equations and in approximation theory. Hermite made use of Padé approximants in his proof of the transcendence of e.

Henri Padé was a nineteenth-century French engineer who developed these approximants in the course of his work. Interest in them intensified in the early twentieth century when the French mathematician **Émile Borel** made extensive use of them in his work on analysis.

Example If $f : x \mapsto e^x = \sum_{i=0}^{\infty} \frac{1}{i!} x^i$ then the function $g_1 : x \mapsto \frac{x^2+4x+6}{6-2x}$ is a Padé approximant to f of type $2/1$ and the function $g_2 : x \mapsto \frac{x^2+6x+12}{x^2-6x+12}$ is a Padé approximant to f of type $2/2$.

If we are given an analytic f as above, how do we calculate Padé approximants to it? One way is by using determinants. First of all, define $c_{-i} = 0$ for all positive integers i. Then, given positive integers k and n, define the matrices $P_{k/n}(X), Q_{k/n}(X) \in \mathcal{M}_{(n+1)\times(n+1)}(\mathbb{R}[X])$ by setting:

$$P_{k/n}(X) = \begin{bmatrix} c_{k-n+1} & c_{k-n+2} & \cdots & c_{k+1} \\ c_{k-n+2} & c_{k-n+3} & \cdots & c_{k+2} \\ \vdots & \vdots & \ddots & \vdots \\ c_k & c_{k+1} & \cdots & c_{k+n} \\ \sum_{i=0}^{k-n} c_i X^{n+i} & \sum_{i=0}^{k-n+1} c_i X^{n+i-1} & \cdots & \sum_{i=0}^{k} c_i X^i \end{bmatrix}$$

and

$$Q_{k/n}(X) = \begin{bmatrix} c_{k-n+1} & c_{k-n+2} & \cdots & c_{k+1} \\ c_{k-n+2} & c_{k-n+3} & \cdots & c_{k+2} \\ \vdots & \vdots & \ddots & \vdots \\ c_k & c_{k+1} & \cdots & c_{k+n} \\ X^n & X^{n-1} & \cdots & 1 \end{bmatrix}.$$

Then the polynomials $p(X) = |P_{k/n}(X)|$ and $q(X) = |Q_{k/n}(X)|$ are of the desired size, and our approximant is given by $x \mapsto p(x)q(x)^{-1}$.

Exercises

Exercise 637

Calculate $\begin{vmatrix} \sin(a) & \cos(a) \\ \sin(b) & \cos(b) \end{vmatrix}$ and $\begin{vmatrix} \cos(a) & \sin(a) \\ \sin(b) & \cos(b) \end{vmatrix}$ for real numbers a and b.

Exercise 638

Calculate $\begin{vmatrix} 1 & i & 1+i \\ -i & 1 & 0 \\ 1-i & 0 & 1 \end{vmatrix} \in \mathbb{C}$.

Exercise 639

Calculate $\begin{vmatrix} a-6 & 0 & 0 & -8 \\ 5 & a-4 & 0 & 12 \\ -1 & 3 & a-2 & -6 \\ 0 & -\frac{1}{2} & 1 & 1 \end{vmatrix}$ for any $a \in \mathbb{R}$.

Exercise 640
Find the image of the function f from $\mathbb{R}$ to itself defined by

$$f : t \mapsto \begin{vmatrix} 1 & 0 & -t \\ 1 & 1 & -1 \\ t & 0 & -1 \end{vmatrix}.$$

Exercise 641
For real numbers a, b, c, and d, show that

$$\begin{vmatrix} a^2 & (a+1)^2 & (a+2)^2 & (a+3)^2 \\ b^2 & (b+1)^2 & (b+2)^2 & (b+3)^2 \\ c^2 & (c+1)^2 & (c+2)^2 & (c+3)^2 \\ d^2 & (d+1)^2 & (d+2)^2 & (d+3)^2 \end{vmatrix} = 0.$$

Exercise 642
Let n be a positive integer and let c be a fixed real number. Calculate the determinant of the matrix $A = [a_{ij}] \in \mathcal{M}_{n \times n}(\mathbb{R})$ defined by

$$a_{ij} = \begin{cases} c & \text{if } i < j, \\ i & \text{if } i = j, \\ 0 & \text{if } i > j. \end{cases}$$

Exercise 643

For $a, b \in \mathbb{R}$, calculate $\begin{vmatrix} a & b & a+b \\ b & a+b & a \\ a+b & a & b \end{vmatrix}$.

Exercise 644

Let $F = GF(2)$. Does there exist a matrix $A \in \mathcal{M}_{2\times2}(F)$ other than I satisfying the condition that $|A| = |A^T A| = 1$?

Exercise 645

If n is a positive integer, we define the nth *Hankel matrix* $H_n \in \mathcal{M}_{n\times n}(\mathbb{R})$ to be the matrix $[a_{ij}]$ satisfying

$$a_{ij} = \begin{cases} 0 & \text{if } i + j - 1 > n, \\ i + j - 1 & \text{otherwise.} \end{cases}$$

Calculate $|H_n|$.

The nineteenth-century German mathematician **Hermann Hankel** was among the first to recognize and popularize the work of Grassmann.

Exercise 646

Let $p(X) = a_0 + a_1 X + a_2 X^2$ and $q(X) = b_0 + b_1 X + b_2 X^2$ be polynomials in $\mathbb{C}[X]$. Show that there exists a complex number c satisfying $p(c) = q(c) = 0$ if and only if
$$\begin{vmatrix} a_0 & a_1 & a_2 & 0 \\ 0 & a_0 & a_1 & a_2 \\ b_0 & b_1 & b_2 & 0 \\ 0 & b_0 & b_1 & b_2 \end{vmatrix} = 0.$$

Exercise 647

Find the set of all pairs (a, b) of real numbers such that

$$\begin{vmatrix} a+1 & 3a & b+3a & b+1 \\ 2b & b+1 & 2-b & 1 \\ a+2 & 0 & 1 & a+3 \\ b-1 & 1 & a+2 & a+b \end{vmatrix} = 0.$$

Exercise 648

For $a, b, c \in \mathbb{R}$, show that $\begin{vmatrix} 0 & (a-b)^2 & (a-c)^2 \\ (b-a)^2 & 0 & (b-c)^2 \\ (c-a)^2 & (c-b)^2 & 0 \end{vmatrix} \geq 0.$

Exercise 649
Let n be a positive even integer and let $c, d \in \mathbb{Q}$. Let $A = [a_{ij}] \in \mathcal{M}_{n \times n}(\mathbb{Q})$ be the matrix defined by

$$a_{ij} = \begin{cases} c & \text{if } i = j, \\ d & \text{if } i + j = n + 1, \\ 0 & \text{otherwise.} \end{cases}$$

Calculate $|A|$.

Exercise 650
Let n be a positive integer and let $A = [a_{ij}] \in \mathcal{M}_{n \times n}(\mathbb{Q})$ be the tridiagonal matrix defined by

$$a_{ij} = \begin{cases} 1 & \text{if } |i - j| \leq 1, \\ 0 & \text{otherwise.} \end{cases}$$

Show that

$$|A| = \begin{cases} -1 & \text{if } n = 3k, \\ 1 & \text{if } n = 3k + 1, \\ 0 & \text{if } n = 3k + 2 \end{cases}$$

for some nonnegative integer k.

Exercise 651
Find $a, b, c \in \mathbb{Z}$ for which $\begin{vmatrix} a+b & c & c \\ a & b+c & a \\ b & b & a+c \end{vmatrix}$ is divisible by 8.

Exercise 652
Let n be a positive integer and let $A \in \mathcal{M}_{n \times n}(\mathbb{Q})$ be a nonsingular matrix satisfying the condition that all of the entries of A and of A^{-1} are integers. Show that $|A| = \pm 1$.

Exercise 653
For elements a, b, and c of a field F, calculate $\begin{vmatrix} -2a & a+b & a+c \\ a+b & -2b & b+c \\ a+c & b+c & -2c \end{vmatrix}$.

Exercise 654
We know that the integers 23028, 31882, 86469, 6327, and 61902 are all divisible by 19. Show that $\begin{vmatrix} 2 & 3 & 0 & 2 & 8 \\ 3 & 1 & 8 & 8 & 2 \\ 8 & 6 & 4 & 6 & 9 \\ 0 & 6 & 3 & 2 & 7 \\ 6 & 1 & 9 & 0 & 2 \end{vmatrix}$ is also divisible by 19.

Exercise 655

Let $q \in \mathbb{Q}$. Show that there are infinitely-many matrices in $\mathcal{M}_{3\times3}(\mathbb{Q})$ of the

form $\begin{bmatrix} 2 & 2 & 3 \\ 3q+2 & 4q+2 & 5q+3 \\ a & b & c \end{bmatrix}$, where $a < b < c$, the determinant of which

equals q.

Exercise 656

Let n be a positive integer and let F be a field. Let $A \in \mathcal{M}_{n\times n}(F)$ be a non-singular matrix which can be written in block form $A = \begin{bmatrix} A_{11} & A_{12} \\ A_{21} & A_{22} \end{bmatrix}$, where

$A_{11} \in \mathcal{M}_{k\times k}(F)$ for some integer $k < n$. Write A^{-1} as $\begin{bmatrix} B_{11} & B_{12} \\ B_{21} & B_{22} \end{bmatrix}$, where

$B_{11} \in \mathcal{M}_{k\times k}(F)$. Show that $|A_{11}| = |A| \cdot |B_{22}|$.

Exercise 657

Find all real numbers a for which $\begin{vmatrix} a & 1 & 1 & 1 \\ 1 & a & 2 & 3 \\ 0 & -1 & 0 & 1 \\ -1 & 1 & 1 & 2 \end{vmatrix} = 0$.

Exercise 658

Let $a_1, a_2, \dots$ be a sequence of real numbers. For each positive integer n, define the *nth continuant* c_n of the sequence to be the determinant of the tridiagonal matrix $A_n = [a_{ij}] \in \mathcal{M}_{n\times x}(\mathbb{R})$ given by

$$a_{ij} = \begin{cases} a_i & \text{if } i = j, \\ -1 & \text{if } i = j - 1, \\ 1 & \text{if } i = j + 1, \\ 0 & \text{otherwise.} \end{cases}$$

Show that $c_n = a_n c_{n-1} + c_{n-2}$ for all $n > 2$.

Exercise 659

Let $n > 1$ be an integer, let d be a real number, and let $A = [a_{ij}] \in \mathcal{M}_{n\times n}(\mathbb{R})$ be the matrix defined as follows:

$$a_{ij} = \begin{cases} 0 & \text{if } i = j, \\ 1 & \text{if } i > 1 \text{ and } j = 1 \text{ or } i = 1 \text{ and } j > 1, \\ d & \text{otherwise.} \end{cases}$$

Show that $|A| = (-1)^{n-1}(n-1)d^{n-2}$.

Exercise 660

Let $b_1, \ldots, b_n$ be nonzero real numbers and let $A = [a_{ij}] \in \mathcal{M}_{n \times n}(\mathbb{R})$ be the matrix defined as follows:

$$a_{ij} = \begin{cases} 1 + b_j & \text{if } i = j, \\ 1 & \text{otherwise.} \end{cases}$$

Calculate $|A|$.

Exercise 661

Let $A = [a_{ij}] \in \mathcal{M}_{4 \times 4}(\mathbb{Q})$ be a matrix each entry of which is either -2 or 3. Show that $|A|$ is an integer multiple of 125.

Exercise 662

Let $a, b, c,$ and d be real numbers not all of which are equal to 0. Show that the

matrix $\begin{bmatrix} a & b & c & d \\ b & -a & d & -c \\ c & -d & -a & b \\ d & c & -b & -a \end{bmatrix} \in \mathcal{M}_{4 \times 4}(\mathbb{R})$ is nonsingular.

Exercise 663

Does there exist a rational number a satisfying the condition that the matrix $\begin{bmatrix} 1 & a & 0 \\ a & 1 & 1 \\ -1 & a & -1 \end{bmatrix} \in \mathcal{M}_{3 \times 3}(\mathbb{Q})$ is nonsingular?

Exercise 664

Find all matrices $I \neq A \in \mathcal{M}_{2 \times 2}(\mathbb{R})$ satisfying $A^3 = I$.

Exercise 665

Find all triples (a, b, c) of real numbers satisfying the condition

$$\begin{vmatrix} 1 & a & a^3 \\ 1 & b & b^3 \\ 1 & c & c^3 \end{vmatrix} = (b - c)(c - a)(a - b)(a + b + c).$$

Exercise 666

Let n be a positive integer, let $A = [a_{ij}] \in \mathcal{M}_{n \times n}(\mathbb{C})$ and let $B = [b_{ij}] \in \mathcal{M}_{n \times n}(\mathbb{C})$ be defined by $b_{ij} = \bar{a}_{ji}$ for each $1 \leq i, j \leq n$. Show that $|AB|$ is a nonnegative real number.

Exercise 667

Calculate $\begin{vmatrix} 1 & \log_b a \\ \log_a b & 1 \end{vmatrix}$ for given positive real numbers a and b.

Exercise 668

Let F be a field. Calculate $\begin{vmatrix} 1 & a & a^2 & a^3 \\ a^3 & a^2 & a & 1 \\ 1 & 2a & 3a^2 & 4a^3 \\ 4a^3 & 3a^2 & 2a & 1 \end{vmatrix}$ for any $a \in F$.

Exercise 669

Calculate $\begin{vmatrix} \cos(a) & \sin(a) & \cos(a) & \sin(a) \\ \cos(2a) & \sin(2a) & 2\cos(2a) & 2\sin(2a) \\ \cos(3a) & \sin(3a) & 3\cos(3a) & 3\sin(3a) \\ \cos(4a) & \sin(4a) & 4\cos(4a) & 4\sin(4a) \end{vmatrix}$ for $a \in \mathbb{R}$.

Exercise 670

Let n be a positive integer and let $A = [a_{ij}] \in \mathcal{M}_{n \times n}(\mathbb{R})$ be the matrix defined by

$$a_{ij} = \begin{cases} 0 & \text{if } i = j, \\ 1 & \text{otherwise.} \end{cases}$$

Calculate $|A|$.

Exercise 671

Let $A = \begin{bmatrix} 3 & -1 & 1 \\ 0 & 2 & 4 \\ 1 & -1 & 1 \end{bmatrix} \in \mathcal{M}_{3 \times 3}(\mathbb{R})$. Calculate $\text{adj}(A)$.

Exercise 672

Let $F = \text{GF}(2)$ and let $A = \begin{bmatrix} 1 & 1 & 0 \\ 0 & 1 & 1 \\ 1 & 1 & 1 \end{bmatrix} \in \mathcal{M}_{3 \times 3}(F)$. Calculate $\text{adj}(A)$.

Exercise 673

Let F be a field, let n be a positive integer, and let $A, B \in \mathcal{M}_{n \times n}(F)$. Is it necessarily true that $\text{adj}(AB) = \text{adj}(A) \, \text{adj}(B)$?

Exercise 674

Let F be a field, let n be a positive integer, and let $A \in \mathcal{M}_{n \times n}(F)$. Is it necessarily true that $\text{adj}(A^T) = \text{adj}(A)^T$?

Exercise 675

Let F be a field, let n be a positive integer, and let the matrices $A, B \in \mathcal{M}_{n \times n}(F)$ be nonsingular. Show that $\text{adj}(B^{-1}AB) = B^{-1} \, \text{adj}(A)B$.

Exercise 676

Let F be a field, let n be a positive integer, and let $A, B \in \mathcal{M}_{n \times n}(F)$ be matrices satisfying $B \neq O$ and $AB = O$. Show that $|A| = 0$.

Exercise 677

Let $A = \begin{bmatrix} 1 & 2 & 3 \\ 1 & 3 & 4 \\ 1 & 4 & 3 \end{bmatrix} \in \mathcal{M}_{3\times3}(\mathbb{R})$. Use the adjoint of A to calculate A^{-1}.

Exercise 678

Let F be a field, let n be a positive integer, and let $A = [a_{ij}] \in \mathcal{M}_{n\times n}(F)$. Let $B = [b_{ij}] \in \mathcal{M}_{n\times n}(F)$ defined by $b_{ij} = (-1)^{i+j} a_{ij}$ for all $1 \leq i, j \leq n$. Show that $|A| = |B|$.

Exercise 679

Let F be a field, let n be a positive integer, and let $A = [a_{ij}] \in \mathcal{M}_{n\times n}(F)$. Let $B = [b_{ij}] \in \mathcal{M}_{n\times n}(F)$ defined by $b_{ij} = (-1)^{i+j+1} a_{ij}$ for all $1 \leq i, j \leq n$. Show that $(-1)^n |A| = |B|$.

Exercise 680

Let n be a positive integer and let $\pi \in S_n$. Let $A \in \mathcal{M}_{n\times n}(\mathbb{Q})$ be the permutation matrix defined by π. Calculate $|A|$.

Exercise 681

Is the set of all permutation matrices in $\mathcal{M}_{n\times n}(\mathbb{Q})$ closed under multiplication? Is the inverse of a permutation matrix a permutation matrix?

Exercise 682

Let $A = [a_{ij}] \in \mathcal{M}_{3\times3}(\mathbb{R})$ be a matrix in which $a_{i2} \neq 0$ for all $1 \leq i \leq 3$. Denote the minor of a_{ij} for all $1 \leq i, j \leq n$ by A_{ij}. Show that

$$|A| = \frac{1}{a_{12}} \begin{vmatrix} A_{21} & A_{23} \\ A_{31} & A_{32} \end{vmatrix} + \frac{1}{a_{22}} \begin{vmatrix} A_{11} & A_{13} \\ A_{31} & A_{33} \end{vmatrix} + \frac{1}{a_{32}} \begin{vmatrix} A_{11} & A_{13} \\ A_{21} & A_{22} \end{vmatrix}.$$

Exercise 683

Let F be a field, let n be a positive integer, and let $A = [a_{ij}] \in \mathcal{M}_{n\times n}(F)$ be nonsingular. Show that $\mathrm{adj}(\mathrm{adj}(A)) = |A|^{n-2} A$.

Exercise 684

Let a and b be real numbers and let n be an integer greater than 2. Let $D = [d_{ij}] \in \mathcal{M}_{n\times n}(\mathbb{R})$ be the matrix defined by $d_{ij} = \sin(ia + jb)$ for all $1 \leq i, j \leq n$. Show that $|D| = 0$.

Exercise 685

Let F be a field and let $a, b, c, d, e, f, g \in F$. Show that

$$\begin{vmatrix} a & b & b \\ c & d & e \\ f & g & g \end{vmatrix} + \begin{vmatrix} a & b & b \\ e & c & d \\ f & g & g \end{vmatrix} + \begin{vmatrix} a & b & b \\ d & e & c \\ f & g & g \end{vmatrix} = 0.$$

Exercise 686

Let F be a field, let n be a positive integer, and let $A = [a_{ij}] \in M_{n \times n}(F)$. Let $B = [b_{ij}] \in M_{n \times n}(F)$ be the matrix defined by

$$b_{ij} = \begin{cases} a_{ij} + a_{i,j+1} & \text{if } j < n, \\ a_{in} & \text{otherwise.} \end{cases}$$

Show that $|B| = |A|$.

Exercise 687

Let k and n be integers greater than 1. Let F be a field and let $A = [a_{ij}]$ be a matrix in $M_{k \times n}(F)$, the upper row of which contains at least one nonzero entry. For each $2 \le i \le k$ and each $2 \le j \le n$, let $d_{ij} = \begin{vmatrix} a_{11} & a_{1j} \\ a_{i1} & a_{ij} \end{vmatrix}$. Show that the rank of the matrix $D = \begin{bmatrix} d_{22} & \cdots & d_{2n} \\ \vdots & \ddots & \vdots \\ d_{k2} & \cdots & d_{kn} \end{bmatrix} \in M_{(k-1) \times (n-1)}(F)$ is $r - 1$, where r is the rank of A.

Exercise 688

Let F be a field, let $a \ne b$ be elements of F, and let $A, B \in M_{2 \times 2}(F)$ be matrices satisfying the condition that $|A + hB| \in \{a, b\}$ for $h = 1, 2, 3, 4, 5$. Show that $|A + 9B| \in \{a, b\}$.

Exercise 689

Let F be a field and let $a, b, c \in F$. Make use of the matrix $\begin{bmatrix} b & c & 0 \\ a & 0 & c \\ 0 & a & b \end{bmatrix}$ in order to calculate the determinant of the matrix $\begin{bmatrix} b^2 + c^2 & ab & ac \\ ab & a^2 + c^2 & bc \\ ac & bc & a^2 + b^2 \end{bmatrix}$.

Exercise 690

Let $A \in M_{n \times n}(\mathbb{C})$ be a nonsingular matrix, which we will write in the form $B + iC$, where $B, C \in M_{n \times n}(\mathbb{R})$. Show that there is a real number d such that the matrix $B + dC \in M_{n \times n}(\mathbb{R})$ is nonsingular.

Exercise 691

Let F be a field and let n be a positive integer. Let $A \in M_{n \times n}(F)$ be a matrix having the property that the sum of all even-numbered columns (considered as vectors in F^n) of A equals the sum of all odd-numbered columns of A. What is $|A|$?

Exercise 692

Let $V = \mathbb{R}^2$ and let $f : V^3 \to \mathbb{R}$ be the function defined as follows: if $v_i = \begin{bmatrix} a_i \\ b_i \end{bmatrix}$ for $i = 1, 2, 3$, then $f : \begin{bmatrix} v_1 \\ v_2 \\ v_3 \end{bmatrix} \mapsto \begin{vmatrix} a_1 & b_1 & 1 \\ a_2 & b_2 & 1 \\ a_3 & b_3 & 1 \end{vmatrix}$. Show that $f(v_1, v_2, v_3) = f(v_4, v_2, v_3) + f(v_1, v_4, v_3) + f(v_1, v_2, v_4)$ for all $v_1, v_2, v_3, v_4 \in V$.

Exercise 693

Let F be a field and let n be a positive integer. Let $D = [d_{ij}] \in \mathcal{M}_{n \times n}(F)$ be the matrix defined by $d_{ij} = 1$ for all $1 \leq i, j \leq n$. Show that for any matrix $A \in \mathcal{M}_{n \times n}(F)$ precisely one of the following conditions holds: (1) There is a unique scalar $a \in F$ such that $A + aD$ is singular; (2) $A + aD$ is singular for all scalars $a \in F$; (3) $A + aD$ is nonsingular for all scalars $a \in F$.

Exercise 694

Let $A, B, C, D \in \mathcal{M}_{2 \times 2}(\mathbb{R})$ and let M be the matrix $\begin{bmatrix} A & B \\ C & D \end{bmatrix} \in \mathcal{M}_{4 \times 4}(\mathbb{R})$. If all of the "formal determinants" $AD - BC, AD - CB, DA - BC$, and $DA - CB$ are nonsingular, is M necessarily nonsingular?

Exercise 695

Let $A, B, C, D \in \mathcal{M}_{2 \times 2}(\mathbb{R})$ and let M be the matrix $\begin{bmatrix} A & B \\ C & D \end{bmatrix} \in \mathcal{M}_{4 \times 4}(\mathbb{R})$. If M is a nonsingular matrix, is at least one of the "formal determinants" $AD - BC$, $AD - CB, DA - BC$, and $DA - CB$ also nonsingular?

Exercise 696

Let $n > 1$ be an integer and let $A = [a_{ij}] \in \mathcal{M}_{n \times n}(\mathbb{Q})$ be a matrix satisfying the condition that each a_{ij} is either equal to 1 or to -1. Show that $|A|$ is an integer multiple of 2^{n-1}.

Exercise 697

If a, b, c, d, e, f are nonzero elements of a field F, show that

$$\begin{vmatrix} 0 & a^2 & b^2 & c^2 \\ a^2 & 0 & f^2 & e^2 \\ b^2 & f^2 & 0 & d^2 \\ c^2 & e^2 & d^2 & 0 \end{vmatrix} = \begin{vmatrix} 0 & ad & be & cf \\ ad & 0 & cf & be \\ be & cf & 0 & ad \\ cf & be & ad & 0 \end{vmatrix}.$$

Exercise 698

Let n be a positive integer and let $c_1, \ldots, c_n$ be distinct real numbers transcendental over $\mathbb{Q}$. For $1 \leq h \leq n$, let $p_h(X) = \sum_{i=0}^{h-1} a_i X^i \in \mathbb{Q}[X]$ be a polynomial of degree $h - 1$. Let $A = [p_i(c_j)] \in \mathcal{M}_{n \times n}(\mathbb{R})$. Show that $|A| = (a_0 \cdots a_{n-1}) \prod_{i<j} (c_j - c_i)$.

Exercise 699

Let n be a positive integer and let $c_1, \ldots, c_n$ be distinct real numbers transcendental over $\mathbb{Q}$. For each $1 \le i, j \le n$, set $d_{ij} = c_i^j - c_i^{-j}$ and let $A = [d_{ij}] \in \mathcal{M}_{n \times n}(\mathbb{R})$. Show that $|A|$ equals $(c_1 \cdots c_n)^{-n} \prod_{i<j}[(c_i - c_j)(1 - c_i c_j)] \prod_{i=1}^{n}(c_i^2 - 1)$.

Exercise 700

Let a, b, c, d be elements of a field F. Solve the equation

$$\begin{vmatrix} a & b & c & d \\ b & c & d & a \\ c & d & a & b \\ d & a & b & c \end{vmatrix} = X \begin{vmatrix} 0 & 1 & -1 & 1 \\ 1 & c & d & a \\ 1 & d & a & b \\ 1 & a & b & c \end{vmatrix}.$$

Exercise 701

Let F be a field. Does there exist a matrix A in $\mathcal{M}_{3 \times 3}(F)$ satisfying the condition that the rank of $\mathrm{adj}(A)$ equals 2?

Exercise 702

Let a, b, and c be nonzero real numbers. Under which conditions does the equation $\begin{vmatrix} 0 & a-X & b-X \\ -a-X & 0 & c-X \\ -b-X & -c-X & 0 \end{vmatrix} = 0$ have more than one solution?

Exercise 703

Use determinants to show that there is no matrix $A \in \mathcal{M}_{4 \times 4}(\mathbb{Q})$ satisfying the condition that $A^4 = \begin{bmatrix} 1 & 0 & 0 & 0 \\ 0 & 2 & 0 & 0 \\ 0 & 0 & 1 & 0 \\ 0 & 0 & 0 & 1 \end{bmatrix}$.

Exercise 704

Let $A = [a_{ij}] \in \mathcal{M}_{n \times n}(\mathbb{R})$ be a matrix satisfying the condition $|a_{ii}| > \sum_{j \ne i} |a_{ij}|$ for all $1 \le i \le n$. Such matrices are called *strictly diagonally dominant*. Show that $|A| \ne 0$.

Exercise 705

Let F be a field and let $a, b, c \in F$. Is it true that

$$\begin{vmatrix} a & b & c & 0 \\ b & a & 0 & c \\ c & 0 & a & b \\ 0 & c & b & a \end{vmatrix} = \begin{vmatrix} -a & b & c & 0 \\ b & -a & 0 & c \\ c & 0 & -a & b \\ 0 & c & b & -a \end{vmatrix}?$$

Exercise 706

Let F be a field and let $n > 2$ be an integer. Give an example of a matrix $A \in M_{n \times n}(F)$ all of the entries in which are nonzero, satisfying $\mathrm{adj}(A) = O$.

Exercise 707

Let F be a field and let $n > 2$ be an integer. Show that $|\mathrm{adj}(A)| = |A|^{n-1}$ for all $A \in M_{n \times n}(F)$.

Exercise 708

Is the function $\mathrm{adj} : M_{2 \times 2}(\mathbb{R}) \to M_{2 \times 2}(\mathbb{R})$ epic?

Exercise 709

For real numbers s and t, let $A(s, t) = \begin{bmatrix} s & 0 & t \\ 1 & 1 & 1 \\ t & 0 & 1 \end{bmatrix}$ and let $B(s, t) = \mathrm{adj}(A(s, t))$.

Find the set of all real numbers s satisfying the condition that $|A(s, t)| \neq |B(s, t)|$ for all $t \in \mathbb{R}$.

Exercise 710

Let n be a positive integer and for all $1 \leq j \leq n$, let m_j be a positive integer. Define the matrix $A = [a_{ij}] \in M_{n \times n}(\mathbb{Q})$ by setting $a_{ij} = \binom{m_j + i - 1}{j - 1}$ for all $1 \leq i, j \leq n$. Calculate $|A|$.

Exercise 711

Let a and b be distinct elements of a field F and let n be a positive integer. Let $A(n) = [a_{ij}] \in M_{n \times n}(F)$ be the matrix defined by

$$a_{ij} = \begin{cases} a & \text{if } i = j, \\ b & \text{otherwise.} \end{cases}$$

Use induction on n to prove that $|A(n)| = [a + (n-1)b](a-b)^{n-1}$.

Exercise 712

Let n be a positive integer and pick integers $1 \leq h, k \leq n$. Let $f, g \in \mathbb{R}^{\mathbb{R}}$ be the functions defined by

$$f : c \mapsto \begin{cases} |E_{h,c}| & \text{if } c \neq 0, \\ 0 & \text{if } c = 0 \end{cases}$$

and $g : c \mapsto |E_{hk;c}|$. Are these functions continuous?

Exercise 713

Let n be a positive integer and let $A \in M_{n \times n}(\mathbb{Q})$ be a nonsingular matrix the entries of which are integers and the determinant of which is ± 1. Show that all of the entries of A^{-1} are integers.

Exercise 714

Let F be a field and let $A \in M_{2\times2}(F)$. Show that the matrix $A^2 + |A|I$ belongs to the subspace of $M_{2\times2}(F)$ generated by $\{A\}$.

Exercise 715

Let $A = [a_{ij}] \in M_{3\times3}(\mathbb{Q})$ be a matrix all of the entries of which are nonnegative one-digit integers. Let d be a positive integer dividing the three-digit integers $a_{11}a_{12}a_{13}$, $a_{21}a_{22}a_{23}$, and $a_{31}a_{32}a_{33}$. Show that d divides $|A|$.

Exercise 716

Let n be a positive integer and let F be a field. Let $A \in M_{n\times n}(F)$ be a matrix satisfying the condition that $|A + B| = |A|$ for all $B \in M_{n\times n}(F)$. Show that $A = O$.

Exercise 717

Let n be an odd positive integer let $A \in M_{n\times n}(\mathbb{R})$. Show that there exists a diagonal matrix B the diagonal entries of which are ±1 such that $A + B$ is nonsingular.

Exercise 718

Let $n > 1$ be an integer and let F be a field. Show that there exist subspaces W and Y of $M_{n\times n}(F)$ satisfying $M_{n\times n}(F) = W \oplus Y$ such that the restrictions of the determinant function δ_n to W and to Y are linear transformations.

Exercise 719

Let $n > 1$ be an integer and let B be the set of all of the nonsingular matrices in $M_{n\times n}(\mathbb{R})$ all of the entries of which are either 1 or 0. Show that in every matrix in B there are at least $n - 1$ entries which are equal to 0 and that there exists a matrix in B in which there are precisely $n - 1$ entries equal to 0.

Exercise 720

Let A be a matrix formed by permuting the rows or columns of an Hadamard matrix. Is A necessarily an Hadamard matrix?

Exercise 721

Let V be a vector space of finite dimension n over a field F and let $\{v_1, \ldots, v_n\}$ be a given basis for V. Let U be the subset of V consisting of all vectors of the form $y_a = \sum_{i=1}^{n} a^{i-1}v_i$, for $0 \neq a \in F$. Show that any subset of U having n elements is a basis for V.

Exercise 722

Let F be a field and let n be an even positive integer. Let $A \in M_{n\times n}(F)$ be a matrix which can be written in block form as $[A_{ij}]$, where $A_{ij} = \begin{bmatrix} 0 & c_i \\ -c_i & 0 \end{bmatrix}$ if $i = j$, and $A_{ij} = \begin{bmatrix} 0 & 0 \\ 0 & 0 \end{bmatrix}$ otherwise. Calculate the Pfaffian of A.

Exercise 723

Let F be a field and let n be an even positive integer. Let $c \in F$ and let $A \in \mathcal{M}_{2 \times n}(F)$ be a skew-symmetric matrix having Pfaffian d. What is the Pfaffian of cA?

Exercise 724

Let $c = \frac{1}{2}(1 + i\sqrt{3})$. Find the set of all real numbers a such that

$$\begin{vmatrix} a & 1 & 1 \\ 1 & c & c^2 \\ 1 & c^2 & c \end{vmatrix} \in \mathbb{R}.$$

Exercise 725

For elements a, b, c, d of a field F, calculate the value of $\begin{vmatrix} a & b & c & d \\ b & a & d & c \\ c & d & a & b \\ d & c & b & a \end{vmatrix}$.

Exercise 726

Let F be a field and let n be a positive integer. Set $V = \mathcal{M}_{n \times n}(F)$ and let $A, B \in V$ satisfy the condition that $|AB| = 1$. Then the function $\alpha : C \mapsto ACB$ is an endomorphism of V satisfying $|C| = |\alpha(C)|$ for all $C \in V$. Find an endomorphism of V satisfying the same condition, which is not of this form.

Exercise 727

Let F be a field and let n be a positive integer. For $A, B, C, D \in \mathcal{M}_{n \times n}(F)$, show that $\begin{vmatrix} A & B \\ C & D \end{vmatrix} = \begin{vmatrix} -C & -D \\ A & B \end{vmatrix}$.

Exercise 728

Let F be a field and let $A, B, C, D \in \mathcal{M}_{n \times n}(F)$ for some positive integer n. If $CD = DC$ and $|D| \neq 0$, show that $\begin{vmatrix} A & B \\ C & D \end{vmatrix} = |AD - BC|$.

Exercise 729

If F is a field and $A = [a_{ij}] \in \mathcal{M}_{n \times n}(F)$ for some positive integer n, then we define the *permanent* of A to be

$$\sum_{\pi \in S_n} a_{\pi(1),1} \bullet a_{\pi(2),2} \bullet \cdots \bullet a_{\pi(n),n}.$$

(i) Show that the permanent of A is the coefficient of $X_1 \cdots X_n$ in the polynomial $\prod_{i=1}^{n}(a_{i1}X_1 + \cdots + a_{in}X_n) \in F[X_1, \ldots, X_n]$.

(ii) If A is a permutation matrix, what is its permanent?

Exercise 730

Does there exist a matrix $A \in \mathcal{M}_{5 \times 5}(\mathbb{Q})$ the permanent of which equals 120?

Exercise 731

Let F be a field. For any matrix $A = [a_{ij}] \in \mathcal{M}_{2 \times 2}(F)$, let $U(A)$ be the set of all matrices $B \in \mathcal{M}_{2 \times 2}(F)$ satisfying $|A + B| = |A| + |B|$. Is $U(A)$ a subspace of $\mathcal{M}_{2 \times 2}(F)$?

Exercise 732

Let F be a field and let $A \in \mathcal{M}_{2 \times 2}(F)$. Find a necessary and sufficient condition for $|I + A| = 1 + |A|$ to hold.

Exercise 733

Let F be a field and let n be a positive integer. If $A, B, C, D \in \mathcal{M}_{n \times n}(F)$ with D nonsingular, show that $\left| \begin{bmatrix} A & B \\ C & D \end{bmatrix} \right| = |AD - BD^{-1}CD|$.

Exercise 734

Let $a, b, c \in \mathbb{Z}$. Find a positive integer n such that

$$\left| \begin{bmatrix} 2c & a+b+c & a+b+c \\ a+b+c & na & a+b+c \\ a+b+c & a+b+c & 2b \end{bmatrix} \right|$$

is divisible by abc.

Exercise 735

Let n be a positive integer and let $A \in \mathcal{M}_{n \times n}(\mathbb{Q})$ be a matrix all entries of which are integers and satisfying $|A| = \pm 1$. Show that all entries of A^{-1} are integers.

Exercise 736

Find the Padé approximant to $x \mapsto e^x$ of type 2/4.

Exercise 737

Let $a, b,$ and c be elements of a field F. Find an element x of F such that

$$\begin{vmatrix} 1 & a & a^2 \\ 1 & x & b^2 \\ 1 & c & c^2 \end{vmatrix} = \begin{vmatrix} 1 & -b-c & bc \\ 1 & -c-a & ca \\ 1 & -a-b & ab \end{vmatrix}.$$

Eigenvalues and Eigenvectors

One of the central problems in linear algebra is this: given a vector space V finitely generated over a field F, and given an endomorphism α of V, is there a way to select a basis B of V so that the matrix $\Phi_{BB}(\alpha)$ is as nice as possible? In this chapter, we will begin by defining some basic notions which will help us address this problem.

Let V be a vector space over a field F and let $\alpha \in \text{End}(V)$. A scalar $c \in F$ is an *eigenvalue* of α if and only if there exists a vector $v \neq 0_V$ satisfying $\alpha(v) = cv$. Such a vector is called an *eigenvector*[1] of α associated with the eigenvalue c. Thus we see that a nonzero vector $v \in V$ is an eigenvector of α if and only if the subspace Fv of V is invariant under α. Every eigenvector of α is associated with a unique eigenvalue of α but any eigenvalue has, as a rule, many eigenvectors associated with it. The set of all eigenvalues of α is called the *spectrum* of α and is denoted by $\text{spec}(\alpha)$. Thus, $c \in \text{spec}(\alpha)$ if and only if the endomorphism $c\sigma_1 - \alpha$ of V is not monic.

Example If V is a vector space over $\mathbb{R}$ and if $\alpha \in \text{End}(V)$ satisfies $\alpha^2 = -\sigma_1$, then $\text{spec}(\alpha) = \varnothing$. To see this, note that if v is an eigenvector corresponding to an eigenvalue c then $-v = \alpha^2(v) = c^2 v$ and so $(c^2 + 1)v = 0_V$, implying that $c^2 = -1$, which is impossible for a real number c. In particular, if $\alpha \in \text{End}(\mathbb{R}^2)$ is defined by $\alpha : \begin{bmatrix} a \\ b \end{bmatrix} \mapsto \begin{bmatrix} -b \\ a \end{bmatrix}$ then $\text{spec}(\alpha) = \varnothing$.

---

[1] The terms "eigenvalue" and "eigenvector" are due to Hilbert. Eigenvalues and eigenvectors are sometimes called *characteristic values* and *characteristic vectors*, respectively, based on terminology used by Cauchy. Sylvester coined the term "latent values" since, as he put it, such scalars are "latent in a somewhat similar sense as vapor may be said to be latent in water or smoke in a tobacco-leaf".

J.S. Golan, *The Linear Algebra a Beginning Graduate Student Ought to Know*, DOI 10.1007/978-94-007-2636-9_12, © Springer Science+Business Media B.V. 2012

Example Let $\alpha \in \text{End}(\mathbb{R}^2)$ be defined by $\alpha : \begin{bmatrix} a \\ b \end{bmatrix} \mapsto \begin{bmatrix} b \\ a \end{bmatrix}$. Then $c \in \text{spec}(\alpha)$ if and

only if there exists a vector $\begin{bmatrix} a \\ b \end{bmatrix}$ satisfying $\begin{bmatrix} b \\ a \end{bmatrix} = \begin{bmatrix} ca \\ cb \end{bmatrix}$. Therefore, we see that

$\text{spec}(\alpha) = \{-1, 1\}$, where $\begin{bmatrix} a \\ -a \end{bmatrix}$ is an eigenvector of α associated with -1 and

$\begin{bmatrix} a \\ a \end{bmatrix}$ is an eigenvector of α associated with 1, for any $0 \neq a \in \mathbb{R}$.

Example Let V be the vector space of all infinitely-differentiable functions from $\mathbb{R}$ to itself and let δ be the endomorphism of V which assigns to each such function its derivative. Then a function f, which is not the 0-function, is an eigenvector of δ if and only if there exists a scalar $c \in \mathbb{R}$ such that $\delta(f) = cf$. For any real number c, there is indeed such a function in V, namely the function $x \mapsto e^{cx}$. Thus $\text{spec}(\delta) = \mathbb{R}$. The set of all eigenvectors of δ associated with c is $\{ae^{cx} \mid a \neq 0\}$. This fact has important applications in the theory of differential equations.

The first use of eigenvalues to study differential equations is due to the French mathematician **Jean d'Alembert**, one of the foremost researchers of the eighteenth century. Important solutions of eigenvalue problems for second-order differential equations were obtained in the nineteenth century by Swiss mathematician **Charles-François Sturm** and French mathematician **Joseph Liouville**.

Let α be an endomorphism of a vector space V of a field F having an eigenvalue c. If $\beta \in \text{Aut}(V)$ then c is also an eigenvalue of $\beta\alpha\beta^{-1}$. Indeed, if v is an eigenvector of α associated with c then $\beta\alpha\beta^{-1}(\beta(v)) = \beta\alpha(v) = \beta(cv) = c\beta(v)$ and $\beta(v) \neq 0_V$ since β is an automorphism. Therefore, $\beta(v)$ is an eigenvector of $\beta\alpha\beta^{-1}$ associated with c.

Similarly, let $p(X) = \sum_{i=0}^n b_i X^i \in F[X]$. If $v \in V$ is an eigenvector of α associated with an eigenvalue c, then v is also an eigenvector of $p(\alpha) \in \text{End}(V)$ associated with the eigenvalue $p(c)$, since $p(\alpha)v = \sum_{i=0}^n b_i \alpha^i(v) = \sum_{i=0}^n b_i c^i v = p(c)v$. In particular, we see that, for any positive integer n, the vector v is an eigenvector of α^n associated with the eigenvalue c^n.

Let V be a vector space over a field F and let α be an endomorphism of V. A vector $v \in V$ is a *fixed point* of α if and only if $\alpha(v) = v$. It is clear that 0_V is a fixed point of every endomorphism of V and a nonzero vector v is a fixed point of α if and only if $1 \in \text{spec}(\alpha)$ and v is an eigenvalue of α associated with 1.

Proposition 12.1 *Let V be a vector space over a field F and let α be an endomorphism of V having an eigenvalue c. The subset W composed of 0_V and all eigenvectors of α associated to c is a subspace of V.*

Proof If $w, w' \in W$ and $a \in F$ then $\alpha(w + w') = \alpha(w) + \alpha(w') = cw + cw' = c(w + w')$ and $\alpha(aw) = a\alpha(w) = a(cw) = c(aw)$ and so $w + w', aw \in W$, proving that W is a subspace of V. □

Let V be a vector space over a field F and let α be an endomorphism of V having an eigenvalue c. The subset W composed of 0_V and all eigenvectors of α associated with c, which we know by Proposition 12.1 is a subspace of V, is called the *eigenspace* of α associated with c. In particular, if 1 is an eigenvalue of α then the *fixed space* of α is the eigenspace associated with 1. If $1 \notin \mathrm{spec}(\alpha)$ then the fixed space of α is taken to be $\{0_V\}$.

Example Define $\alpha \in \mathrm{End}(\mathbb{R}^3)$ by $\alpha : \begin{bmatrix} a \\ b \\ c \end{bmatrix} \mapsto \begin{bmatrix} a \\ 0 \\ c \end{bmatrix}$. Then $1 \in \mathrm{spec}(\alpha)$ and the

eigenspace of α associated with 1 (namely the fixed space of α) is $\mathbb{R} \left\{ \begin{bmatrix} 1 \\ 0 \\ 0 \end{bmatrix}, \begin{bmatrix} 0 \\ 0 \\ 1 \end{bmatrix} \right\}$.

Example Small errors in recording data may lead to considerable errors in the calculation of eigenspaces, even if the eigenvalues are calculated correctly. For example, let $a, b, c, e \in \mathbb{R}$ and let α and β be the endomorphisms of $\mathbb{R}^3$ represented with respect to some fixed basis by the matrices $\begin{bmatrix} a & 0 & 0 \\ 0 & b & 0 \\ 0 & 0 & c \end{bmatrix}$ and $\begin{bmatrix} a & e & e \\ 0 & b & e \\ 0 & 0 & c \end{bmatrix}$, respectively. Then $\mathrm{spec}(\alpha) = \mathrm{spec}(\beta) = \{a, b, c\}$. The eigenspaces of α associated with a,

b, c are $\mathbb{R} \begin{bmatrix} 1 \\ 0 \\ 0 \end{bmatrix}$, $\mathbb{R} \begin{bmatrix} 0 \\ 1 \\ 0 \end{bmatrix}$, and $\mathbb{R} \begin{bmatrix} 0 \\ 0 \\ 1 \end{bmatrix}$. The eigenspaces of β associated with a, b,

c are $\mathbb{R} \begin{bmatrix} 1 \\ 0 \\ 0 \end{bmatrix}$, $\mathbb{R} \begin{bmatrix} e \\ b-a \\ 0 \end{bmatrix}$, and $\mathbb{R} \begin{bmatrix} e(e+c-b) \\ e(c-a) \\ (c-a)(c-b) \end{bmatrix}$.

Example Let $V = C(0, 1)$ and let α be the endomorphism of V defined by $\alpha(f) : x \mapsto \int_0^1 \cos(\pi[x - t]) f(t)\, dt$ for all $f \in V$. To find the eigenvalues of α, recall the trigonometric identity

$$\cos(\pi[x - t]) = \cos(\pi x)\cos(\pi t) + \sin(\pi x)\sin(\pi t).$$

Using this identity, we see that if $f \in V$ then

$$\alpha(f) : x \mapsto \left[\int_0^1 \cos(\pi t) f(t) \, dt \right] \cos(\pi x) + \left[\int_0^1 \sin(\pi t) f(t) \, dt \right] \sin(\pi x)$$

and so the image of α is contained in the subspace $W = \mathbb{R}\{g_1, g_2\}$ of V, where $g_1 : x \mapsto \cos(\pi x)$ and $g_2 : x \mapsto \sin(\pi x)$. It is easy to see that $\alpha(g_1) = \frac{1}{2} g_1$ and $\alpha(g_2) = \frac{1}{2} g_2$, so both of these functions are eigenvectors of α associated with the eigenvalue $\frac{1}{2}$. Moreover, $\{g_1, g_2\}$ is linearly independent. Thus we see that $\mathrm{spec}(\alpha) = \{\frac{1}{2}\}$ and the eigenspace associated with this sole eigenvalue is W.

Proposition 12.2 *Let V be a vector space finitely generated over a field F and let α be an endomorphism of V. Then the following conditions on a scalar c are equivalent:*
(1) *c is an eigenvalue of α;*
(2) *$c \sigma_1 - \alpha \notin \mathrm{Aut}(V)$;*
(3) *If $A = \Phi_{BB}(\alpha)$ for some basis B of V, then $|cI - A| = 0$.*

Proof (1) $\Leftrightarrow$ (2): Condition (1) is satisfied if and only if there exists a nonzero vector $v \in V$ satisfying $\alpha(v) = cv$, i.e., if and only if $(c\sigma_1 - \alpha)(v) = 0_V$. This is true if and only if $\ker(c\sigma_1 - \alpha) \neq \{0_V\}$. Since V is finitely generated, by Proposition 7.3, we know that this is true if and only if condition (2) holds.

(2) $\Leftrightarrow$ (3): This is a direct consequence of the fact that a matrix is nonsingular if and only if its determinant is nonzero. $\square$

From Proposition 12.2, we see how to define eigenvalues of square matrices over a field: if F is a field and n is a positive integer, then $c \in F$ is an *eigenvalue* of a matrix $A \in \mathcal{M}_{n \times n}(F)$ if and only if $|cI - A| = 0$, namely if and only if the matrix $cI - A$ is singular. The set of all eigenvalues of A will be denoted by $\mathrm{spec}(A)$. In particular, we observe that a matrix A is nonsingular if and only if $0 \notin \mathrm{spec}(A)$.

A vector $\begin{bmatrix} 0 \\ \vdots \\ 0 \end{bmatrix} \neq v \in F^n$ is an *eigenvector* of A associated with the eigenvalue c if

and only if $Av = cv$. The subset of F^n consisting of $\begin{bmatrix} 0 \\ \vdots \\ 0 \end{bmatrix}$ and all eigenvectors of

A associated with c is a subspace of F^n called the *eigenspace* associated with c. In the case that F equals $\mathbb{R}$ or $\mathbb{C}$, the number $\rho(A) = \max\{|c| \mid c \in \mathrm{spec}(A)\}$ is called the *spectral radius* of the matrix A, and plays a very important part in the numerical analysis of matrices. Note that if $F = \mathbb{C}$, then $\rho(A)$ is just the radius of the smallest circle in the complex plane, centered at the origin, containing $\mathrm{spec}(A)$. Moreover, since $\mathrm{spec}(A)$ consists precisely of the poles of the function $z \mapsto |zI - A|^{-1}$, this observation allows the use of powerful techniques of complex analysis in the study of the spectra of complex matrices.

Calculating the spectra of matrices is a critical tool in many applications of mathematics. Thus, for example, in statistics one learns that finding the spectrum of covariance matrices is an integral part of several data analysis techniques.

Example It is not necessarily true that $\rho(AB) = \rho(A)\rho(B)$ for square matrices A and B. For example, if $A = \begin{bmatrix} 0 & 2 \\ 0 & 0 \end{bmatrix}$ and $B = \begin{bmatrix} 0 & 0 \\ 2 & 0 \end{bmatrix}$ in $\mathcal{M}_{2\times2}(\mathbb{R})$, then $\rho(A) = 0 = \rho(B)$, whereas $\rho(AB) = 4$.

Given a matrix $A \in \mathcal{M}_{n\times n}(F)$, we note that $|cI - A| = |(cI - A)^T| = |cI - A^T|$ and so $\text{spec}(A) = \text{spec}(A^T)$. However, for each such common eigenvalue, the associated eigenvectors may be different.

Example Let $A = \begin{bmatrix} 1 & 1 & -2 \\ -1 & 2 & 1 \\ 0 & 1 & -1 \end{bmatrix} \in \mathcal{M}_{3\times3}(\mathbb{R})$. Then $\text{spec}(A) = \{-1, 1, 2\}$ and so this is also $\text{spec}(A^T)$.

(1) The eigenspace of A associated with -1 is $\mathbb{R}\begin{bmatrix} 1 \\ 0 \\ 1 \end{bmatrix}$ and the eigenspace of A^T associated with -1 is $\mathbb{R}\begin{bmatrix} 1 \\ 2 \\ -7 \end{bmatrix}$;

(2) The eigenspace of A associated with 1 is $\mathbb{R}\begin{bmatrix} 3 \\ 2 \\ 1 \end{bmatrix}$ and the eigenspace of A^T associated with 1 is $\mathbb{R}\begin{bmatrix} -1 \\ 0 \\ 1 \end{bmatrix}$;

(3) The eigenspace of A associated with 2 is $\mathbb{R}\begin{bmatrix} 1 \\ 3 \\ 1 \end{bmatrix}$ and the eigenspace of A^T associated with 2 is $\mathbb{R}\begin{bmatrix} -1 \\ 1 \\ 1 \end{bmatrix}$.

It is interesting to note the following. Let F be a field and let n be a positive integer. If $v, w \in F^n$, then $v \wedge w = vw^T \in \mathcal{M}_{n\times n}(F)$ and $v \odot w = v^T w \in F$. Direct calculation then yields $(v \wedge w)v = (v \odot w)v$, showing that v is an eigenvector of $v \wedge w$ associated with the eigenvalue $v \odot w$.

Example Let n be a positive integer and let $A = [a_{uj}]$ be an $n \times n$ Markov matrix, which we will consider as an element of $\mathcal{M}_{n\times n}(\mathbb{C})$. We claim that $\rho(A) \leq 1$. Indeed, let $c \in \text{spec}(A)$ and let $v = \begin{bmatrix} b_1 \\ \vdots \\ b_n \end{bmatrix} \in \mathbb{C}^n$ be an eigenvector associated with c. Let

$1 \leq h \leq n$ satisfy the condition that $|b_i| \leq |b_h|$ for all $1 \leq i \leq n$. Then $Av = cv$ implies, in particular, that $\sum_{j=1}^{n} a_{hj}b_j = cb_h$ and so

$$|c| \cdot |b_h| = |cb_h| = \left| \sum_{j=1}^{n} a_{hj}b_j \right| \leq \sum_{j=1}^{n} a_{hj}|b_j| \leq \left(\sum_{j=1}^{n} a_{hj} \right) |b_h| = |b_h|.$$

Hence $|c| \leq 1$, as claimed.

Example Let n be a positive integer and let $A \in \mathcal{M}_{n \times n}(\mathbb{R})$ be a skew-symmetric matrix. We claim that $\text{spec}(A) \subseteq \{0\}$, with equality when n is odd. Indeed, let $c \in \text{spec}(A)$ and let $v \in \mathbb{R}^n$ be an eigenvector of A associated with c. Then $-A^T v = Av = cv$ and so $-A^T(Av) = -A^T(cv) = c(-A^T v) = c^2 v$. Therefore, $-(Av \odot Av) = -v^T A^T A v = c^2 v^T v = c^2(v \odot v)$. But if $y = \begin{bmatrix} b_1 \\ \vdots \\ b_n \end{bmatrix}$ is any vector

in $\mathbb{R}^n$, then $y \odot y = \sum_{i=1}^{n} b_i^2 \geq 0$, with equality if and only if $y = \begin{bmatrix} 0 \\ \vdots \\ 0 \end{bmatrix}$. Since

v is nonzero, we conclude that we must have $c^2 = 0$ and so $c = 0$. Therefore, $\text{spec}(A) \subseteq \{0\}$. If n is odd then, by the remark after Proposition 11.7, we know that A is singular and so $0 \in \text{spec}(A)$, establishing equality.

Example Let n be a positive integer and let $A \in \mathcal{M}_{n \times n}(\mathbb{C})$. If c is a nonzero eigenvalue of A and if $v \in \mathbb{C}^n$ is an eigenvector associated with c then, by Proposition 11.13, we know that $|A|v = \text{adj}(A)Av = c[\text{adj}(A)]v$ and so $[\text{adj}(A)]v = c^{-1}|A|v$. Thus v is also an eigenvector of $\text{adj}(A)$ associated with the eigenvalue $c^{-1}|A|$.

If F is a field, if n is a positive integer, and if $A \in \mathcal{M}_{n \times n}(F)$ is a matrix having eigenvalue c, then $|cI - A| = 0$ and so, by Proposition 11.13, we have $(cI - A)\text{adj}(cI - A) = O$, whence $A[\text{adj}(cI - A)] = c[\text{adj}(cI - A)]$. From this we conclude that each of the columns of $\text{adj}(cI - A)$ must belong to the eigenspace of A associated with c.

Example Let $A = \begin{bmatrix} 0 & 1 & 0 \\ 0 & 0 & 1 \\ 4 & -17 & 8 \end{bmatrix} \in \mathcal{M}_{3 \times 3}(\mathbb{R})$. Then one can calculate that

$\text{spec}(A) = \{2 - \sqrt{3}, 2 + \sqrt{3}, 4\}$. Moreover, $\text{adj}(4I - A) = \begin{bmatrix} 1 & -4 & 1 \\ 4 & -16 & 4 \\ 16 & -64 & 16 \end{bmatrix}$ and it is easy to check that the columns of this matrix are indeed eigenvectors of A associated with 4.

Proposition 12.3 *If V is a vector space finitely generated over a field F and if $\alpha, \beta \in \mathrm{End}(V)$ then $\mathrm{spec}(\alpha\beta) = \mathrm{spec}(\beta\alpha)$.*

Proof Let $c \in \mathrm{spec}(\alpha\beta)$. If $c = 0$, this means that $\alpha\beta \notin \mathrm{Aut}(V)$. Therefore, either α or β is not an automorphism of V, and so $\beta\alpha \notin \mathrm{Aut}(V)$ as well. Therefore, we can assume $c \neq 0$. Let v be an eigenvector of $\alpha\beta$ associated with c and let $w = \beta(v)$. Then $\alpha(w) = \alpha\beta(v) = cv \neq 0_V$ and so $w \neq 0_V$. Moreover, $\beta\alpha(w) = \beta\alpha\beta(v) = \beta(cv) = c\beta(v) = cw$ and so w is an eigenvector of $\beta\alpha$ associated with c. Thus $\mathrm{spec}(\alpha\beta) \subseteq \mathrm{spec}(\beta\alpha)$. A similar argument shows the reverse inclusion, and so we have equality.   □

In particular, as a consequence of Proposition 12.3, we see that if F is a field, if n is a positive integer, and if $A, B \in \mathcal{M}_{n \times n}(F)$ then $\mathrm{spec}(AB) = \mathrm{spec}(BA)$.

As we noted at the beginning of the chapter, if we are given a vector space V finitely generated over a field F and an endomorphism α of V, we would like to find, to the extent possible, a basis B of V such that the matrix $\Phi_{BB}(\alpha)$ is nice, in the sense that it is amenable to quick and accurate calculations. Let V be a vector space over a field F (not necessarily finitely generated) and let $\alpha \in \mathrm{End}(V)$. Then α is *diagonalizable* if and only if there exists a basis B of V composed of eigenvectors of α.

Example We have already seen that the set B of all functions in $\mathbb{R}^{\mathbb{R}}$ of the form $x \mapsto e^{ax}$, for some $a \in \mathbb{R}$, is linearly independent. Therefore, $W = \mathbb{R}B$ is a subspace of $\mathbb{R}^{\mathbb{R}}$ which is not finitely generated, and B is a basis for W. Let α be the endomorphism of W which assigns to each $f \in W$ its derivative. Since each element of B is an eigenvector of α, we see that α is diagonalizable.

The following result characterizes the diagonalizable endomorphisms of finitely-generated vector spaces.

Proposition 12.4 *Let V be a vector space finitely generated over a field F and let $\alpha \in \mathrm{End}(V)$. Then the following conditions on a basis $B = \{v_1, \ldots, v_n\}$ are equivalent:*
(1) v_i is an eigenvector of α for each $1 \leq i \leq n$;
(2) $\Phi_{BB}(\alpha)$ is a diagonal matrix.

Proof (1) $\Rightarrow$ (2): By (1), we know that for each $1 \leq i \leq n$ there exists a scalar c_i satisfying $\alpha(v_i) = c_i v_i$ and so, by definition, $\Phi_{BB}(\alpha)$ is the diagonal matrix $[a_{ij}]$

given by

$$a_{ij} = \begin{cases} c_i & \text{if } i = j, \\ 0 & \text{otherwise.} \end{cases}$$

$(2) \Rightarrow (1)$: If $\Phi_{BB}(\alpha) = [a_{ij}]$ is a diagonal matrix then for each $1 \le i \le n$ we have $\alpha(v_i) = a_{ii}v_i$ and so v_i is an eigenvector of α for each $1 \le i \le n$. $\qquad \square$

Let V be a vector space over a field F and let $\alpha \in \text{End}(V)$. If B is a basis of V made up of eigenvectors of α then, as we have seen above, the elements of B are also eigenvectors of $p(\alpha)$ for any polynomial $p(X) \in F[X]$. We need not stick to polynomials: suppose that each $v \in B$ is an eigenvector of α associated with an eigenvalue c_v of α. Given any function whatsoever $f : \text{spec}(\alpha) \to F$, we can define the endomorphism $f(\alpha)$ of V by setting $f(\alpha) : \sum_{v \in B} a_v v \mapsto \sum_{v \in B} a_v f(c_v) v$ and the elements of B are also eigenvectors of $f(\alpha)$. We note that if f and g are functions from $\text{spec}(\alpha)$ to F then $f(\alpha)g(\alpha) = g(\alpha)f(\alpha)$.

Now assume that V is finitely generated over F and that $B = \{v_1, \dots, v_n\}$ is a basis of V made up of eigenvectors of $\alpha \in \text{End}(V)$. For each $1 \le i \le n$, let c_i be the eigenvalue of α associated with v_i. We have already seen that for each such i there exists a polynomial $p_i(X)$, namely the Lagrange interpolation polynomial, satisfying the condition that

$$p_i(c_j) = \begin{cases} 1 & \text{if } i = j, \\ 0 & \text{otherwise.} \end{cases}$$

Thus, given a function $f : \text{spec}(\alpha) \to F$, the polynomial $p(X) = \sum_{i=1}^n f(c_i)p_i(X)$ satisfies $p(c_i) = f(c_i)$ for all $1 \le i \le n$, and so $p(\alpha) = f(\alpha)$. Thus, for finitely-generated vector spaces, the above generalization does not in fact contribute anything new; it is important, however, in the case of vector spaces which are not finitely generated.

We now show that the size of the spectrum of an endomorphism of a finitely-generated vector space is limited.

> **Proposition 12.5** *Let V be a vector space over a field F and let $\alpha \in \text{End}(V)$. If $c_1, \dots, c_k$ are distinct eigenvalues of α and if v_i is an eigenvector of α associated with c_i for each $1 \le i \le k$, then the set $\{v_1, \dots, v_k\}$ is linearly independent.*

Proof Assume that the set $\{v_1, \dots, v_k\}$ is linearly dependent. Since $v_1 \ne 0_V$, we know that the set $\{v_1\}$ is linearly independent. Thus there exists an integer $1 \le t < k$ such that the set $\{v_1, \dots, v_t\}$ is linearly independent but $\{v_1, \dots, v_{t+1}\}$ is linearly dependent. In other words, there exist scalars $a_1, \dots, a_{t+1}$, not all of which are equal to 0, such that $\sum_{i=1}^{t+1} a_i v_i = 0_V$ and so $0_V = c_{t+1}(\sum_{i=1}^{t+1} a_i v_i) = \sum_{i=1}^{t+1} a_i c_{t+1} v_i$. On the other hand, $0_V = \alpha(\sum_{i=1}^{t+1} a_i v_i) = \sum_{i=1}^{t+1} a_i \alpha(v_i) = \sum_{i=1}^{t+1} a_i c_i v_i$. Therefore, $0_V = \sum_{i=1}^{t+1} a_i c_i v_i - \sum_{i=1}^{t+1} a_i c_{t+1} v_i = \sum_{i=1}^{t} a_i (c_i - c_{t+1}) v_i$. But the set $\{v_1, \dots, v_t\}$

is linearly independent and so $a_i(c_i - c_{t+1}) = 0$ for all $1 \leq i \leq t$. Since, by assumption, $c_i \neq c_{t+1}$ for all $1 \leq i \leq t$, we have $a_i = 0$ for all $1 \leq i \leq t$ and hence $a_{t+1} = 0$ as well, which is a contradiction. Thus $\{v_1, \ldots, v_k\}$ must be linearly independent. $\square$

Thus we see that if F is a field and if $A \in M_{n \times n}(F)$, then $\mathrm{spec}(A)$ can have at most n elements. In particular, if F has more than n elements, then there exists an element $c \in F \smallsetminus \mathrm{spec}(A)$, and so $(cI - A)v \neq \begin{bmatrix} 0 \\ \vdots \\ 0 \end{bmatrix}$ for all $v \neq \begin{bmatrix} 0 \\ \vdots \\ 0 \end{bmatrix}$. This implies that $cI - A$ is nonsingular.

From Proposition 12.5, we see that if α is a an endomorphism of a vector space V over a field F having distinct eigenvalues $c_1, \ldots, c_k$, and if W_i is the eigenspace associated with c_i for all $1 \leq i \leq t$, then the collection $\{W_1, \ldots, W_k\}$ of subspaces of V is independent. Moreover, if V is finitely generated over F then the number of elements in $\mathrm{spec}(\alpha)$ is no greater than $\dim(V)$.

Proposition 12.6 *Let V be a vector space of finite dimension n over a field F. Then any endomorphism α of V having n distinct eigenvalues is diagonalizable.*

Proof This is a direct consequence of Proposition 12.4 and Proposition 12.5. $\square$

Example Let $\alpha \in \mathrm{End}(\mathbb{R}^2)$ be defined by $\alpha : \begin{bmatrix} a \\ b \end{bmatrix} \mapsto \begin{bmatrix} 3a - b \\ 3b - a \end{bmatrix}$. Then $\alpha\left(\begin{bmatrix} 1 \\ 1 \end{bmatrix}\right) = \begin{bmatrix} 2 \\ 2 \end{bmatrix}$ and so $\begin{bmatrix} 1 \\ 1 \end{bmatrix}$ is an eigenvector of α associated with the eigenvalue 2. Also, $\alpha\left(\begin{bmatrix} 1 \\ -1 \end{bmatrix}\right) = \begin{bmatrix} 4 \\ -4 \end{bmatrix}$ and so $\begin{bmatrix} 1 \\ -1 \end{bmatrix}$ is an eigenvector of α associated with the eigenvalue 4. Thus $B = \left\{ \begin{bmatrix} 1 \\ 1 \end{bmatrix}, \begin{bmatrix} 1 \\ -1 \end{bmatrix} \right\}$ is a basis for $\mathbb{R}^2$ and $\Phi_{BB}(\alpha) = \begin{bmatrix} 2 & 0 \\ 0 & 4 \end{bmatrix}$.

Example Let $\alpha \in \mathrm{End}(\mathbb{R}^2)$ be defined by $\alpha : \begin{bmatrix} a \\ b \end{bmatrix} \mapsto \begin{bmatrix} a+b \\ b \end{bmatrix}$. If $\begin{bmatrix} a \\ b \end{bmatrix} \neq \begin{bmatrix} 0 \\ 0 \end{bmatrix}$ and $\alpha\left(\begin{bmatrix} a \\ b \end{bmatrix}\right) = c\begin{bmatrix} a \\ b \end{bmatrix}$ then $cb = b$ and $a + b = ca$, and this can happen only when $b = 0$ and $c = 1$. Thus $\mathrm{spec}(\alpha) = \{1\}$ and the eigenspace associated with this sole eigenvalue is $\mathbb{R}\begin{bmatrix} 0 \\ 1 \end{bmatrix}$. Since this is not all of $\mathbb{R}^2$, we know that there is no basis of $\mathbb{R}^2$ made up of eigenvectors of α, and hence α is not diagonalizable.

Note that the converse of Proposition 12.6 is false, as we easily see by taking $\alpha = \sigma_1$.

From the above, we know that if $A \in \mathcal{M}_{n \times n}(\mathbb{R})$ then the matrix has at most n distinct eigenvalues. However, it may have many fewer than that. If we assume that the entries of this matrix were chosen independently and randomly from a standard normal distribution, how many distinct eigenvalues should we expect? American mathematicians Alan Edelman, Eric Kostlan, and Michael Shub have shown that if ε_n denotes the mathematical expectancy for the number of eigenvalues of such a matrix in $\mathbb{R}$, then $\lim_{n \to \infty} \frac{1}{\sqrt{n}} \varepsilon_n = \sqrt{\frac{2}{\pi}}$. The situation over the complex numbers is quite different. Given a matrix $A \in \mathcal{M}_{n \times n}(\mathbb{C})$ one can, with probability 1, pick a matrix $B \in \mathcal{M}_{n \times n}(\mathbb{C})$ as near to A as we wish, which has n distinct eigenvalues in $\mathbb{C}$.

If F is a field, if n is a positive integer, and if $A \in \mathcal{M}_{n \times n}(F)$, then we can consider the matrix of polynomials $XI - A \in \mathcal{M}_{n \times n}(F[X])$. The determinant of this matrix, $|XI - A|$, is a polynomial in $F[X]$ called the *characteristic polynomial* of A. Note that this polynomial is always monic and of degree n.

Example The characteristic polynomial of $\begin{bmatrix} 1 & -1 & 0 \\ 2 & 1 & 5 \\ 4 & 2 & 1 \end{bmatrix} \in \mathcal{M}_{3 \times 3}(\mathbb{R})$ is $X^3 - 3X^2 - 5X + 27$.

Example The characteristic polynomial of $A = \begin{bmatrix} 1 & 2 & 1 & 2 \\ 0 & 1 & 2 & 3 \\ 3 & 2 & 1 & 1 \\ 1 & 1 & 2 & 0 \end{bmatrix} \in \mathcal{M}_{4 \times 4}(\mathbb{R})$ is $X^4 - 3X^3 - 11X^2 - 25X - 15$. If we sketch the graph of the polynomial function $t \mapsto t^4 - 3t^3 - 11t^2 - 25t - 15$, we see that it has real roots in the neighborhoods of -0.8 and 5.8. (More precisely, they are approximately equal to -0.8062070604 and 5.7448832706.) These are the only real eigenvalues of the matrix A.

Example Let $F = GF(3)$. The characteristic polynomial of $A = \begin{bmatrix} 1 & 1 & 1 & 1 \\ 2 & 0 & 1 & 0 \\ 0 & 1 & 1 & 0 \\ 1 & 1 & 1 & 0 \end{bmatrix} \in \mathcal{M}_{4 \times 4}(F)$ equals $X^4 + X^3 + 1 = (X + 2)(X^3 + 2X^2 + 2X + 2)$ and so A has only one eigenvalue, namely 1.

Example The characteristic polynomial of $A = \begin{bmatrix} 5 & 4 & 2 \\ 4 & 5 & 2 \\ 2 & 2 & 2 \end{bmatrix}$ in $\mathcal{M}_{3 \times 3}(\mathbb{Q})$ is $(X - 10)(X - 1)^2$ and so $\text{spec}(A) = \{1, 10\}$. The eigenspace of A associated with 10 is $\mathbb{Q} \begin{bmatrix} 2 \\ 2 \\ 1 \end{bmatrix}$, while the eigenspace of A associated with 1 is $\mathbb{Q} \left\{ \begin{bmatrix} -1 \\ 1 \\ 0 \end{bmatrix}, \begin{bmatrix} -1 \\ 0 \\ 2 \end{bmatrix} \right\}$.

Example Let $F = \mathrm{GF}(2)$ and let $A = \begin{bmatrix} 0 & 1 \\ 1 & 1 \end{bmatrix} \in \mathcal{M}_{2 \times 2}(F)$. The characteristic polynomial of A is $p(X) = X^2 + X + 1$ and, since $p(0) = p(1) = 1$ we see that $\mathrm{spec}(A) = \varnothing$. In fact, it is possible to show that for every prime integer p there is a symmetric 2×2 matrix A over $\mathrm{GF}(p)$ satisfying $\mathrm{spec}(A) = \varnothing$. Later, we will show that any symmetric matrix over $\mathbb{R}$ must have an eigenvalue.

Example Let α be the endomorphism of $\mathbb{C}^2$ represented with respect to the canonical basis by the matrix $A = \begin{bmatrix} 1 + i & 1 \\ 1 & 1 - i \end{bmatrix} \in \mathcal{M}_{2 \times x}(\mathbb{C})$. The characteristic polynomial of A is $(X - 1)^2$ and so $\mathrm{spec}(A) = 1$ The eigenspace associated with it is $\mathbb{C}\begin{bmatrix} i \\ 1 \end{bmatrix}$, which has dimension 1. Therefore, α is not diagonalizable.

Proposition 12.7 *Let F be a field and let n be a positive integer. If $A \in M_{n \times n}(F)$ has characteristic polynomial $p(X) = \sum_{i=0}^{n} a_i X^i$, then $|A| = (-1)^n a_0$.*

Proof We note that $a_0 = p(0) = |0I - A| = |-A| = (-1)^n |A|$ and so $|A| = (-1)^n a_0$. $\qquad \square$

The speed with which we can compute the characteristic polynomial of a matrix depends on the speed with which we can multiply two matrices. In 1985, Swiss computer scientist Walter Keller-Gehrig showed that if we can multiply two $n \times n$ matrices over a field F in an order of n^c operations, then we can calculate the characteristic polynomial of an $n \times n$ matrix over F in an order of $n^c \log(n)$ operations. In 2007, French mathematician Clément Pernet and German/Canadian computer scientist Arne Storjohann constructed a new algorithm with an expected cost on the order of n^c, provided that the field F has at least $2n^2$ elements. If one has the use of a computer with n^3 parallel processors, then much faster computation times can be obtained.

Any monic polynomial in $F[X]$ of positive degree is the characteristic polynomial of some square matrix over F. To see this, consider a polynomial $p(X) = \sum_{i=0}^{n} a_i X^i$, for $n > 0$. If $p(X)$ is monic, define the *companion matrix* of $p(X)$, denoted by $\mathrm{comp}(p) \in \mathcal{M}_{n \times n}(F)$, to be the matrix $[a_{ij}]$ given by

$$a_{ij} = \begin{cases} 1 & \text{if } i = j + 1 \text{ and } j < n, \\ -a_{i-1} & \text{if } j = n, \\ 0 & \text{otherwise.} \end{cases}$$

Otherwise, define $\mathrm{comp}(p)$ to be $\mathrm{comp}(a_n^{-1} p)$.

Companion matrices were first studied at the beginning of the twentieth century by German mathematician **Alfred Loewy**. The term was first introduced by the twentieth-century American mathematician **Cyrus Macduffee**.

Proposition 12.8 *Let F be a field and let n be a positive integer. If $p(X) = \sum_{i=0}^{n} a_i X^i \in F[X]$ is monic, then $p(X)$ is the characteristic polynomial of $\mathrm{comp}(p) \in M_{n \times n}(F)$.*

Proof We will proceed by induction on n. For $n = 1$, the result is immediate. If $n = 2$ and if $p(X) = X^2 + a_1 X + a_0$, then $\mathrm{comp}(p) = \begin{bmatrix} 0 & -a_0 \\ 1 & -a_1 \end{bmatrix}$ and so the characteristic polynomial of $\mathrm{comp}(p)$ is $\begin{vmatrix} X & a_0 \\ -1 & X + a_1 \end{vmatrix} = p(X)$ and we are done. Assume now that $n > 2$ and the result has been established for $n - 1$. Then the characteristic polynomial of $\mathrm{comp}(p)$ is $\begin{vmatrix} X & 0 & \cdots & a_0 \\ -1 & X & \cdots & a_1 \\ & \ddots & \ddots & \vdots \\ 0 & \cdots & -1 & X + a_{n-1} \end{vmatrix}$. By Proposition 11.11, this equals $X|\mathrm{comp}(q)| + a_0(-1)^{n-1}|B|$, where $q(X) = \sum_{i=0}^{n-1} a_{i+1} X^i$ and where $B \in M_{(n-1) \times (n-1)}(F)$ is an upper-triangular matrix with diagonal entries all equal to -1. Thus $|B| = (-1)^{n-1}$ and, by the induction hypothesis, $|\mathrm{comp}(q)| = q(X)$. Thus the characteristic polynomial of $\mathrm{comp}(p)$ is $Xq(X) + a_0 = p(X)$, as desired. $\qquad \square$

Let F be a field and let n be a positive integer. Every nonsingular matrix $P \in M_{n \times n}(F)$ defines a function ω_P from $M_{n \times n}(F)$ to itself given by $\omega_P : A \mapsto P^{-1} A P$. In fact, $\omega_P \in \mathrm{Aut}(M_{n \times n}(F))$, where $\omega_P^{-1} = \omega_{P^{-1}}$. This is an automorphism of F-algebras and, indeed, it can be shown that every automorphism of unital F-algebras in $\mathrm{Aut}(M_{n \times n}(F))$ is of this form. Therefore, the set of all automorphisms of the form ω_P is a group of automorphisms of $M_{n \times n}(F)$ and so defines an equivalence relation $\sim$ by setting $A \sim B$ if and only if $B = P^{-1} A P$. In this case, we say that the matrices A and B are *similar*. From what we have already seen, two matrices in $M_{n \times n}(F)$ are similar if and only if they represent the same endomorphism of an n-dimensional vector space over F with respect to different bases. One of the problems before us is to decide, given two square matrices of the same size, if they are similar or not.

Note that if a matrix $A \in M_{n \times n}(F)$ is similar to O, then it must equal O. Indeed, if $P^{-1} A P = O$ then $A = (PP^{-1})A(PP^{-1}) = P(P^{-1}AP)P^{-1} = POP^{-1} = O$.

Example In $\mathcal{M}_{3\times3}(\mathbb{Q})$, the matrices

$$A = \begin{bmatrix} 20 & 10 & 10 \\ 10 & 0 & 10 \\ 10 & 10 & 10 \end{bmatrix} \quad \text{and} \quad B = \begin{bmatrix} 80 & 130 & 100 \\ 10 & 10 & 10 \\ -50 & -80 & -60 \end{bmatrix}$$

are similar, since $B = P^{-1}AP$, where $P = \begin{bmatrix} 1 & 2 & 1 \\ 1 & 0 & 1 \\ 2 & 3 & 3 \end{bmatrix}$. Thus we note that a symmetric matrix may be similar to a matrix which is not symmetric.

Example The matrices $A = \begin{bmatrix} 1 & 0 & 0 \\ -1 & 1 & 1 \\ -1 & 0 & 2 \end{bmatrix}$ and $B = \begin{bmatrix} 1 & 1 & 0 \\ 0 & 1 & 0 \\ 0 & 0 & 2 \end{bmatrix}$ in $\mathcal{M}_{3\times3}(\mathbb{Q})$ are not similar since, were they similar, the matrices $A - I$ and $B - I$ would also be similar, and thus have the same rank. But it is easy to see that the rank of $A - I$ equals 1, while the rank of $B - I$ equals 2.

Example If matrices $A, B \in \mathcal{M}_{n\times n}(F)$ are similar, it does not follow that they commute. For example, let $A = \begin{bmatrix} 1 & 0 & -1 \\ 2 & 3 & 0 \\ -1 & 0 & -2 \end{bmatrix} \in \mathcal{M}_{3\times3}(\mathbb{R})$. Then $P = \begin{bmatrix} 1 & 0 & 0 \\ 0 & 1 & 0 \\ 0 & 1 & 1 \end{bmatrix}$ is nonsingular and so $B = PAP^{-1} = \begin{bmatrix} 1 & 1 & -1 \\ 2 & 3 & 0 \\ 1 & 5 & -2 \end{bmatrix}$ is similar to A. However, $AB \neq BA$.

Example Let F be a field and, for each $1 \leq h \leq t$, let A_h be a square matrix over F, which is similar to a square matrix B_h over F. That is to say, there exists a nonsingular square matrix P_h such that $B_h = P_h A_h P_h^{-1}$. Let A be the matrix in block form $\begin{bmatrix} A_1 & O & \cdots & O \\ O & A_2 & \cdots & O \\ \vdots & \vdots & \ddots & \vdots \\ O & O & \cdots & A_t \end{bmatrix}$ in which all blocks not on the diagonal are equal to O, and

let $B = \begin{bmatrix} B_1 & O & \cdots & O \\ O & B_2 & \cdots & O \\ \vdots & \vdots & \ddots & \vdots \\ O & O & \cdots & B_t \end{bmatrix}$. Then A is similar to B, since $B = PAP^{-1}$, where

$P = \begin{bmatrix} P_1 & O & \cdots & O \\ O & P_2 & \cdots & O \\ \vdots & \vdots & \ddots & \vdots \\ O & O & \cdots & P_t \end{bmatrix}$. We will make us of this fact in the next chapter.

Proposition 12.9 *Let F be a field and let $k < n$ be positive integers. Let $A \in M_{n \times n}(F)$ be a matrix which can be written in block form as $A = \begin{bmatrix} A_{11} & A_{12} \\ O & A_{22} \end{bmatrix}$, where $A_{11} \in M_{k \times k}(F)$ and $A_{22} \in M_{(n-k) \times (n-k)}(F)$. Then $\mathrm{spec}(A) = \mathrm{spec}(A_{11}) \cup \mathrm{spec}(A_{22})$.*

Proof Let $c \in \mathrm{spec}(A)$ and let $v \in F^n$ be an eigenvector associated with c. Write $v = \begin{bmatrix} v_1 \\ v_2 \end{bmatrix}$, where $v_1 \in F^k$ and $v_2 \in F^{n-k}$. Then

$$\begin{bmatrix} A_{11}v_1 + A_{12}v_2 \\ A_{22}v_2 \end{bmatrix} = \begin{bmatrix} A_{11} & A_{12} \\ O & A_{22} \end{bmatrix} \begin{bmatrix} v_1 \\ v_2 \end{bmatrix} = Av = cv = \begin{bmatrix} cv_1 \\ cv_2 \end{bmatrix}.$$

From this we see immediately that if $v_2 \neq \begin{bmatrix} 0 \\ \vdots \\ 0 \end{bmatrix}$ then $c \in \mathrm{spec}(A_{22})$, while if $v_2 = \begin{bmatrix} 0 \\ \vdots \\ 0 \end{bmatrix}$ then $c \in \mathrm{spec}(A_{11})$. Therefore, $\mathrm{spec}(A)$ is contained in $\mathrm{spec}(A_{11}) \cup \mathrm{spec}(A_{22})$.

Conversely, let $c \in \mathrm{spec}(A_{11})$ and let $v_1 \in F^k$ be an eigenvector associated with c. Then $A \begin{bmatrix} v_1 \\ O \end{bmatrix} = \begin{bmatrix} A_{11}v_1 \\ O \end{bmatrix} = c \begin{bmatrix} v_1 \\ O \end{bmatrix}$, proving that $c \in \mathrm{spec}(A)$. Now assume that $d \in \mathrm{spec}(A_{22}) \smallsetminus \mathrm{spec}(A_{11})$ and let $v_2 \in F^{n-k}$ be an eigenvector associated with d. Since $d \notin \mathrm{spec}(A_{11})$, we know that the matrix $B = A_{11} - dI \in M_{k \times k}(F)$ is nonsingular. Set $v_1 = B^{-1}A_{12}(-v_2)$. Then $(A - dI) \begin{bmatrix} v_1 \\ v_2 \end{bmatrix} = \begin{bmatrix} Bv_1 + A_{12}v_2 \\ (A_{22} - dI)v_2 \end{bmatrix} = \begin{bmatrix} 0 \\ \vdots \\ 0 \end{bmatrix}$, showing that $d \in \mathrm{spec}(A)$. Therefore, $\mathrm{spec}(A_{11}) \cup \mathrm{spec}(A_{22}) \subseteq \mathrm{spec}(A)$, proving equality. $\qquad\square$

Example Let $A = \begin{bmatrix} 1 & 1 & 5 & 6 \\ -1 & 1 & 7 & 3 \\ 0 & 0 & 2 & 1 \\ 0 & 0 & -4 & 3 \end{bmatrix} \in M_{4 \times 4}(\mathbb{C})$. Then

$$\mathrm{spec}(A) = \mathrm{spec}\left(\begin{bmatrix} 1 & 1 \\ -1 & 1 \end{bmatrix}\right) \cup \mathrm{spec}\left(\begin{bmatrix} 2 & 1 \\ -4 & 3 \end{bmatrix}\right)$$

$$= \{1 \pm i\} \cup \left\{\frac{1}{2}[5 \pm i\sqrt{15}]\right\}.$$

Proposition 12.10 *Similar matrices in $M_{n \times n}(F)$, where F is a field and where n is a positive integer, have identical characteristic polynomials.*

Proof If $A, B \in M_{n \times n}(F)$ satisfy $B = P^{-1}AP$ then

$$|XI - B| = |XI - P^{-1}AP| = |P^{-1}(XI - A)P|$$
$$= |P|^{-1}|XI - A||P| = |XI - A|,$$

as required. □

Example The converse of Proposition 12.10 is false. Indeed, the matrices $\begin{bmatrix} 1 & 1 \\ 0 & 1 \end{bmatrix}$ and $\begin{bmatrix} 1 & 0 \\ 0 & 1 \end{bmatrix}$ are not similar, despite the fact that both of them have the same characteristic polynomial, $(X - 1)^2$.

A generalization of Proposition 12.10 tells us that if $P, Q \in M_{n \times n}(F)$ are nonsingular matrices satisfying $|PQ| = 1$, then the endomorphism α of $M_{n \times n}(F)$ given by $\alpha_{PQ} : A \mapsto PAQ$ satisfies the condition that A and $\alpha(A)$ always have identical characteristic polynomials. The same goes for the linear transformation $\beta_{PQ} : A \mapsto PA^T Q$. Frobenius proved that any endomorphism of $M_{n \times n}(\mathbb{C})$ which preserves characteristic polynomials must be of one of these two forms. Note that endomorphisms of the form α_{PQ} or β_{PQ} are in fact automorphisms of $M_{n \times n}(F)$. They also satisfy the property that $\alpha_{PQ}(A)$ is singular if and only if A is singular, and similarly $\beta_{PQ}(A)$ is singular if and only if A is singular. Indeed, Dieudonné has shown that, for any field F, an endomorphism of $M_{n \times n}(F)$ satisfying this condition must be of one of these two forms.

With kind permission of the Archives of the Mathematisches Forschungsinstitut Oberwolfach.

The twentieth-century French mathematician **Jean Dieudonné** was one of the founders of the influential group who wrote under the collective name of Nicholas Bourbaki.

Example If A and B are square matrices over a field F, then we know that the matrices AB and BA are not necessarily equal. They are also not necessarily similar. For example, if $A = \begin{bmatrix} 1 & 0 \\ 1 & 0 \end{bmatrix}$ and $B = \begin{bmatrix} 0 & 0 \\ 1 & 1 \end{bmatrix}$ then $AB = O \neq BA$, and so AB and BA are not similar. Nonetheless, by Proposition 12.3, we see that $\operatorname{spec}(AB) = \operatorname{spec}(BA)$.

Proposition 12.10 can be used to facilitate computation, as the following example shows.

Example Let n be a positive integer, let F be a field, and let $A = [a_{ij}] \in \mathcal{M}_{n \times n}(F)$ be a symmetric tridiagonal matrix. That is to say, the entries of A satisfy the condition that $a_{ij} = a_{ji}$ when $|i - j| = 1$ and $a_{ij} = 0$ when $|i - j| > 1$. Set $p_0(X) = 0$ and, for each $1 \leq k \leq n$, let $p_k(X)$ be the characteristic polynomial of the $k \times k$ submatrix of A consisting of the first k rows and first k columns of the matrix $XI - A \in F[X]$. Then $p_n(X)$ is the characteristic polynomial of A and we have $p_1(X) = X - a_{11}$ and $p_k(X) = (X - a_{kk})p_{k-1}(X) - a_{ij}^2 p_{k-2}(X)$ for each $2 \leq k \leq n$. This recursion relation allows us to compute the characteristic polynomial of A quickly. Therefore, if A is any symmetric matrix, a good strategy is to try and find a symmetric tridiagonal matrix similar to it and then compute its characteristic polynomial.

Let α be an endomorphism of a vector space V finitely generated over a field F and let $c \in \mathrm{spec}(\alpha)$. The *algebraic multiplicity* of c is the largest integer k such that $(X - c)^k$ divides the characteristic polynomial of α. The *geometric multiplicity* of c is the dimension of the eigenspace of α associated with c. The geometric multiplicity of c is not greater than its algebraic multiplicity, but these two numbers need not be equal, as the following examples show. If these two multiplicities are equal, we say that c is a *semisimple eigenvalue* of α; an eigenvalue which is not semisimple is *defective*. In particular, if the algebraic multiplicity of c is 1 then the same must be true for its geometric multiplicity. In that case, we say that c is a *simple eigenvalue* of α. If at least one eigenvalue of α has geometric multiplicity greater than 1, then α is *derogatory*; otherwise, it is *nonderogatory*.

Example If $\alpha \in \mathrm{End}(\mathbb{R}^2)$ is defined by $\alpha : \begin{bmatrix} a \\ b \end{bmatrix} \mapsto \begin{bmatrix} a+b \\ b \end{bmatrix}$ then $c = 1$ is an eigenvalue of α with associated eigenspace $\mathbb{R} \begin{bmatrix} 0 \\ 1 \end{bmatrix}$ and so the geometric multiplicity of c is 1. On the other hand, α is represented with respect to the canonical basis by the matrix $\begin{bmatrix} 1 & 1 \\ 0 & 1 \end{bmatrix}$, so its characteristic polynomial is $(X - 1)^2$, implying that the algebraic multiplicity of c is 2.

Example Let $\alpha \in \mathrm{End}(\mathbb{R}^3)$ be the endomorphism represented with respect to the canonical basis by the matrix $\begin{bmatrix} 2 & 3 & 1 \\ 3 & 2 & 4 \\ 0 & 0 & -1 \end{bmatrix}$. The characteristic polynomial of α is $(X - 5)(X + 1)^2$ and so $\mathrm{spec}(\alpha) = \{-1, 5\}$, where the algebraic multiplicity of -1 equals 2 and the algebraic multiplicity of 5 equals 1. The eigenspace associated with -1 is $\mathbb{R} \begin{bmatrix} -1 \\ 1 \\ 0 \end{bmatrix}$ and the eigenspace associated with 5 is $\mathbb{R} \begin{bmatrix} 1 \\ 1 \\ 0 \end{bmatrix}$. Thus both eigenvalues have geometric multiplicity 1. Hence, 5 is a simple eigenvalue of α whereas -1 is defective.

Let n be a positive integer. If $\alpha \in \text{End}(\mathbb{R}^n)$ is represented with respect to a given basis of $\mathbb{R}^n$ by a matrix all entries in which are positive, then Perron, using analytic methods, showed that the eigenvalue of largest absolute value of α is simple and positive, and has an associated eigenvector all entries of which are positive. This result has many important applications in statistics and economics, especially in input–output analysis. It was also used by Thurston in his classification of surface diffeomorphisms in topology. Perron's results were later extended by Frobenius to certain matrices all entries in which are nonnegative, and later by Karlin to certain endomorphisms of spaces which are not finite-dimensional. In 1948, Philip Stein and R.L. Rosenberg used Frobenius' extension of Perron's results to compare the convergence rates of the Jacobi and Gauss–Seidel iteration methods for solution of systems of linear equations. Their results have since been considerably extended.

With kind permission of the Archives of the Mathematisches Forschungsinstitut Oberwolfach (Perron, Frobenius, Thurston).

The twentieth-century German mathematician **Oskar Perron** worked in many areas of algebra and geometry. Fellow German mathematician **Georg Frobenius** is known for his important work in group theory and his work on bilinear forms. He was also the first to consider the rank of a matrix. **William Thurston** is a contemporary American geometer; the twentieth-century American applied mathematician **Samuel Karlin** published extensively in probability and statistics, as well as mathematical biology.

Proposition 12.11 *Let V be a vector space finitely generated over a field F and let α be an endomorphism of V satisfying the condition that the characteristic polynomial of α is completely reducible. Then α is diagonalizable if and only if every eigenvalue of α is semisimple.*

Proof Let $\text{spec}(\alpha) = \{c_1, \ldots, c_k\}$. First of all, we will assume that there exists a basis D of V such that $\Phi_{DD}(\alpha)$ is a diagonal matrix. For each $1 \leq j \leq k$, denote by $m(j)$ the number of times that c_j appears on the diagonal of $\Phi_{DD}(\alpha)$. Then $\sum_{j=1}^{k} m(j) = n$ and, by Proposition 12.4, we know that for each $1 \leq j \leq k$ there exists a subset of D, having $m(j)$ elements, which is a basis for the eigenspace of α associated with c_j. Moreover, the characteristic polynomial of α is $\prod_{j=1}^{k} (X - c_j)^{m(j)}$

and so $m(j)$ equals both the algebraic multiplicity and the geometric multiplicity of c_j for each $1 \leq j \leq k$, proving that each such c_j is semisimple. Conversely, assume that each c_j is semisimple, and for each $1 \leq j \leq k$ let $m(j)$ be the algebraic (and geometric) multiplicity of c_j. Let D_j be a basis for the eigenspace of α associated with c_j, and let $D = \bigcup_{j=1}^{k} D_j$. Then D is a linearly-independent subset of V having n elements, and so is a basis of V over F. The result then follows from Proposition 7.5. $\qquad\square$

Example The condition in Proposition 12.11 that the characteristic polynomial of α be completely reducible is essential. To see this, consider the endomorphism α of $\mathbb{R}^3$ represented with respect to the canonical basis by the matrix $A = \begin{bmatrix} 0 & 1 & 0 \\ -1 & 0 & 0 \\ 0 & 0 & 1 \end{bmatrix}$. The characteristic polynomial of α is $(X-1)(X^2+1) \in \mathbb{R}[X]$ and so $\mathrm{spec}(\alpha) = \{1\}$, where 1 is a simple eigenvalue of α and so it is surely semisimple. The eigenspace of α associated with this eigenvalue is $\mathbb{R}\begin{bmatrix} 1 \\ 0 \\ 0 \end{bmatrix}$ and so its dimension is 1. Hence α is not diagonalizable.

Example Consider the endomorphism α of $\mathbb{R}^3$ represented with respect to the canonical basis by the matrix $\begin{bmatrix} -1 & -1 & -2 \\ 8 & -11 & -8 \\ -10 & 11 & 7 \end{bmatrix}$ and let β be the endomorphism of $\mathbb{R}^3$ represented with respect to the canonical basis by the matrix $\begin{bmatrix} 1 & -4 & -4 \\ 8 & -11 & -8 \\ -8 & 8 & 5 \end{bmatrix}$. These two endomorphisms have the same characteristic polynomial $X^3 + 5X^2 + 3X - 9 = (X-1)(X+3)^2$. Thus the algebraic multiplicity of the eigenvalue 1 equals 1 and the algebraic multiplicity of the eigenvalue -3 equals 2. But for α, the geometric multiplicity of -2 equals 1, so α is not diagonalizable. On the other hand, for β the geometric multiplicity of -2 equals 2, and so β is diagonalizable.

Let F be a field and let $(K, \bullet)$ be an associative unital F-algebra. If $v \in K$ and if $p(X) = \sum_{i=0}^{k} c_i X^i \in F[X]$, then $p(v) = \sum_{i=0}^{k} c_i v^i \in K$. For any polynomial $q(X) \in F[X]$ we have $p(v) \bullet q(v) = q(v) \bullet p(v)$. In particular, $v \bullet p(v) = p(v) \bullet v$. It is clear that $\mathrm{Ann}(v) = \{p(X) \in F[X] \mid p(v) = 0_K\}$ is a subspace of $F[X]$. If $p(v) = 0_K$, we say that v *annihilates* the polynomial $p(X)$.

In particular, we note that all of the above is true for the associative unital F-algebra $\mathcal{M}_{n \times n}(F)$, where n is a positive integer. We note that if $A \sim B$ in $\mathcal{M}_{n \times n}(F)$ then there is a nonsingular matrix P such that $B = P^{-1}AP$ and so $p(B) = P^{-1}p(A)P$ so that if $p(A) = O$ then $p(B) = O$. Thus we see that $\mathrm{Ann}(A) = \mathrm{Ann}(B)$ whenever the matrices A and B are similar.

Example Let $A = \begin{bmatrix} 2 & 1 \\ 1 & 0 \end{bmatrix} \in \mathcal{M}_{2\times 2}(\mathbb{R})$ and let $p(X) = X^2 - X + 2 \in \mathbb{R}[X]$. Then

$p(A) = A^2 - A + 2I = \begin{bmatrix} 5 & 1 \\ 1 & 3 \end{bmatrix}$. If $q(X) = X^2 - 2X - 1$ then $q(A) = O$ so $q(X) \in$ Ann(A).

Proposition 12.12 *Let F be a field and let $(K, \bullet)$ be an associative unital F-algebra finitely generated over F. Then* Ann(v) *is nontrivial for each $v \in K$.*

Proof Let $\dim(V) = n$. If $v \in K$ then $\{v^0, v^1, \dots, v^n\}$ cannot be a linearly independent set and so there exist scalars $a_0, \dots, a_n$, not all equal to 0, such that $\sum_{i=0}^{n} a_i v^i = 0_K$. In other words, there exists a nonzero polynomial $p(X) = \sum_{i=0}^{n} a_i X^i$ in Ann(v). $\qquad\square$

We now show why one cannot define "three-dimensional complex numbers".

Proposition 12.13 *If n is an odd integer greater than 1 then there is no way of defining on $\mathbb{R}^n$ the structure of an $\mathbb{R}$-algebra which is also a field.*

Proof Assume that we can define an operation on $\mathbb{R}^n$ (which we will denote by concatenation) which turns it into an $\mathbb{R}$-algebra which is also a field, and let v_1 be the identity element for this operation. Then $V \neq \mathbb{R}v_1$ since $\dim(V) > 1$. Pick an element $y \in V \smallsetminus \mathbb{R}v_1$ and let $\alpha \in \mathrm{End}(V)$ be given by $\alpha : v \mapsto yv$, which is represented with respect to the canonical basis of $\mathbb{R}^n$ by a matrix A. The characteristic polynomial $p(X)$ of A belongs to $\mathbb{R}[X]$ and has odd degree; therefore, it has a root c in $\mathbb{R}$. Thus $p(X) = (X - c)^k q(X)$ for some $k \geq 1$ and some $q(X) \in \mathbb{R}[X]$ satisfying $q(c) \neq 0$. Let $\beta \in \mathrm{End}(V)$ be given by $\beta : v \mapsto (y - cv_1)^k v$. Then $\beta \neq \sigma_0$ since $y \notin \mathbb{R}v_1$ and $0_V \neq q(c) = q(cv_1)$. But then $(y - cv_1)^k q(cv_1) = 0_V$, contradicting Proposition 2.3(12). $\qquad\square$

Let F be a field and let $(K, \bullet)$ be an associative unital F-algebra. If $v \in K$ satisfies the condition that Ann(v) is nontrivial then Ann(v) must contain a polynomial $p(X) = \sum_{i=0}^{n} a_i X^i$ of minimal degree. This means, in particular, that $a_n \neq 0$ and so the monic polynomial $a_n^{-1} p(X)$ also belongs to Ann(v). We claim that it is the unique monic polynomial of minimal degree in Ann(v). Indeed, if $q(X)$ is a monic polynomial of degree n belonging to Ann(v) not equal to $a_n^{-1} p(X)$, then $r(X) = q(X) - a_n^{-1} p(X) \in$ Ann(v). But $\deg(r) < n$, contradicting the minimality of the degree n of $p(X)$. Thus we see that Ann(v), if nonempty, contains a unique monic polynomial of minimal positive degree, which we call the *minimal polynomial* of v over F and denote by $m_v(X)$.

Example We know that $\mathbb{C}$ is an associative unital $\mathbb{R}$-algebra. If $c = a + bi \in \mathbb{C} \smallsetminus \mathbb{R}$, then its minimal polynomial over $\mathbb{R}$ is $(X - c)(X - \overline{c}) = X^2 - 2aX + (a^2 + b^2)$.

In particular, if F is a field and if n is a positive integer, then any matrix $A \in \mathcal{M}_{n \times n}(F)$ has a minimal polynomial, which we denote by $m_A(X)$. If A and B are similar matrices, then $m_A(X) = m_B(X)$. Similarly, if V is a vector space finitely generated over a field F, and if $\alpha \in \mathrm{End}(V)$ then α has a minimal polynomial $m_\alpha(X)$, and this equals the minimal polynomial of $\Phi_{DD}(\alpha)$ for any basis D of V. If $f(X) \in F[X]$ then it is easy to see that $f(X) = m_{\mathrm{comp}(f)}(X)$ and so every polynomial is the minimal polynomial of some matrix.

Example Let $(K, \bullet)$ be an associative unital entire $\mathbb{R}$-algebra. Assume that $v \in K$ has a minimal polynomial $m_v(X) \in \mathbb{R}[X]$. By Proposition 4.4, we know that $m_v(X) = \prod_{i=1}^{t} p_i(X)$, where the $p_i(X)$ are irreducible polynomials of degree at most 2. But then $\prod_{i=1}^{t} p_i(v) = 0_K$ and, since K is entire, there is some index h such that $p_h(v) = 0_K$. By minimality, this means that $m_v(X) = p_h(X)$. We thus conclude that any element of v having a minimal polynomial has one of degree at most 2.

Proposition 12.14 *Let F be a field and let $(K, \bullet)$ be an associative F-algebra finitely generated over F. If $v \in K$ satisfies the condition that $\mathrm{Ann}(v)$ is nontrivial and if $p(X) \in \mathrm{Ann}(v)$, then there is a polynomial $q(X) \in F[X]$ satisfying $p(X) = m_v(X)q(X)$.*

Proof If $p(X)$ is the 0-polynomial, pick $u(X)$ to be the 0-polynomial, and we are done. Therefore, assume that $\deg(p) \geq 0$. From Proposition 4.2, we know that we can write $p(X) = m_v(X)q(X) + r(X)$, where $q(X), r(X) \in F[X]$, with $\deg(r) < \deg(m_v)$. Since $p(v) = 0_K$, we see that $0_K = m_v(v) \bullet q(v) + r(v) = r(v)$. Since $\deg(v) < \deg(m_A)$, we must have $\deg(v) = -\infty$, and so $p(X) = m_v(X)q(X)$. $\quad\square$

With kind permission of the Archives of the Mathematisches Forschungsinstitut Oberwolfach.

This fundamental result was first established at the beginning of the twentieth century by the German mathematician **Kurt Hensel**.

Proposition 12.15 *Let F be a field and let $(K, \bullet)$ be an associative unital F-algebra with multiplicative identity e. If $v \in K$ has a minimal polynomial $m_v(X) = \sum_{i=0}^{n} a_i X^i$ then:*
(1) *v is a unit of K if and only if $a_0 \neq 0$; and*
(2) *If v is a unit of K then $v^{-1} = g(v)$, where $g(X) = \sum_{i=1}^{n}(-a_0^{-1}a_i)X^{i-1} \in F[X]$.*

Proof If $a_0 \neq 0$ then $m_v(v) = 0_K$ implies that $e = a_0^{-1}[-\sum_{i=1}^{n} a_i v^i] = a_0^{-1}[-\sum_{i=1}^{n} a_i v^{i-1}] \bullet v = g(v) \bullet v = v \bullet g(v)$ and so v is a unit and $v^{-1} = g(v)$. Conversely, assume that v is a unit. Had we $a_0 = 0$, we would have $0_V = m_v(v) = v \bullet [\sum_{i=1}^{n} a_i v^{i-1}]$ and so $0_K = v^{-1}m_v(v) = \sum_{i=1}^{n} a_i v^{i-1}$. Thus $\sum_{i=1}^{n} a_i X^{i-1} \in \text{Ann}(v)$, contradicting the minimality of the degree $m_v(X)$. Hence $a_0 \neq 0$. $\qquad\square$

It is important to note that the minimal polynomial of a matrix over a field need not equal its characteristic polynomial. For example, if we consider $I \in \mathcal{M}_{n \times n}(F)$ for any field F and any integer $n > 1$, then the characteristic polynomial of I is $(X - 1)^n$ whereas its minimal polynomial is $X - 1$.

Example Let F be a field. The matrix $A = \begin{bmatrix} 1 & 0 \\ 0 & 0 \end{bmatrix} \in \mathcal{M}_{2 \times 2}(F)$ annihilates the polynomial $X(X - 1)$, and this is in fact its minimal polynomial. It is also the characteristic polynomial of A. Thus we see that the minimal polynomial of a matrix does not have to be irreducible. Notice too that the rank of A equals 1, but the degree of its minimal polynomial is 2. Thus the degree of the minimal polynomial of a matrix may be larger than its rank.

Example Let F be a field. The matrix $A = \begin{bmatrix} 1 & 0 & 0 \\ 0 & 0 & 0 \\ 0 & 0 & 0 \end{bmatrix} \in \mathcal{M}_{3 \times 3}(F)$ annihilates the polynomial $X(X - 1)$, and this is in fact its minimal polynomial. The characteristic polynomial of A is $X^2(X - 1)$.

Example One can check that $\begin{bmatrix} 1 & 0 & 0 \\ 0 & 3 & 0 \\ 0 & 0 & 3 \end{bmatrix}, \begin{bmatrix} 1 & 0 & 0 \\ 0 & 1 & 0 \\ 0 & 0 & 3 \end{bmatrix} \in \mathcal{M}_{3 \times 3}(\mathbb{Q})$ are not similar, but they both have the same minimal polynomial, namely $(X - 1) \cdot (X - 3)$.

Example Proposition 12.15 can be used to calculate the inverse of a nonsingular matrix, though it is rarely the most efficient method of doing so. For example, the matrix $A = \begin{bmatrix} 2 & -2 & 4 \\ 2 & 3 & 2 \\ -1 & 1 & -1 \end{bmatrix}$ has minimal polynomial $X^3 - 4X^2 + 7X - 10 = 0$

so

$$A^{-1} = \frac{1}{10}(A^2 - 4A + 7I) = \frac{1}{10}\begin{bmatrix} -5 & 2 & -16 \\ 0 & 2 & 4 \\ 5 & 0 & 10 \end{bmatrix}.$$

Proposition 12.16 (Cayley–Hamilton Theorem) *Let F be a field and let n be a positive integer. Then every matrix in $M_{n \times n}(F)$ annihilates its characteristic polynomial.*

Proof Let A be a matrix in $M_{n \times n}(F)$ having minimal polynomial $p(X) = X^n + \sum_{i=0}^{n-1} a_i X^i$. Let us look at the matrix $[g_{ij}(X)] = \mathrm{adj}(XI - A) \in M_{n \times n}(F[X])$, where each $g_{ij}(X)$ is a polynomial of degree at most $n - 1$. Then we can write this matrix in the form $\sum_{i=1}^{n} B_i X^{n-i}$, where the B_i are matrices in $M_{n \times n}(F)$. Moreover, we know that

$$p(X)I = |XI - A|I = (XI - A)\,\mathrm{adj}(XI - A) = (XI - A)\left(\sum_{i=1}^{n} B_i X^{n-i}\right).$$

Equating coefficients of the various powers of X, we thus see

$$B_1 = I,$$
$$B_2 - AB_1 = a_{n-1}I,$$
$$B_3 - AB_2 = a_{n-2}I,$$
$$\vdots$$
$$B_n - AB_{n-1} = a_1 I,$$
$$-AB_n = a_0 I.$$

For $1 \le h \le n$, multiply both sides of the hth equation above on the left by A^{n+1-h} and then sum both sides, to obtain $O = p(A)$. $\square$

We see from Proposition 12.14 and Proposition 12.16 that the minimal polynomial of any $n \times n$ matrix over a field divides its characteristic polynomial and so the degree of the minimal polynomial is at most n.

Let V be a vector space finitely generated over a field F and let $\sigma_0 \neq \alpha \in \mathrm{End}(V)$. In Proposition 12.4, we saw that α is diagonalizable if and only if there is a basis that is composed of eigenvectors of V. Moreover, if $\mathrm{spec}(\alpha) = \{c_1, \dots, c_k\}$ and if, for each $1 \le i \le k$, we denote the eigenspace of α associated with c_i by W_i, then for each $1 \le i \le k$ we have a projection $\pi_i \in \mathrm{End}(V)$ satisfying the following conditions:

(1) $\mathrm{im}(\pi_i) = W_i$;
(2) $\pi_1 + \cdots + \pi_n = \sigma_1$;

(3) $\pi_i \pi_j = \sigma_0$ whenever $i \neq j$;

(4) $\alpha = c_1 \pi_1 + \cdots + c_k \pi_k$.

For each $1 \leq h \leq k$, let $p_h(X)$ be the hth Lagrange interpolation polynomial determined by $c_1, \ldots, c_k$. Then we can check that $\pi_h = p_h(\alpha)$ for each h, since $p_h(X)(X - c_h)$ is just a scalar multiple of the minimal polynomial of α.

Is it possible to simultaneously diagonalize two distinct endomorphisms of V? Indeed, let V be a vector space finitely generated over a field F and let α and β be distinct elements of $\mathrm{End}(V) \smallsetminus \{\sigma_0\}$. There exists a basis D of V such that both $\Phi_{DD}(\alpha)$ and $\Phi_{DD}(\beta)$ are diagonal matrices if and only if the elements of D are eigenvectors of α as well as of β. Suppose that we have in hand such a basis $D = \{u_1, \ldots, u_k\}$. Since diagonal matrices commute with each other, we see that $\Phi_{DD}(\alpha\beta) = \Phi_{DD}(\alpha)\Phi_{DD}(\beta) = \Phi_{DD}(\beta)\Phi_{DD}(\alpha) = \Phi_{DD}(\beta\alpha)$ and so $\alpha\beta = \beta\alpha$. Therefore, a necessary condition for both endomorphisms of V to be represented by diagonal matrices with respect to the same basis is that they form a commuting pair.

We also note that if D is a basis for a vector space V over a field F then the set of all endomorphisms α of V satisfying the condition that $\Phi_{DD}(\alpha)$ is a diagonal matrix is a subspace of $\mathrm{End}(V)$. Indeed, this is an immediate consequence of the fact that the set of all diagonal $n \times n$ matrices is a subspace of $\mathcal{M}_{n \times n}(F)$.

Proposition 12.17 *Let V be a vector space over a field F and let α, β be a commuting pair of endomorphisms of V. Then $p(\alpha)q(\beta) = q(\beta)p(\alpha)$ for any $p(X), q(X) \in F[X]$.*

Proof Initially, we will consider the special case of $q(X) = X$. If $p(X) = \sum_{i=0}^{n} a_i X^i$ then $\beta\alpha^2 = (\beta\alpha)\alpha = (\alpha\beta)\alpha = \alpha(\beta\alpha) = \alpha(\alpha\beta) = \alpha^2\beta$, and, by induction, we similarly have $\beta\alpha^k = \alpha^k \beta$ for every positive integer k. Therefore

$$\beta p(\alpha) = \beta \left(\sum_{i=0}^{n} a_i \alpha^i \right) = \sum_{i=0}^{n} a_i \beta \alpha^i = \sum_{i=0}^{n} a_i \alpha^i \beta = \left(\sum_{i=0}^{n} a_i \alpha^i \right) \beta = p(\alpha)\beta.$$

Now a proof similar to the first part shows that $p(\alpha)\beta^k = \beta^k p(\alpha)$ for every positive integer k and hence, by a proof similar to the second part, we get $p(\alpha)q(\beta) = q(\beta)p(\alpha)$ for any $p(X), q(X) \in F[X]$. $\qquad\square$

As a consequence of this we note that if $\alpha, \beta \in \mathrm{End}(V)$ are commuting projections then $(\alpha\beta)^2 = (\alpha\beta)(\alpha\beta) = \alpha(\beta\alpha)\beta = \alpha(\alpha\beta)\beta = \alpha^2\beta^2 = \alpha\beta$ and so $\alpha\beta$ is a projection as well.

Proposition 12.18 *Let V be a vector space finitely generated over a field F and let $\alpha, \beta \in \mathrm{End}(V)$ be diagonalizable endomorphisms of V. Then there exists a basis of V relative to which both α and β can be represented by diagonal matrices if and only if $\alpha\beta = \beta\alpha$.*

Proof We have already noted that if α and β can both be represented by diagonal matrices with respect to a given basis of V then we must have $\alpha\beta = \beta\alpha$. Conversely, assume that α and β are diagonalizable endomorphisms of V satisfying $\alpha\beta = \beta\alpha$. Then, as we have already seen, there exist distinct scalars $c_1, \ldots, c_k$ and projections $\pi_1, \ldots, \pi_k \in \text{End}(V)$ such that $\pi_1 + \cdots + \pi_k = \sigma_1$, $\pi_i\pi_j = \sigma_0$ for $i \neq j$, and $c_1\pi_1 + \cdots + c_k\pi_k = \alpha$. Similarly, there exist scalars $d_1, \ldots, d_t$ and projections $\eta_1, \ldots, \eta_k \in \text{End}(V)$ such that $\eta_1 + \cdots + \eta_k = \sigma_1$, $\eta_i\eta_j = \sigma_0$ for $i \neq j$, and $d_1\eta_1 + \cdots + d_k\eta_k = \beta$. Therefore, $\alpha = \alpha\sigma_1 = (\sum_{i=1}^{k} c_i\pi_i)(\sum_{j=1}^{t} \eta_j) = \sum_{i=1}^{k} \sum_{j=1}^{t} c_i\pi_i\eta_j$ and $\beta = \beta\sigma_1 = (\sum_{j=1}^{t} d_j\eta_j)(\sum_{i=1}^{k} \pi_i) = \sum_{j=1}^{t} \sum_{i=1}^{k} d_j\eta_j\pi_i$. Since we saw that for each $1 \leq i \leq k$ we have $\pi_i = p_i(\alpha)$ for some $p_i(X) \in F[X]$ and similarly for each $1 \leq j \leq t$ we have $\eta_j = q_j(\beta)$ for some $q_j(X) \in F[X]$, we conclude that $\pi_i\eta_j = \eta_j\pi_i$ for each such i and j. Call this common value θ_{ij}. By the comments after Proposition 12.17, we see that θ_{ij} is also a projection in $\text{End}(V)$.

We note that $\theta_{ij}\theta_{hm} = \pi_i\eta_j\pi_h\eta_m = \pi_i\pi_h\eta_j\eta_m$ and this equals σ_0 when $i \neq j$ or $h \neq m$. Thus $\sum_{i=1}^{k} \sum_{j=1}^{t} \theta_{ij} = (\sum_{i=1}^{k} \pi_i)(\sum_{j=1}^{t} \eta_j) = \sigma_1$. Hence we have shown that α and β are simultaneously diagonalizable, using those projections θ_{ij} which are nonzero (as some of them may be zero). $\square$

We now turn to another classical result.

Proposition 12.19 *Let V be a vector space over a field F and let K be a subalgebra of $\text{End}(V)$ such that there is no nontrivial proper subspace of V which is invariant under every $\alpha \in K$. Suppose that $\beta \in \text{End}(V)$ has a nonempty spectrum and commutes with every element of K. Then $\beta = \sigma_c$ for some $c \in F$.*

Proof Pick $c \in \text{spec}(\beta)$ and let W be the eigenspace of β associated with c. This is a nontrivial subspace of V. If $\alpha \in K$ and $w \in W$ then $\beta\alpha(w) = \alpha\beta(w) = \alpha(cw) = c\alpha(w)$ and so $\alpha(w) \in W$. Thus W is a nontrivial subspace of W invariant under every $\alpha \in K$ and so, by assumption, it cannot be proper. Therefore, $W = V$ and so $\beta = \sigma_c$. $\square$

Recall that if the field F is algebraically closed then any element of $\text{End}(V)$ other than σ_0 has a nonempty spectrum.

Proposition 12.20 *Let V be a vector space over an algebraically-closed field F and let K be a unital subalgebra of $\text{End}(V)$ such that there is no nontrivial proper subspace of V which is invariant under every $\alpha \in K$. Let $v \in V$ and let W be a finitely-generated subspace of V satisfying the condition that if $\alpha \in K$ and $W \subseteq \ker(\alpha)$ then $v \in \ker(\alpha)$. Then $v \in W$.*

Proof We will prove the result by induction on $n = \dim(W)$. If $n = 0$ then W is trivial. Since K is unital, $\sigma_1 \in K$ and $W \subseteq \ker(\sigma_1)$. Therefore, by hypothesis, $v \in \ker(\sigma_1)$ and so $v = 0_V \in W$.

Now assume, inductively, that $n > 1$ and that the result has been established for all subspaces of V of dimension less than n. Pick $0_V \neq w_0 \in W$ and let W_1 be a complement of Fw_0 in W. Set $L = \{\alpha \in K \mid W_1 \subseteq \ker(\alpha)\}$. This set is nonempty since $\sigma_0 \in L$. Moreover, it is in fact a subspace of L as a vector space over F. Moreover, if $\alpha \in L$ and $\beta \in K$ then $\beta\alpha \in L$, so in particular L is a subalgebra of K. Moreover, $Y = \{\alpha(w_0) \mid \alpha \in L\}$ is a subspace of V.

Since $w_0 \notin W_1$, we know that there exists an element α_0 of L satisfying $w_0 \notin \ker(\alpha_0)$ and so Y is nontrivial. However, $\beta(y) \in Y$ for each $y \in Y$ and $\beta \in K$. Thus Y is invariant under every element of K and so, by hypothesis, $Y = V$. Define the function $\theta : V \to V$ by $\theta : \alpha(w_0) \mapsto \alpha(v)$. This function is well-defined for if $\alpha_1(w_0) = \alpha_2(w_0)$ then $w_0 \in \ker(\alpha_1 - \alpha_2)$ and so $W \subseteq \ker(\alpha_1 - \alpha_2)$. Hence, by assumption, $v \in \ker(\alpha_1 - \alpha_2)$, i.e., $\alpha_1(v) = \alpha_2(v)$. It is straightforward to check that in fact $\theta \in \operatorname{End}(V)$.

If $\beta \in K$ then $(\theta\beta)(\alpha(w_0)) = \theta(\beta\alpha(w_0)) = \beta\alpha(v) = \beta(\theta\alpha(w_0)) = (\beta\theta)(\alpha(w_0))$ and so θ commutes with every element of K. By Proposition 12.19, this implies that $\theta = \sigma_c$ for some $c \in \mathbb{C}$. Thus, for any $\alpha \in K$ we have $\alpha(v) = \theta\alpha(w_0) = c\alpha(w_0) = \alpha(cw_0)$ and so $\alpha(v - cw_0) = 0_V$. By the induction hypothesis, this implies that $v - cw_0 \in W_0$ and so $v \in W$, as desired. $\qquad\square$

Proposition 12.21 (Burnside's Theorem) *Let V be a vector space finitely generated over an algebraically-closed field F and let K be a unital subalgebra of $\operatorname{End}(V)$ the elements of which commute with all endomorphisms of the form σ_c for $c \in F$. Assume furthermore that there is no nontrivial proper subspace of V which is invariant under every $\alpha \in K$. Then $K = \operatorname{End}(V)$.*

Proof Pick a basis $\{v_1, \ldots, v_n\}$ for V over F and, for all $1 \leq i, j \leq n$, let θ_{ij} be the endomorphism of V defined by the condition that

$$\theta_{ij} : v_k \mapsto \begin{cases} v_j & \text{if } k = i, \\ 0_V & \text{otherwise.} \end{cases}$$

This is a basis for $\operatorname{End}(V)$ and so it suffices to show that $\theta_{ij} \in K$ for all $1 \leq i, j \leq n$. Fix $i \in \{1, \ldots, n\}$ and let

$$L_i = \{\alpha \in K \mid \alpha(v_h) = 0_V \text{ for all } h \neq i\}.$$

By Proposition 12.20, there is an element α_0 of L_i satisfying $\alpha_0(v_i) \neq 0_V$ and so, as in the proof of that proposition, we see that $\{\alpha(v_h) \mid \alpha \in L_i\}$ equals V. In particular, if $1 \leq j \leq n$ there exists an element β_j of L_i satisfying $\beta_j(v_i) = v_j$. Thus $\theta_{ij} = \beta_j \in K$. $\qquad\square$

The British mathematician **William Burnside** published important works on group theory at the end of the nineteenth century. Burnside's original result has been extensively generalized. The above proof is based on the proof of one such generalization, by the twentieth-century American mathematician **John Tate**.

Proposition 12.21 holds for the case of $F = \mathbb{C}$. If the field F is not algebraically closed, this theorem may not hold.

Example Let $F = \mathbb{R}$ and let α be the endomorphism of $\mathbb{R}^2$ defined by $\alpha : \begin{bmatrix} a \\ b \end{bmatrix} \mapsto \begin{bmatrix} -b \\ a \end{bmatrix}$. Then $\alpha^2 = -\sigma_1$ and so $K = \{c\alpha + c\sigma_1 \mid c \in \mathbb{R}\}$ is a proper subalgebra of $\mathrm{End}(\mathbb{R}^2)$ for which there are no nontrivial proper subspaces of $\mathbb{R}^2$ invariant under every element of K.

Algorithms for the computation of the eigenvalues and eigenvectors of a given matrix are usually very complicated, especially if speed of computation is a major consideration. Therefore, we shall not go into the description of such algorithms in detail. As a rule of thumb, it is best to try to compute eigenvectors directly, and not through finding roots of the characteristic polynomial, since small errors in the computation of eigenvalues may often lead to large errors in the computation of the corresponding eigenvectors. For matrices over $\mathbb{R}$, there are often reasonably efficient iterative methods to find at least some of the eigenvectors. We will bring here one example to find an eigenvector associated with the real eigenvalue of a matrix over $\mathbb{R}$ having greatest absolute value (often called the *dominant eigenvalue*), under assumption that such an eigenvalue indeed exists. The algorithm is based on the observation that if c is an eigenvalue of a matrix $A \in \mathcal{M}_{n \times n}(\mathbb{R})$ then c^k is an eigenvalue of A^k. Hence, if k is sufficiently large, the matrix $A(A^k)$ is approximately equal to cA^k. Therefore, if we select an arbitrary vector $v^{(0)} \in \mathbb{R}^n$ and successively define vectors $v^{(1)}, v^{(2)}, \ldots$ by setting $v^{(i+1)} = Av^{(i)}$ for each $i \geq 0$, then $Av^{(k)} = A^{k+1}v^{(0)}$ and this is roughly equal to $cv^{(k)}$. So, if the circumstances are amenable (and we will not go into the precise conditions necessary for this to happen), the vector $v^{(k)}$ is a reasonable approximation to an eigenvector of A associated with c. Of course, we must always remember that repeated computations lead to accumulating roundoff and truncation errors; one way of combating these is to divide each entry in $v^{(i)}$ by the absolute value of the largest entry, and use this "normalized" vector in the next iteration.

With kind permission of NPL (Wilkinson); with kind permission of the Archives of the Mathematisches Forschungsinstitut Oberwolfach (von Mises).

Of the many numerical analysts who studied computational methods for finding eigenvalues, one of the most important is the British mathematician **James H. Wilkinson**, a former assistant of Alan Turing and one of the major early innovators in numerical linear algebra. The iteration algorithm given here was first studied in the 1920s by the Austrian applied mathematician **Richard von Mises**, who later emigrated to the United States.

After one calculates the dominant eigenvalue of a matrix in $\mathcal{M}_{n \times n}(\mathbb{R})$, there are various techniques, known as *deflation* techniques, for creating a new matrix in $\mathcal{M}_{(n-1) \times (n-1)}(\mathbb{R})$ the eigenvalues of which are the same as all of the eigenvalues of the original matrix, except for the dominant eigenvalue.

Example Consider $A = \begin{bmatrix} 5 & 1 \\ -3 & 1 \end{bmatrix} \in \mathcal{M}_{2 \times 2}(\mathbb{R})$ and let us pick $v^{(0)} = \begin{bmatrix} 1 \\ 1 \end{bmatrix}$, then

$$Av^{(0)} = \begin{bmatrix} 6 \\ -2 \end{bmatrix}, \quad \text{and so we will take} \quad v^{(1)} = \begin{bmatrix} 1 \\ -\frac{1}{3} \end{bmatrix};$$

$$Av^{(1)} = \frac{1}{3}\begin{bmatrix} 14 \\ -10 \end{bmatrix}, \quad \text{and so we will take} \quad v^{(2)} = \begin{bmatrix} 1 \\ -\frac{5}{7} \end{bmatrix};$$

$$Av^{(2)} = \frac{1}{7}\begin{bmatrix} 30 \\ -26 \end{bmatrix}, \quad \text{and so we will take} \quad v^{(3)} = \begin{bmatrix} 1 \\ -\frac{15}{13} \end{bmatrix};$$

$$Av^{(3)} = \frac{1}{13}\begin{bmatrix} 62 \\ -58 \end{bmatrix}, \quad \text{and so we will take} \quad v^{(4)} = \begin{bmatrix} 1 \\ -\frac{29}{31} \end{bmatrix};$$

$$Av^{(4)} = \frac{1}{31}\begin{bmatrix} 126 \\ -122 \end{bmatrix}, \quad \text{and so we will take} \quad v^{(5)} = \begin{bmatrix} 1 \\ -\frac{61}{63} \end{bmatrix}.$$

It seems that this sequence of vectors is converging to $\begin{bmatrix} 1 \\ -1 \end{bmatrix}$ and, indeed, one can check that this is an eigenvector of A associated with the eigenvalue 4.

Again, preconditioning can be used to make iterative methods for finding eigenvalues converge more rapidly.

Example Let n be a positive integer and let $A \in \mathcal{M}_{n \times n}(\mathbb{R})$ be a matrix of the form $[cB + (1-c)D]^T$, where $B \in \mathcal{M}_{n \times n}(\mathbb{R})$ is a Markov matrix, $c \in \mathbb{R}$ satisfies $0 \le c \le$ 1, and $D = \begin{bmatrix} 1 \\ \vdots \\ 1 \end{bmatrix} \wedge \begin{bmatrix} d_1 \\ \vdots \\ d_n \end{bmatrix}$ for nonnegative real numbers d_i satisfying $\sum_{i=1}^{n} d_i = 1$.

Such matrices have been called *Google matrices* since they are needed for the *Page-Rank* algorithm used by the internet search engine *Google*™ to compute an estimate of webpage importance for ranking search results (for these purposes, a typical value for c is 0.85). The value of n can be very large, often far larger than 10^9.

One can show that the eigenvalues $e_1, \ldots, e_n$ of such a matrix satisfy $1 = |e_1| \geq |e_2| \geq \cdots \geq |e_n| \geq 0$, and so the power method mentioned above can be (and is) used by *Google* to rapidly compute an eigenvector associated to e_1. Stanford University researchers Taher Haveliwala and Sepandar Kamvar have shown that for any Google matrix, $|e_2| \leq c$, with equality happening under conditions that hold in the case of those matrices arising in this particular application. Eigenvectors corresponding to this second eigenvalue can be used to detect and combat link spamming on the internet.

One can also consider various generalizations of the eigenvalue problem. Thus, for example, given endomorphisms α and β of a vector space V, one can seek to find all scalars c such that $c\beta - \alpha$ is not monic. Problems of this sort arise naturally, for example, in plasma physics and in the design of control systems. Very often, such problems can be formulated as a matter of minimizing the largest generalized eigenvalue of a pair of symmetric matrices. When β is an automorphism, as is usually the case, such generalized eigenvalue problems can be reduced to the usual eigenvalue problem for the endomorphism $\beta^{-1}\alpha$, but there are often reasons for not wanting to do so. For example, even if both α and β are represented with respect to a given basis by symmetric matrices, the matrix representing $\beta^{-1}\alpha$ may not be symmetric. Therefore, some specialized algorithms have been developed to find solutions of the generalized eigenvalue problem directly.

If V has finite dimension n and the endomorphisms α and β are represented with respect to some basis by matrices A and B, respectively, one can look at the *generalized characteristic polynomial* $|XB - A|$. Problems arise, however, since the degree of this polynomial may be less than n, if the matrix B is singular. In fact, this polynomial may even be the 0-polynomial.

A further generalization of the eigenvalue problem is the following: Given endomorphisms $\alpha_0, \ldots, \alpha_n$ of V, find all scalars c such that the endomorphism $\sum_{i=0}^{n} c^i \alpha_i$ is not monic. Various techniques have been developed to handle this problem directly in special cases. Also, it can sometimes be reduced to the case of $n = 1$. For example, finding a vector $0_V \neq v \in V$ in the kernel of $c^2\alpha_2 + c\alpha_1 + \alpha_0$ is equivalent to finding a nonzero element of the form $\begin{bmatrix} cv \\ v \end{bmatrix}$ in the kernel of $c\beta_1 - \beta_0$, where the β_i are the endomorphisms of V^2 defined by

$$\beta_0 : \begin{bmatrix} x \\ y \end{bmatrix} \mapsto \begin{bmatrix} \alpha_1(x) + \alpha_0(y) \\ \alpha_0(y) \end{bmatrix} \quad \text{and} \quad \beta_1 : \begin{bmatrix} x \\ y \end{bmatrix} \mapsto \begin{bmatrix} -\alpha_2(x) \\ \alpha_0(y) \end{bmatrix}.$$

Exercises

Exercise 738
Let n be a positive integer and let $A \in \mathcal{M}_{n \times n}(\mathbb{Q})$. Let $c_1, \ldots, c_n$ be the list of (not necessarily distinct) eigenvalues of A, considered as a matrix in $\mathcal{M}_{n \times n}(\mathbb{C})$. Show that $\sum_{i=1}^{n} c_i$ and $\prod_{i=1}^{n} c_i$ are rational numbers.

Exercise 739
Let F be a field, let n be a positive integer, and let $A, B \in \mathcal{M}_{n \times n}(F)$. Assume that A and B have the same characteristic polynomial $p(X) \in F[X]$. Is it necessarily true that $p(X)$ is the characteristic polynomial of AB?

Exercise 740
Find infinitely-many matrices in $\mathcal{M}_{3 \times 3}(\mathbb{R})$, all of which have characteristic polynomial $X(X - 1)(X - 2)$.

Exercise 741
Find the characteristic polynomial of $\begin{bmatrix} 3 & 2 & 2 \\ 1 & 4 & 1 \\ 2 & -4 & -1 \end{bmatrix} \in \mathcal{M}_{3 \times 3}(\mathbb{R})$.

Exercise 742
Let $a, b, c \in \mathbb{R}$. Find the characteristic polynomial of the matrix

$$\begin{bmatrix} 0 & 0 & 0 & a \\ a & 0 & 0 & b \\ 0 & b & 0 & c \\ 0 & 0 & c & 0 \end{bmatrix} \in \mathcal{M}_{4 \times 4}(\mathbb{R}).$$

Exercise 743
Let n be a positive integer. Show that every matrix $A \in \mathcal{M}_{n \times n}(\mathbb{R})$ can be written as the sum of two nonsingular matrices.

Exercise 744
Let $F = \mathrm{GF}(3)$ and let n be a positive integer. Let $D = [d_{ij}] \in \mathcal{M}_{n \times n}(F)$ be a nonsingular diagonal matrix and let $A \in \mathcal{M}_{n \times n}(F)$. Show that $1 \notin \mathrm{spec}(DA)$ if and only if $D - A$ is nonsingular.

Exercise 745
Let F be a field of characteristic other than 2. For each positive integer n, let T_n be the set of all diagonal matrices in $\mathcal{M}_{n \times n}(\mathbb{R})$ the diagonal entries of which belong to $\{-1, 1\}$. For any $A \in \mathcal{M}_{n \times n}(\mathbb{R})$, show that there exists a matrix $D \in T_n$ satisfying $1 \notin \mathrm{spec}(DA)$.

Exercise 746

Let n be a positive integer and let $\alpha : \mathcal{M}_{n \times n}(\mathbb{C}) \rightarrow \mathbb{C}^n$ be the function defined

by $\alpha : A \mapsto \begin{bmatrix} a_0 \\ \vdots \\ a_{n-1} \end{bmatrix}$, where $X^n + \sum_{i=0}^{n-1} a_i X^i$ is the characteristic polynomial

of A. Is α a linear transformation?

Exercise 747

Define $\alpha \in \text{End}(\mathbb{R}^3)$ by $\alpha : \begin{bmatrix} a \\ b \\ c \end{bmatrix} \mapsto \begin{bmatrix} a - b \\ a + 2b + c \\ -2a + b - c \end{bmatrix}$. Find the eigenvalues of α

and, for each eigenvalue, find the associated eigenspace.

Exercise 748

Let A is a nonempty set and let V be the collection of all subsets of A, which is a vector space over GF(2). Let B be a fixed subset of A and let $\alpha : V \rightarrow V$ be the endomorphism defined by $\alpha : Y \mapsto Y \cap B$. Find the eigenvalues of α and, for each eigenvalue, find the associated eigenspace.

Exercise 749

Let V be a vector space over a field F and let α be an endomorphism of F. Show that the one-dimensional subspaces of V invariant under α are precisely those of the form Rv, where v is an eigenvector of α.

Exercise 750

Define $\alpha \in \text{End}(\mathbb{R}^4)$ by $\alpha : \begin{bmatrix} a \\ b \\ c \\ d \end{bmatrix} \mapsto \begin{bmatrix} b + c \\ c \\ 0 \\ 0 \end{bmatrix}$. Find the eigenvalues of α. Do

there exist two-dimensional subspaces W and Y of $\mathbb{R}^4$, both invariant under α, such that $\mathbb{R}^4 = W \oplus Y$?

Exercise 751

Let V be a vector space finitely generated over a field F and let α be an endomorphism of V having an eigenvalue c. For any $p(X) \in F[X]$, show that $p(c)$ is an eigenvalue of $p(\alpha)$.

Exercise 752

Let V be the vector space of all functions in $\mathbb{R}^{\mathbb{R}}$ which are infinitely differentiable and let $\alpha : V \rightarrow V$ be the endomorphism of V defined by $\alpha : f \mapsto f''$. If $n > 0$ is an integer, show that the function $f : x \mapsto \sin(nx)$ is an eigenvector of α^2 and find the associated eigenvalue.

Exercise 753
Let F be a field and let $V = F^{(\mathbb{Z})}$. Let α be the endomorphism of V defined by $\alpha(f) : i \mapsto f(i + 1)$ for all $i \in \mathbb{Z}$. Determine whether $\mathrm{spec}(\alpha)$ is nonempty or not.

Exercise 754
Let V be the vector space composed of all polynomial functions from $\mathbb{R}$ to itself, let $a \in \mathbb{R}$, and let α be the endomorphism of V defined by $\alpha(p) : x \mapsto (x - a)[p'(x) + p'(a)] - 2[p(x) - p(a)]$, where p' denotes the derivative of p. Find the eigenvalues of α and for each such eigenvalue, find the associated eigenspace.

Exercise 755
Let α be the endomorphism of $\mathcal{M}_{2\times2}(\mathbb{R})$ defined by

$$\alpha : \begin{bmatrix} a & b \\ c & d \end{bmatrix} \mapsto \begin{bmatrix} d & -b \\ -c & a \end{bmatrix}.$$

Find the eigenvalues of α and for each such eigenvalue, find the associated eigenspace.

Exercise 756
Let V be a vector space over $\mathbb{Q}$ and let $\alpha \in \mathrm{End}(V)$ be a projection. Show that $\mathrm{spec}(\alpha) \subseteq \{0, 1\}$. Is the converse true?

Exercise 757
Let V be a vector space of dimension $n > 0$ over a field F. Let α be an endomorphism of V for which there exists a set A of $n + 1$ distinct eigenvectors satisfying the condition that every subset of A of size n is a basis for V. Show that all of the eigenvectors in V are associated with the same eigenvalue c of α and that $\alpha = c\sigma_1$.

Exercise 758
For $a, b \in \mathbb{R}$, let $A = \begin{bmatrix} a & b & 0 \\ b & a & b \\ 0 & b & a \end{bmatrix} \in \mathcal{M}_{3\times3}(\mathbb{R})$. Find the eigenvalues of A.

Exercise 759
Let $A \in \mathcal{M}_{2\times2}(\mathbb{R})$ be a matrix of the form $\begin{bmatrix} a & b \\ c & a \end{bmatrix}$, where $a > 0$ and $bc > 0$. Show that A has two distinct eigenvalues in $\mathbb{R}$.

Exercise 760
Find the eigenvalues of the matrix $\begin{bmatrix} 5 & 6 & -3 \\ -1 & 0 & 1 \\ 2 & 2 & -1 \end{bmatrix} \in \mathcal{M}_{3\times3}(\mathbb{R})$ and, for each such eigenvalue, find the associated eigenspace.

Exercise 761

Find the eigenvalues of the matrix $\begin{bmatrix} 0 & 2 & 1 \\ -2 & 0 & 3 \\ -1 & -3 & 0 \end{bmatrix} \in \mathcal{M}_{3\times3}(\mathbb{C})$ and, for each

such eigenvalue, find the associated eigenspace.

Exercise 762

Let W be the subspace of $\mathbb{R}^\infty$ consisting of all convergent sequences and let α be the endomorphism of W defined by

$$\alpha : [a_1, a_2, \ldots] \mapsto \left[\left(\lim_{i\to\infty} a_i \right) - a_1, \left(\lim_{i\to\infty} a_i \right) - a_2, \ldots \right].$$

Find all eigenvalues of α and, for each eigenvalue, find the corresponding eigenspace.

Exercise 763

Find the eigenvalues of the matrix $\begin{bmatrix} 1 & -1 & 1 \\ 1 & 0 & 0 \\ 0 & 1 & 0 \end{bmatrix}$ in $\mathcal{M}_{3\times3}(\mathbb{C})$ and, for each

such eigenvalue, find the associated eigenspace.

Exercise 764

Does there exist a real number a such that

$$\text{spec} \left(\begin{bmatrix} 1 & -1 & 0 \\ 0 & a & -1 \\ -6 & 11 & -5 \end{bmatrix} \right) = \{-2, -1, 0\}?$$

Exercise 765

Let α be an endomorphism of a vector space V over a field F and let v and w be eigenvectors of α. If $v + w \neq 0_V$, show that $v + w$ is an eigenvector of α if and only if both v and w correspond to the same eigenvalue.

Exercise 766

Show that the matrix $A = \begin{bmatrix} 1 & 0 & a \\ a & a & a \\ a & 0 & -1 \end{bmatrix}$ has three distinct eigenvalues for any

real number a.

Exercise 767

Let n be a positive integer and let t be a nonzero real number. Let $A \in \mathcal{M}_{n\times n}(\mathbb{R})$ be the matrix all of the entries of which equal t. Find the eigenvalues of A and, for each such eigenvalue, find the associated eigenspace.

Exercise 768

Let n be a positive integer and let F be a field. Let A be a nonsingular matrix in $\mathcal{M}_{n\times n}(F)$. Given the eigenvalues of A, find the eigenvalues of A^{-1}.

Exercise 769
Let n be a positive integer and let F be a field. Let $A = [a_{ij}] \in \mathcal{M}_{n \times n}(F)$, and let $c \in \text{spec}(A)$. If $b, d \in F$, show that $bc + d \in \text{spec}(bA + dI)$.

Exercise 770
Let $A = \begin{bmatrix} a & b \\ c & d \end{bmatrix} \in \mathcal{M}_{2 \times 2}(\mathbb{R})$. If $t \in \mathbb{R}$ is a root of the polynomial $bX^2 + (a - d)X - c \in \mathbb{R}[X]$, show that $\begin{bmatrix} 1 \\ t \end{bmatrix}$ is an eigenvector of A associated with the eigenvalue $a + bt$.

Exercise 771
Let $A \in \mathcal{M}_{2 \times 2}(\mathbb{C})$ be a matrix having two distinct eigenvalues. Show that there are precisely four distinct matrices $B \in \mathcal{M}_{2 \times 2}(\mathbb{C})$ satisfying $B^2 = A$.

Exercise 772
Find all $a \in \mathbb{R}$ such that $\begin{bmatrix} a & 0 & 0 \\ 2a & 2a & 2a \\ 0 & 0 & a \end{bmatrix}$ has a unique eigenvalue.

Exercise 773
Find a real number a such that the only eigenvalue of the matrix

$$\begin{bmatrix} a & 1 & 0 \\ -1 & 0 & -1 \\ 0 & 1 & -a \end{bmatrix} \in \mathcal{M}_{3 \times 3}(\mathbb{R})$$

is 0.

Exercise 774
For each $1 \le i \le 3$ and $2 \le j \le 3$, find a real number a_{ij} such that $\begin{bmatrix} 1 \\ -1 \\ 0 \end{bmatrix}$, $\begin{bmatrix} 1 \\ 0 \\ -1 \end{bmatrix}$, and $\begin{bmatrix} 1 \\ 1 \\ 1 \end{bmatrix}$ are all eigenvectors of the matrix $\begin{bmatrix} 1 & a_{12} & a_{13} \\ 1 & a_{22} & a_{23} \\ 1 & a_{32} & a_{33} \end{bmatrix} \in \mathcal{M}_{3 \times 3}(\mathbb{R})$.

Exercise 775
Let $0 \ne r \in \mathbb{C}$ and let n and m be positive integers. Let $A = [a_{ij}] \in \mathcal{M}_{n \times n}(\mathbb{C})$ be given and let $B = [b_{ij}] \in \mathcal{M}_{n \times n}(\mathbb{C})$ be the matrix defined by $b_{ij} = r^{m+i-j} a_{ij}$ for all $1 \le i, j \le n$. Show that if $d \in \mathbb{C}$ is an eigenvalue of A then $r^m d$ is an eigenvalue of B.

Exercise 776
Let n be a positive integer and let F be a field. A matrix $A \in \mathcal{M}_{n \times n}(F)$ is a *magic matrix* if and only if there exists a scalar $c \in F$ such that the sum of the

entries in each row and each column is c. Characterize magic matrices in terms of their eigenvalues.

Exercise 777

Let $A \in M_{2\times 2}(\mathbb{C})$ be a matrix having distinct eigenvalues $a \neq b$. Show that, for all $n > 0$,

$$A^n = \frac{a^n}{a-b}(A - bI) + \frac{b^n}{b-a}(A - aI).$$

Exercise 778

Let $A \in M_{2\times 2}(\mathbb{C})$ be a matrix having a unique eigenvalue c. Show that $A^n = c^{n-1}[nA - (n-1)cI]$ for all $n > 0$.

Exercise 779

Let n be a positive integer and let $A \in M_{n\times n}(\mathbb{C})$. Show that every eigenvector of A is also an eigenvector of $\mathrm{adj}(A)$.

Exercise 780

Let n be a positive integer. Let G be the set of all matrices $A \in M_{n\times n}(\mathbb{C})$ satisfying the condition that $\mathbb{C}^n$ has a basis composed of eigenvectors of A. Is G closed under taking sums? Is it closed under taking products?

Exercise 781

Let $p(X) \in \mathbb{C}[X]$ and let $A \in M_{n\times n}(\mathbb{C})$ for some positive integer n. Calculate the determinant of the matrix $p(A)$ using the eigenvalues of A.

Exercise 782

Let $-1 \neq a \in \mathbb{R}$ and let $A = \begin{bmatrix} 1 - a + a^2 & 1 - a \\ a - a^2 & a \end{bmatrix} \in M_{2\times 2}(\mathbb{R})$. Calculate A^n for all $n \geq 1$.

Exercise 783

Let n be a positive integer. Given a matrix $A \in M_{n\times n}(\mathbb{Q})$, find infinitely-many distinct matrices having the same eigenvalues as A.

Exercise 784

Let $c \in \mathbb{R}$. Find the spectral radius of $A = \begin{bmatrix} 1 & 0 & 0 & 0 \\ 0 & 0 & c & 0 \\ 0 & -c & 0 & 0 \\ 0 & 0 & 0 & 0 \end{bmatrix} \in M_{4\times 4}(\mathbb{C})$.

Exercise 785

Let $A \in M_{n\times n}(\mathbb{R})$ be a matrix all entries in which are positive and let c be a positive real number greater than the spectral radius of A. Show that $|cI - A| > 0$.

Exercise 786
Let n be a positive integer. Show that 1 is an eigenvalue of any Markov matrix in $\mathcal{M}_{n \times n}(\mathbb{R})$.

Exercise 787
Let n be a positive integer. Let $A = [a_{ij}] \in \mathcal{M}_{n \times n}(\mathbb{R})$ satisfy the condition that $\sum_{j=1}^{n} |a_{ij}| \leq 1$ for all $1 \leq j \leq n$. Show that $|c| \leq 1$ for all $c \in \mathrm{spec}(A)$.

Exercise 788
Let n be a positive integer and let F be a field. Let $A = [a_{ij}] \in \mathcal{M}_{n \times n}(F)$ be a matrix satisfying the condition that the sum of the entries in each row equals 1.

Let $1_F \neq c \in \mathrm{spec}(A)$ and let $\begin{bmatrix} b_1 \\ \vdots \\ b_n \end{bmatrix}$ be an eigenvector of A associated with c.

Show that $\sum_{j=1}^{n} b_j = 0$.

Exercise 789
Give an example of a matrix $A \in \mathcal{M}_{2 \times 2}(\mathbb{R})$ satisfying the condition that $\mathrm{spec}(A) = \varnothing$ but $\mathrm{spec}(A^4) \neq \varnothing$.

Exercise 790
Find an example of matrices $A, B \in \mathcal{M}_{2 \times 2}(\mathbb{R})$ satisfying the condition that every element of $\mathrm{spec}(A) \cup \mathrm{spec}(B)$ is positive but every element of $\mathrm{spec}(AB)$ is negative.

Exercise 791
Find a polynomial $p(X) \in \mathbb{C}[X]$ of degree 2 satisfying the condition that all matrices in $\mathcal{M}_{2 \times 2}(\mathbb{C})$ of the form $\begin{bmatrix} 1-a & 1 \\ p(a) & a \end{bmatrix}$, for $a \in \mathbb{C}$, have the same characteristic polynomial.

Exercise 792
Let F be a field and let n be an even positive integer. Let $A, B \in \mathcal{M}_{n \times n}(F)$ be matrices satisfying $A = B^2$. Let $p(X)$ be the characteristic polynomial of A and let $q(X)$ be the characteristic polynomial of B. Show that $p(X^2) = q(X)q(-X)$.

Exercise 793
Let F be a field. Characterize the matrices in $\mathcal{M}_{2 \times 2}(F)$ having the property that their characteristic polynomial is not equal to their minimal polynomial.

Exercise 794
Let $(K, \bullet)$ be an associative unital F-algebra, let $v \in K$, and let $\alpha : K \to K$ be a homomorphism of F-algebras. Show that $\mathrm{Ann}(v) \subseteq \mathrm{Ann}(\alpha(v))$.

Exercise 795

Are the matrices $\begin{bmatrix} 1 & 2 \\ 3 & 4 \end{bmatrix}$ and $\begin{bmatrix} 4 & 2 \\ 3 & 1 \end{bmatrix}$ in $\mathcal{M}_{2\times 2}(\mathbb{R})$ similar?

Exercise 796

Are the matrices $\begin{bmatrix} 1 & i & 0 \\ i & 2 & -1 \\ 0 & i & 1 \end{bmatrix}$ and $\begin{bmatrix} 1+i & 7 & 2 \\ 0 & 1 & 9 \\ 0 & 0 & 2-i \end{bmatrix}$ in $\mathcal{M}_{3\times 3}(\mathbb{C})$ similar?

Exercise 797

Let n be a positive integer and let $A, B \in \mathcal{M}_{n\times n}(\mathbb{R})$. Show that if A and B are similar when considered as elements of $\mathcal{M}_{n\times n}(\mathbb{C})$, they are also similar in $\mathcal{M}_{n\times n}(\mathbb{R})$.

Exercise 798

Find a diagonal matrix in $\mathcal{M}_{3\times 3}(\mathbb{R})$ similar to the matrix $\begin{bmatrix} 1 & 0 & 1 \\ 0 & 1 & 0 \\ 1 & 0 & 1 \end{bmatrix}$.

Exercise 799

Find a diagonal matrix in $\mathcal{M}_{3\times 3}(\mathbb{R})$ similar to the matrix $\begin{bmatrix} 0 & 0 & 1 \\ 0 & 0 & 0 \\ 1 & 0 & 0 \end{bmatrix}$.

Exercise 800

Is there a diagonal matrix in $\mathcal{M}_{3\times 3}(\mathbb{R})$ similar to the matrix

$$\begin{bmatrix} 8 & 3 & -3 \\ -6 & -1 & 3 \\ 12 & 6 & -4 \end{bmatrix}?$$

Exercise 801

Show that every matrix in the subspace of $\mathcal{M}_{2\times 2}(\mathbb{R})$ generated by

$$\left\{ \begin{bmatrix} 0 & 1 \\ 1 & 0 \end{bmatrix}, \begin{bmatrix} 1 & 0 \\ 0 & -1 \end{bmatrix} \right\}$$

is similar to a diagonal matrix.

Exercise 802

Determine if the matrices $\begin{bmatrix} 0 & 1 & 0 \\ 0 & 0 & 1 \\ 0 & 0 & 0 \end{bmatrix}$ and $\begin{bmatrix} 0 & 0 & 0 \\ 1 & 0 & 0 \\ 0 & 1 & 0 \end{bmatrix}$ in $\mathcal{M}_{3\times 3}(GF(5))$ are similar.

Exercise 803

Is there a diagonal matrix in $\mathcal{M}_{3\times 3}(\mathbb{R})$ similar to the matrix

$$\begin{bmatrix} 1 & -1 & 1 \\ -2 & 1 & 2 \\ -2 & -1 & 4 \end{bmatrix}?$$

Exercise 804

Let $A = \begin{bmatrix} 1 & 0 & 0 \\ 0 & 1 & 1 \\ 0 & 1 & 1 \end{bmatrix} \in \mathcal{M}_{3\times 3}(\mathbb{R})$. Find a nonsingular matrix $P \in \mathcal{M}_{3\times 3}(\mathbb{R})$

such that $P^{-1}AP$ is a diagonal matrix.

Exercise 805

Show that the matrix $A = \begin{bmatrix} 1 & i \\ i & -1 \end{bmatrix} \in \mathcal{M}_{2\times 2}(\mathbb{C})$ is not similar to a diagonal

matrix.

Exercise 806

Are the matrices $\begin{bmatrix} 1 & -1 & 0 \\ 0 & 2 & 5 \\ 0 & 0 & 3 \end{bmatrix}$ and $\begin{bmatrix} 2 & 0 & 0 \\ -1 & 4 & 0 \\ 0 & 3 & 7 \end{bmatrix}$ in $\mathcal{M}_{3\times 3}(\mathbb{Q})$ similar?

Exercise 807

Let k and n be positive integers and let F be a field. Let $A \in \mathcal{M}_{k\times n}(F)$ and
$B \in \mathcal{M}_{n\times k}(F)$. Is the matrix $\begin{bmatrix} AB & O \\ B & O \end{bmatrix}$ similar to the matrix $\begin{bmatrix} O & O \\ B & BA \end{bmatrix}?$

Exercise 808

Let F be a field and let $A = \begin{bmatrix} a & 1 & 0 \\ 0 & a & 1 \\ 0 & 0 & a \end{bmatrix} \in \mathcal{M}_{3\times 3}(F)$. For any $p(X) \in F[X]$,

show that $p(A) = \begin{bmatrix} p(a) & p'(a) & \frac{1}{2}p''(a) \\ 0 & p(a) & p'(a) \\ 0 & 0 & p(a) \end{bmatrix}$, where $p'(X)$ denotes the formal

derivative of the polynomial $p(X)$ and $p''(X)$ is the formal derivative of $p'(X)$.

Exercise 809

Find the characteristic and minimal polynomials of the matrix

$$\begin{bmatrix} 7 & 4 & -4 \\ 4 & -8 & -1 \\ -4 & -1 & -8 \end{bmatrix} \in \mathcal{M}_{3\times 3}(\mathbb{R}).$$

Exercise 810

Let n be a positive integer. Let V be the vector space over $\mathbb{R}$ consisting of all polynomial functions from $\mathbb{R}$ to itself having degree at most n. Let α be the endomorphism of V which assigns to each $f \in V$ its derivative, and let A be a matrix representing α with respect to some basis of V. Find the minimal polynomial of A.

Exercise 811

Find six distinct matrices in $\mathcal{M}_{2\times2}(\mathbb{R})$ which annihilate the polynomial $X^2 - 1$.

Exercise 812

Let n be a positive integer and let c be an element of a field F. Find a matrix $A \in \mathcal{M}_{n\times n}(F)$ having minimal polynomial $(X - c)^n$.

Exercise 813

Use the Cayley–Hamilton Theorem to find the inverse of

$$\begin{bmatrix} 5 & 1 & -1 \\ -6 & 0 & 2 \\ 0 & 0 & 2 \end{bmatrix} \in \mathcal{M}_{3\times3}(\mathbb{R}).$$

Exercise 814

Let n be a positive integer and let $A \in \mathcal{M}_{n\times n}(F)$ be a matrix of rank h. Show that the degree of the minimal polynomial of A is at most $h + 1$.

Exercise 815

Let n be a positive integer and let F be a field. Show that a matrix $A \in \mathcal{M}_{n\times n}(F)$ is nonsingular if and only if $m_A(0) \neq 0$.

Exercise 816

Find the eigenvalues of the matrix $\begin{bmatrix} 0 & 0 & 1 & 1 \\ 0 & 0 & 1 & 0 \\ 0 & 1 & 0 & 0 \\ 1 & 1 & 0 & 0 \end{bmatrix} \in \mathcal{M}_{4\times4}(\mathbb{Q})$ and determine the algebraic multiplicity of each.

Exercise 817

Find the minimal polynomial of the matrix $\begin{bmatrix} 0 & 1 & 0 \\ 0 & 0 & 1 \\ 1 & 3 & -3 \end{bmatrix}$ in $\mathcal{M}_{3\times3}(\mathbb{R})$.

Exercise 818

Let α and β be the endomorphisms of $\mathbb{Q}^4$ represented with respect to the canonical basis by the matrices

$$\begin{bmatrix} 1 & 0 & -1 & 0 \\ 0 & 1 & 0 & -1 \\ 1 & 0 & -1 & 0 \\ 0 & 1 & 0 & -1 \end{bmatrix} \text{ and } \begin{bmatrix} 4 & -1 & -1 & 0 \\ -1 & 4 & 0 & -1 \\ 1 & 0 & 2 & -1 \\ 0 & 1 & -1 & 2 \end{bmatrix}, \text{ re-}$$

spectively. Does there exist a basis of $\mathbb{Q}^4$ with respect to which both of them can be represented by diagonal matrices?

Exercise 819

Let α be the endomorphisms of $\mathbb{R}^3$ represented with respect to the canonical basis by the matrix $\begin{bmatrix} -6 & 2 & -5 \\ 4 & 4 & -2 \\ 10 & -3 & 8 \end{bmatrix}$. Calculate the algebraic and geometric multiplicities of each of the eigenvalues of α.

Exercise 820

Let $A = [a_{ij}] \in \mathcal{M}_{n \times n}(\mathbb{C})$ be a symmetric tridiagonal matrix having an eigenvalue c with algebraic multiplicity k. Show that $a_{i-1,i} = 0$ for at least $k - 1$ values of i.

Exercise 821

Let α be the endomorphisms of $\mathbb{R}^3$ represented with respect to the canonical basis by the matrix $\begin{bmatrix} -8 & -13 & -14 \\ -6 & -5 & -8 \\ 14 & 17 & 21 \end{bmatrix}$. Does there exist a basis of $\mathbb{R}^3$ with respect to which α can be represented by a diagonal matrix?

Exercise 822

Let $A = \begin{bmatrix} 17 & -8 & -12 & 14 \\ 46 & -22 & -35 & 41 \\ -2 & 1 & 4 & -4 \\ 4 & -2 & -2 & 3 \end{bmatrix} \in \mathcal{M}_{4 \times 4}(\mathbb{Q})$. Find the minimal polynomial A.

Exercise 823

For each $t \in \mathbb{R}$, set $A(t) = \begin{bmatrix} \cos^2(t) & \cos(t)\sin(t) \\ \cos(t)\sin(t) & \sin^2(t) \end{bmatrix} \in \mathcal{M}_{2 \times 2}(\mathbb{R})$. Show that all of these matrices have the same characteristic and minimal polynomials.

Exercise 824

Let $a, b, c \in \mathbb{C}$. Find a necessary and sufficient condition for the minimal polynomial of $\begin{bmatrix} 2 & 0 & 0 \\ a & 2 & 0 \\ b & c & 1 \end{bmatrix} \in \mathcal{M}_{3 \times 3}(\mathbb{C})$ to be equal to $(X - 1)(X - 2)$.

Exercise 825

Let $A = \begin{bmatrix} 1 & a & 0 \\ a & a & 1 \\ a & a & -1 \end{bmatrix} \in \mathcal{M}_{3\times 3}(\mathbb{R})$. Find the set of all real numbers a for which

the minimal and characteristic polynomials of A are equal.

Exercise 826

Let $F = GF(5)$. For which values of $a, b \in F$ are the characteristic polynomial

and minimal polynomial of the 5×5 matrix $\begin{bmatrix} a & b & 4 & 2 & 0 \\ b & b & b & 3 & 3 \\ 3 & 4 & 2b & 1 & 3 \\ 0 & 0 & 0 & 0 & 1 \\ 0 & 0 & 0 & 3b & 0 \end{bmatrix}$ equal? What

if $F = GF(7)$?

Exercise 827

Let F be a field and let $O \neq A \in \mathcal{M}_{3\times 3}(F)$ be a matrix satisfying $A^k = O$ for some positive integer k. Show that $A^3 = O$.

Exercise 828

Let F be a field and let $A \in \mathcal{M}_{3\times 3}(F)$ be a matrix which can be written in the form BC, where B and C are involutory matrices in $\mathcal{M}_{3\times 3}(F)$. Show that A is nonsingular and similar to A^{-1}.

Exercise 829

Let $A \in \mathcal{M}_{n\times n}(F)$ be written in the form $A = PB$, where $P, B \in \mathcal{M}_{n\times n}(F)$ and P is nonsingular. Show that A is similar to BP.

Exercise 830

Let $A \in \mathcal{M}_{3\times 3}(\mathbb{Q})$ be a matrix satisfying the condition that $A^5 = I$. Show that $A = I$.

Exercise 831

Let n be a positive integer and let α be an endomorphism of $\mathcal{M}_{n\times n}(\mathbb{C})$, considered as a vector space over $\mathbb{C}$, which satisfies the condition that $\alpha(A)$ is nonsingular if and only if A is nonsingular. Show that α is an automorphism.

Exercise 832

Let F be a field and let n be a positive integer. Let $A \in \mathcal{M}_{n\times n}(F)$ be a matrix having characteristic polynomial $p(X) = X^n + \sum_{i=0}^{n-1} c_i X^i$. Show that, for each $k \geq n$, we have $A^k = \sum_{j=0}^{n-1} b_j(k)A^j$, where
(1) $b_j(n) = -c_j$ for all $0 \leq j \leq n - 1$;
(2) $b_{-1}(k) = 0$ for all $k \geq n$;
(3) $b_j(k+1) = b_{j-1}(k) - a_j b_{n-1}(k)$ for all $k \geq n$ and all $0 \leq j \leq n - 1$.

Exercise 833
Show that there is no matrix $A \in \mathcal{M}_{2 \times 2}(\mathbb{R})$ satisfying the condition that $A^2 = \begin{bmatrix} -1 & 0 \\ 0 & -c \end{bmatrix}$, where $c \neq 1$.

Exercise 834
Let V be a vector space finitely generated over $\mathbb{C}$ and let $\alpha \in \mathrm{End}(V)$ be diagonalizable. If W is a nontrivial subspace of V invariant under α, is the restriction of α to W necessarily diagonalizable?

Exercise 835
Let V be a vector space finitely generated over a field F and let $\alpha \in \mathrm{End}(V)$. Show that α is diagonalizable if and only if the sum of all of its eigenspaces equals V.

Exercise 836
Find all rational numbers a satisfying the condition the endomorphism of $\mathbb{Q}^3$ represented with respect to some basis by the matrix $\begin{bmatrix} 1 & 0 & 0 \\ 1 & a & 0 \\ 0 & 0 & 1 \end{bmatrix}$ is diagonalizable.

Exercise 837
Give an example of an endomorphism α of $\mathbb{R}^3$ having nullity 2 which is not diagonalizable.

Exercise 838
Let n be a positive integer and let $B \in \mathcal{M}_{n \times n}(\mathbb{R})$ be a matrix all entries of which are positive. Let $r > \rho(B)$. Show that
(1) The matrix $A = rI - B$ is nonsingular;
(2) All nondiagonal entries of A are nonpositive;
(3) All entries of A^{-1} are nonnegative; and
(4) If $a + bi \in \mathbb{C}$ is an eigenvalue of A, then $a > 0$.

Exercise 839
Let V be a vector space of finite odd dimension over $\mathbb{R}$ and let $\alpha_1, \dots, \alpha_k$ be distinct mutually-commuting endomorphisms of V, for some $k > 1$. Show that these endomorphisms have a common eigenvector.

Exercise 840
Let $A \in \mathcal{M}_{3 \times 3}(\mathbb{Q})$ have characteristic polynomial $X^3 - bX^2 + cX - d$. For all $n \geq 3$, show that

$$A^n = t_{n-1}A + t_{n-2}\,\mathrm{adj}(A) + (t_n - bt_{n-1})I,$$

where $t_n = \sum_{2i+3j \leq n} (-1)^i \binom{i+j}{j}$.

Exercise 841

Show that the endomorphisms $\alpha : \begin{bmatrix} a \\ b \\ c \\ d \end{bmatrix} \mapsto \begin{bmatrix} 2b \\ 2a \\ 2d \\ 2c \end{bmatrix}$ and $\beta : \begin{bmatrix} a \\ b \\ c \\ d \end{bmatrix} \mapsto \begin{bmatrix} c \\ d \\ a \\ b \end{bmatrix}$ of $\mathbb{Q}^4$

are diagonalizable and commute. Find a basis of $\mathbb{Q}^4$ relative to which both α and β are represented by diagonal matrices.

Exercise 842

Let $A = [a_{ij}] \in \mathcal{M}_{n \times n}(\mathbb{R})$ be a Markov matrix all entries of which are positive. If $c \in \mathbb{C}$ is an eigenvalue of A satisfying $|c| = 1$, show that $c = 1$.

Exercise 843

Let F be a finite field. Show that there exists a symmetric matrix in $\mathcal{M}_{2 \times 2}(F)$ having no eigenvalues.

Exercise 844

Does there exist a square matrix A over $\mathbb{R}$ which is not idempotent but satisfies the condition that $\text{spec}(A) = \{1\}$?

Exercise 845

Let n be a positive integer and let $A \in \mathcal{M}_{n \times n}(\mathbb{Q})$ be a matrix all entries of which are integers. Let k be an integer which is an eigenvalue of A. Show $|A|$ is and integer and that k divides $|A|$.

Exercise 846

Let V be a vector space of dimension 3 over a field F and let $\alpha \in \text{End}(V)$ have nullity 2. Show that the characteristic polynomial of α is of the form $X^2(X - c)$ for some $c \in F$.

Krylov Subspaces

13

Let V be a vector space over a field F and let $\alpha \in \operatorname{End}(V)$. If $0_V \neq v_0 \in V$ then the subspace $F\{v_0, \alpha(v_0), \alpha^2(v_0), \ldots\}$ of V is called the *Krylov subspace* of V defined by α and v_0. The elements of this subspace are precisely those vectors in V of the form $p(\alpha)(v_0)$, where $p(X) \in F[X]$, and so it is natural to denote it by $F[\alpha]v_0$. It is clear that $F[\alpha]v_0$ is invariant under α.

Alexei Nikolaevich Krylov was a Russian applied mathematician who at the end of the nineteenth century developed many of the methods mentioned here in connection with the solution of differential equations.

Proposition 13.1 *Let V be a vector space over a field F, let $\alpha \in \operatorname{End}(V)$, and let $0_V \neq v_0 \in V$.*

(1) *$F[\alpha]v_0$ is the intersection of all subspaces of V containing v_0 and invariant under α;*

(2) *v_0 is an eigenvector of α if and only if $\dim(F[\alpha]v_0) = 1$.*

Proof (1) Since $F[\alpha]v_0$ contains v_0 and invariant under α, it certainly contains the intersection of all such subspaces of V. Conversely, if W is a subspace of V which contains v_0 and invariant under α, then $p(\alpha)(v_0) \in W$ for all $p(X) \in F[X]$ and so $F[\alpha]v_0 \subseteq W$. Thus we have the desired equality.

(2) If v_0 is an eigenvector of α associated with an eigenvalue c then for each $p(X) = \sum_{j=0}^{k} a_j X^j \in F[X]$ we have $p(\alpha)(v_0) = \sum_{j=0}^{k} a_j \alpha^j(v_0) = \sum_{j=0}^{k} a_j c^j v_0 \in Fv_0$, proving that $F[\alpha]v_0 = Fv_0$ and so $\dim(F[\alpha]v_0) = 1$. Conversely, assume

J.S. Golan, *The Linear Algebra a Beginning Graduate Student Ought to Know*,
DOI 10.1007/978-94-007-2636-9_13, © Springer Science+Business Media B.V. 2012

that $\dim(F[\alpha]v_0) = 1$. Then $F[\alpha]v_0 = Fv_0$ since Fv_0 is a one-dimensional subspace of $F[\alpha]v_0$. In particular, $\alpha(v_0) \in Fv_0$ and so there exists a scalar c such that $\alpha(v_0) = cv_0$, which proves that v_0 is an eigenvector of α. $\square$

Since the set $\{v_0, \alpha(v_0), \alpha^2(v_0), \ldots\}$ is a generating set for $F[\alpha]v_0$ over F, Proposition 13.1(2) suggests that $\dim(F[\alpha]v_0)$ can be used to measure how far v_0 is from being an eigenvector of α.

As a first example of the use to which we can put Krylov subspaces, we will see how to use the minimal polynomial to solve systems of linear equations. Let V be a vector space over a field F and let V^∞ be the space of all infinite sequences of elements of V. Every polynomial $p(X) = \sum_{j=0}^{k} a_j X^j \in F[X]$ defines an endomorphism θ_p of V^∞ by $\theta_p : [v_0, v_1, \ldots] \mapsto [\sum_{j=0}^{k} a_j v_j, \sum_{j=0}^{k} a_j v_{j+1}, \sum_{j=0}^{k} a_j v_{j+2}, \ldots]$. Note that if $p(X) = c$ is a polynomial of degree no greater than 0, then $\theta_p = \sigma_c$. It is also easy to verify that $\theta_{pq} = \theta_p \theta_q = \theta_q \theta_p$ for all $p(X), q(X) \in F[X]$.

A sequence $y \in V^\infty$ is *linearly recurrent* if and only if there exists a polynomial $p(X) \in F[X]$ with $y \in \ker(\theta_p)$. In this case, we say that $p(X)$ is a *characteristic polynomial* of y. If $p(X) \in F[X]$ is a characteristic polynomial of $y \in V^\infty$ and if $q(X) \in F[X]$ is a characteristic polynomial of $z \in V^\infty$ then $\theta_{pq}(y+z) = \theta_q \theta_p(y) + \theta_p \theta_q(z) = [0, 0, \ldots]$ and so $p(X)q(X)$ is a characteristic polynomial of $y + z$. It is also clear that $p(X)$ is a characteristic polynomial of cy for all $c \in F$. Thus we see that the set of all linearly recurrent sequences in V^∞ is a subspace of V^∞, which we will denote by $\mathrm{LR}(V)$. If $y \in \mathrm{LR}(V)$, there is precisely one characteristic polynomial which is monic and of minimal degree. This polynomial will be called the *minimal polynomial* of y. The degree of the minimal polynomial of y is the *order of recurrence* of y.

Linearly recurrent sequences in $\mathbb{R}^\infty$ were considered by the seventeenth-century French-born mathematician **Abraham de Moivre**, who spent most of his life in exile in England and was one of the fathers of the theory of probability.

Example Let F be a field and let n be a positive integer. If $A \in \mathcal{M}_{n \times n}(F)$, a polynomial $p(X) \in F[X]$ is a characteristic (resp., minimal) polynomial of the sequence $[I, A, A^2, \ldots]$ if and only if it is the characteristic (resp., minimal) polynomial of the matrix A.

Example Let $V = F = \mathbb{Q}$ and let $y = [a_0, a_1, \ldots] \in V^\infty$ be the sequence defined by $a_0 = 0$, $a_1 = 1$, and $a_{i+2} = a_{i+1} + a_i$ for all $i \geq 0$. This sequence is called the *Fibonacci sequence*. Its minimal polynomial is $X^2 - X - 1$. The roots of this polynomial are $\frac{1}{2}(1 \pm \sqrt{5})$. The number $\frac{1}{2}(1 + \sqrt{5})$ is called the *golden ratio* and artists consider rectangles the sides of which are related by the golden ratio to be of high aesthetic value. This ratio—which appears in ancient Egyptian and Babylonian texts—appears in nature and is basic in the analysis of certain patterns of growth in nature (such as the spirals of a snail shell or a sunflower), of Greek architecture, of Renaissance painting, and even such modern designs as the ratio of the dimensions of a credit card or of A4 paper. Notice that $X^2 - X - 1$ is also the characteristic polynomial of the matrix $\begin{bmatrix} 1 & 1 \\ 1 & 0 \end{bmatrix} \in \mathcal{M}_{2 \times 2}(\mathbb{R})$, and so the eigenvalues of this matrix are also precisely $\frac{1}{2}(1 \pm \sqrt{5})$. The eigenspace associated with $\frac{1}{2}(1 + \sqrt{5})$ is $\mathbb{R} \begin{bmatrix} \frac{1}{2}(1 + \sqrt{5}) \\ 1 \end{bmatrix}$ and the eigenspace associated with $\frac{1}{2}(1 - \sqrt{5})$ is $\mathbb{R} \begin{bmatrix} \frac{1}{2}(1 - \sqrt{5}) \\ 1 \end{bmatrix}$.

Leonardo Fibonacci was born in Italy in the twelfth century and educated in Tunis, bringing back the fruits of Arab mathematics to Europe. His book *Liber Abaci*, written in 1202, contained the first new mathematical research in Christian Europe in over 1000 years. In 1509, **Fra Luca Pacioli**, one of the most important Renaissance mathematicians, wrote a book, *The Divine Proportion*, illustrated by his friend Leonardo da Vinci, about the golden ratio.

We note that if $V = F$ and if $y \in \mathrm{LR}(F)$ is a sequence having order of recurrence at most n, then there exist algorithms, which are essentially extensions of the Euclidean algorithm, to calculate the coefficients of the minimal polynomial of y in an order of n^2 arithmetic operations in F.

Now let V be a vector space of finite dimension n over a field F and let α be an automorphism of V having minimal polynomial $p(X) \in F[X]$. If $w \in V$ then the sequence $y = [w, \alpha(w), \alpha^2(w), \ldots]$ belongs to $\ker(\theta_p)$ and hence to $\mathrm{LR}(V)$. Therefore, this sequence has a minimal polynomial $q(X) = \sum_{j=0}^d c_j X^j$, which divides the polynomial $p(X)$ in $F[X]$. Since α is an automorphism, we can assume that $c_0 \neq 0$ and so we see that if $u = -c_0^{-1} \sum_{j=1}^d c_j \alpha^{j-1}(w)$ then $\alpha(u) = w$ and so $u = \alpha^{-1}(w)$. In particular, if $V = F^n$ for some positive integer n and if α is represented by a matrix A with respect to the canonical basis, then $u = -c_0^{-1} \sum_{j=1}^d c_j A^{j-1} w$ is the unique solution of the system of linear equations $AX = w$. If we set $q^*(X) = -c_0^{-1} \sum_{j=1}^d c_j X^{j-1}$ then $u = q^*(A)w$, and this could be computed quickly were we to already know $q(X)$.

How does one calculate $q^*(X)$ in practice? One method used is basically probabilistic: we randomly choose a vector $u \in F^n$ and compute the minimal polynomial

$q_u(X)$ of $y_u = [u \odot w, u \odot (Aw), u \odot A^2w), \ldots] \in F^\infty$, something which can be done, as we have already observed, in an order of n^2 arithmetic operations in F. After that, we check whether the minimal polynomial of y_u is also the minimal polynomial of y. In general, it will not be so, but it will divide the minimal polynomial of y and so after a reasonable number of such attempts we will, usually, have enough information on hand to reconstruct the minimal polynomial of y.

Example Let $F = \mathrm{GF}(5)$, let $A = \begin{bmatrix} 1 & 4 & 4 \\ 4 & 0 & 3 \\ 1 & 2 & 4 \end{bmatrix} \in M_{3\times3}(F)$, and let $w = \begin{bmatrix} 3 \\ 1 \\ 2 \end{bmatrix} \in$

F^3. The sequence $w, Aw, A^2w, \ldots$ looks like

$$\begin{bmatrix} 3 \\ 1 \\ 2 \end{bmatrix}, \begin{bmatrix} 0 \\ 3 \\ 3 \end{bmatrix}, \begin{bmatrix} 4 \\ 4 \\ 3 \end{bmatrix}, \begin{bmatrix} 2 \\ 0 \\ 4 \end{bmatrix}, \begin{bmatrix} 3 \\ 0 \\ 3 \end{bmatrix}, \begin{bmatrix} 0 \\ 1 \\ 0 \end{bmatrix}, \ldots.$$

If we choose $u = \begin{bmatrix} 1 \\ 0 \\ 0 \end{bmatrix}$ we obtain the sequence $y_u = [3, 0, 4, 2, 3, 0, \ldots]$ in F^∞

and the minimal polynomial $q_u(X)$ of this sequence equals $X^2 + 2X + 2$. Since

$q_u(A)w = \begin{bmatrix} 0 \\ 2 \\ 3 \end{bmatrix}$, we see that this polynomial is not the minimal polynomial of y.

We will try again with $u = \begin{bmatrix} 1 \\ 2 \\ 0 \end{bmatrix}$. For this choice, we get $y_u = [0, 1, 2, 2, 3, 2, \ldots]$

and this has minimal polynomial $X^3 + 3X + 1$. Since the minimal polynomial of y has to be a multiple of this polynomial, and has to be of degree 3, it must equal $X^3 +$

$3X + 1$ and, indeed, $q_u(A)w = \begin{bmatrix} 0 \\ 0 \\ 0 \end{bmatrix}$. Therefore, $q^*(X) = X^2 + 3$ and $q^*(A)w =$

$q_u(A)w = \begin{bmatrix} 2 \\ 3 \\ 1 \end{bmatrix}$.

Now let us return to an important problem which was considered in the previous chapter. Let V be a vector space finitely generated over a field F. Given an endomorphism $\alpha \in \mathrm{End}(V)$, how can we find a basis of V relative to which α is represented by a matrix which is as nice as possible? We have already found out when α is diagonalizable. But what if α is not diagonalizable? Given a vector $0_V \neq w \in V$, there exists a positive integer k such that the set $\{w, \alpha(w), \ldots, \alpha^{k-1}(w)\}$ is linearly independent but the set $\{w, \alpha(w), \ldots, \alpha^k(w)\}$ is linearly dependent. Then $\{w, \alpha(w), \ldots, \alpha^{k-1}(w)\}$ is a basis for the Krylov subspace $F[\alpha]w$ of V, which is called the *canonical basis* of this subspace. The restriction of α to $F[\alpha]w$ is

represented by a matrix of the form
$$\begin{bmatrix} 0 & 0 & \cdots & 0 & 0 & c_1 \\ 1 & 0 & \cdots & 0 & 0 & c_2 \\ 0 & 1 & \cdots & 0 & 0 & c_3 \\ \vdots & \vdots & \ddots & \vdots & \vdots & \vdots \\ 0 & 0 & \cdots & 1 & 0 & c_{k-1} \\ 0 & 0 & \cdots & 0 & 1 & c_k \end{bmatrix}$$ with respect to
the canonical basis, where the scalars $c_1, \ldots, c_k$ satisfy $\alpha^k(w) = \sum_{i=1}^{k} c_i \alpha^{i-1}(w)$. This, of course, is just the companion matrix of the polynomial $X^k - \sum_{i=1}^{k} c_i X^{i-1}$.

Krylov subspaces are also the basis for a family of non-stationary iterative algorithms, known as *Krylov algorithms*, used for approximating solutions to systems of equations of the form $AX = w$, where $A \in \mathcal{M}_{n \times n}(\mathbb{R})$ or $A \in \mathcal{M}_{n \times n}(\mathbb{C})$. Similarly, Krylov subspaces are a basis for a family of non-stationary iterative algorithms, known as *Lanczos algorithms*, used for approximating eigenvalues of sufficiently-nice (e.g., symmetric) matrices. Such algorithms work even under the assumption that we don't even have direct access to the entries of A but do have a "black box" ability to compute Av or $A^T v$ for any given vector $v \in \mathbb{R}^n$. Of course, they do not work for all matrices, but when they work they tend to be fairly efficient and rapid, and are especially good for large sparse matrices. Moreover, they are also amenable to implementation on parallel computers. Parallel Lanczos algorithms have also been developed for solving generalized eigenvalue problems, if the matrices involved are symmetric, Lanczos algorithms can be adapted to work for matrices over finite fields. However, in this case there are also other algorithms available. In particular, one should mention the *Wiedemann algorithm* to solve systems of linear equations of the form $AX = w$, where A is a large nonsingular matrix over a finite field. Such problems arise in the computation of discrete logarithms and in other modes of attack on various encryption methods for transmission of data over the internet. They have also been used to factor large integers. The Wiedemann algorithm,

With kind permission of the Archives of the Mathematisches Forschungsinstitut Oberwolfach (Hestenes, Stiefel); With kind permission of Andrew Odlyzko (Odlyzko).

Hungarian-born applied mathematician **Cornelius Lanczos** developed many important numerical methods for computers in the period after World War II, while working at the US National Bureau of Standards and, later, at the University of Dublin in Ireland. Other major researchers of Krylov algorithms were the American numerical analyst **Magnus Hestenes** and the Swiss numerical analyst **Eduard Stiefel**. A major innovator in the use of Wiedemann and similar algorithms over finite fields has been the contemporary American mathematician **Andrew Odlyzko**.

which is based on computing the minimal polynomial of a certain linearly-recurrent sequence, works especially well for sparse matrices, and is amenable to parallel computation.

A nonzero element a of an associative algebra $(K, \bullet)$ is *nilpotent* if and only if there exists a positive integer k satisfying $a^k = 0_K$. The smallest such integer k, if one exists, is called the *index of nilpotence* of a. In particular, if V is a vector space over a field F, then $\alpha \in \text{End}(V)$ is *nilpotent* if and only if there exists a positive integer k satisfying $\alpha^k = \sigma_0$.

Example Let F be a field and let α be the endomorphism of F^3 defined by

$\alpha : \begin{bmatrix} a \\ b \\ c \end{bmatrix} \mapsto \begin{bmatrix} 0 \\ a \\ b \end{bmatrix}$. Then α is a nilpotent endomorphism, having index of nilpo-

tence 3. The endomorphism β of F^3 defined by $\beta : \begin{bmatrix} a \\ b \\ c \end{bmatrix} \mapsto \begin{bmatrix} -a + 2b + c \\ 0 \\ -a + 2b + c \end{bmatrix}$ is a

nilpotent endomorphism, having index of nilpotence 2.

Example Let F be a field and let α and β be the endomorphisms of F^2 defined by $\alpha : \begin{bmatrix} a \\ b \end{bmatrix} \mapsto \begin{bmatrix} b \\ 0 \end{bmatrix}$ and $\beta : \begin{bmatrix} a \\ b \end{bmatrix} \mapsto \begin{bmatrix} 0 \\ a \end{bmatrix}$. Both endomorphisms are nilpotent, but $\alpha + \beta$ is clearly not nilpotent.

If α is a nilpotent endomorphism of a vector space V and $w \in V \smallsetminus \ker(\alpha)$ then the restriction of α to $F[\alpha]w$ is represented with respect to the canonical basis of $F[\alpha]w$ by a matrix of the form

$$\begin{bmatrix} 0 & 0 & 0 & \cdots & 0 & 0 \\ 1 & 0 & 0 & \cdots & 0 & 0 \\ 0 & 1 & 0 & \cdots & 0 & 0 \\ \vdots & \vdots & \vdots & \ddots & \vdots & \vdots \\ 0 & 0 & 0 & \cdots & 0 & 0 \\ 0 & 0 & 0 & \cdots & 1 & 0 \end{bmatrix}.$$

Proposition 13.2 *Let V be a vector space over a field F and let α be a nilpotent endomorphism of V having index of nilpotence k. Then there exists a vector $w \in V$ satisfying the condition that $\dim(F[\alpha]w) = k$.*

Proof We know that $\alpha^k = \sigma_0$ but not that there exists a vector $0_V \neq w \in V$ such that $\alpha^{k-1}(w) \neq 0_V$. We will have proven the theorem should we are able to show that the set $\{w, \alpha(w), \ldots, \alpha^{k-1}(w)\}$ is linearly independent. And, indeed, assume that we have scalars $a_0, \ldots, a_{k-1} \in F$ satisfying $\sum_{i=0}^{k-1} a_i \alpha^i(w) = 0_V$. Let t be the smallest index such that $a_t \neq 0$. Then if we apply the endomorphism α^{k-t-1} to

$\sum_{i=0}^{k-1} a_i \alpha^i(w)$ we get $0_V = a_t \alpha^{k-1}(w) + a_{t+1} \alpha^k(w) + \cdots + a_{k-2} \alpha^{2k-t-2}(w)$ and so $a_t = 0$, which is a contradiction. Therefore, we conclude that $a_i = 0$ for all i, and so the set is linearly independent, as required. $\qquad \square$

In particular, we see that if V is a vector space of finite dimension over a field F and if α is a nilpotent endomorphism of V, then the index of nilpotence of α is no greater than $\dim(V)$.

Proposition 13.3 *Let V be a vector space finitely generated over a field F and let α be a nilpotent endomorphism of V having index of nilpotence k. If $w \in V$ satisfies the condition that $\dim(F[\alpha]w) = k$ then the subspace $F[\alpha]w$ of V has a complement in V which is invariant under α.*

Proof We will proceed by induction on k. If $k = 1$ then $\alpha = \sigma_0$ and so $F[\alpha]w = Fw$. Then there is a subset B of $V \smallsetminus \{Fw\}$ such that $B \cup \{w\}$ is a basis for V, and B is a basis for a complement of Fw in V. Assume that $k > 1$ and that the result has been established for any vector space finitely generated over F and any nilpotent endomorphism of that space having index of nilpotence less than k.

We know that $\mathrm{im}(\alpha)$ is invariant under α and that the restriction of α to $\mathrm{im}(\alpha)$ is nilpotent, having index of nilpotence $k - 1$. We know that the set $\{w, \alpha(w), \ldots, \alpha^{k-1}(w)\}$ forms a basis for $F[\alpha]w$ and so the set $\{\alpha(w), \ldots, \alpha^{k-1}(w)\}$ forms a basis for the image U of $F[\alpha]w$ under α. Therefore, $U = F[\alpha]\alpha(w)$ is a subspace of $\mathrm{im}(\alpha)$ and, by the induction hypothesis, it has a complement W_2 in $\mathrm{im}(\alpha)$ invariant under α.

Let $W_0 = \{v \in V \mid \alpha(v) \in W_2\}$. This is a subspace of V containing W_2, since W_2 is invariant under α. But $\alpha(v) \in W_2 \subseteq W_0$ for all $v \in W_0$ and so W_0 is also invariant under α. Our first assertion is that $V = F[\alpha]w + W_0$. And, indeed, if $x \in V$ then $\alpha(x) \in \mathrm{im}(\alpha) = U \oplus W_2$ and so $\alpha(x) = u + w_2$, where $u \in U$ and $w_2 \in W_2$. But $u = \alpha(y)$ for some $y \in F[\alpha]w$ and $x = y + (x - y)$. The first summand belongs to $F[\alpha]w$, whereas, as to the second, we have $\alpha(x - y) = \alpha(x) - \alpha(y) = \alpha(x) - u = w_2 \in W_2$ and so $x - y \in W_0$, proving the assertion.

Our second assertion is that $F[\alpha]w \cap W_0 \subseteq U$. Indeed, if $x \in F[\alpha]w \cap W_0$ then $\alpha(x) \in U \cap W_2 = \{0_V\}$ and so $x \in \ker(\alpha)$. Since $x \in F[\alpha]w$, we know that there exist scalars $a_0, \ldots, a_{k-1}$ such that $x = \sum_{i=0}^{k-1} a_i \alpha^i(w)$ and hence $0_V = \alpha(x) = \sum_{i=0}^{k-2} a_i \alpha^{i+1}(w)$, which implies that $a_0 = \cdots = a_{k-2} = 0$. Therefore, $x = a_{k-1} \alpha^{k-1}(w) \in U$, proving the second assertion.

In particular, from what we have seen. we deduce that the subspaces W_2 and $F[\alpha]w \cap W_0$ are disjoint. Therefore, $W_2 \oplus (F[\alpha]w \cap W_0)$ is a subspace of W_0. This subspace has a complement W_1 in W_0. Thus we have $W_0 = W_1 \oplus W_2 \oplus (F[\alpha]w \cap W_0)$.

Our third assertion is that $W = W_1 \oplus W_2$ is a complement of $F[\alpha]w$ in V which is invariant under α, and should we prove this, we will have proven the proposition. Indeed, we immediately note that $\alpha(W) \subseteq \alpha(W_0) \subseteq W_2 \subseteq W$ and so W is surely invariant under α. Moreover, $F[\alpha]w \cap W = \{0_V\}$ since this subspace is contained in the intersection of W and $F[\alpha]w \cap W_0$, which, by the choice of W, equals $\{0_V\}$. Finally,

$$V = F[\alpha]w + W_0 = F[\alpha]w + \left[W_1 + W_2 + (F[\alpha]w \cap W_0)\right]$$
$$= F[\alpha]w + W_1 + W_2 = F[\alpha]w + W,$$

and so $V = F[\alpha]w \oplus W$.   □

Proposition 13.4 (Rational Decomposition Theorem) *Let V be a vector space of finite dimension n over a field F let α be a nilpotent endomorphism of V having index of nilpotence k. Then there exist natural numbers $k = k_1 \geq \cdots \geq k_t$ satisfying $k_1 + \cdots + k_t = n$, and there exist vectors $v_1, \ldots, v_t$ in V such that $\{v_1, \alpha(v_1), \ldots, \alpha^{k_1-1}(v_1), v_2, \alpha(v_2), \ldots, \alpha^{k_2-1}(v_2), \ldots, v_t, \alpha(v_t), \ldots, \alpha^{k_t-1}(v_t)\}$ forms a basis for V. The matrix which represents α with respect to this basis is of the form*
$$\begin{bmatrix} A_1 & O & \cdots & O \\ O & A_2 & \cdots & O \\ \vdots & \vdots & \ddots & \vdots \\ O & O & \cdots & A_t \end{bmatrix}, \text{ where each } A_i \text{ is of}$$
the form
$$\begin{bmatrix} 0 & 0 & \cdots & 0 & 0 \\ 1 & 0 & \cdots & 0 & 0 \\ 0 & 1 & \cdots & 0 & 0 \\ \vdots & \vdots & \ddots & \vdots & \vdots \\ 0 & 0 & \cdots & 1 & 0 \end{bmatrix} \text{ in } \mathcal{M}_{k_i \times k_i}(F).$$

Proof Choose $k_1 = k$ and choose $v_1 \notin \ker(\alpha^{k_1-1})$. Then $U_1 = F[\alpha]v_1$ has a basis $\{v_1, \alpha(v_1), \ldots, \alpha^{k_1-1}(v_1)\}$. It is invariant under α and of dimension k_1. By Proposition 13.3, $V = U_1 \oplus W_1$, where W_1 is a subspace of V invariant under α. The restriction of α to W_1 is a nilpotent endomorphism of W_1 with index of nilpotence $k_2 \leq k_1$. We now repeat the above procedure for W_1. Pick $v_2 \in W_1 \setminus \ker(\alpha^{k_2-1})$. Then $U_2 = F[\alpha]v_2$ of W_1 has a basis $\{v_2, \alpha(v_2), \ldots, \alpha^{k_2-1}(v_2)\}$. It is invariant under α and of dimension k_2. Moreover, we can write $W_1 = U_2 \oplus W_2$, where W_2 is invariant under α. Continuing in this manner, we end up with a decomposition $V = U_1 \oplus \cdots \oplus U_t$, where each U_i is a subspace of V invariant under α having a basis of the form $\{v_i, \alpha(v_i), \ldots, \alpha^{k_i-1}(v_i)\}$ as above. This proves the first contention of the proposition. The second one follows since $U_i = F[\alpha]v_i$ for all i, which leads to a matrix of the desired form.   □

A matrix of the form given in Proposition 13.4 is called a representation of the nilpotent endomorphism α in *Jordan canonical form*. Let V be a vector space over a field F and let α be an endomorphism of V having an eigenvalue c. A vector $0_V \neq v \in V$ is a *generalized eigenvector* of α associated with c of *degree* $k > 0$ if and only if v is in $\ker((\alpha - c\sigma_1)^k) \setminus \ker((\alpha - c\sigma_1)^{k-1})$. Thus, in particular, the eigenvectors of α associated with c, in the previous sense, are just the generalized eigenvectors of α of degree 1 associated with c.

The nineteenth-century French mathematician **Camille Jordan** made major contributions to linear algebra, group theory, the theory of finite fields, and the beginnings of topology.

Example Let α be the endomorphism of $\mathbb{R}^4$ represented with respect to the canonical basis by the matrix $\begin{bmatrix} 2 & -2 & 1 & 1 \\ 0 & 1 & 1 & 1 \\ 0 & 0 & 2 & 1 \\ 0 & 0 & 0 & 2 \end{bmatrix}$. This endomorphism has an eigenvector $\begin{bmatrix} 2 \\ 1 \\ 0 \\ 0 \end{bmatrix}$ associated with the eigenvalue 1 and an eigenvector $\begin{bmatrix} 1 \\ 0 \\ 0 \\ 0 \end{bmatrix}$ associated with the eigenvalue 2. It also has a generalized eigenvector $\begin{bmatrix} 0 \\ -1 \\ -1 \\ 0 \end{bmatrix}$ of degree 2 associated with the eigenvalue 2 and a generalized eigenvector $\begin{bmatrix} 0 \\ -1 \\ -1 \\ -1 \end{bmatrix}$ of degree 3 associated with the eigenvalue 2.

We now prove a generalization of Proposition 12.1.

Proposition 13.5 *Let V be a vector space over a field F and let $\alpha \in \mathrm{End}(V)$ have an eigenvalue c. Then the set of all generalized eigenvectors of α (of all degrees) associated with c, together with 0_V, forms a subspace of V which is invariant under any endomorphism of V which commutes with α.*

Proof Let $a \in F$ and let $v, w \in V$ be generalized eigenvectors of α associated with c, of degrees k and h, respectively. Then both v and w belong to $\ker(\alpha - c\sigma_1)^{h+k}$ and hence the same is true for $v + w$ and av. This means that there exist positive integers $s, t \leq h + k$ such that $v + w \in \ker((\alpha - c\sigma_1)^s) \smallsetminus \ker((\alpha - c\sigma_1)^{s-1})$ and $av \in \ker((\alpha - c\sigma_1)^t) \smallsetminus \ker((\alpha - c\sigma_1)^{t-1})$, proving that we have a subspace.

If β is an endomorphism of V which commutes with α and if v is a generalized eigenvector of α associated with c such that $v \in \ker((\alpha - c\sigma_1)^k)$, then $(\alpha - c\sigma_1)^k \beta(v) = \beta(\alpha - c\sigma_1)^k(v) = 0_V$ so $\beta(v)$ is also a generalized eigenvector of α associated with c, proving invariance. $\square$

Let V be a vector space over a field F and let $\alpha \in \mathrm{End}(V)$ have an eigenvalue c. The subspace of V defined in Proposition 13.5 is called the *generalized eigenspace* of α associated with c.

Proposition 13.6 *Let V be a vector space over a field F and let $\alpha \in \mathrm{End}(V)$ have an eigenvalue c. Let v be a generalized eigenvector of degree k associated with c. Then the set of vectors $\{v, (\alpha - c\sigma_1)(v), \ldots, (\alpha - c\sigma_1)^{k-1}(v)\}$ is linearly independent.*

Proof Set $\beta = \alpha - c\sigma_1$ and, for each $1 \le j \le k$, let $v_j = \beta^{k-j}(v)$. Assume that there exist scalars $c_1, \ldots, c_k \in F$ satisfying $\sum_{j=1}^{k} c_j v_j = 0_V$. Then $0_V = \beta^{k-1}(\sum_{j=1}^{k} c_j v_j) = \beta^{k-1}(c_k v_k) = c_k \beta^{k-1}(v_k)$ and so, since $\beta^{k-1}(v_k) \ne 0_V$, we conclude that $c_k = 0$. We work backwards in this manner to see that $c_j = 0$ for all $1 \le j \le k$, and so the given set is linearly independent. $\square$

In particular, let V be a vector space of finite dimension n over a field F and let $\alpha \in \mathrm{End}(V)$. If v is a generalized eigenvector of α of degree k associated to an eigenvalue c of α, then we must have $k \le n$. Thus we see that $\dim(V)$ is an upper bound to the degree of generalized eigenvalue of α and we see that the generalized eigenspace of α associated to an eigenvalue c is just $\ker((\alpha - c\sigma_1)^n)$.

Proposition 13.7 *Let V be a vector space of finite dimension n over a field F and let $\alpha \in \mathrm{End}(V)$ satisfy the condition that the characteristic polynomial $p(X)$ of α is completely reducible, say $p(X) = \prod_{j=1}^{m}(X - c_j)^{n_j}$, where $\mathrm{spec}(\alpha) = \{c_1, \ldots, c_m\}$. Then there exist subspaces $U_1, \ldots, U_m$ of V, each of which invariant under α, such that:*
(1) $V = U_1 \oplus \cdots \oplus U_m$;
(2) $\dim(U_h) = n_h$ for each $1 \le h \le m$;
(3) For each $1 \le h \le m$, the restriction of α to U_h is of the form $c_h \tau_h + \beta_h$, where $\beta_h \in \mathrm{End}(U_h)$ is nilpotent and τ_h is the restriction of σ_1 to U_h.

Proof For each $1 \le h \le m$, consider the endomorphism $\beta_h = \alpha - c_h \sigma_1$ of V, and let U_h be the generalized eigenspace of α associated with c_h. Then U_h is a subspace of V invariant under β_h and also invariant under α since for all $v \in U_h$ we have $\beta_h^n \alpha(v) = \alpha \beta_h^n(v) = \alpha(0_V) = 0_V$. We claim that there exists a positive integer k, independent of h, such that all elements of U_h are generalized eigenvectors of α of degree at most k. Indeed, we see that $\ker(\beta_h) \subseteq \ker(\beta_h^2) \subseteq \ker(\beta_h^3) \subseteq \cdots$ and since V is finitely-generated, there are at most a finite number of proper containments. Thus there exists a k such that $\ker(\beta_h^k) = \ker(\beta_h^{k+1}) = \cdots$. From here it is clear that $\ker(\beta_h^k) = U_h$, proving the claim.

In particular, this claim shows that the restriction of β_h to U_h is a nilpotent endomorphism having index of nilpotence k. More than that, the restriction of α to U_h equals $c_h \tau_h + \beta_h$, proving (3). We now notice that if $t \neq h$ then U_t is invariant under β_h. We claim that the restriction of β_h to U_t is an automorphism. Since U_t is finite-dimensional, it is sufficient to prove that it is a monomorphism. Indeed, suppose that $v \in U_t \cap \ker(\beta_h)$. Then there exists a positive integer k such that $\beta_t^k(v) = 0_V$ and so $0_V = \beta_t^k(v) = [\beta_h + (c_h - c_t)]^k(v) = (c_h - c_t)^k(v)$ and, since $c_h - c_t \neq 0$, we must have $v = 0_V$, proving the claim.

The next step is to show that the collection $\{U_1, \ldots, U_m\}$ of subspaces of V is independent. Indeed, let $1 \leq h \leq m$ and let $Y = U_h \cap \sum_{j \neq h} U_j$. Then Y is a subspace of V invariant under β_h on which β_h is monic (since $Y \subseteq \sum_{j \neq h} U_j$) and nilpotent (since $Y \subseteq U_h$), which is possible only if $Y = \{0_V\}$. This proves independence, and we will set $U = U_1 \oplus \cdots \oplus U_m$. We want to show that $U = V$. Let $v \in U$. By the Cayley–Hamilton Theorem (Proposition 12.16), we see that α annihilates its characteristic polynomial $p(X)$ and so $[\prod_{i=1}^m \beta_i^{n_i}](v) = 0_V \in U$. Suppose that $\beta_t^{n_1}(v) \in U$, say that it is equal to $\sum_{i=1}^m u_i$, where $u_h \in U_h$ for all $1 \leq h \leq m$. Since $\beta_t^{n_1}$ is epic when restricted to U_h, for each $h \neq 1$, we can find an elements w_h of U_h for each $1 < h \leq m$, such that $u_h = \beta_t^{n_1}(w_h)$ for all such h. Therefore, $\beta_t^{n_1}(u - \sum_{h=2}^m w_h) = u_1 \in U_1$. By definition of U_1, it follows that $w_1 = u - \sum_{h=2}^m w_h \in U_1$. Therefore, $v = \sum_{h=1}^m w_h \in U$. If, on the other hand, $\beta_t^{n_1}(v) \notin U$, then let t be the smallest element of $\{2, \ldots, m\}$ satisfying the condition that $[\prod_{i=1}^t \beta_i^{n_i}](v) \in U$ and $[\prod_{i=1}^{t-1} \beta_i^{n_i}](v) \notin U$. A similar argument to the preceding then shows that we must have $v \in U$.

We are left to show that $\dim(U_h) = n_h$ for all $1 \leq h \leq m$. Pick a basis for V which is a union of bases of the U_h. With respect to this basis, the endomorphism

α is represented by a matrix of the form $\begin{bmatrix} A_1 & O & \cdots & O \\ O & A_2 & \cdots & O \\ \vdots & \vdots & \ddots & \vdots \\ O & O & \cdots & A_m \end{bmatrix}$, where each A_h

is a matrix representing the restriction of α to U_h. By Proposition 11.12, the characteristic polynomial of α is therefore of the form $|XI - A| = \prod_{h=1}^m |XI - A_h|$. From this decomposition and from the fact that each β_h restricts to an automorphism of U_t for all $t \neq h$, it follows that the only eigenvalue of the restriction of α to U_h is c_h, and the algebraic multiplicity of this eigenvalue is at most n_h. Since $\sum_{h=1}^m \dim(U_h) = \sum_{h=1}^m n_h$, it then follows that $\dim(U_h) = n_h$ for each h. $\qquad \square$

Proposition 13.7 shows that when conditions are right—for example, when the field F is algebraically closed—and when we are given an endomorphism α of a finite-dimensional vector space V, it is possible to choose a basis for V relative to which α is represented in a particularly simple form. We do this in two steps.

(I) Write V as a direct sum $U_1 \oplus \cdots \oplus U_m$ as above. By choosing a basis for V which is a union of bases of the U_h, we get a matrix representing α composed of blocks strung out along the main diagonal, each representing the restriction of α to one of the subspaces U_h.

(II) For each h, we have $\alpha = c_h \tau_h + \beta_h$, where β_h is a nilpotent endomorphism of U_h. We now choose a basis of U_h relative to which β_h is represented in Jordan canonical form.

Thus, in the end, we have a representation of α by a matrix of the form

$$\begin{bmatrix} A_1 & O & \cdots & O \\ O & A_2 & \cdots & O \\ \vdots & \vdots & \ddots & \vdots \\ O & O & \cdots & A_m \end{bmatrix}, \text{ where each block } A_h \text{ is a matrix with blocks of the form}$$

$$\begin{bmatrix} c_h & 0 & 0 & \cdots & 0 \\ 1 & c_h & 0 & \cdots & 0 \\ 0 & 1 & c_h & \cdots & 0 \\ & & \ddots & \ddots & \\ 0 & 0 & \cdots & 1 & c_h \end{bmatrix} \text{ on its diagonal (these may be } 1 \times 1\text{!) and all other en-}$$

tries equal to 0. A matrix of this form is called the *Jordan canonical form* of α. By Proposition 13.7, we see that if V is a vector space finitely generated over a field F and if α is an endomorphism of V having a completely reducible characteristic polynomial in $F[X]$, then there is a basis of V relative to which α can be represented by a matrix in Jordan canonical form. Thus, this can always be done if the field F is algebraically closed. If F is not algebraically closed then it is always possible to extend the field F to a larger field K such that the characteristic polynomial of α is completely reducible in $K[X]$.

Example Consider $\alpha \in \mathrm{End}(\mathbb{R}^4)$ represented with respect to the canonical basis by $A = \begin{bmatrix} 0 & 1 & 0 & 0 \\ 0 & 0 & 1 & 0 \\ 0 & 0 & 0 & 1 \\ -1 & 4 & -6 & 4 \end{bmatrix}$. The characteristic polynomial of A is $X^4 - 4X^3 + 6X^2 - 4X + 1 = (X - 1)^4$ and so its only eigenvalue is 1. Then A is similar to $B = \begin{bmatrix} 1 & 0 & 0 & 0 \\ 1 & 1 & 0 & 0 \\ 0 & 1 & 1 & 0 \\ 0 & 0 & 1 & 1 \end{bmatrix}$ in Jordan canonical form. Indeed, $B = PAP^{-1}$, where $P = \begin{bmatrix} -1 & 3 & -3 & 1 \\ 1 & -2 & 1 & 0 \\ -1 & 1 & 0 & 0 \\ 1 & 0 & 0 & 0 \end{bmatrix}$.

Example Consider $\alpha \in \text{End}(\mathbb{R}^5)$ represented with respect to the canonical basis

by $A = \begin{bmatrix} 3 & 0 & 0 & 0 & 0 \\ 0 & 4 & 2 & -1 & 4 \\ 0 & 0 & 2 & 0 & 0 \\ 0 & 1 & 3 & 2 & 1 \\ 0 & 0 & 0 & 0 & 2 \end{bmatrix}$. The characteristic polynomial of A is $(X - 3)^3 \cdot$

$(X - 2)^2$ and A is similar to $B = \begin{bmatrix} 2 & 0 & 0 & 0 & 0 \\ 0 & 3 & 0 & 0 & 0 \\ 0 & 1 & 3 & 0 & 0 \\ 0 & 0 & 0 & 2 & 0 \\ 0 & 0 & 0 & 0 & 3 \end{bmatrix}$ in Jordan canonical form. In-

deed,

$B = PAP^{-1}$, where $P = \begin{bmatrix} 0 & 0 & -3 & 0 & -4 \\ 0 & 1 & -1 & -1 & 3 \\ 2 & 1 & 3 & 0 & 1 \\ 0 & 0 & 0 & 0 & -1 \\ -1 & 0 & 0 & 0 & 0 \end{bmatrix}$.

Example Consider $\alpha \in \text{End}(\mathbb{C}^4)$ which is represented with respect to the canon-

ical basis by $A = \begin{bmatrix} 0 & 0 & 2i & 0 \\ 1 & 0 & 0 & 2i \\ -2i & 0 & 0 & 0 \\ 0 & -2i & 1 & 0 \end{bmatrix}$. The characteristic polynomial of A

is $(X - 2)^2(X + 2)^2$ and A is similar to $B = \begin{bmatrix} -2 & 0 & 0 & 0 \\ 1 & -2 & 0 & 0 \\ 0 & 0 & 2 & 0 \\ 0 & 0 & 1 & 2 \end{bmatrix}$. Indeed,

$B = PAP^{-1}$, where $P = \dfrac{1}{2}\begin{bmatrix} 1 & 0 & -i & 0 \\ 0 & 1 & 0 & -i \\ 1 & 0 & i & 0 \\ 0 & 1 & 0 & i \end{bmatrix}$.

We now use Jordan canonical forms to prove a result interesting in its own right.

Proposition 13.8 *Let n be a positive integer and let $A \in M_{n \times n}(F)$, where F is an algebraically-closed field. Then A can be written as a product of two symmetric matrices.*

Proof By Proposition 13.7, we know that A is similar to a matrix B in Jordan canonical form. In other words, there exists a nonsingular matrix $Q \in M_{n \times n}(F)$ satisfying $A = QBQ^{-1}$. If we can write $B = CD$, where both C and D are symmetric, then $A = QBDQ^{-1} = (QCQ^T)((Q^T)^{-1}DQ^{-1}) = (QCQ^T)((Q^{-1})^T DQ^{-1})$, where both QCQ^T and $(Q^{-1})^T DQ^{-1}$ are symmetric. Therefore, without loss of generality, we can assume that A is in Jordan canonical form, say

$$A = \begin{bmatrix} A_1 & O & \cdots & O \\ O & A_2 & \cdots & O \\ \vdots & \vdots & \ddots & \vdots \\ O & O & \cdots & A_m \end{bmatrix},$$

where each block A_h is of the form

$$\begin{bmatrix} a_h & 0 & 0 & \cdots & 0 \\ 1 & a_h & 0 & \cdots & 0 \\ 0 & 1 & a_h & \cdots & 0 \\ & & \ddots & \ddots & \\ 0 & 0 & \cdots & 1 & a_h \end{bmatrix} \in \mathcal{M}_{n_h \times n_h}(F).$$

Define the matrix $D_h \in \mathcal{M}_{n_h \times n_h}(F)$ to be $[d_{ij}]$, where

$$d_{ij} = \begin{cases} 1 & \text{if } i + j = n_h + 1, \\ 0 & \text{otherwise.} \end{cases}$$

Then D_h is a symmetric matrix satisfying $D_h^{-1} = D_h$. Moreover, the matrix

$$D = \begin{bmatrix} D_1 & O & \cdots & O \\ O & D_2 & \cdots & O \\ \vdots & \vdots & \ddots & \vdots \\ O & O & \cdots & D_m \end{bmatrix} \in \mathcal{M}_{n \times n}(F)$$

is also symmetric and satisfies $D^{-1} = D$. Furthermore, the matrix

$$C = \begin{bmatrix} A_1 D_1 & O & \cdots & O \\ O & A_2 D_2 & \cdots & O \\ \vdots & \vdots & \ddots & \vdots \\ O & O & \cdots & A_m D_m \end{bmatrix} \in \mathcal{M}_{n \times n}(F)$$

is also symmetric and $A = CD$, as required. $\square$

Another interesting result is the following.

Proposition 13.9 *If $A \in \mathcal{M}_{n \times n}(F)$, where F is an algebraically-closed field, then A is similar to its transpose.*

Proof Since F is algebraically closed, we know that A is similar to a matrix B in Jordan canonical form, and to show that B is similar to its transpose it suffices to show that each Jordan block is similar to its transpose. This is surely true of blocks

of size 1×1. If $C = \begin{bmatrix} c & 0 & 0 & \cdots & 0 \\ 1 & c & 0 & \cdots & 0 \\ 0 & 1 & c & \cdots & 0 \\ & & \ddots & \ddots & \\ 0 & 0 & \cdots & 1 & c \end{bmatrix}$ is a Jordan block of size $h \times h$ for

$h > 1$, then we note that $PCP = C^T$, where $P = [p_{ij}]$ is the involutory matrix
defined by the condition that

$$p_{ij} = \begin{cases} 1 & \text{if } i + j = n + 1, \\ 0 & \text{otherwise} \end{cases}$$

thus proving the result.   □

Contemporary American mathematician Richard Brualdi and Chinese mathematicians Pei Pei and Xingzhi Zhan have shown that the Jordan canonical form
of a matrix in $\mathcal{M}_{n \times n}(\mathbb{C})$ is the best one can get in terms of sparseness, namely they
proved that among all the matrices that are similar to a given matrix in $\mathcal{M}_{n \times n}(\mathbb{C})$,
the Jordan canonical form has the greatest number of off-diagonal zero entries.

Exercises

Exercise 847
Find endomorphisms α and β of $\mathbb{R}^3$ satisfying the condition that $\alpha\beta$ is not nilpotent but $c\alpha + d\beta$ is nilpotent for all $c, d \in \mathbb{R}$.

Exercise 848
Let V be a vector space over a field F and let $\alpha \in \text{End}(V)$ be nilpotent, having
index of nilpotence $k > 0$. Show that $\sigma_1 + \alpha \in \text{Aut}(V)$.

Exercise 849
Let V be a vector space finitely-generated over $\mathbb{C}$. Do there exist endomorphisms
α and β of V satisfying the condition that $\sigma_1 + \alpha\beta - \beta\alpha$ is nilpotent?

Exercise 850
Let F a field and let B be a given basis of F^3. Let $a \in F$ and let α be the
endomorphism of F^3 satisfying $\Phi_{BB}(\alpha) = \begin{bmatrix} -a & a & a \\ 0 & 0 & 0 \\ -a & a & a \end{bmatrix}$. For which values of
a is this endomorphism nilpotent?

Exercise 851
Let F be a field. Give an example of a nilpotent endomorphism of F^5 having
index of nilpotence 3.

Exercise 852

Let α be the endomorphism of $V = \mathbb{R}^4$ represented with respect to a basis B of

V by the matrix $\begin{bmatrix} 2 & -8 & 12 & -60 \\ 2 & -5 & 9 & -48 \\ 6 & -17 & 29 & -152 \\ 1 & -3 & 5 & -26 \end{bmatrix}$. Show that α is nilpotent and find its

index of nilpotence.

Exercise 853

Let V be a vector space over a field F and let $\alpha \in \mathrm{End}(V)$ be nilpotent. Does $\beta\alpha$ have to be nilpotent for all $\beta \in \mathrm{End}(V)$?

Exercise 854

Let V be a vector space over a field F and let $\alpha \in \mathrm{End}(V)$ be nilpotent. Find $\mathrm{spec}(\alpha)$.

Exercise 855

Let V be a vector space finitely generated over $\mathbb{C}$ and let $\alpha \in \mathrm{End}(V)$ satisfy $\mathrm{spec}(\alpha) = \{0\}$. Show that α is nilpotent.

Exercise 856

Let V be a vector space over a field F and let $\alpha \in \mathrm{End}(V)$ be nilpotent, having index of nilpotence k. Find the minimal polynomial of α.

Exercise 857

Let V be a vector space finitely generated over a field F and let $\alpha \in \mathrm{End}(V)$ satisfy the condition that for each $v \in V$ there exists a positive integer $n(v)$ satisfying $\alpha^{n(v)}(v) = 0_V$. Show that α is nilpotent. Does the same result hold if V is not assumed to be finitely generated over F?

Exercise 858

Let F be a field and let α be an endomorphism of F^3 represented with respect to

a basis B of F^3 by the matrix $\begin{bmatrix} 0 & a & 0 \\ 0 & 0 & b \\ 0 & 0 & 0 \end{bmatrix}$, where a and b are nonzero scalars.

Does there exist a endomorphism β of F^3 satisfying $\beta^2 = \alpha$?

Exercise 859

Let α be a nilpotent endomorphism of a vector space V over a field F having characteristic 0. Show that there exists an endomorphism β of V belonging to $F[\alpha]$ and satisfying $\beta^2 = \sigma_1 + \alpha$.

Exercise 860

Let $\alpha \in \mathrm{End}(\mathbb{R}^3)$ be represented with respect to the canonical basis by the matrix
$\begin{bmatrix} 1 & 2 & -2 \\ 3 & 0 & 3 \\ 1 & 1 & -2 \end{bmatrix}$. Calculate $\mathbb{R}[\alpha] \begin{bmatrix} 1 \\ 0 \\ 0 \end{bmatrix}$.

Exercise 861

Let V be the space of all infinitely-differentiable functions from $\mathbb{R}$ to itself. Let δ be the endomorphism of V which assigns to each function its derivative. What is $\mathbb{R}[\delta]\sin(x)$?

Exercise 862

Define $\alpha \in \mathrm{End}(\mathbb{R}^3)$ by $\alpha : \begin{bmatrix} a \\ b \\ c \end{bmatrix} \mapsto \begin{bmatrix} a+c \\ b-a \\ b \end{bmatrix}$. Find $\mathbb{R}[\alpha] \begin{bmatrix} 1 \\ 1 \\ 1 \end{bmatrix}$.

Exercise 863

Let V be a vector space over a field F and let $\alpha \in \mathrm{End}(V)$. Let $v \in V$ be a vector satisfying $F[\alpha^2]v = V$. Show that $F[\alpha]v = V$.

Exercise 864

Given $a \in \mathbb{R}$, let $\alpha_a \in \mathrm{End}(\mathbb{R}^4)$ be represented with respect to the canonical basis by the matrix $\begin{bmatrix} 0 & a & 1 & 0 \\ 1 & -2 & 1 & 1 \\ 0 & 0 & 1 & 0 \\ 0 & 1 & 0 & -2 \end{bmatrix}$. For which values of a is the dimension of $\mathbb{R}[\alpha_a] \begin{bmatrix} 0 \\ 0 \\ 0 \\ -1 \end{bmatrix}$ equal to 3?

Exercise 865

Let $\alpha \in \mathrm{End}(\mathbb{R}^3)$ be represented with respect to the canonical basis by the matrix $\begin{bmatrix} 1 & 1 & 1 \\ 0 & 1 & 0 \\ 0 & 0 & 2 \end{bmatrix}$. Find the eigenvalues of α and the generalized eigenspace associated with each.

Exercise 866

Let V be a vector space finitely generated over $\mathbb{C}$ and let $\alpha \in \text{End}(V)$. Show that α is diagonalizable if and only if every generalized eigenvector of α is an eigenvector of α.

Exercise 867

Let $B = \{v_1, v_2, v_3\}$ be the canonical basis of $\mathbb{R}^3$ and let α be the endomorphism of $\mathbb{R}^3$ satisfying $\Phi_{BB}(\alpha) = \begin{bmatrix} 1 & 0 & 2 \\ 0 & 1 & 0 \\ 0 & 0 & 1 \end{bmatrix}$. Show that $W = \mathbb{R}\{v_1, v_3\}$ and $Y = \mathbb{R}v_2$ are complements of each other in $\mathbb{R}^3$ and that each of these spaces is invariant under α.

Exercise 868

Let $A = \begin{bmatrix} 1 & 2 & 0 \\ 0 & 2 & 0 \\ 2 & -2 & -1 \end{bmatrix} \in \mathcal{M}_{3\times 3}(\mathbb{R})$. Find the Jordan canonical form of A.

Exercise 869

Let $A = \begin{bmatrix} -2 & 8 & 6 \\ -4 & 10 & 6 \\ 4 & -8 & -4 \end{bmatrix} \in \mathcal{M}_{3\times 3}(\mathbb{R})$. Find the Jordan canonical form of A.

Exercise 870

Let $O \neq A \in \mathcal{M}_{3\times 3}(\mathbb{C})$ be of the form $\begin{bmatrix} 0 & a & -b \\ -a & 0 & c \\ b & -c & 0 \end{bmatrix}$, where a, b, and c are real numbers. What is the Jordan canonical form of A?

Exercise 871

Let $A \in \mathcal{M}_{5\times 5}(\mathbb{Q})$ be a matrix in Jordan canonical form having minimal polynomial $(X - 3)^2$. What does A look like?

Exercise 872

Give an example of a matrix in $\mathcal{M}_{4\times 4}(\mathbb{R})$ which is not similar to a matrix in Jordan canonical form.

Exercise 873

Let V be a vector space finitely generated over a field F and let α and β be nilpotent endomorphisms of V represented with respect to some given basis by matrices A and B, respectively. If the matrices A and B are similar, does the index of nilpotence of α have to equal that of β?

Exercise 874

Let F be a field and let $A = \begin{bmatrix} a & 0 & 0 \\ 1 & a & 0 \\ 0 & 1 & a \end{bmatrix} \in \mathcal{M}_{3 \times 3}(F)$. Show that

$$A^k = \begin{bmatrix} a^k & 0 & 0 \\ ka^{k-1} & a^k & 0 \\ \frac{1}{2}k(k-1)a^{k-2} & ka^{k-1} & a^k \end{bmatrix} \text{ for all } k > 0.$$

Exercise 875

Let n be a positive integer and, for all $1 \le i, j \le n$, let $p_{ij}(X) \in \mathbb{C}[X]$. Let $\varphi : \mathbb{C} \to \mathcal{M}_{n \times n}(\mathbb{C})$ be the function defined by $\varphi : z \mapsto [p_{ij}(z)]$. Furthermore, let us assume that $\varphi(z)$ is nonsingular for each $z \in \mathbb{C}$. Show that there exists a nonzero complex number d such that $|\varphi(z)| = d$ for all $z \in \mathbb{C}$.

Exercise 876

For each $t \in \mathbb{C}$, let α_t be the endomorphism of $\mathbb{C}^3$ represented with respect to the canonical basis by the matrix $\begin{bmatrix} 0 & 1 & 0 \\ 0 & 0 & 0 \\ t & 0 & 0 \end{bmatrix}$. Is the representation of α_t in Jordan canonical form dependent on t?

Exercise 877

Let $A = \begin{bmatrix} 0 & 0 & 4 \\ 0 & 0 & 0 \\ 0 & 0 & 0 \end{bmatrix} \in \mathcal{M}_{3 \times 3}(\mathbb{C})$. Find the set of all $c \in \mathbb{C}$ satisfying the condition that cA is similar to A.

Exercise 878

Find the Jordan canonical form of $A = \dfrac{1}{2}\begin{bmatrix} -1 & 1 & 0 \\ 1 & -1 & 0 \\ 0 & 0 & 0 \end{bmatrix} \in \mathcal{M}_{3 \times 3}(\mathbb{R})$.

Exercise 879

Let $a \in \mathbb{R}$ be positive. Find the Jordan canonical form of

$$\begin{bmatrix} a & 0 & 0 & 1 \\ 0 & 1 & 1 & 0 \\ 0 & 1 & 1 & 0 \\ 0 & 0 & 0 & 1 \end{bmatrix} \in \mathcal{M}_{4 \times 4}(\mathbb{R}).$$

Exercise 880

Let $A \in \mathcal{M}_{n \times n}(\mathbb{R})$ differ from I and O. If A is idempotent, show that its Jordan canonical form is a diagonal matrix.

Exercise 881

Let $I \neq A \in \mathcal{M}_{n \times n}(\mathbb{R})$ be an involutory matrix. Show that the Jordan canonical form of A is a diagonal matrix.

The Dual Space

14

Let V be a vector space over a field F. A linear transformation from V to F (considered as a vector space over itself) is a *linear functional* on V. The space $\mathrm{Hom}(V, F)$ of all such linear functionals is called the *dual space* of V and will be denoted by $D(V)$. Note that $D(V)$ is a vector space over F, the identity element of which for addition is the 0-*functional*, $v \mapsto 0$. Since $\dim(F) = 1$, we immediately see that every linear functional other than the 0-functional must be an epimorphism.

With kind permission of the Archives of the Mathematisches Forschungsinstitut Oberwolfach.

Linear functionals were first studied systematically by the French mathematician **Jacques Hadamard**, whose long life ranged from the mid nineteenth century to the mid twentieth century, and by his student **Maurice Fréchet**. Their work on functionals turned them into a major tool in analysis.

Example Let F be a field and let n be a positive integer. Any $v \in F^n$ defines a linear functional in $D(F^n)$ by $w \mapsto v \odot w$.

Example Let V be a vector space over a field F and let B be a basis for V. Each $u \in B$ defines a function $f_u \in F^B$ defined by

$$f_u : u' \mapsto \begin{cases} 1 & \text{if } u' = u, \\ 0 & \text{otherwise}, \end{cases}$$

and by Proposition 6.2 we know that this function in turn defines a linear functional $\delta_u \in D(V)$. In particular, if $V = F^n$ and if $B = \{u_1, \dots, u_n\}$ is the canonical basis for V, then $\delta_{u_h} : \begin{bmatrix} a_1 \\ \vdots \\ a_n \end{bmatrix} \mapsto a_h$ for each $1 \le h \le n$.

J.S. Golan, *The Linear Algebra a Beginning Graduate Student Ought to Know*,
DOI 10.1007/978-94-007-2636-9_14, © Springer Science+Business Media B.V. 2012

Example Suppose that $V = C(a, b)$ and that $g_0 \in V$. Then the function $\eta : V \to \mathbb{R}$ defined by $\eta : f \mapsto \int_a^b f(x)g_0(x)\,dx$ belongs to $D(V)$. Hadamard's initial work on linear functionals concerned those of the form $f \mapsto \lim_{n \to \infty} \int_a^b f(x)g_n(x)\,dx$ for suitable sequences $g_1, g_2, \ldots$ in V.

Example Let V be the subspace of $\mathbb{R}^\mathbb{R}$ consisting of all infinitely-differentiable functions f satisfying the condition that there exist real numbers $a \le b$ such that $f(x) = 0$ if $x \notin [a, b]$. Then the function $f \mapsto \int_{-\infty}^\infty f(x)\,dx$ belongs to $D(V)$. Elements of $D(V)$ are known as *distributions* and play an important role in analysis and theoretical physics.

Note that the linear functional $\mathrm{tr} : \mathcal{M}_{n \times n}(F) \to F$ is not a homomorphism of F-algebras whenever $n > 1$. If $(K, \bullet)$ is an algebra over a field F then a nonzero linear functional $\delta \in D(K)$ which is also a homomorphism of F-algebras is called a *weight function* on K and an algebra having a weight function is called a *baric algebra*.[1] Nonassociative baric algebras are an important context for mathematical models in genetics.

Let F be a field and let n be a positive integer. Then there exists a linear functional $\mathrm{tr} : \mathcal{M}_{n \times n}(F) \to F$ which assigns to each matrix the sum of the elements of its diagonal, i.e., $\mathrm{tr} : [a_{ij}] \mapsto \sum_{i=1}^n a_{ii}$. This linear functional is called the *trace*. This functional will play an important part in our later discussion. Note that $v \odot w = \mathrm{tr}(v \wedge w)$ for all $v, w \in F^n$.

If $A = [a_{ij}]$ and $B = [b_{ij}]$ are matrices in $\mathcal{M}_{n \times n}(F)$, then it is easy to see that $\mathrm{tr}(AB) = \sum_{i=1}^n \sum_{h=1}^n a_{ih}b_{hi} = \mathrm{tr}(BA)$. We also notice that $\mathrm{tr}(I) = n$, where I is the identity matrix of $\mathcal{M}_{n \times n}(F)$. If the characteristic of the field F does not divide n, we claim that these conditions uniquely characterize the trace.

With kind permission of the Clarke University Archives.

The trace of a matrix was first defined by the nineteenth-century American mathematician **Henry Taber**.

Proposition 14.1 *Let n be a positive integer and let F be a field the characteristic of which does not divide n. Let δ be a linear functional on $\mathcal{M}_{n \times n}(F)$ satisfying the conditions that $\delta(AB) = \delta(BA)$ for all $A, B \in \mathcal{M}_{n \times n}(F)$ and that $\delta(I) = n$. Then $\delta = \mathrm{tr}$.*

---

[1] Such structures were first studied by the twentieth-century Scottish mathematician **I.M.H. Etherington**, who formulated the Mendelian laws algebraically.

Proof If $\{H_{ij} \mid 1 \le i, j \le n\}$ is the canonical basis of $\mathcal{M}_{n\times n}(F)$, then it suffices to show that $\delta(H_{ij}) = \text{tr}(H_{ij})$ for all $1 \le i, j \le n$. In particular, if $1 \le i, j \le n$ then $\delta(H_{ii}) = \delta(H_{ij}H_{ji}) = \delta(H_{ji}H_{ij}) = \delta(H_{jj})$. Since $I = \sum_{i=1}^{n} H_{ii}$, this implies that $n = \delta(I) = \sum_{i=1}^{n} \delta(H_{ii})$ and so $\delta(H_{ii}) = 1 = \text{tr}(H_{ii})$ for all $1 \le i \le n$. If $i \ne j$ then $H_{1j}H_{i1} = O$ and so $\delta(H_{ij}) = \delta(H_{i1}H_{1j}) = \delta(H_{1j}H_{i1}) = \delta(O) = 0 = \text{tr}(H_{ij})$, and we are done. $\qquad\square$

By the above, we see that if F is a field, if n is a positive integer, and if $A, B \in \mathcal{M}_{n\times n}(F)$, then $\text{tr}(A \bullet B) = \text{tr}(AB) - \text{tr}(BA) = 0$, where $\bullet$ denotes the Lie product on $\mathcal{M}_{n\times n}(F)$. In fact, over fields of characteristic 0 the converse is also true. In order to establish this fact, we first need a technical result.

Proposition 14.2 *Let F be a field of characteristic 0 and let n be a positive integer. Let $A \in \mathcal{M}_{n\times n}(F)$ have the property that it is similar to no matrix in $\mathcal{M}_{n\times n}(F)$ having a 0 for its $(1, 1)$-entry. Then A is a scalar matrix.*

Proof Clearly, A is not O and so there exists a vector $w \in F^n$ satisfying $A^T w \ne \begin{bmatrix} 0 \\ \vdots \\ 0 \end{bmatrix}$. Assume that there exists a vector $v \in F^n$ satisfying the condition that $w \odot v = 1$ and $(A^T w) \odot v = 0$. Let $\delta \in D(F^n)$ be given by $\delta : y \mapsto w \odot y$. Then the nullity of δ is $n - 1$ and we can pick a basis $\{y_2, \dots, y_n\}$ for $\ker(\delta)$. Since $v \notin \ker(\delta)$, we see that the set $\{v, y_2, \dots, y_n\}$ is linearly independent. Therefore, the matrix P the columns of which are $v, y_2, \dots, y_n$ is nonsingular, and w^T is the first row of P^{-1}. Moreover, the $(1, 1)$-entry of the matrix $P^{-1}AP$ is $(A^T w) \odot v = 0$, contradicting the assumption on A. This means that there is no vector v satisfying the given conditions and so $A^T w = c_w w$ for some scalar $c_w \in F$. Thus we conclude that if w is any vector in F^n then $A^T w$ is either the 0-vector or a scalar multiple of w and so, for any nonsingular matrix $Q \in \mathcal{M}_{n\times n}(F)$ we see that $Q^{-1}AQ = [b_{ij}]$ is a diagonal matrix. If $b_{hh} \ne b_{kk}$ and if y is the difference between the hth and kth rows of Q^{-1}, then $A^T y$ cannot be of the form $c_y y$, which is again a contradiction. Thus A must in fact be a scalar matrix. $\qquad\square$

Proposition 14.3 *Let F be a field of characteristic 0, let n be a positive integer, and let $C \in \mathcal{M}_{n\times n}(F)$. Then $\text{tr}(C) = 0$ if and only if C is the Lie product of matrices $A, B \in \mathcal{M}_{n\times n}(F)$.*

Proof We have already seen that if C is the Lie product of two matrices in $\mathcal{M}_{n\times n}(F)$ then $\text{tr}(C) = 0$. We will prove the converse by induction on n. The result is clearly true if $n = 1$ so we can assume, inductively, that $n > 1$ and that the result has been established for matrices in $\mathcal{M}_{k\times k}(F)$ for any $k < n$. Moreover, if $C = O$, take $A = B = O$ and we are done. Hence we can assume that $C \ne O$.

Since F has characteristic 0 and $\text{tr}(C) = 0$, we know that C is not a scalar matrix. By Proposition 14.2, this means that there is a nonsingular matrix P such that $P^{-1}CP$ has a 0 for its $(1, 1)$-entry. That is to say, we can write $P^{-1}CP$ in block form as $\begin{bmatrix} 0 & x^T \\ z & C' \end{bmatrix}$, where $x, z \in F^{n-1}$ and $C' \in \mathcal{M}_{(n-1)\times(n-1)}(F)$. Moreover, $\text{tr}(D) = \text{tr}(P^{-1}CP) = \text{tr}(C) = 0$ and so, by the induction hypothesis, there exist matrices $A', B' \in \mathcal{M}_{(n-1)\times(n-1)}(F)$ satisfying $C' = A'B' - B'A'$. If A' is singular, then we can replace A' by $A' - c'I$ for any scalar $c' \notin \text{spec}(A')$ (and such an element c' exists since F has characteristic 0 and hence is infinite). Therefore, without loss of generality, we can assume that A' is nonsingular. Then $P^{-1}CP = A''B'' - B''A''$, where $A'' = \begin{bmatrix} 0 & O \\ O & A' \end{bmatrix}$ and $B'' = \begin{bmatrix} 0 & -x^T A'^{-1} \\ A'^{-1}z & B' \end{bmatrix}$. Thus $C = (PA''P^{-1})(PB''P^{-1}) - (PB''P^{-1})(PA''P^{-1})$. □

The first of many proofs of this result was given by **Kenjiro Shoda**, one of the major figures in the twentieth-century Japanese algebra. The proof given here is due to Kahan.

Example Note that this result may be false if the field F has positive characteristic. For example, if $F = \text{GF}(2)$ then $\text{tr}\left(\begin{bmatrix} 1 & 0 \\ 0 & 1 \end{bmatrix}\right) = 0$ but there are no matrices A and B in $\mathcal{M}_{2\times 2}(F)$ satisfying $AB - BA = \begin{bmatrix} 1 & 0 \\ 0 & 1 \end{bmatrix}$.

Thus, if F has characteristic 0 then the set of all matrices $C \in \mathcal{M}_{n\times n}(F)$ satisfying $\text{tr}(C) = 0$ forms a subalgebra of the general Lie algebra $\mathcal{M}_{n\times n}(F)^-$, called the *special Lie algebra* defined by F^n.

If n is a positive integer, F is a field, and P is a nonsingular matrix in $\mathcal{M}_{n\times n}(F)$, then $\text{tr}(PAP^{-1}) = \text{tr}(PP^{-1}A) = \text{tr}(A)$ and so similar matrices have identical traces. In general, if B and C are fixed matrices in $\mathcal{M}_{n\times n}(F)$ then the functions $A \mapsto \text{tr}(BA)$ and $A \mapsto \text{tr}(AC)$ belong to $D(\mathcal{M}_{n\times n}(F))$.

The following result shows that traces essentially define all linear functionals on spaces of square matrices.

Proposition 14.4 *Let F be a field, let n be a positive integer, and let $\delta \in D(\mathcal{M}_{n\times n}(F))$. Then there exists a matrix $C \in \mathcal{M}_{n\times n}(F)$ satisfying $\delta : A \mapsto \text{tr}(AC)$ for all $A \in \mathcal{M}_{n\times n}(F)$.*

Proof For each $1 \leq i, j \leq n$, let H_{ij} be the matrix having (i, j)-entry equal to 1 and all of the other entries equal to 0. Then we know that the the set of all such matrices is a basis for $\mathcal{M}_{n \times n}(F)$. Let $C = [c_{ij}] \in \mathcal{M}_{n \times n}(F)$ be the matrix defined by $c_{ij} = \delta(H_{ji})$ for all $1 \leq i, j \leq n$. Then for each matrix $A = [a_{ij}] \in \mathcal{M}_{n \times n}(F)$ we have $\delta(A) = \delta(\sum_{i=1}^{n} \sum_{j=1}^{n} a_{ij} H_{ij}) = \sum_{i=1}^{n} \sum_{j=1}^{n} a_{ij} \delta(H_{ij}) = \sum_{i=1}^{n} \sum_{j=1}^{n} a_{ij} c_{ji} = \text{tr}(AC)$. $\quad\square$

Proposition 14.5 *Let F be a field, let n be a positive integer, and let $\delta \in D(\mathcal{M}_{n \times n}(F))$ be a linear functional satisfying $\delta(AB) = \delta(BA)$ for all $A, B \in \mathcal{M}_{n \times n}(F)$. Then there exists a scalar $c \in F$ such that $\delta(A) = c \cdot \text{tr}(A)$ for all $A \in \mathcal{M}_{n \times n}(F)$.*

Proof Again, for each $1 \leq i, j \leq n$, let H_{ij} be the matrix having (i, j)-entry equal to 1 and all of the other entries equal to 0. If $1 \leq i \neq j \leq n$ then $\delta(H_{ij}) = \delta(H_{ii} H_{ij}) = \delta(H_{ij} H_{ii}) = \delta(O) = 0$. Moreover, for all $1 \leq j, k \leq n$ we have $\delta(H_{jj}) = \delta(H_{jk} H_{kj}) = \delta(H_{kj} H_{jk}) = \delta(H_{kk})$. Thus we see that there exists a $c \in F$ such that $\delta(H_{jj}) = c$ for all $1 \leq j \leq n$ and from Proposition 14.4 we conclude that $\delta(A) = \text{tr}(A \cdot cI) = c \cdot \text{tr}(A)$ for all $A \in \mathcal{M}_{n \times n}(F)$. $\quad\square$

Proposition 14.6 (Taber's Theorem) *Let F be a field, let n be a positive integer, and let $A \in \mathcal{M}_{n \times n}(F)$ be a matrix the characteristic polynomial of which is completely reducible. Then $\text{tr}(A)$ is the sum of the eigenvalues of A (with the appropriate multiplicities).*

Proof Let $p(X) = \sum_{i=0}^{n} c_i X^i$ be the characteristic polynomial of A. We know that this polynomial is completely reducible, say $p(X) = \prod_{i=1}^{n} (X - b_i)$, and after multiplying this out, we see that $c_{n-1} = - \sum_{i=1}^{n} b_i$. But from the definition of the characteristic polynomial, we also see that $c_{n-1} = -\text{tr}(A)$. Thus we see that, for any such matrix, $\text{tr}(A)$ is the sum of the eigenvalues of A (with the appropriate multiplicities). $\quad\square$

Example Let F be a field, let Ω be a nonempty set, and let $V = F^{\Omega}$. For each $a \in \Omega$ there exists a linear functional $\delta_a \in D(V)$ defined by evaluation: $\delta_a : f \mapsto f(a)$. In the case that F is $\mathbb{R}$ and Ω is the unit interval of the real line, this functional is known to physicists as the *Dirac functional*. In analysis, evaluation functionals are often used to establish boundary conditions on classes of functions being studied.

Paul Dirac, the Nobel-prize-winning twentieth-century British physicist, built the first accepted model of quantum mechanics, in which linear functionals played a fundamental part.

Example Let n be a positive integer and let c and d be complex numbers. Can we find all matrices $A \in \mathcal{M}_{n \times n}(\mathbb{C})$ having the property that c is an eigenvalue of A having geometric multiplicity $n - 1$ and d is an eigenvalue of A having geometric multiplicity 1? (Certainly one such matrix always exists, namely a diagonal matrix with c appearing $n - 1$ times on the diagonal and d once.) In general, in order for c to be an eigenvalue of A of geometric multiplicity $n - 1$, the eigenspace associated with it has to be of dimension $n - 1$. In other words, the nullity of the matrix $A - cI$ must equal $n - 1$. From this we see that the dimension of the column space of $A - cI$ must equal 1, and so there must exist nonzero vectors $u = \begin{bmatrix} b_1 \\ \vdots \\ b_n \end{bmatrix}$ and $v = \begin{bmatrix} e_1 \\ \vdots \\ e_n \end{bmatrix}$ in $\mathbb{C}^n$ such that $A - cI = u \wedge v$, whence $A = u \wedge v + cI$. Conversely, if A is a matrix of the form $u \wedge v + cI$, then c is an eigenvalue of A having multiplicity at least $n - 1$. Note that $\mathrm{tr}(A) = \sum_{i=1}^{n} b_i e_i + nc$. But, as we just noted, $\mathrm{tr}(A)$ is also the sum of the eigenvalues of A, counted by multiplicity, and so we want it to equal $d + (n - 1)c$. Thus we are reduced to finding vectors u and v as above satisfying the condition that $\sum_{i=1}^{n} b_i e_i = d - c$. This is easy to do in concrete cases.

The following proposition shows that there always enough linear functionals to enable us to distinguish between vectors.

Proposition 14.7 *Let V be a vector space of a field F. If $v \neq w$ are elements of V then there exists a linear functional $\delta \in D(V)$ satisfying $\delta(v) \neq \delta(w)$.*

Proof Since the set $\{v - w\}$ is linearly independent, it can be completed to a basis B of V. By Proposition 6.2, there exists a linear functional $\delta \in D(V)$ satisfying $\delta(v - w) = 1$ and $\delta(u) = 0$ for all $u \in B \setminus \{v - w\}$. This is the linear functional we want. $\qquad\square$

In particular, if $0_V \neq v \in V$ then there is a linear functional $\delta \in D(V)$ satisfying $\delta(v) \neq 0$.

Proposition 14.8 *Let V be a vector space of a field F. Then $D(V) \cong F^B$. In particular, if V is finitely generated over F then $D(V) \cong V$.*

Proof We have a function $\alpha : D(V) \to F^B$ given by restriction, and it is straight-forward to check that this function is an R-homomorphism. Since every element of $D(V)$ is totally defined by its action on a basis, this function is monic and, by Proposition 6.2, it is epic. Therefore, it is an isomorphism. If B is finite, then $V \cong F^{(B)} = F^B$. $\qquad\square$

In particular, we see the important relationship between $F^{(\Omega)}$ and F^{Ω} for any nonempty set Ω, namely that F^{Ω} is isomorphic to the dual space of $F^{(\Omega)}$. Note too that $D(V) \not\cong V$ whenever the vector space V is not finitely generated over F since it can be shown, using the arithmetic of transfinite cardinals, that $F^{(\Omega)}$ and F^{Ω} are never isomorphic when Ω is infinite.

Let us consider the idea inherent in Proposition 14.8. Let V be a vector space over a field F and let B be a given basis for V. For each $v \in B$, let $\delta_v \in D(V)$ satisfy $\delta_v(v) = 1$ and $\delta_v(u) = 0$ for all $v \neq u \in B$. We claim that $E = \{\delta_v \mid v \in B\}$ is a linearly-independent subset of $D(V)$. Indeed, if $c_1, \ldots, c_n$ are scalars in F and $u_1, \ldots, u_n$ are elements of B satisfying the condition that $\sum_{i=1}^{n} c_i \delta_{u_i}$ is the 0-functional. Then for all $1 \le h \le n$ we have $0 = (\sum_{i=1}^{n} c_i \delta_{u_i})(u_h) = \sum_{i=1}^{n} c_i \delta_{u_i}(u_h) = c_h$. This establishes the claim. If V is finitely-generated then B is finite and so E is a basis for $D(V)$, since it is easy to check that $\delta = \sum_{u \in B} \delta(u) \delta_u$ for all $\delta \in D(V)$. Such a basis E for $D(V)$ is called the *dual basis* of the basis B for V. If V is not finitely generated, then FE is a subspace of $D(V)$ composed of all those linear functionals $\delta \in D(V)$ satisfying the condition that $\delta(u) \neq 0$ for at most finitely-many elements u of B. This subspace is called the *weak dual space* of V.

Example Let V be the vector space of all polynomial functions in $\mathbb{R}^{\mathbb{R}}$ having degree at most 4. Suppose that $B = \{a_1, \ldots, a_5\}$ is a set of distinct positive real numbers and, for each $1 \le i \le 5$, let $\delta_i \in D(V)$ be the linear transformation defined by $\delta_i : p(t) \longmapsto \int_0^{a_i} p(t)\,dt$. We claim that $B = \{\delta_1, \ldots, \delta_5\}$ is a basis for $D(V)$. Indeed, since we know by Proposition 14.8 that $\dim(D(V)) = 5$, all we have to show is that the set B is linearly independent. That is to say, we must show that if there exist real numbers $b_1, \ldots, b_5$ satisfying the condition that $\sum_{i=1}^{5} b_i \delta_i$ is the 0-functional, then $b_i = 0$ for all $1 \le i \le 5$. Since $\sum_{i=1}^{5} b_i \delta_i(t^h) = \sum_{i=1}^{5} [\frac{1}{h+1} a_i^{h+1}] b_i$ for all $0 \le h \le 4$, we must show that

$$\begin{vmatrix} a_1 & \frac{1}{2}a_1^2 & \frac{1}{3}a_1^3 & \frac{1}{4}a_1^4 & \frac{1}{5}a_1^5 \\ a_2 & \frac{1}{2}a_2^2 & \frac{1}{3}a_2^3 & \frac{1}{4}a_2^4 & \frac{1}{5}a_2^5 \\ \vdots & \vdots & \vdots & \vdots & \vdots \\ a_5 & \frac{1}{2}a_5^2 & \frac{1}{3}a_5^3 & \frac{1}{4}a_5^4 & \frac{1}{5}a_5^5 \end{vmatrix}$$

is nonsingu-lar, which is the case since this is just a nonzero scalar multiple of a Vandermonde matrix.

Example Let $a < b$ be real numbers and let $t_1, \ldots, t_n$ be distinct real numbers in the closed interval $[a, b]$ of the real line and let W be the subspace of $\mathbb{R}^{\mathbb{R}}$ consisting of all polynomial functions of degree less than n. Then $\dim(W) = n$. For each $1 \le i \le n$, let $\delta_i \in D(W)$ be the linear functional defined by $\delta_i : p \mapsto p(t_i)$. We claim that the subset $B = \{\delta_1, \ldots \delta_n\}$ of $D(W)$ is linearly independent. Indeed, if $\sum_{i=1}^{n} c_i \delta_i$ is the 0-functional and if $1 \le h \le n$, then $0 = (\sum_{i=1}^{n} c_i \delta_i) \prod_{j \ne h} (X - t_j) = c_h \prod_{j \ne h} (t_h - t_j)$, which implies that $c_h = 0$ since the t_i are distinct. Therefore, by Proposition 14.8, B is a basis for $D(W)$. Since the function $p \mapsto \int_a^b p(x)\, dx$ also belongs to $D(W)$, we conclude that there exist real numbers $c_1, \ldots, c_n$ satisfying $\int_a^b p(x)\, dx = \sum_{i=1}^{n} c_i p(t_i)$ for any $p \in W$.

Let V and W be vector spaces over a field F and let $\alpha \in \mathrm{Hom}(V, W)$. If $\delta \in D(W)$ then $\delta\alpha \in D(V)$. Moreover, if $\delta_1, \delta_2 \in D(W)$ and if $v \in V$ then $[(\delta_1 + \delta_2)\alpha](v) = (\delta_1 + \delta_2)\alpha(v) = \delta_1\alpha(v) + \delta_2\alpha(v) = [\delta_1\alpha + \delta_2\alpha](v)$ and so $(\delta_1 + \delta_2)\alpha = \delta_1\alpha + \delta_2\alpha$. Similarly, if $c \in F$ and if $\delta \in D(W)$ then $c(\delta\alpha) = (c\delta)\alpha$. Therefore, we see that α defines a linear transformation $D(\alpha) \in \mathrm{Hom}(D(W), D(V))$ by setting $D(\alpha) : \delta \mapsto \delta\alpha$. If V, W, and Y are vector spaces over F and if $\alpha \in \mathrm{Hom}(V, W)$ and $\beta \in \mathrm{Hom}(W, Y)$ then it is straightforward to show that $D(\beta\alpha) = D(\alpha)D(\beta)$. If α is an isomorphism, then $D(\alpha)$ is also an isomorphism, where $D(\alpha)^{-1} = D(\alpha^{-1})$.

Proposition 14.9 *Let F be a field and let V and W be vector spaces finitely generated over F. Let $B = \{v_1, \ldots, v_k\}$ be a basis for V, the dual basis of which is $C = \{\delta_1, \ldots, \delta_k\}$, and let $D = \{w_1, \ldots, w_n\}$ be a basis for W, the dual basis of which is $E = \{\eta_1, \ldots, \eta_n\}$. If $\alpha : V \to W$ is a linear transformation then $\Phi_{EC}(D(\alpha)) = \Phi_{BD}(\alpha)$.*

Proof Let $\Phi_{BD}(\alpha) = [a_{ij}]$. For each $1 \le i \le k$ we have $\alpha(v_i) = \sum_{h=1}^{n} a_{hi} w_h$ and so for all $1 \le i \le k$ and all $1 \le j \le n$ we have $[D(\alpha)(\eta_j)](v_i) = \eta_j\alpha(v_i) = \sum_{h=1}^{n} a_{hi}\eta_j(w_h) = a_{ji}$. But each $\delta \in D(V)$ satisfies $\delta = \sum_{i=1}^{k} \delta(v_i)\delta_i$ and so, in particular, $D(\alpha)(\eta_j) = \sum_{i=1}^{k} [D(\alpha)(\eta_j)](v_i)\delta_i = \sum_{i=1}^{k} a_{ji}\delta_i$, which gives the desired result. $\qquad \square$

We have already seen that, given a vector space V over a field F, we can build the dual space $D(V)$. Since this too is a vector space over F, we can go on to built its dual space, $D^2(V) = D(D(V))$. What do some elements of this space look like? Each $v \in V$ defines a function $\theta_v : D(V) \to F$ by setting $\theta_v : \delta \mapsto \delta(v)$. This is indeed a linear functional and so is an element of $D^2(V)$, which we call the *evaluation functional* at v.

> **Proposition 14.10** *Let V be a vector space over a field F. The function $v \mapsto \theta_v$ is a monomorphism from V to $D^2(V)$, which is an isomorphism in the case V is finitely generated.*

Proof We first have to show that this function is a linear transformation. And, indeed, if $v, w \in V$, if $a \in F$, and if $\delta \in D(V)$, then as a direct consequence of the definitions we obtain $\theta_{v+w}(\delta) = \delta(v + w) = \delta(v) + \delta(w) = \theta_v(\delta) + \theta_w(\delta) = [\theta_v + \theta_w](\delta)$ and so $\theta_{v+w} = \theta_v + \theta_w$. Similarly, $\theta_{av}(\delta) = \delta(av) = a\delta(v) = a\theta_v(\delta)$ and so $\theta_{av} = a\theta_v$. Thus we have shown that we do indeed have a homomorphism. If v belongs to the kernel of this function then $\theta_v(\delta) = \delta(v)$ for all $\delta \in D(V)$ and so, by Proposition 14.5, we know that $v = 0_V$. Thus it is a monomorphism. Finally, if V is finitely generated then, by Proposition 14.6, we see that $\dim(D^2(V)) = \dim(D(V)) = \dim(V)$ and so any monomorphism from V to $D^2(V)$ has to be an isomorphism. $\square$

We should note that the importance of Proposition 14.10 lies not in the existence of an isomorphism between V and $D^2(V)$, which could be inferred from dimension arguments alone, but in finding a specific, natural, such isomorphism.

A proper subspace W of a vector space V over a field F is a *maximal subspace* if and only if there is no subspace of V properly contained in V and properly containing W. By the Hausdorff Maximum Principle, we know that any nontrivial vector space contains a maximal subspace. The maximal subspaces of finitely-generated vector spaces are usually called *hyperplanes* of the space. We will now use linear functionals in order to characterize these subspaces of V.

> **Proposition 14.11** *A subspace W of a vector space V over a field F is maximal if and only if there exists a linear functional $\delta \in D(V)$ which is not the 0-functional, with kernel W.*

Proof Let us assume that $W = \ker(\delta)$, where δ is a linear functional which is not the 0-functional, and assume that there exists a proper subspace Y of V which properly contains W. Pick $y \in Y \smallsetminus W$ and $x \in V \smallsetminus Y$. These two vectors have to be nonzero and the set $\{x, y\}$ is linearly independent by Proposition 5.3, since $Fy \subseteq Y$ and $x \notin Y$. Set $U = F\{x, y\}$. Then $\ker(\delta)$ and U are disjoint, so the restriction of δ to U is a monomorphism, which is impossible since $\dim(U) = 2$ and $\dim(F) = 1$. Therefore, W must be a maximal subspace of V. Conversely, let W be a maximal subspace of V and let $y \in V \smallsetminus W$. Then $Fy \cap W = \{0_V\}$ and $Fy + W = V$ by the maximality of W. Therefore, $V = Fy \oplus W$ and so every vector in V can be written in the form $ay + w$, where $a \in F$ and $w \in W$. The function $\delta : ay + w \mapsto a$ is a linear functional in $D(V)$ the kernel of which equals W. $\square$

Proposition 14.12 *Let V be a vector space over a field F and let $\delta, \delta_1, \ldots, \delta_n$ be elements of $D(V)$. Then $\delta \in F\{\delta_1, \ldots, \delta_n\}$ if and only if $\bigcap_{i=1}^{n} \ker(\delta_i) \subseteq \ker(\delta)$.*

Proof Assume that $\delta \in F\{\delta_1, \ldots, \delta_n\}$. Then there exist scalars $a_1, \ldots, a_n$ such that $\delta = \sum_{i=1}^{n} a_i \delta_i$. If $v \in \bigcap_{i=1}^{n} \ker(\delta_i)$ then $\delta_i(v) = 0$ for all $1 \le i \le n$ and so $\delta(v) = \sum_{i=1}^{n} a_i \delta_i(v) = 0$. Thus $v \in \ker(\delta)$. Conversely, suppose that $\bigcap_{i=1}^{n} \ker(\delta_i) \subseteq \ker(\delta)$. We will proceed by induction on n. First, assume that $n = 1$. If δ is the 0-functional, then surely we are done. Thus let us assume that this is not the case and let $v \in V \smallsetminus \ker(\delta)$. Since $\ker(\delta_1) \subseteq \ker(\delta)$, this means that $\delta_1(v) \ne 0$. Set $a = \delta_1(v)^{-1}\delta(v)$. Then $\delta(v) = a\delta_1(v) = (a\delta_1)(v)$ and so $v \in \ker(\delta - a\delta_1)$. But $\ker(\delta_1) \subseteq \ker(\delta - a\delta_1)$, and so this containment is again proper. By Proposition 14.11, $\ker(\delta_1)$ is a maximal subspace of V and so $\ker(\delta - a\delta_1) = V$, which shows that $\delta = a\delta_1$.

Now let us assume that we have prove the result for a given n and assume we have linear functionals $\delta, \delta_1, \ldots, \delta_{n+1}$ in $D(V)$ satisfying $\bigcap_{i=1}^{n+1} \ker(\delta_i) \subseteq \ker(\delta)$. Set $W = \ker(\delta_{n+1})$ and for each $1 \le i \le n$ let β_i be the restriction of δ_i to W. Also, let β be the restriction of δ to W. Then $\bigcap_{i=1}^{n} \ker(\beta_i) \subseteq \ker(\beta)$ and so, by the induction hypothesis, we know that there exist scalars $a_1, \ldots, a_n$ such that $\beta = \sum_{i=1}^{n} a_i \beta_i$. Therefore, $\ker(\delta_{n+1}) \subseteq \ker(\delta - \sum_{i=1}^{n} a_i \delta_i)$ and, as in the case $n = 1$, it follows that there exists a scalar a_{n+1} such that $\delta - \sum_{i=1}^{n} a_i \delta_i = a_{n+1}\delta_{n+1}$, proving that $\delta = \sum_{i=1}^{n+1} a_i \delta_i$. $\qquad\square$

In the context of functional analysis, the following consequence of Proposition 14.11, taken together with the Riesz Representation Theorem (Proposition 16.14), is known as the *Fredholm alternative*, and has many important applications.

The Swedish mathematician **Ivar Fredholm** was active in the late nineteenth century and studied the solvability of integral equations.

Proposition 14.13 *Let V and W be vector spaces over a field F, let $\alpha \in \mathrm{Hom}(V, W)$, and let $w \in W$. Then $w \in \mathrm{im}(\alpha)$ if and only if $w \in \ker(\delta)$ for any $\delta \in D(W)$ satisfying $\mathrm{im}(\alpha) \subseteq \ker(\delta)$.*

Proof If $w \in \mathrm{im}(\alpha)$ then the given condition clearly holds. Conversely, assume that $w \notin \mathrm{im}(\alpha)$ and let B be a basis for $\mathrm{im}(\alpha)$. By Proposition 5.3, the set $\{w\} \cup B$

is linearly independent, and so there exists a subset B' of W containing B such that $\{w\} \cup B'$ is a basis for W. Then FB' is a maximal subspace of W and so, by Proposition 14.11, there exists a $\delta \in D(W)$ satisfying $\delta(w) \neq 0$ and $\mathrm{im}(\alpha) \subseteq FB' = \ker(\delta)$. $\square$

Exercises

Exercise 882
Let $V = C(0, 1)$. From calculus we know that for each $f \in V$ there exists a maximal element a_f of $\{f(t) \mid 0 \leq t \leq 1\}$. Is the function $f \mapsto a_f$ a linear functional on V?

Exercise 883
Let W be a subspace of $\mathbb{Q}[X]$ generated by a countably-infinite linearly-independent set $\{p_1(X), p_2(X), \ldots\}$ of polynomials. Let $\delta : W \to \mathbb{Q}$ be the function defined by $\delta : \sum_{i=1}^{\infty} a_i p_i(X) \mapsto \sum_{i=1}^{\infty} a_i \deg(p_i)$ (where only finitely-many of the a_i are nonzero). Does δ belong to $D(W)$?

Exercise 884
Let $F = \mathrm{GF}(2)$ and let $\delta : F^3 \to F$ be the function which assigns to each vector $v = \begin{bmatrix} a \\ b \\ c \end{bmatrix}$ the value (0 or 1) appearing in the majority of entries of v. Is δ a linear functional?

Exercise 885
Find a linear functional $\delta \in D(\mathbb{R}^3)$ which is not the 0-functional but which satisfies $\delta \left(\begin{bmatrix} 3 \\ 2 \\ -1 \end{bmatrix} \right) = \delta \left(\begin{bmatrix} 3 \\ 2 \\ 1 \end{bmatrix} \right) = 0$.

Exercise 886
Let $V = \mathbb{Q}[X]$ and to each vector $v = [b_1, b_2, \ldots] \in \mathbb{Q}^{\infty}$ assign a linear functional $\delta_v \in D(V)$ defined by $\delta_v : \sum_{n=0}^{\infty} a_n X^n \mapsto \sum_{n=0}^{\infty} n! a_n b_{n+1}$. Is the function $\alpha : \mathbb{Q}^{\infty} \to D(V)$ defined by $v \mapsto \delta_v$ an isomorphism?

Exercise 887
Let V be a vector space over a field F and let $\alpha, \beta \in D(V)$ satisfy the condition that $\ker(\beta) \subseteq \ker(\alpha)$. Show that $\alpha \in F\beta$.

Exercise 888
Let F be a field and let $0 \neq a \in F$. Let $\alpha : F[X] \to F$ be the function defined by $\alpha : p(X) \mapsto p(a) - p(0)$. Is α a linear functional?

Exercise 889
Let F be a field of characteristic 0 and let n be a positive integer. Show that any matrix $A \in \mathcal{M}_{n \times n}(F)$ is similar to a matrix all diagonal entries of which are equal to 0.

Exercise 890
Let $V = \mathbb{R}^3$ and consider the linear functionals

$$\delta_1 : \begin{bmatrix} a \\ b \\ c \end{bmatrix} \mapsto 2a - b + 3c, \quad \delta_2 : \begin{bmatrix} a \\ b \\ c \end{bmatrix} \mapsto 3a - 5b + c, \quad \text{and}$$

$$\delta_3 : \begin{bmatrix} a \\ b \\ c \end{bmatrix} \mapsto 4a - 7b + c$$

on V. Is $\{\delta_1, \delta_2, \delta_3\}$ a basis for $D(V)$?

Exercise 891
Let V be a vector space finitely generated over a field F and let W be a subspace of V having a complement Y in V. Show that $D(V) = W' \oplus Y'$, where W' is a subspace of $D(V)$ isomorphic to W and Y' is a subspace of $D(V)$ isomorphic to Y.

Exercise 892
Let n be a positive integer and let V be the vector space of all polynomial functions from $\mathbb{R}$ to itself of degree no more than n. For all $0 \le k \le n$, let $\delta_n : V \to \mathbb{R}$ be the function defined by $\delta_k : p \mapsto \int_{-1}^{1} t^k p(t)\, dt$. Show that $\{\delta_1, \ldots, \delta_n\}$ is a basis of $D(V)$.

Exercise 893
Let $B = \left\{ \begin{bmatrix} 0 \\ 3 \\ -2 \end{bmatrix}, \begin{bmatrix} 0 \\ 1 \\ -1 \end{bmatrix}, \begin{bmatrix} 1 \\ -1 \\ 3 \end{bmatrix} \right\} \subseteq \mathbb{R}^3$. Find the dual basis of B.

Exercise 894
Let n be a positive integer and let V be a vector space of dimension n over a field F. Let $B = \{\delta_1, \ldots, \delta_n\}$ be a subset of $D(V)$ and assume that there exists a vector $0_V \ne v \in V$ satisfying $\delta_i(v) = 0$ for all $0 \le i \le n$. Show that B is linearly dependent.

Exercise 895
Let V be a vector space over a field F. For every subspace W of V, let $E(W) = \{\delta \in D(V) \mid \ker(\delta) \supseteq W\}$. Show that $E(W)$ is a subspace of $D(V)$. Moreover, if $\{W_i \mid i \in \Omega\}$ is a family of subspaces of V, show that $E(\sum_{i \in \Omega} W_i) = \bigcap_{i \in \Omega} E(W_i)$.

Exercise 896

Let V be a vector space finitely generated over a field F and let W be a subspace of V. For $E(W) = \{\delta \in D(V) \mid \ker(\delta) \supseteq W\}$, show that $\dim(W) + \dim(E(W)) = \dim(V)$.

Exercise 897

Let V be a vector space finitely generated over a field F, let W be a subspace of V, and let Y be a subspace of $D(V)$. Are the following conditions equivalent:
(1) $Y = \{\delta \in D(V) \mid \ker(\delta) \supseteq W\}$;
(2) $W = \bigcap_{\delta \in Y} \ker(\delta)$?

Exercise 898

Let $A, B \in \mathcal{M}_{2 \times 2}(\mathbb{R})$. Show that $\mathrm{tr}(AB) = \mathrm{tr}(A) \cdot \mathrm{tr}(B)$ if and only if $|A + B| = |A| + |B|$.

Exercise 899

Let n be a positive integer and let U be a finite subset of $\mathcal{M}_{n \times n}(\mathbb{C})$ which is closed under multiplication of matrices. Show that there exists a matrix A in U satisfying $\mathrm{tr}(A) \in \{1, \dots, n\}$.

Exercise 900

Let n be a positive integer and let F be a field. For any matrix $A = [a_{ij}] \in \mathcal{M}_{n \times n}(F)$, define the *antitrace* of A to be $\mathrm{antitr}(A) = \sum_{i=1}^{n} a_{i,n+1-i}$. Is the function $A \mapsto \mathrm{antitr}(A)$ a linear functional on $\mathcal{M}_{n \times n}(F)$?

Exercise 901

Let F be a field and let $A \in \mathcal{M}_{2 \times 2}(F)$ be a matrix satisfying $\mathrm{tr}(A) = \mathrm{tr}(A^2) = 0$. Is it necessarily true that $A = O$?

Exercise 902

Let k and n be positive integers. If $O \neq A \in \mathcal{M}_{k \times n}(\mathbb{R})$, does there necessarily exist a matrix $B \in \mathcal{M}_{n \times k}(\mathbb{R})$ satisfying $\mathrm{tr}(AB) \neq 0$?

Exercise 903

Let F be a field and let $k \neq n$ be positive integers. Let $A \in \mathcal{M}_{k \times n}(F)$ and $B \in \mathcal{M}_{n \times k}(F)$. Are $\mathrm{tr}(AB)$ and $\mathrm{tr}(BA)$ necessarily equal?

Exercise 904

Show that the matrices $\begin{bmatrix} 1 & 2-i & 1+i \\ 4+i & 1+i & 0 \\ 1+i & 1 & 1 \end{bmatrix}$ and $\begin{bmatrix} 1 & 1+i & 2-i \\ 3-i & 1+i & 0 \\ 1 & 27 & 1-i \end{bmatrix}$ in $\mathcal{M}_{3 \times 3}(\mathbb{C})$ are not similar.

Exercise 905

Let n be a positive integer and let V be the subspace of $\mathbb{R}[X]$ composed of all polynomials of degree at most n. What is the dual basis of $\{1, X, \dots, X^n\}$?

Exercise 906
Let n be a positive integer. If B and C are elements of $\mathcal{M}_{n \times n}(\mathbb{R})$ satisfying $\mathrm{tr}(B) \leq \mathrm{tr}(C)$, and if $A \in \mathcal{M}_{n \times n}(\mathbb{R})$, is it necessarily true that $\mathrm{tr}(AB) \leq \mathrm{tr}(AC)$?

Exercise 907
For a matrix $A \in \mathcal{M}_{3 \times 3}(\mathbb{R})$, find a positive integer c satisfying

$$|A| = \frac{1}{c} \begin{vmatrix} \mathrm{tr}(A) & 1 & 0 \\ \mathrm{tr}(A^2) & \mathrm{tr}(A) & 2 \\ \mathrm{tr}(A^3) & \mathrm{tr}(A^2) & \mathrm{tr}(A) \end{vmatrix}.$$

Exercise 908
Let k and n be positive integers and let F be a field. Define a function $\alpha : \mathcal{M}_{kn \times kn}(F) \to \mathcal{M}_{n \times n}(F)$ as follows: if $A \in \mathcal{M}_{kn \times kn}(F)$, write $A = [A_{ij}]$, where each A_{ij} is a $(k \times k)$-block. Then set $\alpha(A) = [b_{ij}] \in \mathcal{M}_{n \times n}(F)$, where $b_{ij} = \mathrm{tr}(A_{ij})$ for each $1 \leq i, j \leq n$. Is α a linear transformation? Is it a homomorphism of unital F-algebras?

Exercise 909
Let A be a nonempty set and let V be the collection of all subsets of A, which is a vector space over GF(2). Is the characteristic function of $\varnothing \neq D \subseteq A$ a linear functional on V?

Exercise 910
For each integer $n > 1$, find a nonsingular matrix $A \in \mathcal{M}_{n \times n}(\mathbb{Q})$ satisfying $\mathrm{tr}(A) = 0$.

Exercise 911
Let $n > 1$ be an integer and let $A \in \mathcal{M}_{n \times n}(\mathbb{R})$. Does there necessarily exist a symmetric matrix $B \in \mathcal{M}_{n \times n}(\mathbb{R})$ satisfying $\mathrm{tr}(A) = \mathrm{tr}(B)$?

Exercise 912
Let V be a vector space finitely generated over $\mathbb{Q}$ and let $\alpha \in \mathrm{End}(V)$ be a projection. Show that there is a basis D of V satisfying the condition that the rank of α equals $\mathrm{tr}(\Phi_{DD}(\alpha))$.

Exercise 913
Let W be a proper subspace of a vector space V over a field F and let $v \in V \smallsetminus W$. Show that there is a linear functional $\delta \in D(V)$ satisfying $\delta(v) \neq 0$ but $\delta(w) = 0$ for all $w \in W$.

Exercise 914
Let V be a vector space finitely generated over a field F and let W_1 and W_2 be proper subspaces of V satisfying $V = W_1 \oplus W_2$. Show that $D(V) = E_1 \oplus E_2$, where $E_j = \{\delta \in D(V) \mid W_j \subseteq \ker(\delta)\}$ for $j = 1, 2$.

Exercise 915
Let F be a field and let $A \in \mathcal{M}_{2 \times 2}(F)$. Show that we always have $A^2 - \operatorname{tr}(A)A + |A|I = O$.

Exercise 916
Let V be the subspace of $\mathbb{R}^\infty$ consisting of all sequences $[a_1, a_2, \ldots] \in \mathbb{R}^\infty$ satisfying the condition that $\lim_{i \to \infty} a_i$ exists in $\mathbb{R}$. Define linear functionals $\delta_1, \delta_2, \ldots, \delta_\infty \in D(V)$ by setting $\delta_h : [a_1, a_2, \ldots] \mapsto a_h$ for each $h = 1, 2, \ldots$ and $\delta_\infty : [a_1, a_2, \ldots] \mapsto \lim_{i \to \infty} a_i$. Is the subset $\{\delta_1, \delta_2, \ldots, \delta_\infty\}$ of $D(V)$ necessarily linearly independent?

Exercise 917
Let F be a field and, for each $a \in F$, let $\varepsilon_a : F[X] \to F$ be the linear functional defined by $\varepsilon_a : p(X) \mapsto p(a)$. Show that the subset $\{\varepsilon_a \mid a \in F\}$ of $D(F[X])$ is linearly independent.

Exercise 918
Let V be a vector space over a field F and let $\delta_1, \delta_2 \in D(V)$ be linear functionals satisfying the condition that $\delta_1(v)\delta_2(v) = 0$ for all $v \in V$. Show that one of the δ_i must be the 0-functional.

Exercise 919
Let $n > 1$ be an integer and let $f : \mathbb{R}^n \to \mathbb{R}$ be a continuous function which maps the 0-vector to 0 and which satisfies the condition that $f(v + w) + f(v - w) = 2f(v)$ for all $v, w \in \mathbb{R}^n$. Show that $f \in D(\mathbb{R}^n)$.

Exercise 920
Let $a \in \mathbb{R}$, let n be a positive integer, and let $A, B \in \mathcal{M}_{n \times n}(\mathbb{R})$. Does there necessarily exist a matrix $C \in \mathcal{M}_{n \times n}(\mathbb{R})$ satisfying $AC + \operatorname{tr}(C)A = B$?

Exercise 921
Let F be a field and let n be a positive integer. Let $\delta : \mathcal{M}_{n \times n}(F) \to F$ be the linear functional given by $\delta : [a_{ij}] \mapsto \sum_{i=1}^{n} \sum_{j=1}^{n} a_{ij}$. Find an endomorphism α of $\mathcal{M}_{n \times n}(F)$ satisfying the condition that $\delta(A) = a \cdot \operatorname{tr}(\alpha(A))$ for all $A \in \mathcal{M}_{n \times n}(F)$.

Exercise 922
Let F be a field and let n be a positive integer. Let $A, B \in \mathcal{M}_{n \times n}(F)$ be matrices satisfying $A^2 + B^2 = I$ and $AB + BA = O$. Show that $\operatorname{tr}(A) = \operatorname{tr}(B) = 0$.

Exercise 923
Let F be a field and let n be a positive integer. Given a positive integer k, is it necessarily true that $\operatorname{tr}((AB)^k) = \operatorname{tr}(A^k)\operatorname{tr}(B^k)$ for all $A, B \in \mathcal{M}_{n \times n}(F)$?

Exercise 924
Let V be a vector space finitely generated over a field F and let $\alpha \in \text{End}(V)$. Show that α and $D(\alpha)$ have identical minimal polynomials.

Exercise 925
Let V be a vector space over a field F and let n be a positive integer. Let $v_1, \dots, v_n$ be distinct vectors in V and assume that there exist $\alpha \in \text{End}(V)$ and $\delta \in D(V)$ such that the matrix $[\delta\alpha^{i-1}(v_j)] \in \mathcal{M}_{n \times n}(F)$ is nonsingular. Show that the set $\{v_1, \dots, v_n\}$ is linearly independent.

Exercise 926
Let W be a subspace of a vector space V over a field F. Show that W is a maximal subspace of V if and only if every complement of W in V has dimension 1.

Exercise 927
Let F be a field and let n be a positive integer. For matrices $A, B \in \mathcal{M}_{n \times n}(F)$, calculate $\text{tr}([AB - BA][AB + BA])$.

Exercise 928
Let n be a positive integer. Can we find matrices $A, B \in \mathcal{M}_{n \times n}(\mathbb{C})$ satisfying the condition that all eigenvalues of A and of B are positive real numbers, but not all eigenvalues of $A + B$ are positive real numbers?

Exercise 929
Let k and n be positive integers, let F be a field, and let $O \neq A \in \mathcal{M}_{k \times n}(F)$. Does there necessarily exist a matrix $B \in \mathcal{M}_{n \times k}(F)$ satisfying $\text{tr}(AB) \neq 0$.

Exercise 930
Let V be a vector space of finite dimension n over a field F. A nonempty finite collection $\{W_1, \dots, W_k\}$ of hyperplanes of V is *co-independent* if and only if $\dim(\bigcap_{i=1}^{k} W_i) = n - k$. Is a nonempty subcollection of a co-independent collection of hyperplanes necessarily co-independent?

Exercise 931
If V is a vector space over $\mathbb{R}$ then the complexification of $D(V)$ is isomorphic to $\text{Hom}_{\mathbb{R}}(V, \mathbb{C})$ as vector spaces over $\mathbb{C}$.

Exercise 932
Let F be a field and let k and n be positive integers. If $A \in \mathcal{M}_{k \times n}(F)$, are $\text{tr}(AA^T)$ and $\text{tr}(A^T A)$ necessarily equal?

Exercise 933
Let F be a field of characteristic other than 2. Show that any matrix $A \in \mathcal{M}_{2 \times 2}(F)$ can be written in the form $cI + B$, where $c \in F$ and $\text{tr}(B) = 0$.

Inner Product Spaces

15

In this chapter, we will have to restrict the set of fields over which we work. A subfield F of $\mathbb{R}$ is *real Euclidean* if and only if for each $0 \le c \in F$ there exists an element $d \in F$ satisfying $d^2 = c$ and a subfield K of $\mathbb{C}$ is *Euclidean* if and only if there exists a real Euclidean field F such that $K = \{a + bi \mid a, b \in F\}$. It is immediately clear that if K is a Euclidean field and $c \in K$, then $\bar{c} \in K$. Being a Euclidean field is intimately tied in with the constructibility of elements of the complex plane by straightedge and compass constructions, and in fact every real Euclidean field must contain all those real numbers which are then lengths of line segments obtainable from the unit line segment by straightedge and compass construction methods. Clearly, $\mathbb{R}$ itself is real Euclidean, while $\mathbb{Q}$, as we have already noted, is not; the set real numbers algebraic over $\mathbb{Q}$ is real Euclidean and properly contained in $\mathbb{R}$. The field $\mathbb{C}$ is Euclidean, and the set of all algebraic numbers is Euclidean and properly contained in $\mathbb{C}$.

Let V be a vector space over a Euclidean field F. A function μ from $V \times V$ to F is an *inner product* on V if and only if:

(1) For each $w \in V$, the function $v \mapsto \mu(v, w)$ from V to F is a linear functional;
(2) If $v, w \in V$ then $\mu(v, w) = \overline{\mu(w, v)}$;
(3) If $v \in V$ then $\mu(v, v)$ is a nonnegative real number, which equals 0 if and only if $v = 0_V$.

Note that, in the above situation, if $v, w \in V$ then, as a consequence of (2), $\mu(v, w) + \mu(w, v) = 2\operatorname{Re}(\mu(v, w))$ is also always a real number, though it may, of course, be negative.

In general, once we have fixed an inner product on a space, we will write $\langle v, w \rangle$ instead of $\mu(v, w)$. A vector space over a Euclidean subfield F of $\mathbb{C}$ on which we have an inner product defined is called an *inner product space*. Another term for such a space, coming from functional analysis, is a *pre-Hilbert space*. Abstract inner product spaces were first studied in an axiomatic manner by von Neumann. While inner product spaces over general Euclidean fields may prove to be interesting in the future, at the moment the study of such spaces is almost universally restricted to spaces over $\mathbb{R}$ or $\mathbb{C}$, and so from now on we will do the same and consider only these as possible fields of scalars. When we talk about an inner product space

J.S. Golan, *The Linear Algebra a Beginning Graduate Student Ought to Know*,
DOI 10.1007/978-94-007-2636-9_15, © Springer Science+Business Media B.V. 2012

without specifying the field of scalars, we will always assume that it is one of these two fields.

Example Let n be a positive integer and let F be either $\mathbb{R}$ or $\mathbb{C}$. We define an inner product on F^n, called the *dot product*, as follows: if $v = \begin{bmatrix} a_1 \\ \vdots \\ a_n \end{bmatrix}$ and $w = \begin{bmatrix} b_1 \\ \vdots \\ b_n \end{bmatrix}$, then we set $v \cdot w = \sum_{i=1}^{n} a_i \overline{b}_i$. Note that if $F = \mathbb{R}$, then this product just coincides with the interior product $v \odot w$ which defined earlier. However, that is not true for the case $F = \mathbb{C}$, so we must be very careful to distinguish between the two products. This modification of the definition is necessary since, over $\mathbb{C}$, we have $\begin{bmatrix} 1 \\ i \end{bmatrix} \odot \begin{bmatrix} 1 \\ i \end{bmatrix} = 0$, even though $\begin{bmatrix} 1 \\ i \end{bmatrix} \neq \begin{bmatrix} 0 \\ 0 \end{bmatrix}$. Hence the interior product $\odot$ is not an inner product as we have defined it in this chapter.

With kind permission of the Archives of the Mathematisches Forschungsinstitut Oberwolfach (Artin).

The problem arises because, in $\mathbb{C}$, 0 can be written as the sum of squares of nonzero elements. A field F in which 0 cannot be the sum of squares of nonzero elements of F is *formally real*; so $\mathbb{R}$ is formally real while $\mathbb{C}$ is not. The theory of formally real fields was developed in the 1920s by the Austrian mathematicians **Emil Artin** and **Otto Schreier**.

We can generalize the previous example. If F is either $\mathbb{R}$ or $\mathbb{C}$, and if $D = [d_{ij}]$ is a nonsingular matrix in $\mathcal{M}_{n \times n}(F)$, we can define an inner product on F^n by setting $\left\langle \begin{bmatrix} a_1 \\ \vdots \\ a_n \end{bmatrix}, \begin{bmatrix} b_1 \\ \vdots \\ b_n \end{bmatrix} \right\rangle = \begin{bmatrix} a_1 & \cdots & a_n \end{bmatrix} D D^H \begin{bmatrix} \overline{b}_1 \\ \vdots \\ \overline{b}_n \end{bmatrix}$, where $D^H = [\overline{d}_{ij}]^T$. The matrix D^H is called the *conjugate transpose* or *Hermitian transpose* of D, and it again belongs to $\mathcal{M}_{n \times n}(F)$. Conjugate transposes of matrices over $\mathbb{C}$ will play an important part in the following discussion; of course, $D^H = D^T$ for any matrix $D \in \mathcal{M}_{n \times n}(\mathbb{R})$.

The properties of the conjugate transpose are very much like those of the transpose. Indeed, we note that if $A, B \in \mathcal{M}_{n \times n}(\mathbb{C})$ and $c \in \mathbb{C}$, then $(A + B)^H = A^H + B^H$, $(cA)^H = \overline{c} A^H$, $A^{HH} = A$, and $(AB)^H = B^H A^H$. In particular, if A is nonsingular then $I = I^H = (AA^{-1})^H = (A^{-1})^H A^H$, proving that $(A^{-1})^H = (A^H)^{-1}$.

Example If we are given positive real numbers $c_1, \ldots, c_n$ and consider the diagonal matrix $D = [d_{ij}] \in \mathcal{M}_{n \times n}(\mathbb{R})$ the diagonal entries of which are given by $d_{ii} = \sqrt{c_i}$ for $1 \le i \le n$, then, by the above, we have an inner product on $\mathbb{C}^n$ given by

$$\left\langle \begin{bmatrix} a_1 \\ \vdots \\ a_n \end{bmatrix}, \begin{bmatrix} b_1 \\ \vdots \\ b_n \end{bmatrix} \right\rangle = \sum_{i=1}^{n} c_i a_i \overline{b}_i.$$ Such a product is called a *weighted dot product*.

Weighted dot products are extremely important in statistics and data analysis, where we often want to emphasize the values of certain parameters and de-emphasize others.

Example Let $a < b$ be real numbers and let $V = C(a, b)$. This is, as we have seen, a vector space over $\mathbb{R}$, on which we can define an inner product $\langle f, g \rangle = \int_a^b f(x)g(x)\, dx$. Continuity is important here. The set Y of all functions from $[a, b]$ to $\mathbb{R}$ which are continuous at all but finitely-many points is a subspace of $\mathbb{R}^{[a,b]}$ properly containing $C(a, b)$ but $\langle f, g \rangle = \int_a^b f(x)g(x)\, dx$ is not an inner product on Y. Indeed, if we select a real number c satisfying $a < c < b$ and define the function $f \in Y$ by

$$f : x \mapsto \begin{cases} 1 & \text{if } x = c, \\ 0 & \text{otherwise} \end{cases}$$

then f is a nonzero element of Y but $\langle f, f \rangle = 0$.

Similarly, if V be the set of all continuous complex-valued functions defined on the closed interval $[a, b]$ in $\mathbb{R}$, then V is a vector space over $\mathbb{C}$, on which we can define an inner product $\langle f, g \rangle = \int_a^b f(x)\overline{g(x)}\, dx$.

Example Let F be $\mathbb{R}$ or $\mathbb{C}$, and let $V = M_{n \times n}(F)$, which is a vector space over F. Define an inner product on V by setting $\langle A, B \rangle = \text{tr}(AB^H) = \text{tr}(B^H A)$. If $A = [a_{ij}]$ and $B = [b_{ij}]$, then $\langle A, B \rangle = \sum_{i=1}^{n} \sum_{j=1}^{n} a_{ij} \overline{b}_{ij}$. In particular, $\langle A, A \rangle = \sum_{i=1}^{n} \sum_{j=1}^{n} |a_{ij}|^2$.

Example Let V be the subspace of $\mathbb{C}^\infty$ composed of all those sequences $[c_0, c_1, \dots]$ of complex numbers satisfying $\sum_{i=0}^{\infty} |c_i|^2 < \infty$. This vector space is very important in analysis, and we can define an inner product on it by setting $\langle [c_0, c_1, \dots], [d_0, d_1, \dots] \rangle = \sum_{i=0}^{\infty} c_i \overline{d}_i$.

Let F be $\mathbb{R}$ or $\mathbb{C}$, and let W be a subspace of an inner product space V over F. The restriction of this inner product to a function from $W \times W$ to F is an inner product on W. Thus we can always assume that any subspace of an inner product space V inherits the inner-product-space structure of V.

Example Let V be an inner product space over $\mathbb{R}$ and let K be the set of all matrices of the form $\begin{bmatrix} a & v \\ w & b \end{bmatrix}$, where $a, b \in \mathbb{R}$ and $v, w \in V$. Then K is a vector space over $\mathbb{R}$, where addition and scalar multiplication are defined by

$$\begin{bmatrix} a & v \\ w & b \end{bmatrix} + \begin{bmatrix} a' & v' \\ w' & b' \end{bmatrix} = \begin{bmatrix} a + a' & v + v' \\ w + w' & b + b' \end{bmatrix} \quad \text{and} \quad c \begin{bmatrix} a & v \\ w & b \end{bmatrix} = \begin{bmatrix} ca & cv \\ cw & cb \end{bmatrix}.$$

We create the structure of an $\mathbb{R}$-algebra on K by defining an operation $\bullet$ as follows:

$$
\begin{bmatrix} a & v \\ w & b \end{bmatrix} \bullet \begin{bmatrix} a' & v' \\ w' & b' \end{bmatrix} = \begin{bmatrix} aa' + \langle v, w' \rangle & av' + b'v \\ a'w + bw' & bb' + \langle w, v' \rangle \end{bmatrix}.
$$

This algebra is a division algebra, called the *Cayley algebra*, and it is not associative.

We now look at some properties of general inner product spaces.

Proposition 15.1 *Let V be an inner product space. For $v, w_1, w_2 \in V$ and for a scalar a, we have:*
(1) $\langle v, w_1 + w_2 \rangle = \langle v, w_1 \rangle + \langle v, w_2 \rangle$;
(2) $\langle v, aw_1 \rangle = \overline{a}\langle v, w_1 \rangle$;
(3) $\langle 0_V, w_1 \rangle = \langle v, 0_V \rangle = 0$.

Proof From the definition of the inner product, we have $\langle v, w_1 + w_2 \rangle = \overline{\langle w_1 + w_2, v \rangle} = \overline{\langle w_1, v \rangle + \langle w_2, v \rangle} = \overline{\langle v, w_1 \rangle} + \overline{\langle v, w_2 \rangle} = \langle v, w_1 \rangle + \langle v, w_2 \rangle$, which proves (1). We also have $\langle v, aw_1 \rangle = \overline{\langle aw_1, v \rangle} = \overline{a\langle w_1, v \rangle} = \overline{a}\,\overline{\langle w_1, v \rangle} = \overline{a}\langle v, w_1 \rangle$, which proves (2). Finally, $\langle 0_V, w_1 \rangle = \langle 00_V, w_1 \rangle = 0$, and similarly $\langle v, 0_V \rangle = 0$, proving (3). $\qquad\square$

By Proposition 15.1 we see that if V is an inner product space over $\mathbb{R}$ then for each $v \in V$ the function $w \mapsto \langle v, w \rangle$ from V to F is a linear transformation, but that is not the case for inner product spaces over $\mathbb{C}$.

Let V be a finitely-generated inner product space and let $v_1, \ldots, v_k$ be a list of vectors in V. The *Gram matrix* of this list is the $k \times k$ matrix $G = [g_{ij}]$ defined by $g_{ij} = \langle v_i, v_j \rangle$ for all $1 \le i, j \le k$. Let $B = \{v_1, \ldots, v_n\}$ be a basis for V. Given vectors $v = \sum_{i=1}^{n} a_i v_i$ and $w = \sum_{j=1}^{n} b_j v_j$ in V, we note that

$$
\langle v, w \rangle = \sum_{i=1}^{n} \sum_{j=1}^{n} a_i \overline{b}_j \langle v_i, v_j \rangle = \begin{bmatrix} a_1 & \cdots & a_n \end{bmatrix} G \begin{bmatrix} \overline{b}_1 \\ \vdots \\ \overline{b}_n \end{bmatrix},
$$

where G is the Gram matrix of B.

Jorgen Gram was a Danish mathematician who at the end of the nineteenth century developed computational techniques for inner product spaces in connection with his work for insurance companies.

Example Let V be the subspace of $\mathbb{C}[X]$ consisting of all polynomials of degree at most 5, and let B be the canonical basis for V. Define an inner product on V by setting $\langle f, g \rangle = \int_0^1 f(x)\overline{g(x)}\, dx$. (Note that we are using the same notation for a polynomial and its corresponding polynomial function in $\mathbb{C}^{\mathbb{R}}$.) Then the Gram matrix defined by B is precisely the Hilbert matrix H_6, which we have seen earlier.

Proposition 15.2 (Cauchy–Schwarz–Bunyakovsky Theorem) *Let V be an inner product space. If $v, w \in V$, then $|\langle v, w \rangle|^2 \leq \langle v, v \rangle \langle w, w \rangle$.*

Proof If $v = 0_V$ or $w = 0_V$ then the result is immediate, and so we can assume that both vectors differ from 0_V. Let $a = -\langle w, v \rangle$ and $b = \langle v, v \rangle$. Then $\overline{a} = -\langle v, w \rangle$ and $\overline{b} = b$ so

$$0 \leq \langle av + bw, av + bw \rangle = a\overline{a}\langle v, v \rangle + ab\langle v, w \rangle + b\overline{a}\langle w, v \rangle + b^2\langle w, w \rangle$$

$$= a\overline{a}b - a\overline{b}a - ab\overline{a} + b^2\langle w, w \rangle = b\left[-a\overline{a} + b\langle w, w \rangle\right].$$

Since $v \neq 0_V$, it follows that b is a positive real number and so $a\overline{a} \leq b\langle w, w \rangle$, which is what we want. $\qquad\square$

With kind permission of ETH-Bibliothek Zurich, Image Archive (Schwarz).

Herman Schwarz was a German mathematician who in the late nineteenth century studied spaces of functions and their structure as inner product spaces. **Viktor Yakovlevich Bunyakovsky** was a Russian student of Cauchy who proved this theorem a generation before Schwarz, but since his work was published in an obscure journal, it was not widely recognized until the twentieth century.

Example If $a_1, \ldots, a_n, b_1, \ldots, b_n, c_1, \ldots, c_n$ are real numbers, with $c_i > 0$ for all $1 \leq i \leq n$, then

$$\left|\sum_{i=1}^n c_i a_i b_i\right| \leq \left(\sqrt{\sum_{i=1}^n c_i a_i^2}\right)\left(\sqrt{\sum_{i=1}^n c_i b_i^2}\right).$$

Indeed, this is a consequence of the Cauchy–Schwarz–Bunyakovsky Theorem, using the weighted dot product $\left\langle \begin{bmatrix} a_1 \\ \vdots \\ a_n \end{bmatrix}, \begin{bmatrix} b_1 \\ \vdots \\ b_n \end{bmatrix} \right\rangle = \sum_{i=1}^n c_i a_i b_i$ defined on $\mathbb{R}^n$.

In general, the Cauchy–Schwarz–Bunyakovsky Theorem is an extremely rich source of inequalities between real-valued functions of several real variables. For example, consider the vectors $v = \dfrac{1}{\sqrt{a+b+c}} \begin{bmatrix} \sqrt{a+b} \\ \sqrt{a+c} \\ \sqrt{b+c} \end{bmatrix}$ and $w = \begin{bmatrix} 1 \\ 1 \\ 1 \end{bmatrix}$ in $\mathbb{R}^3$, where $a, b,$ and c are positive. Then, by the Cauchy–Schwarz–Bunyakovsky Theorem, we see that

$$\sqrt{\frac{a+b}{a+b+c}} + \sqrt{\frac{a+c}{a+b+c}} + \sqrt{\frac{b+c}{a+b+c}} = v \cdot w \leq \sqrt{\langle v, v \rangle \langle w, w \rangle} = \sqrt{6}.$$

Similarly, we note that the matrix $D = \begin{bmatrix} \sqrt{3} & 0 \\ 1 & \sqrt{2} \end{bmatrix} \in \mathcal{M}_{2\times 2}(\mathbb{R})$ is nonsingular and so, by a previous example, we have an inner product μ on $\mathbb{R}^2$ defined by

$$\mu\left(\begin{bmatrix} a \\ b \end{bmatrix}, \begin{bmatrix} c \\ d \end{bmatrix} \right) = \begin{bmatrix} a \\ b \end{bmatrix}^T DD^T \begin{bmatrix} c \\ d \end{bmatrix} = 3(ac + bd) + (\sqrt{3})(ad + bc).$$

Applying the Cauchy–Schwarz–Bunyakovsky Theorem, we see that for all real numbers $a, b, c,$ and d we have

$$\left[3(ac + bd) + (\sqrt{3})(ad + bc) \right]^2$$
$$\leq \left[3(a^2 + b^2) + (2\sqrt{3})ab \right]\left[3(c^2 + d^2) + (2\sqrt{3})cd \right].$$

In particular, if we take $b = d = \sqrt{3}$, we see that $(ac + a + c + 3)^2 \leq (a^2 + 2a + 3)(c^2 + 2c + 3)$ for all real numbers a and c.

Let V be an inner product space. The *norm* of a vector $v \in V$ is defined to be the scalar $\|v\| = \sqrt{\langle v, v \rangle}$. A vector v satisfying $\|v\| = 1$ is *normal*.

Example Let $V = \mathbb{R}^n$, and endow V with the dot product. Then

$$\left\| \begin{bmatrix} a_1 \\ \vdots \\ a_n \end{bmatrix} \right\| = \sqrt{\sum_{i=1}^{n} a_i^2}.$$

This norm is known as the *Euclidean norm* on V.

Example Let $V = C(-\pi, \pi)$, on which we have defined the inner product $\langle f, g \rangle = \int_{-\pi}^{\pi} f(x)g(x)\,dx$. For each positive integer k, consider the function $f_k : x \mapsto \sin(kx)$. Then $\|f_k\| = \sqrt{\langle f_k, f_k \rangle} = \sqrt{\int_{-\pi}^{\pi} \sin^2(kx)\,dx} = \sqrt{\pi}$ and so $g_k = \frac{1}{\sqrt{\pi}} f_k$ is a normal vector in this space.

We have seen how the vector space $\mathbb{R}^3$, endowed with the cross product $\times$, is a Lie algebra. It is easy to check that the cross product is related to the dot product on

$\mathbb{R}^3$ by the relations

$$u \times (v \times w) = (u \cdot w)v - (u \cdot v)w \quad \text{and}$$

$$(u \times v) \times w = (u \cdot w)v - (v \cdot w)u$$

for all $u, v, w \in \mathbb{R}^3$. Moreover, we have the following identities:
(1) $v \cdot (v \times w) = 0$ for all $v, w \in \mathbb{R}^3$;
(2) (Lagrange identity) $\|v \times w\|^2 = \|v\|^2 \|w\|^2 - (v \cdot w)^2$.

There are only two possible anticommutative operations on $\mathbb{R}^3$ which turn it into an $\mathbb{R}$-algebra satisfying these two identities, namely $\times$ and the operation $\times'$ given by $v \times' w = -(v \times w)$. Furthermore, if $n > 3$ no such operation can be defined on $\mathbb{R}^n$, except for the case of $n = 7$. In that case, we can define an operation $\times$ as follows: write elements of $\mathbb{R}^7$ in the form $\begin{bmatrix} v \\ a \\ v' \end{bmatrix}$, where $v, v' \in \mathbb{R}^3$ and $a \in \mathbb{R}$ and then set

$$\begin{bmatrix} v \\ a \\ v' \end{bmatrix} \times \begin{bmatrix} w \\ b \\ w' \end{bmatrix} = \begin{bmatrix} aw' - bv' + (v \times w) - (v' \times w') \\ -v \cdot w + v' \cdot w' \\ bv - aw + (v \times w') - (v' \times w) \end{bmatrix}.$$

We also note that if $u = \begin{bmatrix} a_1 \\ a_2 \\ a_3 \end{bmatrix}$, $v = \begin{bmatrix} b_1 \\ b_2 \\ b_3 \end{bmatrix}$, and $w = \begin{bmatrix} c_1 \\ c_2 \\ c_3 \end{bmatrix}$ in $\mathbb{R}^3$ then

$u \cdot (v \times w) = \begin{vmatrix} a_1 & b_1 & c_1 \\ a_2 & b_2 & c_2 \\ a_3 & b_3 & c_3 \end{vmatrix}$. As an immediate consequence, we observe that if

$u, v, w \in \mathbb{R}^3$ then:
(1) $u \cdot (v \times w) = v \cdot (w \times u) = w \cdot (u \times v)$;
(2) $u \cdot (v \times w) = 0$ if and only if two of these vectors are equal or the set $\{u, v, w\}$ is linearly dependent.

The scalar value $u \cdot (v \times w)$ is often called the *scalar triple product* of the vectors u, v, w, to distinguish it from the *vector triple product* $u \times (v \times w)$.

Proposition 15.3 *Let V be an inner product space. If $v, w \in V$ and if a is a scalar, then:*
(1) $\|av\| = |a| \cdot \|v\|$;
(2) $\|v\| \geq 0$, *with equality if and only if* $v = 0_V$;
(3) *(Minkowski's inequality):* $\|v + w\| \leq \|v\| + \|w\|$;
(4) *(Parallelogram law):* $\|v + w\|^2 + \|v - w\|^2 = 2(\|v\|^2 + \|w\|^2)$;
(5) *(Triangle difference inequality):* $\|v - w\| \geq |\|v\| - \|w\||$.

Hermann Minkowski, a German mathematician at the end of the nineteenth century, built an elegant mathematical framework for the theory of relativity, using four-dimensional non-Euclidean geometry.

Proof We see that $\|av\| = \sqrt{\langle av, av \rangle} = \sqrt{a\overline{a}\langle v, v \rangle} = |a| \cdot \|v\|$, proving (1). Inequality (2) follows immediately from the definition. As for (3), note that if $z = a + bi$ then $z + \overline{z} = 2a \leq 2|a| = 2\sqrt{a^2} \leq 2\sqrt{a^2 + b^2} = 2|z|$. As a consequence of the Cauchy–Schwarz–Bunyakovsky Theorem, we see that $|\langle v, w \rangle| = |\langle w, v \rangle| \leq \|v\| \cdot \|w\|$, and so

$$\|v + w\|^2 = \langle v + w, v + w \rangle = \langle v, v \rangle + \langle v, w \rangle + \langle w, v \rangle + \langle w, w \rangle$$

$$\leq \|v\|^2 + 2\|v\| \cdot \|w\| + \|w\|^2 = \big(\|v\| + \|w\|\big)^2,$$

and that proves (3). Moreover, we know that

$$\|v + w\|^2 = \langle v + w, v + w \rangle = \langle v, v \rangle + \langle v, w \rangle + \langle w, v \rangle + \langle w, w \rangle$$

and $\|v - w\|^2 = \langle v - w, v - w \rangle = \langle v, v \rangle - \langle v, w \rangle - \langle w, v \rangle + \langle w, w \rangle$. Adding these two gives us (4).

Finally, by (3), we have $\|w\| = \|w + (v - w)\| \leq \|w\| + \|v - w\|$, and so $\|v - w\| \geq \|v\| - \|w\|$. Interchanging the roles of v and w and using (1), gives us $\|v - w\| = \|w - v\| \geq \|w\| - \|v\|$, and so we have (5).   □

Note that by Proposition 15.3 we see that if $0_V \neq v \in V$ then $\frac{1}{\|v\|}v$ is a normal vector. Moreover, if v is normal and c is a scalar satisfying $|c| = 1$, then cv is again normal.

Example Let V be an inner product space, and let Ω be a nonempty set. A function $f \in V^\Omega$ is *bounded* if and only if there exists a real number b_f satisfying $\|f(i)\| \leq b_f$ for all $i \in \Omega$. If $f, g \in V^\Omega$ are bounded functions then, from Minkowski's inequality, we conclude that $\|(f + g)(i)\| \leq \|f(i)\| + \|g(i)\| \leq b_f + b_g$ for all $i \in \Omega$. If c is a scalar then $\|(cf)(i)\| = |c| \cdot \|f(i)\| \leq |c|b_f$ for all $i \in \Omega$. Thus both $f + g$ and cf are both bounded, and we see that the set of all bounded elements of V^Ω is a subspace of V^Ω.

Example We now return to a previous example. Let p be an integer greater than 1, not necessarily prime, and let $G = \mathbb{Z}/(p)$, on which we have an operation of addition as defined in Chap. 2. Let $V = \mathbb{C}^G$, which is a vector space of dimension p over $\mathbb{C}$. On this space, we can define an inner product by setting $\langle f, g \rangle = \sum_{n \in G} f(n)\overline{g(n)}$. Every element $n \in G$ defines a function $h_n : k \mapsto$

$\cos(\frac{2\pi nk}{p}) + i\sin(\frac{2\pi nk}{p})$ which belongs to V. Given a function $f \in V$, define a function $\widehat{f} \in V$ as $\widehat{f}: n \mapsto \langle f, h_n \rangle = \sum_{k \in G} f(k)h_n(-k)$. This function is called the *discrete Fourier transform* of f of order p. One can show that the function $f \mapsto \widehat{f}$ is in fact an automorphism of V. Moreover, $f(n) = \frac{1}{p}\widehat{\widehat{f}}(-n)$ and $\|f\| = \frac{1}{\sqrt{p}}\|\widehat{f}\|$ for all $f \in V$ and all $n \in G$.

Example There are various generalizations of Theorem 15.2 which, as a rule, require more sophisticated methods of complex analysis to prove. For example, the contemporary Greek mathematicians Manolis Magiropoulos and Dimitri Karayannakis have shown that if V is an inner product space and if u, v, and w are distinct elements of V, then

$$2|\langle u, v \rangle| \cdot |\langle u, w \rangle| \leq \langle u, u \rangle \big[\|v\| \cdot \|w\| + |\langle v, w \rangle| \big].$$

In case the set $\{v, w\}$ is linearly dependent, it is clear that this reduces to the inequality in Proposition 15.2. Inequalities such as these allow us to get better bounds on inner products. For example, let $0 < a < b$ be real numbers and let $V = C(a, b)$, on which we have the inner product $\langle f, g \rangle = \int_a^b f(x)g(x)\,dx$. If u, v, $w \in V$ are given by $u: x \mapsto 1/x$, $v: x \mapsto \sin(x)$, and $w: x \mapsto \cos(x)$ then Proposition 15.2 gives us the bound

$$|\langle u, v \rangle| \cdot |\langle u, w \rangle| \leq \left(\int_a^b \frac{dx}{x^2} \right) \sqrt{\int_a^b \sin^2(x)\,dx} \sqrt{\int_a^b \cos^2(x)\,dx}$$

whereas this result gives us the better upper bound

$$\frac{1}{2}\left(\int_a^b \frac{dx}{x^2} \right) \left[\sqrt{\int_a^b \sin^2(x)\,dx} \sqrt{\int_a^b \cos^2(x)\,dx} + \left| \int_a^b \sin(x)\cos(x)\,dx \right| \right].$$

Proposition 15.4 *Let V be an inner product space and let $\alpha \in \mathrm{End}(V)$ satisfy the condition that there exists a real number $0 < c < 1$ such that $\|\alpha(v)\| \leq c\|v\|$ for all $v \in V$. Then $\sigma_1 + \alpha$ is monic.*

Proof If $0_V \neq v \in V$ then, by Proposition 15.3,

$$\|v\| = \|v + \alpha(v) - \alpha(v)\| \leq \|v + \alpha(v)\| + \|\alpha(v)\|$$
$$= \|(\sigma_1 + \alpha)(v)\| + \|\alpha(v)\|,$$

and so $\|(\sigma_1 + \alpha)(v)\| \geq \|v\| - \|\alpha(v)\| \geq (1 - c)\|v\| > 0$, which shows that $v \notin \ker(\sigma_1 + \alpha)$. Thus $\sigma_1 + \alpha$ is monic. $\qquad\square$

In particular, if V is a finitely-generated inner product space and if $\alpha \in \mathrm{End}(V)$ satisfies the condition that there exists a real number $0 < c < 1$ such that $\|\alpha(v)\| \leq$

$c\|v\|$ for all $v \in V$, then $\sigma_1 + \alpha \in \text{Aut}(V)$. Let $\beta = (\sigma_1 + \alpha)^{-1}$. If $0_V \neq v \in V$ then

$$\|v\| = \|(\sigma_1 + \alpha)\beta(v)\| = \|\beta(v) + \alpha\beta(v)\| \geq \|\beta(v)\| - \|\alpha\beta(v)\|$$
$$\geq \|\beta(v)\| - c\|\beta(v)\| = (1 - c)\|\beta(v)\|.$$

Similarly, $\|v\| \leq \|\beta(v)\| + \|\alpha\beta(v)\| \leq \|\beta(v)\| + c\|\beta(v)\| = (1 + c)\|\beta(v)\|$ and so $\frac{1}{1+c}\|v\| \leq \|\beta(v)\| \leq \frac{1}{1-c}\|v\|$ for all $v \in V$.

Sometimes, however, we need a bit more generality. If V is a vector space over $\mathbb{R}$ or $\mathbb{C}$ then, in general, a function $v \mapsto \|v\|$ satisfying conditions (1)–(3) of Proposition 15.3 is called a *norm* and a vector space on which a fixed norm is defined is called a *normed space* or, in a functional-analysis context, a *pre-Banach space*. An immediate question is whether every norm defined on a vector space comes from an inner product. The answer is negative: if, for example, we define the norm $\| \cdot \|_1$ on $\mathbb{C}^n$ by setting $\left\| \begin{bmatrix} a_1 \\ \vdots \\ a_n \end{bmatrix} \right\|_1 = \sum_{i=1}^{n} |a_i|$, then this cannot come from an inner product since the parallelogram law is not satisfied by this norm. In fact, satisfying the parallelogram law is necessary for a norm to come from an inner product in the following sense: let V be a vector space over $\mathbb{R}$ or $\mathbb{C}$ on which we have a norm $\psi : V \to \mathbb{R}$ satisfying $\psi(v + w)^2 + \psi(v - w)^2 = 2[\psi(v)^2 + \psi(w)^2]$ for all $v, w \in V$, and write $\lambda(v, w) = \frac{1}{4}[\psi(v + w)^2 - \psi(v - w)^2]$. Then it is possible to define an inner product on V relative to which the norm of a vector v is precisely $\psi(v)$. In the case the field of scalars is $\mathbb{R}$, then this inner product is defined by $\langle v, w \rangle = \lambda(v, w)$ and otherwise this inner product is defined by $\langle v, w \rangle = \lambda(v, w) + i\lambda(v, iw)$.

With kind permission of the Archives of the Mathematisches Forschungsinstitut Oberwolfach (Wiener); © Stefan Banach (Banach).

Normed spaces were first studied at the beginning of the twentieth century by the Austrian mathematician **Hans Hahn**, and then by the American mathematician **Norbert Wiener** and the Polish mathematician **Stefan Banach**.

Example Every vector space over $\mathbb{R}$ can be turned into a normed space in at least one way. Indeed, let V be a vector space over $\mathbb{R}$ for which we fix a basis $\{v_i \mid i \in \Omega\}$. Then the function $\psi : V \to \mathbb{R}$ defined by $\psi : \sum_{i \in \Omega} a_i v_i \mapsto \sum_{i \in \Omega} |a_i|$ can easily be seen to be a norm on V.

Example Let $V = C(0, 1)$, which is a vector space over $\mathbb{R}$, and for each positive integer n, let $f_n \in V$ be the function defined by

$$f_n : x \mapsto \begin{cases} 1 - nx & \text{if } 0 \leq x \leq \frac{1}{n}, \\ 0 & \text{otherwise.} \end{cases}$$

Let $\| \cdot \|$ be the norm defined on V by the inner product $\langle f, g \rangle = \int_0^1 f(x)g(x)\,dx$ and let $\| \cdot \|_\infty$ be the norm on V defined by $\|f\|_\infty = \sup\{|f(x)| \mid 0 \leq x \leq 1\}$. Then $\|f_n\| = \frac{1}{\sqrt{3n}}$ for all positive integers n, whereas $\|f_n\|_\infty = 1$ for all positive integers n. Thus there can be no real number c satisfying $\|f\|_\infty \leq c\|f\|$ for all $f \in V$.

Example Let V and W be normed spaces over the same field of scalars F (which is either $\mathbb{R}$ or $\mathbb{C}$). If $\alpha \in \mathrm{Hom}(V, W)$, set

$$\|\alpha\| = \sup\left\{ \frac{\|\alpha(v)\|}{\|v\|} \;\middle|\; 0_V \neq v \in V \right\}$$

where the norm in the numerator is the one defined on W and the norm in the denominator is the one defined on V. (If V is trivial then the only such α is the 0-function, the norm of which we set equal to 0.) Note that the fraction $\|\alpha(v)\|/\|v\|$ is just $\|\alpha(v')\|$, where v' is the normal vector $\frac{1}{\|v\|} v$, so we see that $\|\alpha\|$ is just $\sup\{\|\alpha(v')\|\}$, where the supremum runs over all normal vectors v' in V. In particular, if $\delta \in D(V)$ then we define the norm of δ to be

$$\|\delta\| = \sup\left\{ \frac{|\delta(v)|}{\|v\|} \;\middle|\; 0_V \neq v \in V \right\}.$$

Note that $\|\alpha\|$ may not be finite, though it surely will be if α is bounded. For example, let V be the space of all polynomial functions in $\mathbb{R}^{\mathbb{R}}$ on which we define the norm $\|f\| = \max\{f(t) \mid 0 \leq t \leq 1\}$. Let α be the differentiation endomorphism of V and, for each $h \geq 1$, let $f_h \in V$ be given by $f_h : x \mapsto x^h$. Then

$$\frac{\|\alpha(f_h)\|}{\|f_h\|} = h$$

for each $h \geq 1$, showing that $\|\alpha\|$ is infinite. If V is finitely generated, then we assert that $\|\alpha\|$ is finite for all $\alpha \in \mathrm{Hom}(V, W)$, a claim which we will justify in the next chapter.

We claim that, if $\|\alpha\|$ is finite for all α, then this is a norm defined on $\mathrm{Hom}(V, W)$, called norm *induced* by the respective norms on V and W. Indeed, as an immediate consequence of the definition we see that $\|\alpha\| \geq 0$ for all $\alpha \in \mathrm{Hom}(V, W)$, with equality happening only when α is the 0-function. We also see that if α is not the 0-function then $\|\alpha\|$ is the smallest positive real number c such that $\|\alpha(v)\| \leq c\|v\|$ for all $v \in V$. (We note a subtle point here: the norms on V and W are, of course, different. Therefore, in the case $V = W$, and if we have two different norms defined on V, we may use one in the numerator and another in the denominator, though usually one uses the same norm in both instances.)

Now let $\alpha \in \mathrm{Hom}(V, W)$ and $a \in F$. Then

$$\|a\alpha\| = \sup\left\{ \frac{\|a\alpha(v)\|}{\|v\|} \,\middle|\, 0_V \neq v \in V \right\}$$

$$= \sup\left\{ \frac{|a| \cdot \|\alpha(v)\|}{\|v\|} \,\middle|\, 0_V \neq v \in V \right\} = |a| \cdot \|\alpha\|.$$

Finally, if $\alpha, \beta \in \mathrm{Hom}(V, W)$ then

$$\|\alpha + \beta\| = \sup\left\{ \frac{\|(\alpha + \beta)(v)\|}{\|v\|} \,\middle|\, 0_V \neq v \in V \right\}$$

$$= \sup\left\{ \frac{\|\alpha(v) + \beta(v)\|}{\|v\|} \,\middle|\, 0_V \neq v \in V \right\}$$

$$\leq \sup\left\{ \frac{\|\alpha(v)\| + \|\beta(v)\|}{\|v\|} \,\middle|\, 0_V \neq v \in V \right\} \leq \|\alpha\| + \|\beta\|.$$

If $V = F^n$ and $W = F^k$, endowed with respective dot products and the norms defined by them, then the induced norm on $\mathrm{Hom}(V, W)$ does, in fact, always exist and is called the *spectral norm*. If $A \in \mathcal{M}_{k \times n}(F)$, then the *spectral norm* of A is defined to be the spectral norm of the homomorphism from F^n to F^k given by $v \mapsto Av$.

In 1941, Gelfand showed that if n is a positive integer and $A \in \mathcal{M}_{n \times n}(\mathbb{C})$, then the spectral radius of A satisfies $\rho(A) = \lim_{k \to \infty} \sqrt[k]{\|A^k\|}$, where $\| \cdot \|$ is any norm defined on $\mathcal{M}_{n \times n}(\mathbb{C})$. In other words, we see that, given $A \in \mathcal{M}_{n \times n}(\mathbb{C})$, there exists a sufficiently large k such that $\|A^k\|$ is approximately equal to $\rho(A)^k$.

Example If p is any positive integer, we can define the *Hölder norm* $\| \cdot \|_p$ on $\mathbb{C}^n$ by

setting $\left\| \begin{bmatrix} a_1 \\ \vdots \\ a_n \end{bmatrix} \right\|_p = \left[\sum_{i=1}^{n} |a_i|^p \right]^{1/p}$. For the case $p = 2$, this, of course, reduces to

the norm coming from the dot product. The proof that this is a norm in the general case relies on a generalization of Minkowski's inequality: $\|v + w\|_p \leq \|v\|_p + \|w\|_p$ for all $v, w \in \mathbb{C}^n$ and any positive integer p. This norm can be used to define a norm on $\mathrm{Hom}(\mathbb{C}^n, \mathbb{C}^k)$ for positive integers k and n, by setting

$$\|\alpha\|_p = \sup\left\{ \frac{\|\alpha(v)\|_p}{\|v\|_p} \,\middle|\, 0_V \neq v \in V \right\}$$

for any $\alpha \in \mathrm{Hom}(\mathbb{C}^n, \mathbb{C}^k)$.

General matrix norms were first discussed by the twentieth-century American mathematician **Albert H. Bowker**. The nineteenth-century German algebraist **Otto Hölder** was strongly influenced by the work of Kronecker.

Example Let n be a positive integer. If $A = [a_{ij}] \in \mathcal{M}_{n \times n}(\mathbb{R})$, set $\|A\|_C = \max\{|\sum_{i=1}^{n} \sum_{j=1}^{n} a_{ij} c_i c'_j| \mid c_i, c'_j \in \{0, 1\}\}$. This defines a norm on $\mathcal{M}_{n \times n}(\mathbb{R})$, known as the *cut norm*. This norm has important applications in graph theory and combinatorics, but is hard to calculate. However, efficient methods of approximating the cut norm of a matrix exist, making use of the following remarkable result, known as *Grothendieck's inequality*: there exists a universal constant k_G (not dependent of n) satisfying the condition that any normal vectors $v_1, \ldots, v_n, w_1, \ldots, w_n$ in $\mathbb{R}^n$ and any scalars $e_1, \ldots, e_n, e'_1, \ldots, e'_n \in \{-1, 1\}$ satisfy $\sum_{i=1}^{n} \sum_{j=1}^{n} a_{ij} v_i \cdot w_j \leq k_G \sum_{i=1}^{n} \sum_{j=1}^{n} a_{ij} e_i e'_j$. The precise value of the constant k_G, known as *Grothendieck's constant*, has not been determined, but the French mathematician Jean-Louis Krivine has shown that $1.677 \ldots \leq k_G \leq 1.782 \ldots$.

The French algebraic geometer **Alexandre Grothendieck** is considered one of the most influential of contemporary mathematicians.

Example For positive integers k and n, we define the *Frobenius norm* or *Hilbert–Schmidt norm* of $A = [a_{ij}] \in \mathcal{M}_{k \times n}(\mathbb{C})$ by

$$\|A\|_{\mathfrak{F}} = \sqrt{\mathrm{tr}(AA^H)} = \sqrt{\sum_{i=1}^{k} \sum_{j=1}^{n} |a_{ij}|^2}.$$

This is precisely the norm coming from the inner product on $\mathcal{M}_{k \times n}(\mathbb{C})$ given by $\langle A, B \rangle = \mathrm{tr}(AB^H)$. If $A \in \mathcal{M}_{k \times n}(\mathbb{C})$ has spectral norm $\|A\|$ and Frobenius norm $\|A\|_{\mathfrak{F}}$, then it is straightforward to show that $\|A\| \leq \|A\|_{\mathfrak{F}} \leq (\sqrt{n})\|A\|$.

For vector spaces V finitely generated over $\mathbb{R}$ or $\mathbb{C}$, it does not matter which norm once chooses. To see this, we need the following preliminary result.

Proposition 15.5 *Let* $\{v_1, \ldots, v_n\}$ *be a finite linearly-independent subset of a normed space* V. *Then there exists a positive real number* c *such that* $\| \sum_{i=1}^{n} a_i v_i \| \geq c(\sum_{i=1}^{n} |a_i|)$ *for all scalars* $a_1, \ldots, a_n$.

Proof Let W be the subspace of V generated by $\{v_1, \ldots, v_n\}$ and let Y be the subset of W consisting of all linear combinations $\sum_{i=1}^{n} a_i v_i$ for which $\sum_{i=1}^{n} |a_i| = 1$. Pick $w = \sum_{i=1}^{n} a_i v_i \in W$. If $w = 0_V$, then $a_i = 0$ for each i and so $\sum_{i=1}^{n} |a_i| = 0$. Therefore, any positive real number c will do. Hence we can assume that $w \neq 0_V$ and so $d = \sum_{i=1}^{n} |a_i| > 0$. Moreover, $y = \sum_{i=1}^{n} (a_i d^{-1}) v_i \in Y$ and $\|w\| \geq c(\sum_{i=1}^{n} |a_i|)$ if and only if $\|y\| \geq c$. Therefore, to prove the proposition it suffices to show that there exists a positive real number c satisfying the condition that $\|y\| \geq c$ for all $y \in Y$.

Suppose that this is not the case. Then we can find a sequence $y_1, y_2, \ldots$ of vectors in Y such that $y_h = \sum_{i=1}^{n} b_{ih} v_i$ with $\sum_{i=1}^{n} |b_{ih}| = 1$ and $\lim_{h \to \infty} \|y_h\| = 0$. In particular, we note that $|b_{ih}| \leq 1$ for each $1 \leq i \leq n$ and each $h \geq 1$. Thus, in particular, the sequence $b_{11}, b_{12}, \ldots$ of scalars is bounded. By the Bolzano–Weierstrass Theorem (which holds for both real and complex numbers), this sequence must therefore have a convergent subsequence. Throwing away all of y_h for which b_{1h} is not in that subsequence, we can assume without loss of generality that the sequence $b_{11}, b_{12}, \ldots$ converges to some scalar b_1. Similarly, the sequence $b_{21}, b_{22}, \ldots$ has a convergent subsequence and, throwing away all of the y_h for which b_{2h} is not in that sequence, we can assume that the sequence $b_{21}, b_{22}, \ldots$ converges to some scalar b_2 as well. Continuing in this manner, we finally obtain an infinite sequence $y_1, y_2, \ldots$ of vectors in Y such that, for each $1 \leq i \leq n$, the sequence of scalars $b_{i1}, b_{i2}, \ldots$ converges to some scalar b_i.

Set $y = \sum_{i=1}^{n} b_i v_i$. Clearly, $y \in W$ and so not all of the b_i are equal to 0. In particular, $y \neq 0_V$ and so $\|y\| = r > 0$. On the other hand, for each $h \geq 1$ we have $\|y\| \leq \|y - y_h\| + \|y_h\| = \| \sum_{i=1}^{n} (b_i - b_{ih}) v_i \| + \|y_h\| \leq (\sum_{i=1}^{n} |b_i - b_{ih}| \cdot \|v_i\|) + \|y_h\|$. But $\lim_{h \to \infty} \|y_h\| = 0$ and $\lim_{h \to 0} |b_i - b_{ih}| = 0$ for each $1 \leq i \leq n$, and so there exists an integer h so large that $\|y\| < r$. This is a contradiction, from which the result follows. $\qquad\square$

Norms $\| \cdot \|_a$ and $\| \cdot \|_b$ are defined on the same vector space V are *equivalent* if and only if there exist positive real numbers c and d such that $c\|v\|_a \leq \|v\|_b \leq d\|v\|_a$ for all $v \in V$.

Proposition 15.6 *Any two norms defined on a finitely-generated vector space* V *over* $\mathbb{R}$ *or* $\mathbb{C}$ *are equivalent.*

Proof Let $\{v_1, \ldots, v_n\}$ be a basis for a vector space V over $\mathbb{R}$ or $\mathbb{C}$ on which we have norms $\| \cdot \|_a$ and $\| \cdot \|_b$ defined. By Proposition 15.5, there exists a scalar c such that $\| \sum_{i=1}^{n} a_i v_i \|_b \geq c(\sum_{i=1}^{n} |a_i|)$ for any vector $v = \sum_{i=1}^{n} a_i v_i$ in V. On the other hand, from the triangle inequality, we have $\|v\|_a \le \sum_{i=1}^n |a_i| \cdot \|v_i\|_a \le r \sum_{i=1}^n |a_i|$, where $r = \max\{\|v_1\|_a, \ldots, \|v_n\|_a\} > 0$ and so $(cr^{-1})\|v\|_a \le \|v\|_b$ for each $v \in V$. Interchanging the roles of $\|\cdot\|_a$ and $\|\cdot\|_b$, we repeat this proof to obtain a positive real number d such that $\|v\|_b \le d\|v\|_a$ for each $v \in V$. $\qquad\square$

Proposition 15.7 (Hahn–Banach Theorem) *Let V be a vector space over a field F which is either $\mathbb{R}$ or $\mathbb{C}$ and let $v \mapsto \|v\|$ be a norm defined on V. Moreover, let W be a subspace of V and let $\delta \in D(W)$ satisfy the condition that $|\delta(w)| \le \|w\|$ for all $w \in W$. Then there exists a linear functional $\theta \in D(V)$ which is an extension of δ satisfying $|\theta(v)| \le \|v\|$ for all $v \in V$.*

Proof (1) We first consider the case $F = \mathbb{R}$. Let $\mathcal{C}$ be the set of all pairs (Y, ψ), where Y is a subspace of V containing W and $\psi \in D(Y)$ satisfies the conditions that $\psi(y) \le \|y\|$ for all $y \in Y$ and ψ is an extension of δ. This set is nonempty since $|\delta(w)| \le \|w\|$ surely implies that $\delta(w) \le \|w\|$ and so $(W, \delta) \in \mathcal{C}$. Moreover, we can define a partial order on $\mathcal{C}$ by setting $(Y, \psi) \preccurlyeq (Y', \psi')$ if and only if $Y \subseteq Y'$ and $\psi'(y) = \psi(y)$ for all $y \in Y$. If $((Y_h, \psi_h) \mid h \in \Omega)$ is a chain in $\mathcal{C}$, set $Y = \bigcup_{h \in \Omega} Y_h$ and define $\psi \in D(Y)$ by setting $\psi(y) = \psi_h(y)$ when $y \in Y_h$. This function is well-defined since $\mathcal{C}$ is a chain, and it surely belongs to $\mathcal{C}$. Moreover, it is clear that $(Y_h, \psi_h) \preccurlyeq (Y, \psi)$ for each $h \in \Omega$. Therefore, by the Hausdorff Maximum Principle, $\mathcal{C}$ has a maximal element, which we will denote by (Y_0, θ).

We want to show that $Y_0 = V$. Indeed, assume that this is not the case and let $z \in V \smallsetminus Y_0$. Then $Y_1 = Y_0 + \mathbb{R}z$ properly contains Y_0 and, for any $c_0 \in \mathbb{R}$ we can define the linear functional $\theta_1 \in D(Y_1)$ defined by $\theta_1 : y_0 + az \mapsto \theta(y_0) + ac_0$ which surely is an extension of θ. We will be done if we can pick c_0 in such a manner that $\theta_1(y_1) \le \|y_1\|$ for each $y_1 \in Y_1$, for if we can do that, then we would have $(Y_1, \theta) \in \mathcal{C}$, contradicting the maximality of Y_0. If $y_1, y_2 \in Y_1$ then

$$\theta(y_1) - \theta(y_2) = \theta(y_1 - y_2) \le \|y_1 - y_2\|$$
$$= \|y_1 + z - z - y_2\| \le \|y_1 + z\| + \|-z - y_2\|.$$

This implies that $-\|-z - y_2\| - \theta(y_2) \le \|y_1 + z\| - \theta(y_1)$. Since y_2 does not appear on the right side of this equality nor does y_1 appear on the left side, we see that the real numbers

$$d_1 = \inf\{\|y_1 + z\| - \theta(y_1) \mid y_1 \in Y_1\}$$

and

$$d_2 = \sup\{-\|-z - y_2\| - \theta(y_2) \mid y_2 \in Y_1\}$$

satisfy $d_2 \le d_1$. Now choose c_0 to be any real number satisfying $d_2 \le c_0 \le d_1$.

We claim that $\theta(y_0) + ac_0 \le \|y_0 + az\|$ for all real numbers a. If $a = 0$, we know it is true by the choice of θ. If $a > 0$ we have $c_0 \le d_1 \le \|a^{-1}y_0 + z\| - \theta(a^{-1}y_0)$

and so $ac_0 \leq a\|a^{-1}y_0 + z\| - \theta(y_0) = \|y_0 + az\| - \theta(y_0)$, whence $\theta(y_0) + ac_0$ $\leq \|y_0 + az\|$. If $a < 0$ we have $c_0 \geq d_2 \geq -\|-z - a^{-1}y_0\|$ and so $-ac_0 \geq$ $-a\|-z - a^{-1}y_0\| + \theta(y_0) = -\|az + y_0\| + \theta(y_0)$ and so $-\theta(y_0) - ac_0 \geq$ $-\|az + y_0\|$, whence $\theta(y_0) + ac_0 \leq \|y_0 + az\|$.

Thus we see that $\theta \in D(V)$ satisfies $\theta(v) \leq \|v\|$ for all $v \in V$. If $v \in V$. then $-\theta(v) = \theta(-v) \leq \|-v\| = |(-1)| \cdot \|v\| = \|v\|$ as well as so $|\theta(v)| \leq \|v\|$ for all $v \in V$, proving our result in the case the field of scalars is $\mathbb{R}$.

(2) Now assume that $F = \mathbb{C}$. Since W and V are vector spaces over $\mathbb{C}$, they are also vector spaces over $\mathbb{R}$. Write δ as $\delta : w \mapsto \delta_1(w) + i\delta_2(w)$, where $\delta_1, \delta_2 \in D(W)$, considering W as a vector space over $\mathbb{R}$. Moreover, $\delta_1(w) \leq |\delta(w)|$ for all $w \in W$, since $\mathrm{Re}(z) \leq |z|$ for any $z \in \mathbb{C}$. Therefore, $\delta_1(w) \leq \|w\|$ for all $w \in W$ and, as in the last part of the proof of part (1), we actually have $|\delta_1(w)| \leq \|w\|$ for all $w \in W$. By part (1), we then know that there exists a linear functional $\theta_1 \in D(V)$ satisfying $\theta_1(v) \leq \|v\|$ for all $v \in V$.

But $i[\delta_1(w) + i\delta_2(w)] = i\delta(w) = \delta(iw) = \delta_1(iw) + i\delta_2(iw)$ for all $w \in W$. Since the real parts of both sides must be equal, we see that $\delta_2(w) = -\delta_1(iw)$. Now define the function $\theta : V \to \mathbb{C}$ by setting $\theta : v \mapsto \theta_1(v) - i\theta_1(iv)$. This is a linear functional on V, considered as a vector space over $\mathbb{C}$, since clearly $\theta(v + v') = \theta(v) + \theta(v')$ for all $v, v' \in V$ and for each $a + bi \in \mathbb{C}$ and $v \in V$ we have

$$\theta\big((a + bi)v\big) = \theta_1(av + ibv) - i\theta_1(iav - bv)$$
$$= a\theta_1(v) + b\theta_1(iv) - i\big[a\theta_1(iv) - b\theta_1(v)\big]$$
$$= (a + bi)\big[\theta_1(v) - i\theta_1(iv)\big] = (a + ib)\theta(v).$$

Furthermore, θ is an extension of δ.

We claim that $|\theta(v)| \leq \|v\|$ for all $v \in V$. To begin with, we note that if $\theta(v) = 0$ this holds, since $\|v\| \geq 0$ for all $v \in V$. Now assume that $\|v\| > 0$. Then there exists a real number r such that $\theta(v) = |\theta(v)|e^{ir}$ and so $|\theta(v)| = \theta(v)e^{-ir}$. Since $|\theta(v)|$ is real, this means that $\theta(v)e^{-ir} \in \mathbb{R}$ and so $|\theta(v)| = \theta(v)e^{-ir} = \theta_1(e^{-ir}v) \leq$ $\|e^{-ir}v\| = |e^{-ir}| \cdot \|v\| = \|v\|$. Thus the proposition is proven. $\square$

Proposition 15.8 *Let V be a normed space and let W be nontrivial subspace of V on which we are given a linear functional δ, for which $\|\delta\|$ is finite. Then there exists a linear functional $\theta \in D(V)$ which is an extension of δ satisfying $\|\theta\| = \|\delta\|$.*

Proof For each $w \in W$ we have $|\delta(w)| \leq \|\delta\| \cdot \|w\|$. Moreover, we have a norm $v \mapsto \|v\|^*$ on V by setting $\|v\|^* = \|\delta\| \cdot \|v\|$ for all $v \in V$. Therefore, by Proposition 15.7, we know that there exists a linear functional $\theta \in D(V)$ extending δ and satisfying $|\theta(v)| \leq \|v\|^* = \|\delta\| \cdot \|v\|$, and so $\frac{|\theta(v)|}{\|v\|} \leq \|\delta\|$ for all $0_V \neq v \in V$. Thus

$\|\theta\| \leq \|\delta\|$. On the other hand,

$$\|\delta\| = \sup\left\{ \frac{|\delta(w)|}{\|w\|} \,\middle|\, 0_W \neq w \in W \right\} \leq \sup\left\{ \frac{|\theta(v)|}{\|v\|} \,\middle|\, 0_V \neq v \in V \right\} = \|\theta\|,$$

and so we have the desired equality. $\square$

The norm $\| \cdot \|_1$ defined on $\mathbb{C}^n$ is important in various contexts. Let n be a positive integer and let θ be the function from $\mathcal{M}_{n \times n}(\mathbb{C})$ to $\mathbb{R}$ defined by $\theta : [a_{ij}] \mapsto \max\{\sum_{i=1}^{n} |a_{ij}| \mid 1 \leq j \leq n\}$, which we have already seen when we defined condition numbers. Numerical algorithms that compute the eigenvalues of a matrix, as a rule, make roundoff errors on the order of $c\theta(A)$, where c is a constant determined by the precision of the computer on which the algorithm is running. Since the eigenvalues of similar matrices are identical, it is usually useful, given a square matrix A, to find a matrix B similar to A with $\theta(B)$ small. This can often be done by choosing B of the form PAP^{-1}, where P is a nonsingular diagonal matrix.

Example If $A = \begin{bmatrix} 1 & 0 & 10^{-4} \\ 1 & 1 & 10^{-2} \\ 10^4 & 10^2 & 1 \end{bmatrix}$, then $\theta(A) = 1002$. However, if we choose $P = \begin{bmatrix} 10^2 & 0 & 0 \\ 0 & 1 & 0 \\ 0 & 0 & 10^{-2} \end{bmatrix}$, then $\theta(PAP^{-1}) = 3$.

Let α be the endomorphism of $\mathbb{C}^n$ represented with respect to the canonical basis by a matrix $A \in \mathcal{M}_{n \times n}(\mathbb{C})$. Then for each $v \in \mathbb{C}^n$ we have $\theta(A)\|v\|_1 \geq \|\alpha(v)\|_1$. In particular, if c is an eigenvalue of α associated with an eigenvector v then $\theta(A)\|v\|_1 \geq \|\alpha(v)\|_1 = |c| \cdot \|v\|_1$ and so $\theta(A) \geq |c|$. Thus we see that $\theta(A) \geq \rho(A)$, where $\rho(A)$ is the spectral radius of A. This bound is called the *Gershgorin bound*. In fact, we can sharpen this result.

With kind permission of the Archives of the Mathematisches Forschungsinstitut Oberwolfach (Taussky-Todd).

Semyon Aranovich Gershgorin was a twentieth century Russian mathematician. Gershgorin's theorem was published in a Russian journal in 1931 and was generally ignored, until it was noticed and publicized by the Austrian-born American mathematician **Olga Taussky-Todd**, one of the most important researchers in matrix theory, who worked on the development of numerical linear algebra methods for computers after World War II.

Proposition 15.9 (Gershgorin's Theorem) *Let α be the endomorphism of $\mathbb{C}^n$ represented with respect to the canonical basis by the matrix $A = [a_{ij}] \in M_{n \times n}(\mathbb{C})$ and, for each $1 \leq i \leq n$, let $r_i = \sum_{j \neq i} |a_{ij}|$. Let K_i be the circle in the complex plane with radius r_i and center a_{ii}. Then $\mathrm{spec}(\alpha) \subseteq K = \bigcup_{i \neq j} K_i$.*

Proof Let c be an eigenvalue of α and let $v = \begin{bmatrix} b_1 \\ \vdots \\ b_n \end{bmatrix}$ be an eigenvector of α associated with c. Let h be an index satisfying $|b_h| \geq |b_i|$ for all $1 \leq i \leq n$. Then $b_h \neq 0$ and $Av = cv$ so $(c - a_{hh})b_h = \sum_{j \neq h} a_{hj} b_j$ and hence $|c - a_{hh}||b_h| \leq \sum_{j \neq i} |a_{hj} b_j| \leq |b_h| r_h$. Thus $|c - a_{hh}| \leq r_h$ and so $c \in K_h \subseteq K$, as desired. $\square$

Proposition 15.10 (Diagonal Dominance Theorem[1]) *Let n be a positive integer and let $A = [a_{ij}] \in M_{n \times n}(\mathbb{C})$ satisfy the condition that $|a_{ii}| > \sum_{j \neq i} |a_{ij}|$ for all $1 \leq i \leq n$. Then A is nonsingular.*

Proof The stated condition says that 0 does not belong to any of the circles K_i defined in Gershgorin's Theorem and so it cannot be an eigenvalue of A. Hence A is nonsingular. $\square$

Example Let α be the endomorphism of $\mathbb{C}^4$ represented with respect to the canonical basis by the matrix $A = \begin{bmatrix} 3 & 1 & 2 & 0 \\ 4 & 15 & 0 & -2 \\ -3 & 0 & 0 & -1 \\ 0 & 0 & 3 & 5 \end{bmatrix}$. Then $\mathrm{spec}(A) = \{15.32, 4.49, 1.59 \pm 2.35i\}$. These numbers are found in the union K of the following circles in the complex plane: the circle of radius 3 around the point $(3, 0)$; the circle of radius 6 around the point $(15, 0)$, the circle of radius 4 around the point $(0, 0)$, and the circle of radius 3 around the point $(5, 0)$. We furthermore note that $\mathrm{spec}(A) = \mathrm{spec}(A^T)$ and so, by the same argument, we see that the eigenvalues of α lie in the union K' of the following circles in the complex plane: the circle of radius 7 around the point $(3, 0)$, the circle of radius 1 around the point $(15, 0)$, the circle of radius 5 around the point $(0, 0)$, and the circle of radius 3 around the point $(5, 0)$.

---

[1]This theorem was proven by the French mathematicians L. Lévy and J. Desplanques at the end of the nineteenth century. It was independently rediscovered by several other algebraists, including Hadamard, Minkowski, and Nekrasov.

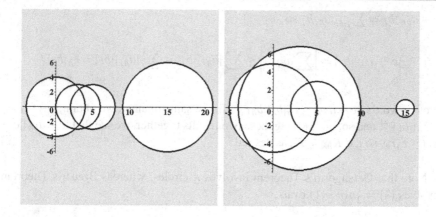

These circles and the location of the eigenvalues can be seen in the figure above. Thus, the eigenvalues of α lie in $K \cap K'$.

Since any polynomial in $\mathbb{C}[X]$ is the characteristic polynomial of a matrix, we can use Gershgorin's Theorem to get a bound on the location of the zeros of any polynomial. However, there are more sophisticated methods available to get much better bounds.

We will not go into the many results explicating Gershgorin's Theorem. One of these, for example, states that if the union of s of the disks in the complex plane defined by Gershgorin circles forms a connected domain which is isolated from the disks defined by the remaining circles, then this domain contains precisely s of the eigenvalues of the given matrix. There are also many generalizations of Gershgorin's Theorem, the best-known of which is the following.

Proposition 15.11 (Brauer's Theorem) *Let α be the endomorphism of $\mathbb{C}^n$ represented with respect to the canonical basis by the matrix $A = [a_{ij}] \in M_{n \times n}(\mathbb{C})$ and, for each $1 \leq i \leq n$, let $r_i = \sum_{j \neq i} |a_{ij}|$. For each $1 \leq i \neq j \leq n$, let K_{ij} be the Cassini oval $\{z \in \mathbb{C} \mid |z - a_{ii}||z - a_{jj}| \leq r_i r_j\}$ in the complex plane. Then $\mathrm{spec}(\alpha) \subseteq K = \bigcup_{i \neq j} K_{ij}$.*

Proof Let c be an eigenvalue of α and let $v = \begin{bmatrix} b_1 \\ \vdots \\ b_n \end{bmatrix}$ be an eigenvector of α asso-

ciated with c. Let h and k be indices such that $|b_h| \geq |b_k| \geq |b_i|$ for all $i \neq h, k$. We know that $b_h \neq 0$, and we can assume, as well, that $b_k \neq 0$ for otherwise we would have $a_{hh} = c$, in which case surely $c \in K$. Since $Av = cv$, we have

$(c - a_{hh})b_h = \sum_{j \neq h} a_{hj}b_j$ so

$$|c - a_{hh}||b_h| = \left|\sum_{j \neq h} a_{hj}b_j\right| \leq \sum_{j \neq h} |a_{hj}||b_j| \leq \sum_{j \neq h} |a_{hj}||b_k| = r_h|b_k|.$$

In other words, $|c - a_{hh}| \leq r_h|b_k||b_h|^{-1}$. In the same manner, we obtain $|c - a_{kk}| \leq r_k|b_h||b_k|^{-1}$ and so, multiplying these two results together, we see that $|c - a_{hh}||c - a_{kk}| \leq r_h r_k$, so $c \in K_{hk} \subseteq K$, as desired.   □

Note that Gershgorin's Theorem involves n circles, whereas Brauer's Theorem involves $\binom{n}{2} = \frac{1}{2}n(n - 1)$ ovals.

With kind permission of the Archives of the Mathematisches Forschungsinstitut Oberwolfach (Brauer).

The twentieth-century German mathematician **Alfred Brauer** emigrated to the United States in 1939; his research was primarily in matrix theory. **Giovanni Domenico Cassini** was a seventeenth-century Italian mathematician and astronomer.

Example It is sometimes useful to consider norms on vector spaces V not over subfields of $\mathbb{C}$, namely functions $v \mapsto \|v\|$ from V to $\mathbb{R}$ satisfying conditions (1)–(3) of Proposition 15.3. For example, let F be a finite field and let $V = F^n$ for some positive integer n. Define $\|v\|$ to be the number of nonzero entries in v, for each $v \in V$. This function is called the *Hamming norm* and is of extreme importance in algebraic coding theory, where one is interested in vector spaces over F in which every nonzero vector has a large Hamming norm. In an example at the beginning of Chap. 5, we showed a vector space of dimension 3 over GF(2), every nonzero element of which has Hamming norm equal to 4.

With kind permission of the Special Collections & Archives, Dudley Knox Library, Naval Postgraduate School.

Richard Hamming, a twentieth-century American mathematician and computer scientist, is best known for his development of the theory of error-detecting and error-correcting codes.

If v and w are vectors in space V over which we have a norm defined, then the *distance* between v and w is defined as $d(v, w) = \|v - w\|$. If $v \in V$ and $\varnothing \neq U \subseteq V$, we define the distance of v from U by $d(v, U) = \inf\{d(v, u) \mid u \in U\}$.

When $V = \mathbb{R}^n$ on which we have the dot product, this just gives us the ordinary notion of Euclidean distance. The ability to define the notion of distance in such spaces is important, since it allows us to measure the degree of error in algorithmic computations by measuring the distance between a computed value and the value predicted by theory. It also allows us to define the notion of convergence.

The following proposition shows that this abstract notion of distance indeed has the geometric properties that one would expect from a notion of distance.

Proposition 15.12 *Let V be a normed space and let $v, w, y \in V$. Then:*
(1) $d(v, w) = d(w, v)$;
(2) $d(v, w) \geq 0$, *where equality exists if and only if $v = w$;*
(3) *(Triangle inequality)* $d(v, w) \leq d(v, y) + d(y, w)$.

Proof This is an immediate consequence of Proposition 15.3.   □

Example Let A be a finite set, and let V be the collection of all subsets of A, which is a vector space over $F = GF(2)$. We have a norm defined on V by letting $\|B\|$ be the number of elements in B. Then the distance between subsets B and C of A is $\|B + C\|$, namely the number of elements in their symmetric difference.

If A and B are nonempty subsets of a space V over which we have a norm defined, then we set $d(A, B) = \inf\{d(v, w) \mid v \in A \text{ and } w \in B\}$. In particular, if $v \in V$ and B is a nonempty subset of V, we set $d(v, B) = d(\{v\}, B)$.

Let n be a positive integer. If $A = [a_{ij}] \in M_{n \times n}(\mathbb{C})$, and if $k > 0$ is an integer, let us define the matrix $P(k) = [p_{ij}^{(k)}]$ to be $I + \sum_{h=1}^{k} \frac{1}{h!} A^h$. We claim that, for each fixed $1 \leq i, j \leq n$, the limit $\lim_{h \to \infty} p_{ij}^{(h)}$ exists in $\mathbb{C}$. Indeed, if $B = [b_{ij}] \in M_{n \times n}(F)$, set $m(B) = \max_{1 \leq i, j \leq n} |b_{ij}|$. Then every entry in the matrix A^2 equals the sum of n products of pairs of entries of A and so, in absolute value, is equal to at most $m(A)^2 n$. Thus we see that $m(A^2) \leq m(A)^2 n$. Similarly, $m(A^3) \leq m(A^2)m(A)n \leq m(A)^3 n^2$ and so forth. Thus, in general,

$$m\left(\frac{1}{k!}A^k\right) \leq \frac{n^{k-1}}{k!}m(A)^k \leq \frac{1}{k!}[m(A)n]^k$$

and so, in particular, $m(P(k)) \leq \sum_{k=0}^{k} \frac{1}{k!}m(A)^k$ for all $k \geq 1$. But from calculus we know that the series $\sum_{h=0}^{\infty} \frac{1}{h!} r^h$ converges absolutely to e^r for each real number r. Therefore, the limit we seek exists, and, at least by analogy, we are justified in denoting the matrix $[\lim_{h \to \infty} p_{ij}^{(h)}]$ by e^A.

Matrix exponentials were explicitly studied by the American mathematician **William Henry Metzler** at the end of the nineteenth century. They appear earlier in the work of Laguerre and Peano.

Proposition 15.13 *If n is a positive integer $A = [a_{ij}] \in M_{n \times n}(F)$ is a diagonal matrix, where F is $\mathbb{R}$ or $\mathbb{C}$, then $e^A = [b_{ij}]$ is a diagonal matrix with $b_{ii} = e^{a_{ii}}$ for all $1 \le i \le n$.*

Proof This is an immediate consequence of the definition.   □

In particular, $e^O = I$. Moreover, this implies that if $B \in M_{n \times n}(F)$ is similar to a diagonal matrix then B and e^B have the same eigenvectors, while the eigenvalues of e^B are the exponentials of the eigenvalues of B.

Actually, we can do a bit better: if $A = \begin{bmatrix} A_1 & O & \dots & O \\ O & A_2 & \dots & O \\ \vdots & \vdots & \ddots & \vdots \\ O & O & \dots & A_m \end{bmatrix}$, where each A_h

is a square matrix, then $e^A = \begin{bmatrix} e^{A_1} & O & \dots & O \\ O & e^{A_2} & \dots & O \\ \vdots & \vdots & \ddots & \vdots \\ O & O & \dots & e^{A_m} \end{bmatrix}$.

Example If $O \ne A \in M_{n \times n}(F)$, where F is either $\mathbb{R}$ or $\mathbb{C}$, is a nilpotent matrix with index of nilpotence k, then $e^A = I + \sum_{h=1}^{k} \frac{1}{h!} A^h$. Thus, for example, if $A = \begin{bmatrix} 0 & 1 & 2 \\ 0 & 0 & -1 \\ 0 & 0 & 0 \end{bmatrix}$, we have $e^A = I + A + \frac{1}{2} A^2 = \begin{bmatrix} 1 & 1 & \frac{3}{2} \\ 0 & 1 & -1 \\ 0 & 0 & 1 \end{bmatrix}$.

Example If $A = \begin{bmatrix} 0 & k \\ 0 & 0 \end{bmatrix} \in M_{2 \times 2}(\mathbb{R})$, then $e^A = \begin{bmatrix} 1 & k \\ 0 & 1 \end{bmatrix}$; if $B = \begin{bmatrix} 0 & r \\ -r & 0 \end{bmatrix}$ then $e^B = \begin{bmatrix} \cos(r) & \sin(r) \\ -\sin(r) & \cos(r) \end{bmatrix}$.

Example If $A = \begin{bmatrix} 1 & 1 & 1 \\ 1 & 1 & 1 \\ 1 & 1 & 1 \end{bmatrix} \in \mathcal{M}_{3\times3}(\mathbb{R})$, then

$$e^A = \begin{bmatrix} \frac{2}{3} + \frac{1}{3}e^3 & \frac{1}{3}e^3 - \frac{1}{3} & \frac{1}{3}e^3 - \frac{1}{3} \\ \frac{1}{3}e^3 - \frac{1}{3} & \frac{2}{3} + \frac{1}{3}e^3 & \frac{1}{3}e^3 - \frac{1}{3} \\ \frac{1}{3}e^3 - \frac{1}{3} & \frac{1}{3}e^3 - \frac{1}{3} & \frac{2}{3} + \frac{1}{3}e^3 \end{bmatrix}.$$

If $P \in \mathcal{M}_{n\times n}(F)$ is nonsingular, where F is $\mathbb{R}$ or $\mathbb{C}$, then

$$P^{-1}\left[I + \sum_{h=1}^{k} \frac{1}{h!}A^h\right]P = I + \sum_{h=1}^{k} \frac{1}{h!}(P^{-1}AP)^h$$

for each k and so $P^{-1}e^A P = e^{P^{-1}AP}$. Thus we see that the exponentials of similar matrices are themselves similar. This is very important in calculations. In particular, if A is diagonalizable there exists a nonsingular matrix P such that $P^{-1}AP$ is a diagonal matrix $D = [d_{ij}]$ and so $P^{-1}e^A P = e^D$ is also a diagonal matrix. Thus e^A is diagonalizable whenever A is.

If A, B is a commuting pair of matrices in $\mathcal{M}_{n\times n}(F)$, then as a direct consequence of the definition we see that $e^A e^B = e^{A+B} = e^B e^A$. But this is not true in general, as the following example shows.

Example If $A = \begin{bmatrix} 0 & 1 \\ 0 & 0 \end{bmatrix}$ and $B = \begin{bmatrix} -1 & 0 \\ 0 & 0 \end{bmatrix}$, then

$$e^A e^B = \begin{bmatrix} 1 & 1 \\ 0 & 1 \end{bmatrix}\begin{bmatrix} e^{-1} & 0 \\ 0 & 1 \end{bmatrix} = \begin{bmatrix} e^{-1} & 1 \\ 0 & 1 \end{bmatrix} \neq \begin{bmatrix} e^{-1} & 1 - e^{-1} \\ 0 & 1 \end{bmatrix} = e^{A+B}.$$

Example The condition that A, B be a commuting pair is sufficient for $e^A e^B = e^{A+B}$ to hold, but is not necessary. Thus, for example, if $A = \begin{bmatrix} 0 & \pi \\ -\pi & 0 \end{bmatrix}$ and $B = \begin{bmatrix} 0 & (7 + 4\sqrt{3})\pi \\ (-7 + 4\sqrt{3})\pi & 0 \end{bmatrix}$, then $AB \neq BA$, but $e^A = e^B = -I$ so $e^A e^B = I = e^{A+B}$.

This fact is significant when it comes to calculating e^A in many cases. For example, suppose that A is an $n \times n$ matrix having a single eigenvalue c of multiplicity n. Then for each scalar t, the matrices ctI and $t(A - cI)$ commute and so $e^{tA} = e^{tcI}e^{t(A-cI)} = (e^{tc}I)\sum_{k=0}^{\infty} \frac{t^k}{k!}(A - cI)^k$ and, from the Cayley–Hamilton Theorem, we know that $(A - cI)^k = O$ for all $k \geq n$. Thus we see that $e^{tA} = (e^{tc}I)\sum_{k=0}^{n-1} \frac{t^k}{k!}(A - cI)^k$ and so there exists a polynomial $p(X) \in F[X]$ satisfying $e^{tA} = p(A)$.

Thus we can put much of what we said together. Given a matrix $A \in \mathcal{M}_{n \times n}(\mathbb{C})$, we know by Proposition 13.7 that it is similar to a matrix in Jordan canonical form. That is to say, there exists a nonsingular matrix P such that $P^{-1}AP$ is of the form

$$\begin{bmatrix} A_1 & O & \cdots & O \\ O & A_2 & \cdots & O \\ \vdots & \vdots & \ddots & \vdots \\ O & O & \cdots & A_m \end{bmatrix}, \text{ where each } A_i \text{ is a square matrix of the form } c_i I + N_i, \text{ for}$$

N_i a nilpotent matrix of a particularly simple form. Thus, for each i, we have $e^{A_i} =$

$$e^{c_i}e^{N_i} \text{ and so } e^A = P \begin{bmatrix} e^{c_1}e^{N_1} & O & \cdots & O \\ O & e^{c_2}e^{N_2} & \cdots & O \\ \vdots & \vdots & \ddots & \vdots \\ O & O & \cdots & e^{c_m}e^{N_m} \end{bmatrix} P^{-1}. \text{ Moreover, each } e^{N_i}$$

is just $p_i(N_i)$ for some polynomial $p_i(X) \in \mathbb{C}[X]$.

We also note that any matrix A commutes with $-A$, so $e^A e^{-A} = e^{A-A} = e^O = I$, proving that e^A is nonsingular and $e^{-A} = (e^A)^{-1}$. Therefore, we have a function $A \mapsto e^A$ from $\mathcal{M}_{n \times n}(F)$ (where F is either $\mathbb{R}$ or $\mathbb{C}$) to the set of all nonsingular matrices in $\mathcal{M}_{n \times n}(F)$, which is not monic. In the case $F = \mathbb{C}$, this function is in fact epic. If $A \in \mathcal{M}_{n \times n}(F)$ then a matrix $B \in \mathcal{M}_{n \times n}(F)$ is a *matrix logarithm* of A if and only if $A = e^B$. From the previous discussion, we see that only nonsingular matrices have logarithms. If $F = \mathbb{C}$, then every nonsingular matrix has a logarithm, but not necessarily a unique one.

Example If $A = \begin{bmatrix} 0 & 0 \\ 0 & 0 \end{bmatrix}$ and $B = \begin{bmatrix} 0 & 2\pi \\ -2\pi & 0 \end{bmatrix}$ then $e^A = e^B = I$. Therefore, both A and B are logarithms of I, which are not even similar.

A similar proof can be used to show that if A has distinct eigenvalues $\{c_1, \ldots, c_n\}$ and if $p_k(X) = \prod_{j \neq k}(c_k - c_j)^{-1}(X - c_j I)$ for all $1 \leq k \leq n$ then for any scalar t we have $e^{tA} = \sum_{k=1}^{n} e^{tc_k} p_k(A)$.

What about, say, $\cos(A)$ and $\sin(A)$? We know that the cosine function has a Maclaurin representation

$$\cos(x) = \sum_{i=0}^{\infty} \frac{(-1)^i}{(2i)!} x^{2i}.$$

For each natural number n, let us consider the polynomial

$$p_n(X) = \sum_{i=0}^{n} \frac{(-1)^i}{(2i)!} X^{2i}.$$

Then we can surely calculate $p_n(A)$ for each n and see whether the sequence of such matrices converges in some sense. However, there is another possibility. We know that for any real or complex number z we have $\cos(z) = \frac{1}{2}[e^{iz} + e^{-iz}]$ and so we can just define $\cos(A)$ to be the matrix $\frac{1}{2}[e^{iA} + e^{-iA}]$, which we know always exists.

Example We see that $\cos(I) = \begin{bmatrix} \cos(1) & 0 & 0 \\ 0 & \cos(1) & 0 \\ 0 & 0 & \cos(1) \end{bmatrix}$ and

$$\cos\left(\begin{bmatrix} 1 & 1 & 1 \\ 1 & 1 & 1 \\ 1 & 1 & 1 \end{bmatrix}\right) = \begin{bmatrix} \frac{2}{3} + \frac{1}{3}\cos(3) & \frac{1}{3}\cos(3) - \frac{1}{3} & \frac{1}{3}\cos(3) - \frac{1}{3} \\ \frac{1}{3}\cos(3) - \frac{1}{3} & \frac{2}{3} + \frac{1}{3}\cos(3) & \frac{1}{3}\cos(3) - \frac{1}{3} \\ \frac{1}{3}\cos(3) - \frac{1}{3} & \frac{1}{3}\cos(3) - \frac{1}{3} & \frac{2}{3} + \frac{1}{3}\cos(3) \end{bmatrix}.$$

Similarly, we know that $\sin(z) = \frac{-i}{2}[e^{iz} - e^{-iz}]$ and so we can define $\sin(A)$ to be $\frac{-i}{2}[e^{iA} - e^{-iA}]$.

Exercises

Exercise 934
Let $V = C(-1, 1)$ and let $a > -\frac{1}{2}$ be a real number. Is the function $\mu : V \times V \to \mathbb{R}$ defined by $\langle f, g \rangle = \int_{-1}^{1} [1 - t^2]^{a-1/2} f(t)g(t)\, dt$ an inner product on V?

Exercise 935
Is the function $\mu : \mathbb{R}^2 \times \mathbb{R}^2 \to \mathbb{R}$ defined by

$$\mu : \left(\begin{bmatrix} a_1 \\ a_2 \end{bmatrix}, \begin{bmatrix} b_1 \\ b_2 \end{bmatrix}\right) \mapsto a_1(b_1 + b_2) + a_2(b_1 + 2b_2)$$

an inner product on $\mathbb{R}^2$?

Exercise 936
Is the function $\mu : \mathbb{R}^2 \times \mathbb{R}^2 \to \mathbb{R}$ defined by

$$\mu : \left(\begin{bmatrix} a_1 \\ a_2 \end{bmatrix}, \begin{bmatrix} b_1 \\ b_2 \end{bmatrix}\right) \mapsto a_1 b_1 - a_1 b_2 - a_2 b_1 + 4a_2 b_2$$

an inner product on $\mathbb{R}^2$?

Exercise 937

Is the function $\mu : \mathbb{R}^3 \times \mathbb{R}^3 \to \mathbb{R}$ defined by

$$\mu : \left(\begin{bmatrix} a_1 \\ a_2 \\ a_3 \end{bmatrix}, \begin{bmatrix} b_1 \\ b_2 \\ b_3 \end{bmatrix} \right) \mapsto a_1 b_1 + 2 a_2 b_2 + 3 a_3 b_3 + a_1 b_2 + a_2 b_1$$

an inner product on $\mathbb{R}^3$?

Exercise 938

Verify whether the function $\mu : \mathbb{R}[X] \times \mathbb{R}[X] \to \mathbb{R}$ defined by $\mu : (f, g) \mapsto \deg(fg)$ is an inner product on $\mathbb{R}[X]$.

Exercise 939

Give an example of a function $\mu : \mathbb{R}^2 \times \mathbb{R}^2 \to \mathbb{R}$ which satisfies the first two conditions of an inner product, which does not satisfy the third, but does satisfy

$$\mu \left(\begin{bmatrix} 1 \\ 0 \end{bmatrix}, \begin{bmatrix} 1 \\ 0 \end{bmatrix} \right) = 1.$$

Exercise 940

Is the function $\mu : \mathbb{R}[X] \times \mathbb{R}[X] \to \mathbb{R}$ defined by

$$\mu : \left(\sum_{i=0}^{\infty} a_i X^i, \sum_{j=0}^{\infty} b_j X^j \right) \mapsto \sum_{i=0}^{\infty} \sum_{j=0}^{\infty} \frac{1}{i+j+1} a_i b_j$$

an inner product on $\mathbb{R}[X]$?

Exercise 941

Let V be the vector space of all continuous functions from $\mathbb{R}$ to itself. Let $\mu : V \times V \to \mathbb{R}$ be the function given by $\mu : (f, g) \mapsto \lim_{t \to \infty} \frac{1}{t} \int_{-t}^{t} f(s) g(s) \, ds$. Is μ an inner product?

Exercise 942

Let V be a vector space over $\mathbb{C}$ and let $\mu : V \times V \to \mathbb{C}$ be a function satisfying the following conditions:
(1) For each $w \in V$, the function $v \mapsto \mu(v, w)$ from V to $\mathbb{C}$ is a linear functional;
(2) If $v, w \in V$ then $\mu(v, w) = \overline{\mu(w, v)}$;
(3) If $v \in V$ satisfies $\mu(v, w) = 0$ for all $w \in V$, then $v = 0_V$.
Is μ an inner product on V?

Exercise 943

Let V be the vector space of all continuously differentiable functions from the interval $[a, b]$ in $\mathbb{R}$ to $\mathbb{R}$. If $f, g \in V$, define the *Sobolev inner product*

$$\langle f, g \rangle = \int_a^b f(t)g(t)\,dt + \int_a^b f'(t)g'(t)\,dt,$$

where f' and g' are the derivatives of f and g, respectively. Verify that this is indeed an inner product on V.

With kind permission of the Archives of the Mathematisches Forschungsinstitut Oberwolfach.

Sergei Lvovich Sobolev was a twentieth-century Russian mathematician who worked primarily in functional analysis.

Exercise 944

Let $\{u, v\}$ be a linearly-dependent subset of an inner product space V. Show that $\|u\|^2 v = \langle v, u \rangle u$.

Exercise 945

Let n be a positive integer and let $v = \begin{bmatrix} c_1 \\ \vdots \\ c_n \end{bmatrix} \in \mathbb{R}^n$. Is the function $\mu_v :$ $\mathbb{R}[X_1, \dots, X_n] \times \mathbb{R}[X_1, \dots, X_n] \to \mathbb{R}$ defined by

$$\mu_v : (p, q) \mapsto p(c_1, \dots, c_n)q(c_1, \dots, c_n)$$

an inner product on $\mathbb{R}[X_1, \dots, X_n]$?

Exercise 946

Let $a < b$ be real numbers and let $V = C(a, b)$. Let $h_0 \in V$ be a function satisfying the condition that $h_0(t) > 0$ for all $a < t < b$. Show that the function $\mu : V \times V \to \mathbb{R}$ defined by $\mu : (f, g) \mapsto \int_a^b f(x)g(x)h_0(x)\,dx$ is an inner product on V.

Exercise 947

Let c and d be given real numbers. Find a necessary and sufficient condition that the function $\mu : \left(\begin{bmatrix} a_1 \\ a_2 \end{bmatrix}, \begin{bmatrix} b_1 \\ b_2 \end{bmatrix} \right) \mapsto ca_1b_1 + da_2b_2$ be an inner product on $\mathbb{R}^2$.

Exercise 948

Is the function $\mu : \mathbb{R}^3 \times \mathbb{R}^3 \to \mathbb{R}$ defined by

$$\mu : \left(\begin{bmatrix} a_1 \\ a_2 \\ a_3 \end{bmatrix}, \begin{bmatrix} b_1 \\ b_2 \\ b_3 \end{bmatrix} \right) \mapsto a_1^2 b_2 + b_1^2 a_2 + (a_3 b_3)^2$$

an inner product on $\mathbb{R}^3$?

Exercise 949

Let V be an inner product space over $\mathbb{R}$ and let $n > 1$ be an integer. For positive real numbers $a_1, \ldots, a_n$, define the function $\mu : V^n \times V^n \to \mathbb{R}$ by

$$\mu : \left(\begin{bmatrix} v_1 \\ \vdots \\ v_n \end{bmatrix}, \begin{bmatrix} w_1 \\ \vdots \\ w_n \end{bmatrix} \right) \mapsto \sum_{i=1}^n a_i \langle v_i, w_i \rangle.$$

Is μ an inner product on V^n?

Exercise 950

Let n be a positive integer and let V be the subspace of $\mathbb{R}[X]$ consisting of all polynomials of degree at most n. Is the function $\mu : V \times V \to \mathbb{R}$ defined by $\mu : (p, q) \mapsto \sum_{i=0}^n p(\frac{i}{n}) q(\frac{i}{n})$ an inner product on V?

Exercise 951

Let $0 < n \in \mathbb{Z}$. Is the function $\mu : \mathbb{C}^n \times \mathbb{C}^n \to \mathbb{C}$ defined by

$$\mu : \left(\begin{bmatrix} a_1 \\ \vdots \\ a_n \end{bmatrix}, \begin{bmatrix} b_1 \\ \vdots \\ b_n \end{bmatrix} \right) \mapsto \sum_{i=1}^n a_i \overline{b}_{n-i+1}$$

an inner product?

Exercise 952

Let $V = \mathbb{C}^2$ on which we have defined the dot product, and let $D = \{v \in V \mid \|v\| = 1\}$. Find $\{\langle Av, v \rangle \mid v \in D\}$, where $A = \begin{bmatrix} 1 & 0 \\ 0 & 0 \end{bmatrix} \in \mathcal{M}_{2 \times 2}(\mathbb{C})$.

Exercise 953

Let n be a positive integer and let $\{v_1, \ldots, v_k\}$ be a set of vectors in $\mathbb{R}^n$ satisfying $v_i \cdot v_j \leq 0$ for all $1 \leq i < j \leq k$. Show that $k \leq 2n$ and give an example in which equality holds.

Exercise 954

Let n be a positive integer and let $A \in \mathcal{M}_{n \times n}(\mathbb{C})$ be idempotent. Is A^H necessarily idempotent?

Exercise 955
Let V be an inner product space and let $v \neq v'$ be vectors in V. Show that there exists a vector $w \in V$ satisfying $\langle v, w \rangle \neq \langle v', w \rangle$.

Exercise 956
Let V be an inner product space finitely generated over its field of scalars, and let $B = \{v_1, \ldots, v_n\}$ be a basis of V. Show that there exists a basis $\{w_1, \ldots, w_n\}$ of V satisfying the condition that

$$\langle v_i, w_j \rangle = \begin{cases} 1 & \text{if } i = j, \\ 0 & \text{otherwise.} \end{cases}$$

Exercise 957
Let W be a subspace of a vector space V over $\mathbb{R}$ and let Y be a complement of W in V. Define an inner product μ on W and an inner product v on Y. Is the function from $V \times V \to \mathbb{R}$ defined by $(w + y, w' + y') \mapsto \mu(w, w') + v(y, y')$ an inner product on V?

Exercise 958
Let $V = C(0, 1)$. Let $A = \{f_1, \ldots, f_n\}$ be a linearly-independent subset of V and define a function $u : \mathbb{R} \times \mathbb{R} \to \mathbb{R}$ by $u : (a, b) \mapsto \sum_{j=1}^{n} f_j(a) \cos^j(b)$. Show that if $h \in V$ and if there exists a function $g \in V$ such that $h(x) = \int_0^1 u(x, y)g(y)\,dy$ for all $x \in \mathbb{R}$, then $h \in \mathbb{R}A$.

Exercise 959
Let V be an inner product space over $\mathbb{R}$. For each real number a, set $U(a) = \{v \in V \mid \langle v, v \rangle \leq a\}$. Given a real number a, find a real number b such that $\langle v + w, v + w \rangle \in U(b)$ for all $v, w \in U(a)$.

Exercise 960
Let V be an inner product space and let $\alpha \in \text{End}(V)$. Show that $\langle \alpha(v), v \rangle \langle v, \alpha(v) \rangle \leq \|\alpha(v)\|^2$ for every normal vector $v \in V$.

Exercise 961
For real numbers $a_1, \ldots, a_n$, show that

$$\sum_{i=1}^{n} a_i \leq \left(\sqrt{\sum_{i=1}^{n} |a_i|^{2/3}} \right) \left(\sqrt{\sum_{i=1}^{n} |a_i|^{4/3}} \right).$$

Exercise 962
(Binet–Cauchy identity) For $u, v, w, y \in \mathbb{R}^3$, show that $(v \times w)(y \times u) = (v \cdot y)(w \cdot u) - (v \cdot u)(y \cdot w)$.

Exercise 963

Let v be a normal vector in $\mathbb{R}^3$. Show that the function $\alpha_v : \mathbb{R}^3 \to \mathbb{R}^3$ defined by $\alpha_v : w \mapsto v \times (v \times w) + w$ is a projection in $\mathrm{End}(\mathbb{R}^3)$.

Exercise 964

Let V be an inner product space and let $\alpha \in \mathrm{End}(V)$ be a projection. Does it necessarily follow that $\|\alpha(v)\| \le \|v\|$ for all $v \in V$?

Exercise 965

For $u, v, w, y \in \mathbb{R}^3$, show that $(u \times v) \cdot (w \times y) = \begin{vmatrix} u \cdot w & u \cdot y \\ v \cdot w & v \cdot y \end{vmatrix}$.

Exercise 966

For nonnegative real numbers a, b, and c, show that

$$(a+b+c)\sqrt{2} \le \sqrt{a^2+b^2} + \sqrt{b^2+c^2} + \sqrt{a^2+c^2}.$$

Exercise 967

For real numbers $0 < a \le b \le c$, show that

$$\sqrt{b^2+c^2} \le (\sqrt{2})a \le \sqrt{(b-a)^2+(c-a)^2}.$$

Exercise 968

Let n be a positive integer and let $A \in \mathcal{M}_{n \times n}(\mathbb{R})$ be a matrix the n Gershgorin circles of which are mutually disjoint. Prove that all of the eigenvalues of A are real.

Exercise 969

Show that $\left[\int_0^1 f(x)\,dx \right]^2 \le \int_0^1 f(x)^2\,dx$ for any $f \in C(0,1)$.

Exercise 970

Let $f : \mathbb{R} \to \mathbb{R}$ be the constant function $x \mapsto 1$. Calculate $\|f\|$ when f is considered as an element of $C(0, \frac{\pi}{2})$ and compare it to $\|f\|$, when f is considered as an element of $C(0, \pi)$.

Exercise 971

Let V be an inner product space over $\mathbb{R}$ and let $v, w \in V$ satisfy $\|v + w\| = \|v\| + \|w\|$. Show that $\|av + bw\| = a\|v\| + b\|w\|$ for all $0 \le a, b \in \mathbb{R}$.

Exercise 972

(Real polarization identity) Let V be an inner product space over $\mathbb{R}$. Show that $\langle u, v \rangle = \frac{1}{4}(\|u+v\|^2 - \|u-v\|^2)$ for all $u, v \in V$.

Exercise 973

(Complex polarization identity) Let V be an inner product space over $\mathbb{C}$. Show that $\langle u, v \rangle = \frac{1}{4}(\|u+v\|^2 - \|u-v\|^2 + i\|u+iv\|^2 - i\|u-iv\|^2)$ for all $u, v \in V$.

Exercise 974

Let $V = C(0, 1)$ on which we have defined the inner product $\langle f, g \rangle = \int_0^1 f(t)g(t)\,dt$. Let W be the subspace of V generated by the function $x \mapsto x^2$. Find all elements of W normal with respect to this inner product.

Exercise 975

Let V be an inner product space over $\mathbb{R}$ and assume that $v, w \in V$ are nonzero vectors satisfying the condition $\langle v, w \rangle = \|v\| \cdot \|w\|$. Show that $\mathbb{R}v = \mathbb{R}w$.

Exercise 976

Let V be a vector space over $\mathbb{R}$ on which we have two inner products, μ and μ' defined, which in turn define distance functions d and d' respectively. If $d = d'$, does it necessarily follow that $\mu = \mu'$?

Exercise 977

(Apollonius' identity) Let V be an inner product space. Show that

$$\|u - w\|^2 + \|v - w\|^2 = \frac{1}{2}\|u - v\|^2 + 2\left\|\frac{1}{2}(u+v) - w\right\|^2$$

for all $u, v, w \in V$.

The Greek geometer **Apollonius of Perga**, who worked in Alexandria in the third century BC, in his famous book *Conics*, was the first to introduce the terms "hyperbola", "parabola", and "ellipse".

Exercise 978

Let n be a positive integer and let $\|\cdot\|$ be a norm defined on $\mathbb{C}^n$. For each $A \in \mathcal{M}_{n\times n}(\mathbb{C})$, let $\|A\|$ be the spectral norm of A. If $A \in \mathcal{M}_{n\times n}(\mathbb{C})$ is nonsingular, show that every singular matrix $B \in \mathcal{M}_{n\times n}(\mathbb{C})$ satisfies $\|A - B\| \geq \|A^{-1}\|^{-1}$. Does there necessarily exist a singular matrix B for which equality holds?

Exercise 979

Let V be an inner product space over $\mathbb{R}$ and consider the function $\theta : V \times V \times V \to \mathbb{R}$ defined by

$$\theta(v, w, y) = \|v + w + y\|^2 + \|v + w + y\|^2 - \|v - w - y\|^2 - \|v - w + y\|^2.$$

Show that, for any $v, w, y \in V$, the value of $\theta(v, w, y)$ does not depend on y.

Exercise 980

Let V be an inner product space over $\mathbb{R}$ and let $n > 2$. Let $\theta : V^n \to V$ be the function defined by $\theta : \begin{bmatrix} v_1 \\ \vdots \\ v_n \end{bmatrix} \mapsto \frac{1}{n} \sum_{i=1}^n v_i$. Show that

$$\sum_{i=1}^n \left\| v_i - \theta\left(\begin{bmatrix} v_1 \\ \vdots \\ v_n \end{bmatrix} \right) \right\|^2 = \sum_{i=1}^n \|v_i\|^2 - n \left\| \begin{bmatrix} v_1 \\ \vdots \\ v_n \end{bmatrix} \right\|^2.$$

Exercise 981

Let V be an inner product space over $\mathbb{R}$ and let v and w be nonzero vectors in V. Show that $|\langle v, w \rangle|^2 = \langle v, v \rangle \langle w, w \rangle$ if and only if the set $\{v, w\}$ is linearly dependent.

Exercise 982

Let V be a finitely-generated inner product space and let $B = \{v_1, \ldots, v_n\}$ be a set of vectors in V. Show that B is linearly dependent if and only if its Gram matrix is singular.

Exercise 983

Let V be a vector space over $\mathbb{R}$ and let $\| \cdot \|$ be a norm defined on V. Show that $\|\|v\| - \|w\|\| \leq \|v - w\|$ for all $v, w \in V$.

Exercise 984

Let V be an inner product space finitely generated over $\mathbb{R}$ and let $\delta \in D(V)$. Pick $v_0 \in V$. Show that for each real number $e > 0$ there exists a real number $d > 0$ such that $|\delta(v) - \delta(v_0)| < e$ whenever $\|v - v_0\| < d$.

Exercise 985

Let V be an inner product space. For any $u, v, w \in V$, show that

$$\begin{vmatrix} 0 & 1 & 1 & 1 \\ 1 & 0 & d(u, v)^2 & d(u, w)^2 \\ 1 & d(u, v)^2 & 0 & d(v, w)^2 \\ 1 & d(u, w)^2 & d(v, w)^2 & 0 \end{vmatrix} \leq 0.$$

Exercise 986

Let $n > 1$. Show that there is no norm $v \mapsto \|v\|$ defined on $\mathbb{C}^n$ satisfying $\|A\|_{\mathfrak{z}} = \sup\{\frac{\|Av\|}{\|v\|} \mid 0_V \neq v \in \mathbb{C}^n\}$ for all $A \in \mathcal{M}_{n \times n}(\mathbb{C})$.

Exercise 987

Let $n > 1$ be an integer. For each $A \in \mathcal{M}_{n \times n}(\mathbb{C})$, let $\|A\| = \rho(A)$, the spectral radius of A. Does this turn $\mathcal{M}_{n \times n}(\mathbb{C})$ into a normed space?

Exercise 988

Let V be the vector space of all continuous functions from the unit interval $[0, 1]$ on the real line to $\mathbb{R}$ and for each $f \in V$, set $\|f\| = \int_0^1 |f(t)| \, dt$. Is $\| \cdot \|$ a norm on V?

Exercise 989

Let $p > 2$ be prime and let n be a positive integer. For each $1 \leq i \leq n$, define $w(i) = \min\{i - 1, p - i + 1\}$. Does the function $\mathrm{GF}(p)^n \to \mathbb{R}$ defined by $\begin{bmatrix} a_1 \\ \vdots \\ a_n \end{bmatrix} \mapsto \sum_{i=1}^n w(i) a_i$ turn $\mathrm{GF}(p)^n$ into a normed space?

Exercise 990

Let $0 < p < 1$ and let $f : \mathbb{R}^n \to \mathbb{R}$ be the function defined by

$$f : \begin{bmatrix} a_1 \\ \vdots \\ a_n \end{bmatrix} \to \left(\sum_{i=1}^n |a_i|^p \right)^{1/p}.$$

Show that this is not a norm but does satisfy the inequality $f(v + w) \leq 2^{(1-p)/p}[f(v) + f(w)]$ for all $v, w \in \mathbb{R}^n$.

Exercise 991

Let $V_1, \ldots, V_n$ be normed spaces over the same field and let $V = \prod_{i=1}^n V_i$. For each $1 \leq i \leq n$, denote the norm defined on V_i by $\| \cdot \|_i$ and define a function $v \mapsto \|v\|$ from V to $\mathbb{R}$ by setting $\left\| \begin{bmatrix} v_1 \\ \vdots \\ v_n \end{bmatrix} \right\| = \sum_{i=1}^n \|v_i\|_i$. Is this a norm on V?

Exercise 992

Let V be a normed space and let $0_V \neq v_0 \in V$. Show that there exists a linear functional $\theta \in D(V)$ satisfying $\|\theta\| = 1$ and $\theta(v_0) = \|v_0\|$.

Exercise 993

Let $n > 1$ and let $A \in \mathcal{M}_{n \times n}(\mathbb{C})$. Show that there are infinitely-many other matrices in $\mathcal{M}_{n \times n}(\mathbb{C})$ having the same Gershgorin circles as A.

Exercise 994
Let $A = [a_{ij}] \in M_{2 \times 2}(\mathbb{C})$ and let K be the Cassini oval defined by A. Show that every point on the boundary of K is an eigenvalue of a matrix $B \in M_{2 \times 2}(\mathbb{C})$ defining the same Cassini oval.

Exercise 995
Let $n > 1$ be an integer and let $A = [a_{ij}] \in M_{n \times n}(\mathbb{R})$. For any $e > 0$, show that there exists a nonsingular matrix $B \in M_{n \times n}(\mathbb{R})$ satisfying $\|A - B\|_{\mathfrak{F}} < e$.

Exercise 996
Let n be a positive integer and let $A = [a_{ij}] \in M_{n \times n}(\mathbb{C})$. Let $f : \mathbb{R} \to M_{n \times n}(\mathbb{C})$ be defined by $f : t \mapsto e^{tA}$. Show that the derivative of f is given by $f' : t \mapsto Ae^{tA}$.

Exercise 997
Let F be a field and let n be a positive integer. Let $\alpha \in \text{End}(F^n)$ and let V be a subspace of F disjoint from $\ker(\alpha)$. If $\| \cdot \|$ denotes the Hamming norm on F^n, is it necessarily true that $\|v\| = \|\alpha(v)\|$ for all $v \in V$?

Exercise 998
Let n be a positive integer and let α be an endomorphism of $\mathbb{C}^n$ represented with respect to the canonical basis by a matrix $A \in M_{n \times n}(\mathbb{C})$. Then the canonical inner product on $\mathbb{C}^n$ defines norms on $\mathbb{C}^n$ and on $\text{End}(\mathbb{C}^n)$. Show that $\rho(A) \leq \|\alpha^k\|^{1/k}$ for any integer $k > 0$.

Exercise 999
Let V and W be normed spaces over $\mathbb{R}$ or $\mathbb{C}$ and let $\alpha : V \to W$ be a linear transformation for which $\|\alpha\|$ exists. Show that $D(\alpha)$ satisfies $\|D(\alpha)\| = \|\alpha\|$.

Exercise 1000
Let V be a vector space finitely-generated over a field F and let L be the set of all subspaces of V. For $W, Y \in L$, define $d(W, Y) = \dim(W + Y) - \dim(W \cap Y)$. Does this function satisfy the conditions of Proposition 15.12?

Exercise 1001
For each real number t, set $A(t) = \begin{bmatrix} t & 1 & 1 \\ 1 & 0 & 1 \\ 1 & 1 & t \end{bmatrix} \in M_{3 \times 3}(\mathbb{R})$. Does there exist a value of t for which $\|A(t)\|_{\mathfrak{F}} = \|A(t)\|_2$?

Exercise 1002
For each $t > 0$, let $f(t) = \left\| \begin{bmatrix} t & 1 & 0 & 0 \\ 1 & t & 0 & 0 \\ 0 & 0 & 1 & t \\ 0 & 0 & t & 1 \end{bmatrix} \right\|_{\mathfrak{F}}$. Calculate $\lim_{t \to \infty} \frac{1}{t} f(t)$.

Exercise 1003

Let V be a vector space over $\mathbb{R}$ and let Y be its complexification. Define a function $\mu : Y \times Y \to \mathbb{C}$ by

$$\mu : \left(\begin{bmatrix} v_1 \\ v_2 \end{bmatrix}, \begin{bmatrix} w_1 \\ w_2 \end{bmatrix} \right) \mapsto \langle v_1, w_1 \rangle + \langle v_2, w_2 \rangle + i \big[\langle v_w, w_1 \rangle - \langle v_1, w_2 \rangle \big].$$

Show that μ is an inner product on Y and calculate $\left\| \begin{bmatrix} v_1 \\ v_2 \end{bmatrix} \right\|$ for each $\begin{bmatrix} v_1 \\ v_2 \end{bmatrix} \in Y$.

Exercise 1004

Let V be a normed space and let $\alpha \in \mathrm{End}(V)$ have an induced norm satisfying $\|\alpha\| < 1$. Show that $\sigma_1 - \alpha \in \mathrm{Aut}(V)$.

Exercise 1005

Let V be the set of all "infinite matrices" $A = [a_{ij}]$, where $a_{ij} \in \mathbb{R}$ for all $i, j \geq 0$, which is a vector space over $\mathbb{R}$ with addition and scalar multiplication defined elementwise. Let $p > 1$ be a real number and let q be a real number satisfying $\frac{1}{p} + \frac{1}{q} = 1$. Let W be the subset of V consisting of all those matrices A satisfying the condition that $\sum_{i=1}^{\infty} [\sum_{j=1}^{\infty} |a_{ij}|^q]^{p/q}$ is finite. Show that W is a subspace of V and that the function $A \mapsto (\sum_{i=1}^{\infty} [\sum_{j=1}^{\infty} |a_{ij}|^q]^{p/q})^{1/p}$ is a norm on W.

Orthogonality

16

Let V be an inner product space and let $0_V \neq v, w \in V$. From Proposition 15.2 we see that

$$-1 \leq \frac{\langle v, w \rangle + \langle w, v \rangle}{2\|v\| \cdot \|w\|} \leq 1,$$

and so there exists a real number $0 \leq t \leq \pi$ satisfying

$$\cos(t) = \frac{\langle v, w \rangle + \langle w, v \rangle}{2\|v\| \cdot \|w\|}.$$

This number t is the *angle* between v and w. Note that if we are working over $\mathbb{R}$, then

$$\cos(t) = \frac{\langle v, w \rangle}{\|v\| \cdot \|w\|}.$$

Example If $V = \mathbb{R}^n$ is endowed with the dot product, and if $0_V \neq v, w \in V$ then, using analytic geometry, it is easy to show that the angle as defined here is indeed the angle between the straight line determined by v and the origin, and the straight line determined by w and the origin. If we define different inner products on V, we build in this manner various non-Euclidean geometries in n-space.

Example Let $V = C(0, 1)$, on which we have defined the inner product $\langle f, g \rangle = \int_0^1 f(x)g(x)\,dx$. In particular, consider the functions $f : x \mapsto 5x^2$ and $g : x \mapsto 3x$. Then $\|f\| = \sqrt{5}$ and $\|g\| = \sqrt{3}$, and the angle t between f and g satisfies $\cos(t) = \sqrt{\frac{1}{15}} \int_0^1 (5x^2)(3x)\,dx = \frac{1}{4}\sqrt{15}$.

Vectors v and w in an inner product space V are *orthogonal* if and only if $\langle v, w \rangle = 0$. In this case, we write $v \perp w$. We note that if $v \perp w$ then $\|v + w\|^2 = \|v\|^2 + \langle v, w \rangle + \langle w, v \rangle + \|w\|^2 = \|v\|^2 + \|w\|^2$. A nonempty subset D of V is a set of *mutually orthogonal* vectors if $v \perp w$ whenever $v \neq w$ in D. If $\{v_1, \dots, v_n\}$ is a mu-

J.S. Golan, *The Linear Algebra a Beginning Graduate Student Ought to Know*,
DOI 10.1007/978-94-007-2636-9_16, © Springer Science+Business Media B.V. 2012

tually orthogonal set of vectors in V then one shows, similarly, that $\|\sum_{i=1}^{n} v_i\|^2 = \sum_{i=1}^{n} \|v_i\|^2$.

Example We have already seen that if $v, w \in \mathbb{R}^3$, then $v \cdot (v \times w) = 0$. This says that a vector v is orthogonal to $v \times w$, for any vector w. The same is also true for w and $v \times w$ and so we see that if $\{v, w\}$ is a linearly-independent subset of $\mathbb{R}^3$ then the set $\{v, w, v \times w\}$ is linearly independent and so is a basis of $\mathbb{R}^3$.

Moreover, as an immediate consequence of the Lagrange identity on $\mathbb{R}^3$, we see that if $v \times w = \begin{bmatrix} 0 \\ 0 \\ 0 \end{bmatrix}$ and $v \cdot w = 0$ then either $v = \begin{bmatrix} 0 \\ 0 \\ 0 \end{bmatrix}$ or $w = \begin{bmatrix} 0 \\ 0 \\ 0 \end{bmatrix}$. If $v, w \in \mathbb{R}^3$, then the angle t between them satisfies the condition that $v \cdot w = (\|v\| \cdot \|w\|) \cos(t)$. Using the Lagrange identity, we see that

$$\|v \times w\|^2 = \|v\|^2 \|w\|^2 - (v \cdot w)^2 = \|v\|^2 \|w\|^2 [1 - \cos^2(t)]$$
$$= \|v\|^2 \|w\|^2 \sin^2(t),$$

and so $\|v \times w\| = (\|v\| \cdot \|w\|)|\sin(t)|$. Thus $|\cos(t)| = \frac{\|v \times w\|}{\|v\| \cdot \|w\|}$.

Example If $V = \mathbb{C}^2$ on which we have the dot product, then it is easy to see that $\begin{bmatrix} 2 + 3i \\ -1 + 5i \end{bmatrix} \perp \begin{bmatrix} 1 + i \\ -i \end{bmatrix}$.

Example Let $V = C(-1, 1)$, on which we have defined the inner product $\langle f, g \rangle = \int_{-1}^{1} f(x)g(x)\,dx$. For all $i \geq 0$, define the functions $p_i \in V$ as follows: $p_0 : x \mapsto 1$; $p_1 : x \mapsto x$; and

$$p_{h+1} : x \mapsto \left(\frac{2h+1}{h+1}\right) x p_h(x) - \left(\frac{h}{h+1}\right) p_{h-1}(x) \quad \text{whenever} \quad h > 1.$$

These polynomial functions are known as *Legendre polynomials*. It is easy to verify that $p_i \perp p_h$ whenever $i \neq h$.

On the same space, we can define another inner product, namely

$$\langle f, g \rangle = \int_{-1}^{1} \frac{f(x)g(x)}{\sqrt{1 - x^2}}\,dx.$$

For each $i \geq 0$, define the function $q_i \in V$ by setting $q_0 : x \mapsto 1$; $q_1 : x \mapsto x$; and $q_{h+1} : x \mapsto 2x q_h(x) - q_{h-1}(x)$ whenever $h > 1$. These polynomial functions are known as *Chebyshev polynomials*. It is again easy to verify that $q_i \perp q_h$ whenever $i \neq h$.

Both of the these products are special instances of a more general construction. For any $-1 < r, s \in \mathbb{R}$, it is possible to define an inner product on $C(-1, 1)$ by setting $\langle f, g \rangle = \int_{-1}^{1} f(x)g(x)(1-x)^r(1+x)^s \, dx$. The set of polynomial functions which are mutually orthogonal with respect to this inner product is called the set of *Jacobi polynomials* of type (r, s). Such polynomials are important in many areas of numerical analysis, and in particular in numerical integration.

With kind permission of the Bibliothèque de l'Institut de France (Legendre).

Adrien-Marie Legendre was one of the first-rate mathematicians who worked in France during the time of the revolution and the generation after it. Among other things, he served on the committee that defined the metric system. **Pafnuty Lvovich Chebyshev**, a nineteenth-century Russian mathematician, made important contributions to both pure and applied mathematics.

Proposition 16.1 *Let V be an inner product space over a field of scalars F.*
(1) *If $v \in V$ satisfies $v \perp w$ for all $w \in V$, then $v = 0_V$.*
(2) *If $\varnothing \neq A \subseteq V$ and if $v \in V$ satisfies the condition that $v \perp w$ for all $w \in A$, then $v \perp w$ for $w \in FA$.*

Proof (1) is an immediate consequence of the fact that if $v \neq 0_V$ then $\langle v, v \rangle \neq 0$. Now assume that $\varnothing \neq A \subseteq V$ and that $v \perp w$ for all $w \in A$. If $y \in FA$ then there exist elements $w_1, \ldots, w_n \in A$ and scalars $a_1, \ldots, a_n$ such that $y = \sum_{i=1}^{n} a_i w_i$ and so $\langle v, y \rangle = \sum_{i=1}^{n} \overline{a}_i \langle v, w_i \rangle = 0$, whence $v \perp y$. $\square$

Proposition 16.2 *Let V be an inner product space and let A be a nonempty set of nonzero mutually-orthogonal vectors in V. Then A is linearly independent.*

Proof Let $\{v_1, \ldots, v_n\}$ be a finite subset of A and assume that there exist scalars $c_1, \ldots, c_n$ such that $\sum_{i=1}^{n} c_i v_i = 0_V$. Then, for $1 \leq h \leq n$, we have $c_h \langle v_h, v_h \rangle = \sum_{i=1}^{n} c_i \langle v_i, v_h \rangle = \langle \sum_{i=1}^{n} c_i v_i, v_h \rangle = \langle 0_V, v_h \rangle = 0$ and hence $c_h = 0$. Thus any finite subset of A is linearly independent, and therefore A is linearly independent. $\square$

If V is an inner product space then any vector $0_V \neq w \in V$ defines a function

$$\pi_w : v \mapsto \frac{\langle v, w \rangle}{\langle w, w \rangle} w$$

from V to itself, which is in fact a projection the image of which is the subspace of V generated by $\{w\}$. This easily-checked remark is the basis for the following theorem.

> **Proposition 16.3 (Gram–Schmidt Theorem)** *Any finitely-generated inner product space V has a basis composed of mutually-orthogonal vectors.*

Proof We will proceed by induction on $\dim(V)$. If $\dim(V) = 1$ the result is immediate. Therefore, we can assume that the proposition is true for any inner product space of dimension k, and assume that $\dim(V) = k + 1$. Let W be a subspace of V of dimension k. By the induction hypothesis, there exists a basis $\{v_1, \ldots, v_k\}$ of W composed of mutually-orthogonal vectors. Let $v \in V \smallsetminus W$ and set $v_{k+1} = v - \sum_{i=1}^{k} \pi_{v_i}(v)$. This vector does not belong to W since $v \notin W$. Therefore, $\{v_1, \ldots, v_{k+1}\}$ is a generating set for V. Moreover, for $1 \le j \le k$, we have

$$\langle v_{k+1}, v_j \rangle = \langle v, v_j \rangle - \sum_{i=1}^{k} \frac{\langle v, v_i \rangle}{\langle v_i, v_i \rangle} \langle v_i, v_j \rangle = \langle v, v_j \rangle - \frac{\langle v, v_j \rangle}{\langle v_j, v_j \rangle} \langle v_j, v_j \rangle = 0$$

and so $v_{k+1} \perp v_j$ for all $1 \le j \le k$. By Proposition 5.3, it follows that the set $\{v_1, \ldots, v_{k+1}\}$ is linearly independent and so is a basis for V. $\qquad\square$

We should note that the proof of Proposition 16.3 is an algorithm, called the *Gram–Schmidt process*, which is easy to implement by a computer program to create a basis composed of mutually-orthogonal vectors of V, when we are given a basis of any sort for the space.

Example Let $v_1 = \begin{bmatrix} 3 \\ 0 \\ 0 \\ 0 \end{bmatrix}$, $\quad v_2 = \begin{bmatrix} 0 \\ 1 \\ 2 \\ 1 \end{bmatrix}$, and $v_3 = \begin{bmatrix} 3 \\ -1 \\ 3 \\ 2 \end{bmatrix}$ be vectors in $\mathbb{R}^4$, on which we have defined the dot product. The set $\{v_1, v_2, v_3\}$ is linearly independent and so generates a three-dimensional subspace W of $\mathbb{R}^4$. Let us use the Gram–Schmidt process to build a basis for W composed of mutually-orthogonal vectors.

Indeed, we define $u_1 = v_1$, $u_2 = v_2 - \pi_{u_1}(v_2) = \begin{bmatrix} 0 \\ 1 \\ 2 \\ 1 \end{bmatrix}$, and $u_3 = v_3 - \pi_{u_1}(v_3) -$

$\pi_{u_2}(v_3) = \frac{1}{6} \begin{bmatrix} 0 \\ -13 \\ 4 \\ 5 \end{bmatrix}$ then $\{u_1, u_2, u_3\}$ is a basis for W, the vectors of which are mutually orthogonal.

Example Let $V = C(-1, 1)$, on which we have defined the inner product $\langle f, g \rangle = \int_{-1}^{1} f(x)g(x)\,dx$. For all $i \ge 0$, let f_i be the polynomial function $f_i : x \mapsto x^i$. Then for each $n > 0$, the set $\{f_0, \ldots, f_n\}$ is linearly independent and so forms a basis for a subspace W of V. We now apply the Gram–Schmidt process to this basis, to obtain a basis $\{p_0, \ldots, p_n\}$ of vectors in V which are mutually orthogonal, where the p_j are precisely the Legendre polynomials we introduced earlier.

Example If α is a diagonalizable endomorphism of a finitely-generated inner product space V then there exists a basis B composed of eigenvectors of α. Applying the Gram–Schmidt process to B will yield a basis of V composed of mutually-orthogonal vectors, but they may no longer be eigenvectors of α. Thus, for example, if $V = \mathbb{R}^2$, if $\alpha : \begin{bmatrix} a \\ b \end{bmatrix} \mapsto \begin{bmatrix} a+b \\ 2b \end{bmatrix}$, and if $B = \left\{ \begin{bmatrix} 1 \\ 0 \end{bmatrix}, \begin{bmatrix} 1 \\ 1 \end{bmatrix} \right\}$, then the Gram–Schmidt process yields $\left\{ \begin{bmatrix} 1 \\ 0 \end{bmatrix}, \begin{bmatrix} 0 \\ 1 \end{bmatrix} \right\}$, where $\begin{bmatrix} 0 \\ 1 \end{bmatrix}$ is not an eigenvector of α.

Actually, the assumption that we have a basis in hand when initiating the Gram–Schmidt process is one of convenience rather than necessity. We could begin with an arbitrary generating set $\{v_1, \ldots, v_n\}$ for the given space. In that case, at the hth stage of the process we would begin by checking whether v_h is a linear combination of the set of mutually-orthogonal vectors $\{u_1, \ldots, u_{h-1}\}$ we have already created. If it is, we just discard it and go on to v_{h+1}.

We should point out that the Gram–Schmidt process is not considered computationally stable—small errors and roundoffs in the computational process accumulate rapidly and can lead at the end to a significant difference between the true solution and the computed solution. There are, fortunately, other more sophisticated methods of constructing a basis composed of mutually-orthogonal vectors from a given basis.

Proposition 16.4 (Hadamard inequality) *Let n be a positive integer, let $A = [a_{ij}] \subset M_{n \times n}(\mathbb{R})$ be a nonsingular matrix, and let $e = |A|$. Then $|e| \le g^n \sqrt{n^n}$, where $g = \max\{|a_{ij}| \mid 1 \le i, j \le n\}$.*

Proof Denote the rows of A by $v_1, \ldots, v_n$. Then $\{v_1, \ldots, v_n\}$ is a basis for $V = \mathbb{R}^n$ and so, using the Gram–Schmidt method, we can find a new basis $\{u_1, \ldots, u_n\}$ for V, on which we consider the dot product, composed of mutually-orthogonal vectors, and defined by setting $u_1 = v_1$ and $u_h = v_h - \sum_{j=1}^{h-1} c_{hj} u_j$, where $c_{hj} = (v_h \cdot u_j)(u_j \cdot u_j)^{-1}$. If $B \in \mathcal{M}_{n \times n}(\mathbb{R})$ is the matrix the rows of which are $u_1, \ldots, u_n$, then $A = CB$, where $C = \begin{bmatrix} 1 & 0 & 0 & \cdots & 0 \\ c_{21} & 1 & 0 & \cdots & 0 \\ c_{31} & c_{32} & 1 & \cdots & 0 \\ \vdots & \vdots & \ddots & \ddots & \vdots \\ c_{n1} & c_{n2} & \cdots & c_{n,n-1} & 1 \end{bmatrix}$. Since C is a lower-triangular matrix, its determinant is the product of the entries on its diagonal, namely 1. Therefore $e = |B|$.

By looking at the Gram–Schmidt method, we see that $\|u_i\| \leq \|v_i\|$ for all $1 \leq i \leq n$. Moreover, since the u_i are mutually orthogonal, we see that $BB^T = D$, where $D = [d_{ij}]$ is the diagonal matrix defined by $d_{ii} = \|u_i\|^2$ for all $1 \leq i \leq n$. Therefore, $e^2 = |BB^T| = |D| = \prod_{i=1}^{n} \|u_i\|^2$. Now let $g = \max\{|a_{ij}| \mid 1 \leq i, j \leq n\}$. Then $\|u_i\| \leq \|v_i\| \leq g\sqrt{n}$ for all $1 \leq i \leq n$ and so $|e| \leq g^n \sqrt{n^n}$, as desired. $\square$

Let V be an inner product space having a subspace W. Let $W^\perp = \{v \in V \mid \langle v, w \rangle = 0$ for all $w \in W\}$. By Proposition 16.1, we know that $W^\perp$ is a subspace of V. Since $\langle v, v \rangle \neq 0$ for all $0_V \neq v \in V$, it is clear that W and $W^\perp$ are disjoint. Also, again by Proposition 16.1, we see that $V^\perp = \{0_V\}$ and $\{0_V\}^\perp = V$. The space $W^\perp$ is called the *orthogonal complement* of W in V, and this name is justified by the following result:

Proposition 16.5 *Let W be a subspace of a finitely-generated inner product space V. Then $V = W \oplus W^\perp$ and $W = (W^\perp)^\perp$. Moreover, if Y is another subspace of V then $W^\perp \cap Y^\perp = (W + Y)^\perp$.*

Proof By Proposition 16.3, we know that it is possible to find a basis $\{v_1, \ldots, v_k\}$ of W which is composed of mutually-orthogonal vectors, and by the construction method used in the proof of this proposition, we see that this can be extended to a basis $\{v_1, \ldots, v_n\}$ of V, the elements of which are still mutually orthogonal. Thus $v_i \in W^\perp$ for all $k < i \leq n$, proving that $V = W + W^\perp$. But we already know that W and $W^\perp$ are disjoint and so we have $W \oplus W^\perp$. Moreover, $\{v_{k+1}, \ldots, v_n\}$ is a basis for $W^\perp$ and so $W = (W^\perp)^\perp$.

Now let Y be another subspace of V. If $v \in W^\perp \cap Y^\perp$ then, for each $w \in Y$ and $y \in Y$ we have $\langle v, w + y \rangle = \langle v, w \rangle + \langle v, y \rangle = 0$ and so $v \in (W + Y)^\perp$. Conversely, if $v \in (W + Y)^\perp$ then $\langle v, w \rangle = \langle v, y \rangle = 0$ for all $w \in W$ and $y \in Y$, so $v \in W^\perp \cap Y^\perp$. $\square$

In particular, if V is an inner product space having a subspace W then we have a natural projection of $W \oplus W^\perp$ onto W, called the *orthogonal projection*. The image of a vector $v \in W \oplus W^\perp$ under this projection is the unique element of W closest to v, according to the distance function defined by the inner product on V, in the sense of the following theorem.

Proposition 16.6 *Let W be a subspace of an inner product space V and let $v = w + y$, where $w \in W$ and $y \in W^\perp$. Then $\|v - w'\| \geq \|v - w\|$ for all $w' \in W$, with equality holding if and only if $w' = w$.*

Proof If $w' \in W$ then

$$
\begin{aligned}
\|v - w'\|^2 = \|w - w' + y\|^2 &= \langle w - w' + y, w - w' + y \rangle \\
&= \langle w - w', w - w' \rangle + \langle y, w - w' \rangle + \langle w - w', y \rangle + \langle y, y \rangle \\
&= \langle w - w', w - w' \rangle + \langle y, y \rangle \\
&= \|w - w'\|^2 + \|y\|^2 = \|w - w'\|^2 + \|v - w\|^2,
\end{aligned}
$$

and from here the result follows immediately. $\qquad\square$

One of the important problems in computational algebra is the following: Given an endomorphism α of a finitely-generated inner product space and a vector $0_V \neq v_0 \in V$, find an efficient procedure to define an orthogonal projection onto the Krylov subspace $F[\alpha]v_0$. One of the first of these is the *Arnoldi process*, a modification of the Gram–Schmidt process. Several variants of this procedure have been devised, depending on special properties of α. This process is not considered as computationally efficient as the Lanczos algorithm mentioned earlier. Arnoldi's process is also the basis for the *GMRES algorithm* (GMRES = generalized minimal residual) for solution of systems of linear equations, devised by Yousef Saad and Martin Schultz in 1986.

© Y. Saad (Saad); © Martin Schultz (Schultz).

Algerian/American **Yousef Saad** and American **Martin H. Schultz** are contemporary computer scientists. Walter Edward Arnoldi was a twentieth century American engineer whose career was mostly spent with United Aircraft Corporation.

Note that Proposition 16.5 is not necessarily true if the space V is not finitely generated, as the following example shows.

Example Let $V = \mathbb{R}^{(\infty)}$. For each $h \geq 0$, let v_h be the sequence in which the hth entry equals 1 and all other entries equal 0. Then $B = \{v_h \mid h \geq 0\}$ is a basis for V composed of mutually-orthogonal vectors. Let $W = \mathbb{R}\{v_0 - v_1, v_1 - v_2, \ldots\}$. This subspace of V is proper since $v_0 \in V \smallsetminus W$. If $0_V \neq y \in W^\perp$ then there exists a nonnegative integer n such that $y = \sum_{i=0}^n a_i v_i$, where the a_i are real numbers and $a_n \neq 0$. But then $a_n = \langle y, v_n - v_{n+1} \rangle = 0$, and that is a contradiction. Therefore, we have shown that $W^\perp = \{0_V\}$, despite the fact that $W \neq V$. Moreover, in this case $V \neq W \oplus W^\perp$ and $(W^\perp)^\perp = V \neq W$.

Let V be an inner product space. A nonempty subset A of V is *orthonormal* if and only if the elements of A are mutually orthogonal, and each of them is normal. Thus, for example, the canonical basis of $\mathbb{R}^n$, equipped with the dot product, is orthonormal.

Example Let $V = C(-\pi, \pi)$, on which we have an inner product defined by $\langle f, g \rangle = \frac{1}{\pi} \int_{-\pi}^{\pi} f(x)g(x)\,dx$. Then we have an orthonormal subset $\{\frac{1}{\sqrt{2}}\} \cup \{\sin(nx) \mid n \geq 1\} \cup \{\cos(nx) \mid n \geq 1\}$ of V.

Example Let V be the subspace of $\mathbb{R}^{\mathbb{R}}$ consisting of all functions f for which $\int_{-\infty}^{\infty} |f(x)|^2\,dx$ is finite, where the norm is taken with respect to the inner product $\langle f, g \rangle = \int_{-\infty}^{\infty} f(x)g(x)\,dx$ defined on V. Let $h \in V$ be the function defined by

$$
h : x \mapsto \begin{cases} 1 & \text{for } 0 < x \leq \frac{1}{2}, \\ -1 & \text{for } \frac{1}{2} < x \leq 1, \\ 0 & \text{otherwise.} \end{cases}
$$

This function is known as the *Haar wavelet*. For each $j, k \in \mathbb{N}$ define the function $h_k^j \in V$ by setting $h_k^j : x \mapsto 2^{j/2} h(2^j x - k)$. Then the subset $\{h_k^j \mid j, k \in \mathbb{N}\}$ of V is orthonormal. Haar wavelets have important applications in image compression.

The twentieth-century Hungarian mathematician **Alfréd Haar** worked primarily in analysis.

Proposition 16.7 *Every finitely-generated inner product space V has an orthonormal basis.*

Proof By Proposition 16.3, we know that V has a basis $\{v_1, \ldots, v_n\}$ the elements of which are mutually orthogonal. For each $1 \leq i \leq n$, let $w_i = \|v_i\|^{-1} v_i$. Then each w_i is normal and $\{w_1, \ldots, w_n\}$ is a basis for V, the elements of which remain mutually orthogonal. $\square$

We can modify the Gram–Schmidt method to provide an algorithm for constructing an orthonormal basis from any given basis of a finitely-generated inner product space V, by normalizing each basis element as it is created. This has the added advantage of tending to reduce accumulated roundoff and truncation errors. The examples after Proposition 16.3 and Proposition 16.6 show that inner product spaces which are not finitely generated may have orthonormal bases as well, but this is not always true. Making use of the Hausdorff Maximum Principle, it is possible to show that every inner product space V has a maximal orthonormal set, which must be linearly independent by Proposition 16.2. Such a subset is called a *Hilbert subset* of V. Clearly, a subset A of V is a Hilbert subset if and only if for every $0_V \neq y \in V$

there exists a $v \in A$ satisfying $\langle v, y \rangle \neq 0$. If V is finitely generated then any Hilbert subset of V is a basis for V, but this is not necessarily true for inner product spaces which are not finitely generated.

Example Let V be the set of all infinite sequences $c_0, c_1, \ldots$ of complex numbers satisfying the condition that $\sum_{i=0}^{\infty} |c_i| < \infty$. We have already seen that this is an inner product space. For each $k \geq 0$, let v_k be the sequence $c_0, c_1, \ldots$ in which

$$c_i = \begin{cases} 1 & \text{if } i = k, \\ 0 & \text{otherwise.} \end{cases}$$

Then $\{v_k \mid k \geq 0\}$ is a Hilbert subset of V which is not a basis for V.

Example It is, of course, possible that a finitely-generated inner product space may have many different orthonormal bases. For example, the canonical basis of $\mathbb{R}^4$ is orthonormal, as is the basis

$$\left\{ \begin{bmatrix} 1 \\ 0 \\ 0 \\ 0 \end{bmatrix}, \begin{bmatrix} 0 \\ \frac{1}{\sqrt{2}} \\ \frac{1}{\sqrt{2}} \\ 0 \end{bmatrix}, \begin{bmatrix} 0 \\ \frac{-1}{\sqrt{2}} \\ \frac{1}{\sqrt{2}} \\ 0 \end{bmatrix}, \begin{bmatrix} 0 \\ 0 \\ 0 \\ 1 \end{bmatrix} \right\}.$$

We can use Proposition 16.7 to verify the assertion made in the previous chapter.

Proposition 16.8 *If V and W are inner product spaces over the same field F, and if V is finitely generated, then $\|\alpha\|$ is finite for all $\alpha \in \mathrm{Hom}(V, W)$.*

Proof Pick an orthonormal basis $\{v_1, \ldots, v_n\}$ for V and let $\alpha \in \mathrm{Hom}(V, W)$. If $v \in \sum_{i=1}^{n} a_i v_i$ is normal, then $1 = \|v\|^2 = \langle v, v \rangle = \sum_{i=1}^{n} \sum_{j=1}^{n} a_i \bar{a}_j \langle v_i, v_j \rangle = \sum_{i=1}^{n} |a_i|^2$ and so $|a_i| \leq 1$ for each $1 \leq i \leq n$. Therefore, $\|\alpha(v)\| \leq \sum_{i=1}^{n} |a_i| \cdot \|\alpha(v_i)\| \leq \sum_{i=1}^{n} \|\alpha(v_i)\|$. Thus $\|\alpha\|$ is finite and less than $c = \sum_{i=1}^{n} \|\alpha(v_i)\|$. $\square$

Example Of course, $\|\alpha\|$ may be finite even when V is not finitely generated. For example, let $V = C(0, 1)$ and define a norm on V by setting $\|f\| = \max\{|f(t)| \mid 0 \leq t \leq 1\}$. Let $g : [0, 1] \times [0, 1] \to \mathbb{R}$ be a continuous function. Let α be the endomorphism of V defined by $\alpha(f) : t \mapsto \int_0^1 g(t, s) f(s) \, ds$. Since g is continuous on a closed subset of $\mathbb{R}^2$, we note that it is bounded there, say $|g(t, s)| \leq c$ for all $0 \leq t, s \leq 1$. Moreover, $|f(s)| \leq \max\{|f(t)| \mid 0 \leq t \leq 1\} = \|f\|$ for all $0 \leq s \leq 1$, and so

$$\|\alpha(f)\| = \max\left\{ \left| \int_0^1 g(t, s) f(s) \, ds \right| \,\middle|\, 0 \leq t \leq 1 \right\}$$

$$\leq \max\left\{ \int_0^1 |g(t, s)| \cdot |f(s)| \, ds \,\middle|\, 0 \leq t \leq 1 \right\} \leq c \|f\|$$

for all $f \in V$.

Proposition 16.9 *Let V be an inner product space having an orthonormal basis B. If $v = \sum_{y \in B} a_y y$ and $w = \sum_{y \in B} b_y y$ are vectors in V (where only finitely-many of the a_y and b_y are nonzero), then $\langle v, w \rangle = \sum_{y \in B} a_y \overline{b}_y$.*

Proof By the properties of the inner product, we have

$$\langle v, w \rangle = \left\langle \sum_{y \in B} a_y y, \sum_{x \in B} b_x x \right\rangle = \sum_{y \in B} \sum_{x \in B} a_y \overline{b}_x \langle y, x \rangle = \sum_{y \in B} a_y \overline{b}_y. \qquad \square$$

Proposition 16.10 *Let V be an inner product space having an orthonormal basis B. Then each $v \in V$ satisfies $v = \sum_{x \in B} \langle v, x \rangle x$.*

Proof We know that there exist scalars $\{a_x \mid x \in B\}$, only finitely-many of which are nonzero, such that $v = \sum_{x \in B} a_x x$. Then for each $x \in B$ we have $\langle v, x \rangle = \langle \sum_{y \in B} a_y y, x \rangle = \sum_{y \in B} a_y \langle y, x \rangle = a_x \langle x, x \rangle = a_x$, which yields the desired result.
$\square$

The coefficients $\langle v, v_i \rangle$ encountered in Proposition 16.10 are called the *Fourier coefficients* of the vector v with respect to the given orthonormal basis.

Example Consider the vector space $C(-1, 1)$ over $\mathbb{R}$, on which we have the inner product $\langle f, g \rangle = \int_{-1}^{1} f(x)g(x)\,dx$. We want to find a polynomial function of degree at most 3 which most closely approximates the function $f : x \mapsto \sin(x)$ on the interval $[-1, 1]$. To do so, consider the subspace V of $C(-1, 1)$ generated by the functions $p_i : x \mapsto x^i$ for $0 \le i \le 3$ and f. Apply the Gram–Schmidt process to the basis $\{p_0, \dots, p_3, f\}$ of V to get an orthonormal basis $\{q_0, \dots, q_3, g\}$, where $q_0 : x \mapsto \frac{1}{2}$; $q_1 : x \mapsto \sqrt{\frac{3}{2}}x$; $q_2 : x \mapsto \sqrt{\frac{5}{2}}(\frac{3}{2}x^2 - \frac{1}{2})$; and $q_3 : x \mapsto \sqrt{\frac{7}{2}}(\frac{5}{2}x^3 - \frac{9}{6}x)$. By Proposition 16.6 and Proposition 16.10, we know that the polynomial function of degree at most 3 which most closely approximates the function f is $\sum_{i=0}^{3} \langle f, q_i \rangle q_i$, where the Fourier coefficients $\langle f, q_i \rangle$ are given by

$$\langle f, q_0 \rangle = \int_{-1}^{1} \frac{1}{2} \sin(x)\,dx = 0;$$

$$\langle f, q_1 \rangle = \int_{-1}^{1} \sqrt{\frac{3}{2}} \sin(x)x \; dx = \sqrt{6}\big(\sin(1) - \cos(1)\big) = 0.738;$$

$$\langle f, q_2 \rangle = \int_{-1}^{1} \sqrt{\frac{5}{2}} \left(\frac{3}{2}x^2 - \frac{1}{2}\right) \sin(x)\,dx = 0;$$

$$\langle f, q_3 \rangle = \int_{-1}^{1} \sqrt{\frac{7}{2}} \left(\frac{5}{2}x^3 - \frac{9}{6}x\right) \sin(x)\,dx$$

$$= 14\sqrt{14}\cos(1) - 9\sqrt{14}\sin(1) = -0.034.$$

Thus the polynomial function we seek is given by $u : x \mapsto -0.315x^3 + 0.998x$.

Proposition 16.11 *Let F be $\mathbb{R}$ or $\mathbb{C}$ and let k and n be positive integers. Let $A \in M_{k \times n}(F)$ be a matrix the columns of which are linearly independent in F^k. Then there exist matrices $Q \in M_{k \times n}(F)$ and $R \in M_{n \times n}(F)$ such that*
(1) $A = QR$;
(2) The columns of Q are orthonormal with respect to the dot product on F^k;
(3) R is nonsingular and upper-triangular.

Proof Let $u_1, \ldots, u_n$ be the columns of A. Apply the Gram–Schmidt process to the set $\{u_1, \ldots, u_n\}$ and then normalize each of the resulting vectors to obtain an orthonormal set $\{v_1, \ldots, v_n\}$ of vectors in F^k. Let $Q \in M_{k \times n}(F)$ be the matrix having columns $v_1, \ldots, v_n$. Then, by Proposition 16.10, we see that $u_i = \sum_{j=1}^{n}(u_i \cdot v_j)v_j$ for all $1 \le i \le n$, and so $A = QR$, where $R = [r_{ij}] \in M_{n \times n}(F)$ is given by $r_{ij} = u_j \cdot v_i$ for all $1 \le i, j \le n$. This matrix is clearly nonsingular. Moreover, we note that the Gram–Schmidt process is such that v_j is orthogonal to $u_1, \ldots, u_{j-1}$ for all $2 \le j \le n$ and so $r_{ij} = 0$ when $i > j$. Therefore, R is also upper-triangular. $\qquad \square$

A factorization of a matrix in the form given by Proposition 16.11 is called a *QR-decomposition*. Such decompositions form a basis of many important numerical algorithms, and are widely used, for example, in computing eigenvalues of large matrices. The use is primarily iterative. If A is an $n \times n$ matrix over $\mathbb{R}$ or $\mathbb{C}$ the eigenvalues of which have distinct absolute values and if we can indefinitely perform the iteration
(1) $A_1 = A$;
(2) If A_i has a QR-decomposition $A_i = Q_i R_i$ then set $A_{i+1} = R_i Q_i$; then, under rather mild conditions on A, the sequence $A_1, A_2, \ldots$ of matrices tends to an upper triangular matrix in which the eigenvalues of A appear in decreasing order of absolute value along the diagonal.

© Walter Gander (Rutishauser); © Vera Kublanovskaya (Kublanovskaya); © Frank Uhlig (Francis).

QR-decompositions were developed independently by the Swiss computer scientist **Heinz Rutishauser**, one of the fathers of ALGOL, by the Russian computer scientist **Vera Kublanovskaya**, and by **John G.F. Francis** of the British computer manufacturer Ferranti Ltd.

One of the major advantages of QR-decompositions is that they are easy to update. If we are given a decomposition $A = QR$, and then the matrix A is altered slightly to obtain a matrix A' by changing a few of its entries, it is relatively easy to alter Q and R to get a QR-decomposition for A'. This is important since many applications of linear algebra involve solving successive systems of linear equations of the form $A^{(i)}X = w^{(i)}$, where $A^{(i+1)}$ and $w^{(i+1)}$ are obtained from $A^{(i)}$ and $w^{(i)}$ by relatively minor modifications, based on data from some external source which is periodically updated.

The following *QR-algorithm* is used to compute a QR-decomposition of a matrix $A \in \mathcal{M}_{k \times n}(F)$ with columns $u_1, \ldots, u_n$:

For $i = 1$ to n do steps (1)–(3):
(1) $v_i = u_i$;
(2) For $j = 1$ to $i - 1$ set $r_{ji} = u_i \cdot v_j$ and $v_i = v_i - r_{ji}v_j$;
(3) Set $r_{ii} = \|v_i\|$ and $v_i = r_{ii}^{-1}v_i$.

Then Q is the matrix with columns $v_1, \ldots, v_n$ and $R = [r_{ij}]$. Note that step (3) presupposes that we have already checked that the set of columns of A is linearly independent. If not, then we have to add an initial check to insure that r_{ii} is nonzero, before we attempt to invert it. As already noted, the Gram–Schmidt method is not numerically stable and hence neither is this algorithm for finding a QR-decomposition. It can be modified to produce a somewhat more stable algorithm by replacing the definition of r_{ji} in step (2) by $r_{ji} = v_i \cdot v_j$.

A variant on this algorithm, called the *QZ algorithm*, has been devised by Moler and Stewart to find solutions for generalized eigenvalue problems.

With kind permission of The MathWorks, Inc. (Moler); © Eric de Sturler (Stewart).

The contemporary American computer scientist **Cleve Moler**, after a distinguished academic career, became chairman and chief scientist of MathWorks, the company that developed MAT-LAB. **G.W. Stewart** is a contemporary American computer scientist.

Proposition 16.12 *Let V be an inner product space having an orthonormal basis B. Then for all $v, w \in V$:*
(1) *(Parseval's identity)* $\langle v, w \rangle = \sum_{y \in B} \overline{\langle y, v \rangle} \langle y, w \rangle$;
(2) *(Bessel's identity)* $\|v\|^2 = \sum_{y \in B} |\langle y, v \rangle|^2$.

Proof Parseval's identity follows from the calculation

$$\langle v, w \rangle = \left\langle \sum_{y \in B} \langle v, y \rangle y, w \right\rangle = \sum_{y \in B} \langle v, y \rangle \langle y, w \rangle = \sum_{y \in B} \overline{\langle y, v \rangle} \langle y, w \rangle;$$

Bessel's identity derives from this in the special case $v = w$.   □

Wilhelm Bessel was a nineteenth century astronomer and a friend of Gauss; his mathematical work came as a result of his research on planetary orbits. The French mathematician Marc-Antoine Parseval published only five short papers at the end of the eighteenth century.

The following results shows that orthogonality can be used to determine the relation between two different inner products defined on a vector space over $\mathbb{R}$.

Proposition 16.13 *Let V be a vector space over $\mathbb{R}$ on which we have defined two inner products, μ_1 and μ_2. For $i = 1, 2$, let $Y_i = \{(v, w) \in V \times V \mid \mu_i(v, w) = 0\}$. Then the following conditions are equivalent:*
(1) *There exists a positive real number c such that $\mu_2 = c^2 \mu_1$;*
(2) $Y_1 = Y_2$;
(3) $Y_1 \subseteq Y_2$.

Proof Since it is clear that (1) implies (2), and (2) implies (3), all we have to prove is that (3) implies (1). Therefore, assume (3). First, let us consider the case $\dim(V) = 1$, i.e., the case in which $V = \mathbb{R}$. Then, for $i = 1, 2$, the scalar $b_i = \mu_i(1, 1)$ is nonzero. Set $d = \mu_2(1, 1)/\mu_1(1, 1)$. If $a, b \in \mathbb{R}$, then $\mu_2(a, b) = ab\mu_2(1, 1) = abd\mu_1(1, 1) = d\mu_1(a, b)$ and so, taking $c = \sqrt{d}$, we have established (1). Thus we can assume that $\dim(V) \geq 2$. For each $i = 1, 2$, and each $v \in V$, let $\|v\|_i = \sqrt{\mu_i(v, v)} > 0$. Without loss of generality, we can assume that there exist elements $v, w \in V$ and positive real numbers $a < b$ such that $\|v\|_2 = a\|v\|_1$ and $\|w\|_2 = b\|w\|_1$, since otherwise we would immediately have (1). Suppose that $w = dv$ for some $0 \neq d \in \mathbb{R}$. Then $\|w\|_2 = |d| \cdot \|v\|_2 = |d|a \cdot \|v\|_1 a\|w\|_1$ and so $a = b$, which is contrary to our assumption that $a < b$. Therefore, we conclude that the set $\{v, w\}$ is linearly independent. Normalizing v and w with respect to μ_1 if necessary, we can furthermore assume that $\|v\|_1 = 1 = \|w\|_1$.

We claim that $(v, w) \notin Y_1$. Indeed, assume otherwise. Then we have $\mu_1(v + w, v - w) = \mu_1(v, v) - \mu_1(w, w) = \|v\|_1^2 - \|w\|_1^2 = 0$, which implies $(v + w, v - w) \in Y_1$. Therefore, by (3), $(v + w, v - w) \in Y_2$. But then we have $\mu_2(v + w, v - w) = \|v\|_2^2 - \|w\|_2^2 = a^2 - b^2 \in \mathbb{R} \setminus \{0\}$, yielding a contradiction and establishing the claim. Set $y = v - \|v\|_1^2 \mu_1(v, w)^{-1} w$. Then $\mu_1(y, v) = \|v\|_1^2 - r\mu_1(v, w) = 0$, where $r = \|v\|_1^2 \mu_1(v, w)^{-1}$, and so $(y, v) \in Y_1 \subseteq Y_2$. Since the set $\{v, w\}$ is linearly independent, we know that $y \neq 0_V$. If we set $y' = \|y\|_1^{-1} y$, then $(y', v) \in Y_1 \subseteq Y_2$ and so, as before, $\mu_1(y' + v, y' - v) = \|y'\|_1^2 - \|v\|_1^2 = 0$. Hence $(y' + v, y' - v) \in Y_1$. Thus $\|y\|_1^2 + \|v\|_1^2 = \| - y + v\|_1^2 = \|rw\|_1^2 = \|v\|_1^4 |\mu_1(v, w)|^{-2} \|w\|_1^2$ and so $\|y\|_1^2 = \|v\|_1^4 |\mu_1(v, w)|^{-2} \|w\|_1^2 - \|v\|_1^2$. Since $\mu_2(y, v) = 0$, we see that $\|y\|_2^2 + \|v\|_2^2 = \| - y + v\|_2^2 = \|rw\|_2^2 = $

$\|v\|_1^4 |\mu_1(v, w)|^{-2} \|w\|_2^2$, Thus

$$\begin{aligned}
\|y\|_2^2 &= \|v\|_1^4 |\mu_1(v, w)|^{-2} \|w\|_2^2 - \|v\|_2^2 \\
&= \|v\|_1^4 |\mu_1(v, w)|^{-2} b^2 \|w\|_1^2 - a^2 \|v\|_1^2 \\
&> a^2 \big(\|v\|_1^4 |\mu_1(v, w)|^{-2} \|w\|_1^2 - \|v\|_1^2 \big) = a^2 \|y\|_1^2.
\end{aligned}$$

Since $\mu_2(y', v) = 0$, this implies that $\mu_2(y' + v, y' - v) = \|y'\|_2^2 - \|v\|_2^2 > a^2 - a^2 = 0$, contradicting (3) and the fact that $\mu_1(y' + v, y' - v) = 0$. From this contradiction, we conclude that there can be no elements a and b as above, so there must exist a positive real number c such that $\|v\|_2^2 = c\|v\|_1^2$ for each nonzero vector $v \in V$. Then for each $v, w \in V$ we have

$$\begin{aligned}
\mu_2(v, w) &= \frac{1}{4}\big[\|v + w\|_2 - \|v - w\|_2 \big] \\
&= \frac{c^2}{4}\big[\|v + w\|_1 - \|v - w\|_1 \big] = c^2 \mu_1(v, w),
\end{aligned}$$

which proves (1).   □

Let V be an inner product space. We have already seen that, for each $w \in V$, the function from V to the field of scalars given by $v \mapsto \langle v, w \rangle$ belongs to $D(V)$. If V is finitely generated, we claim that every element of $D(V)$ is of this form. The following result is actually a special case of a much wider, and more complicated, theorem.

Proposition 16.14 (Riesz Representation Theorem) *Let V be a finitely-generated inner product space. If $\delta \in D(V)$ then there exists a unique vector $y \in V$ satisfying $\delta(v) = \langle v, y \rangle$ for all $v \in V$.*

Proof Let $\{v_1, \dots, v_n\}$ be an orthonormal basis for V and let $y = \sum_{i=1}^n \overline{\delta(v_i)} v_i$. Then for all $1 \le h \le n$ we have

$$\langle v_h, y \rangle = \left\langle v_h, \sum_{i=1}^n \overline{\delta(v_i)} v_i \right\rangle = \sum_{i=1}^n \delta(v_i) \langle v_h, v_i \rangle = \delta(v_h),$$

and so $\langle v, y \rangle = \delta(v)$ for all $v \in V$. The vector y is unique since if $\langle v, x \rangle = \langle v, y \rangle$ for all $v \in V$ then $x = \sum_{i=1}^n \langle x, v_i \rangle v_i = \sum_{i=1}^n \overline{\delta(v_i)} v_i = y$, as desired.   □

The twentieth century Hungarian mathematician **Frigyes Riesz** was one of the founders of functional analysis.

Example Let $n > 1$ be an integer and let V be the subspace of $\mathbb{R}^{\mathbb{R}}$ consisting of all polynomial functions of degree at most n, on which we have an inner product defined by $\langle f, g \rangle = \int_{-1}^{1} f(t)g(t) \, dt$. Let $\delta \in D(V)$ be the linear functional defined by $\delta : f \mapsto f(0)$. By Proposition 16.14, there exists a polynomial function $p \in V$ satisfying the condition $f(0) = \int_{-1}^{1} f(t)p(t) \, dt$ for all $f \in V$. The function p is defined to be $\sum_{i=0}^{n} p_i(0)p_i$, where p_i is the ith Legendre polynomial.

> **Proposition 16.15** *Let V and W be finitely-generated inner product spaces, and let $\alpha : V \rightarrow W$ be a linear transformation. Then there exists a unique linear transformation $\alpha^* : W \rightarrow V$ satisfying the condition $\langle \alpha(v), w \rangle = \langle v, \alpha^*(w) \rangle$ for all $v \in V$ and all $w \in W$.*

Proof Let w be a given vector in W. It is easy to check that the function δ from V to F defined by $\delta : v \mapsto \langle \alpha(v), w \rangle$ is a linear functional. By Proposition 16.14, we know that there exists a unique vector $y_w \in V$ satisfying $\delta(v) = \langle v, y_w \rangle$ for all $v \in V$. Define the function $\alpha^* : W \rightarrow V$ by $\alpha^* : w \mapsto y_w$. We have to prove that this function is indeed a linear transformation. Indeed, if $w_1, w_2 \in W$ then

$$\langle v, \alpha^*(w_1 + w_2) \rangle = \langle \alpha(v), w_1 + w_2 \rangle = \langle \alpha(v), w_1 \rangle + \langle \alpha(v), w_2 \rangle$$
$$= \langle v, \alpha^*(w_1) \rangle + \langle v, \alpha^*(w_2) \rangle = \langle v, \alpha^*(w_1) + \alpha^*(w_2) \rangle,$$

and this is true for all $v \in V$, so we have $\alpha^*(w_1 + w_2) = \alpha^*(w_1) + \alpha^*(w_2)$ for all $w_1, w_2 \in W$. If c is a scalar and if $w \in W$ then

$$\langle v, \alpha^*(cw) \rangle = \langle \alpha(v), cw \rangle = \overline{c} \langle \alpha(v), w \rangle = \overline{c} \langle v, \alpha^*(w) \rangle = \langle v, c\alpha^*(w) \rangle$$

for all $v \in V$, and hence $\alpha^*(cw) = c\alpha^*(w)$. Thus α^* is a linear transformation and, since y_w is uniquely defined, it is also unique. $\square$

Let V and W be inner product spaces and let $\alpha : V \rightarrow W$ be a linear transformation. A linear transformation $\alpha^* : W \rightarrow V$ satisfying the condition $\langle \alpha(v), w \rangle = \langle v, \alpha^*(w) \rangle$ for all $v \in V$ and $w \in W$ is called an *adjoint transformation* of α. If such an adjoint exists, it must be unique. Indeed, assume that $\alpha : V \rightarrow W$ has adjoints α^* and $\alpha^\times$ and that there exists an element $w' \in W$ satisfying $\alpha^*(w') \neq \alpha^\times(w')$. Set

$v' = \alpha^*(w') - \alpha^\times(w')$. Then $\langle v', v' \rangle = \langle v', \alpha^*(w') \rangle - \langle v', \alpha^\times(w') \rangle = \langle \alpha(v'), w' \rangle - \langle \alpha(v'), w' \rangle = 0$ and so $v' = 0_V$, which is a contradiction.

By Proposition 16.15, we know that if V and W are finitely generated then every $\alpha \in \mathrm{Hom}(V, W)$ has an adjoint.

Proposition 16.16 *Let V and W be finitely-generated inner product spaces, having orthonormal bases $B = \{v_1, \ldots, v_n\}$ and $D = \{w_1, \ldots, w_k\}$, respectively. Let $\alpha : V \to W$ be a linear transformation. Then $\Phi_{BD}(\alpha)$ is the matrix $A = [a_{ij}]$, where $a_{ji} = \langle \alpha(v_i), w_j \rangle$ and $\Phi_{DB}(\alpha^*) = A^H$.*

Proof For all $1 \le i \le n$, let $\alpha(v_i) = \sum_{h=1}^{k} a_{hj} w_h$. Then for all $1 \le j \le k$ we have $\langle \alpha(v_i), w_j \rangle = \langle \sum_{h=1}^{k} a_{hj} w_h, w_j \rangle = a_{ji}$ and also $\langle \alpha^*(w_j), v_i \rangle = \overline{\langle v_i, \alpha^*(w_j) \rangle} = \overline{\langle \alpha(v_i), w_j \rangle} = \overline{a}_{ji}$, as needed. $\square$

Example It is, of course, possible that a linear transformation between inner product spaces can have an adjoint even if the spaces are not finitely generated. For example, let $[a, b]$ be a closed interval on the real line and let V be the vector space of all differentiable functions from $[a, b]$ to $\mathbb{R}$. Define an inner product on V by setting $\langle f, g \rangle = \int_a^b f(x)g(x)\,dx$. This is an inner product space which is not finitely generated over $\mathbb{R}$. Let α be the endomorphism of V satisfying $\alpha(f) : x \mapsto \int_a^b e^{-tx} f(t)\,dt$. Then $\langle \alpha(f), g \rangle = \langle f, \alpha(g) \rangle$ for all $f, g \in V$, and so α^* exists, and equals α.

Proposition 16.17 *Let V, W, and Y be inner product spaces. Let α and β be linear transformations from V to W having adjoints, let ζ be a linear transformation from W to Y having an adjoint, and let c be a scalar. Then:*
(1) $(\alpha + \beta)^* = \alpha^* + \beta^*$;
(2) $(c\alpha)^* = \overline{c}\alpha^*$;
(3) $(\zeta\alpha)^* = \alpha^*\zeta^*$;
(4) $\alpha^{**} = \alpha$.

Proof (1) For all $v \in V$ and all $w \in W$, we have

$$\langle v, (\alpha + \beta)^*(w) \rangle = \langle (\alpha + \beta)(v), w \rangle = \langle \alpha(v) + \beta(v), w \rangle$$
$$= \langle \alpha(v), w \rangle + \langle \beta(v), w \rangle = \langle v, \alpha^*(w) \rangle + \langle v, \beta^*(w) \rangle$$
$$= \langle v, (\alpha^* + \beta^*)(w) \rangle,$$

and so by the uniqueness of the adjoint we get $(\alpha + \beta)^* = \alpha^* + \beta^*$.

(2) For all $v \in V$ and all $w \in W$, we have

$$\langle v, (c\alpha)^*(w)\rangle = \langle ((c\alpha)(v), w\rangle = \langle c(\alpha(v)), w\rangle = c\langle \alpha(v), w\rangle$$
$$= c\langle v, \alpha^*(w)\rangle = \langle v, \bar{c}\alpha^*(w)\rangle = \langle v, (\bar{c}\alpha^*)(w)\rangle,$$

and so $(c\alpha)^* = \bar{c}\alpha^*$.

(3) For all $v \in V$ and all $y \in Y$, we have $\langle v, (\zeta\alpha)^*(y)\rangle = \langle (\zeta\alpha)(v), y\rangle = \langle \alpha(v), \zeta^*(y)\rangle = \langle v, \alpha^*\zeta^*(y)\rangle$, and so $(\zeta\alpha)^* = \alpha^*\zeta^*$.

(4) For all $v \in V$ and all $w \in W$, we have $\langle w, \alpha^{**}(v)\rangle = \langle \alpha^*(w), v\rangle = \overline{\langle v, \alpha^*(w)\rangle} = \overline{\langle \alpha(v), w\rangle} = \langle w, \alpha(v)\rangle$, and so $\alpha^{**} = \alpha$. $\qquad\square$

If $(K, \bullet)$ is an algebra over a field F, then a function $a \mapsto a^*$ from K to itself is an *involution* of K if and only if the following additional conditions are satisfied:
(1) $(a + b)^* = a^* + b^*$ and $(a \bullet b)^* = b^* \bullet a^*$ for all $a, b \in K$;
(2) $a^{**} = a$ for all $a \in K$.
Note that $0^* = (0 + 0)^* = 0^* + 0^*$ and so $0^* = 0$. This means that if $0 \neq a \in K$ then $0 \neq a^*$, by (2). If K is unital, then $1 = 1^{**} = (1 \bullet 1^*)^* = 1^{**} \bullet 1^* = 1 \bullet 1^* = 1^*$. If this case, if a is a unit of K then $(a^{-1})^* a^* = (aa^{-1})^* = 1^* = 1$ and similarly $a^*(a^{-1})^* = 1$, so $(a^{-1})^* = (a^*)^{-1}$.

An element a of K is *symmetric with respect to* $*$ if and only if $a^* = a$. If $b \in K$ is a unit symmetric with respect to $*$ then it is straightforward to verify that the function $a \mapsto b^{-1}a^*b$ is also an involution of K.

Example If V is a finitely-generated inner product space, then we see that the function $\alpha \mapsto \alpha^*$ is an involution of the F-algebra $\text{End}(V)$. Another involution we have already seen is the function $A \mapsto A^T$ of the F-algebra $\mathcal{M}_{n\times n}(F)$, for any field F. Of course, in the case $F = \mathbb{R}$, the relation between these two involutions can be seen from Proposition 16.16. We have also seen that the function $A \mapsto A^H$ is an involution on $\mathcal{M}_{n\times n}(\mathbb{C})$, and its relation to the involution $\alpha \mapsto \alpha^*$ is also immediate from Proposition 16.15.

Example Let F be a field and let $(K, \bullet)$ be an F-algebra. Define an operation $\diamond$ on K^2 by setting $\begin{bmatrix} a \\ b \end{bmatrix} \diamond \begin{bmatrix} c \\ d \end{bmatrix} = \begin{bmatrix} a \bullet c \\ d \bullet b \end{bmatrix}$. Then $(K^2, \diamond)$ is an F-algebra and the function $\begin{bmatrix} a \\ b \end{bmatrix} \mapsto \begin{bmatrix} b \\ a \end{bmatrix}$ is an involution of this algebra.

Proposition 16.18 *Let $\alpha : V \to W$ be a linear transformation between finitely-generated inner product spaces. Then:*
(1) $\ker(\alpha^*) = \text{im}(\alpha)^\perp$;
(2) $\ker(\alpha) = \text{im}(\alpha^*)^\perp$;
(3) $\text{im}(\alpha) = \ker(\alpha^*)^\perp$;
(4) $\text{im}(\alpha^*) = \ker(\alpha)^\perp$.

Proof (1) We note that

$$\ker(\alpha^*) = \{w \in W \mid \alpha^*(w) = 0_V\}$$
$$= \{w \in W \mid \langle v, \alpha^*(w) \rangle = 0 \text{ for all } v \in V\}$$
$$= \{w \in W \mid \langle \alpha(v), w \rangle = 0 \text{ for all } v \in V\} = \text{im}(\alpha)^\perp.$$

(2) This follows from the same argument as (1), replacing α by α^*.

(3) By (1) and Proposition 16.6, we have $\text{im}(\alpha) = (\text{im}(\alpha)^\perp)^\perp = \ker(\alpha^*)^\perp$.

(4) This follows from (2) in the way (3) follows from (1). $\square$

Proposition 16.19 *If α is an endomorphism of a finitely-generated inner product space V then $\text{null}(\alpha) = \text{null}(\alpha^*)$.*

Proof By Proposition 6.10 and Proposition 16.18, we see that $\text{null}(\alpha) = \dim(\text{im}(\alpha^*)^\perp) = \dim(V) - \dim(\text{im}(\alpha^*)) = \text{null}(\alpha^*)$. $\square$

Example Proposition 16.19 is not necessarily true for inner product spaces which are not finitely generated. For example, let $V = \mathbb{R}^{(\infty)}$ with the inner product $\langle [a_0, a_1, \ldots], [b_0, b_1, \ldots] \rangle = \sum_{i=0}^{\infty} a_i b_i$. Let $\alpha \in \text{End}(V)$ be given by $\alpha : [a_0, a_1, \ldots] \mapsto [0, a_0, a_1, \ldots]$. Then α^* exists and is given by $\alpha^* : [a_0, a_1, \ldots] \mapsto [a_1, a_2, \ldots]$. Clearly, $\ker(\alpha)$ is trivial but $\ker(\alpha^*)$ is not.

Proposition 16.20 *Let $\alpha : V \to W$ be a linear transformation between finitely-generated inner product spaces. Then*
(1) *If α is a monomorphism then $\alpha^*\alpha$ is an automorphism of V;*
(2) *If α is an epimorphism then $\alpha\alpha^*$ is an automorphism of W.*

Proof (1) It suffices to prove that the linear transformation $\alpha^*\alpha$ is monic. And, indeed, if $v \in V$ satisfies $\alpha^*\alpha(v) = 0_V$ then $\langle \alpha(v), \alpha(v) \rangle = \langle \alpha^*\alpha(v), v \rangle = \langle 0_V, v \rangle = 0$ and so $\alpha(v) = 0_W$. Since α is a monomorphism, $v = 0_V$ and so we have shown that $\alpha^*\alpha$ is monic, as we needed.

(2) First of all, we will show that α^* is a monomorphism. Indeed, if $w_1, w_2 \in W$ are vectors satisfying $\alpha^*(w_1) = \alpha^*(w_2)$ then for all $v \in V$ we have $\langle \alpha(v), w_1 - w_2 \rangle = \langle v, \alpha^*(w_1) - \alpha^*(w_2) \rangle = 0$ and since α is an epimorphism, we conclude that $\langle w, w_1 - w_2 \rangle = 0$ for all $w \in W$. This implies that $w_1 - w_2 = 0_V$ and so $w_1 = w_2$, showing that α^* is indeed monic. Now we will show that $\alpha\alpha^*$ is also monic, which will suffice to prove (2). Indeed, if $\alpha\alpha^*(w) = 0_W$ then $\langle \alpha^*(w), \alpha^*(w) \rangle = \langle \alpha\alpha^*(w), w \rangle = 0$ and so $\alpha^*(w) = 0_V$, proving that $w = 0_W$. $\square$

> **Proposition 16.21** *Let* $\alpha : V \to W$ *be an isomorphism between finitely-generated inner produce spaces. Then* $(\alpha^*)^{-1} = (\alpha^{-1})^*$.

Proof Let $\beta = (\alpha^{-1})^*$. Then for all $v_1, v_2 \in V$ we have $\langle v_1, v_2 \rangle = \langle \alpha^{-1}\alpha(v_1), v_2 \rangle = \langle \alpha(v_1), \beta(v_2) \rangle = \langle v_1, \alpha^*\beta(v_2) \rangle$ and so $\alpha^*\beta(v_2) = v_2$ for all $v_2 \in V$, which means that $\alpha^*\beta$ is the identity map on V. Thus $\beta = (\alpha^*)^{-1}$. $\qquad\square$

Finally, we mention a few consequences of some of the above results with which we will not deal at length, but have extensive and interesting discussions in the mathematical literature.

(1) In an inner product space over $\mathbb{R}$, we can also project onto affine subsets and not just onto subspaces. Indeed, if V is an inner product space over $\mathbb{R}$ then any element v of V defines a linear functional $\delta_v \in D(V)$ given by $\delta_v : w \mapsto \langle w, v \rangle$. If $0 \neq c \in \mathbb{R}$, then $\delta_c^{-1}(c)$ is an affine subset of V. Define a function $\theta_v : V \to V$ by setting

$$\theta_v : y \mapsto y + \left[\frac{c - \delta_v(y)}{\|v\|^2} \right] v.$$

Then for all $y \in V$ we have $\delta_v \theta_v(y) = c$ and so we see that $\theta_v(y) \in \delta_c^{-1}(c)$, so that $\operatorname{im}(\theta_v) \subseteq \delta_c^{-1}(c)$. Moreover, $\theta_v^2 = \theta_v$. We call the function θ_v the *projection* on the affine set $\delta_c^{-1}(c)$. Such projections have many applications, such as the algebraic reconstruction technique (ART), which is very important in computerized imaging.

(2) From Proposition 16.5, we see that the rule which assigns to each subspace W of a finitely-generated inner product space V the orthogonal projection of V onto W is an embedding of the set of all subspaces of V into the algebra $\operatorname{End}(V)$. This observation has many ramifications, of which we mention but one. Let V be a finitely-generated inner product space. For subspaces W and Y of V, we can define the *gap* between W and Y to be $g(W, Y) = \|\pi_W - \pi_Y\|$, where π_W and π_Y are the orthogonal projections of V onto W and Y, respectively. This allows us to measure the distance between subspaces of V in a natural way. One immediately sees that $g(W, Y) = g(W^\perp, Y^\perp)$ and that $g(W, Y) \leq 1$ for all such W and Y. Since the gap is a distance function—in the sense of Proposition 15.12—it turns the set of all subspaces of V into a metric space, the topological properties of which can be studied. For example, one can show that this space is compact and, as a result, also complete, meaning that every sequence $W_1, W_2, \ldots$ of subspaces of V satisfying $\lim_{i,j \to \infty} g(W_i, W_j) = 0$ is convergent. It also makes sense to talk about continuous families of subspaces of V. The analysis of the topological space of all subspaces of a finite-dimensional inner product space has proven to be an extremely important tool.

Exercises

Exercise 1006

Let $A = \begin{bmatrix} 2 & 1 \\ 1 & 2 \end{bmatrix} \in \mathcal{M}_{2\times2}(\mathbb{R})$. Does there exist a matrix $B \in \mathcal{M}_{2\times2}(\mathbb{R})$ of the

form $\begin{bmatrix} \cos(t) & -\sin(t) \\ \sin(t) & \cos(t) \end{bmatrix}$ such that the columns of $A + B$ are orthogonal, when

considered as elements of the space $\mathbb{R}^2$, endowed with the dot product?

Exercise 1007

Calculate the angle between the vectors $\begin{bmatrix} 2 \\ 1 \\ 1 \end{bmatrix}$ and $\begin{bmatrix} 3 \\ 1 \\ -2 \end{bmatrix}$ in the space $\mathbb{R}^3$, en-

dowed with the dot product.

Exercise 1008

Calculate the angle between the vectors $\begin{bmatrix} 1 \\ 1 \\ 1 \\ 2 \\ 1 \end{bmatrix}$ and $\begin{bmatrix} 1 \\ -1 \\ 1 \\ -1 \\ 1 \end{bmatrix}$ in the space $\mathbb{R}^5$, en-

dowed with the dot product.

Exercise 1009

Let n be a positive integer. A matrix $A = [a_{ij}] \in \mathcal{M}_{n\times n}(\mathbb{C})$ is a *complex Hadamard matrix* if and only if $|a_{hj}| = 1$ for all $1 \leq h, j \leq n$ and any pair of distinct rows of A, considered as vectors in $\mathbb{C}^n$, is orthogonal. For each n, find a complex number d such that $A = [d^{hj}]$ is a complex Hadamard matrix. For $n = 6$, find a complex Hadamard matrix which is not of this form.

Exercise 1010

Let A and B be nonempty subsets of $\mathbb{R}^3$ which satisfy the condition that $u \times v \in B$ whenever $u \in A$ and $v \in B$. Is it true that $u \times w \in B^{\perp}$ whenever $u \in A$ and $w \in B^{\perp}$.

Exercise 1011

Let $V = C(0, 1)$ on which we have defined the inner product $\langle f, g \rangle = \int_0^1 f(x)g(x)\,dx$. Calculate $\| \cos(t) \|$.

Exercise 1012

Let $V = \mathcal{M}_{2\times2}(\mathbb{R})$ and define an inner product on V by setting

$$\left\langle \begin{bmatrix} a_{11} & a_{12} \\ a_{21} & a_{22} \end{bmatrix}, \begin{bmatrix} b_{11} & b_{12} \\ b_{21} & b_{22} \end{bmatrix} \right\rangle = \sum_{i=1}^{2}\sum_{j=1}^{2} a_{ij}b_{ij}.$$

Find the angle between the matrices $\begin{bmatrix} 1 & 1 \\ 1 & 1 \end{bmatrix}$ and $\begin{bmatrix} 1 & 1 \\ -1 & 1 \end{bmatrix}$.

Exercise 1013

Let V be the space $\mathbb{R}^4$, together with the dot product. Find a normal vector in V which is orthogonal to each of the vectors $\begin{bmatrix} 1 \\ 1 \\ 1 \\ 1 \end{bmatrix}$, $\begin{bmatrix} 1 \\ -1 \\ -1 \\ 1 \end{bmatrix}$, and $\begin{bmatrix} 2 \\ 1 \\ 1 \\ 3 \end{bmatrix}$.

Exercise 1014

Let $V = \mathbb{R}^3$ on which some inner product is defined. Does there exist a vector $0_V \neq v \in V$ which is orthogonal to each of the vectors $\begin{bmatrix} 2 \\ 1 \\ 3 \end{bmatrix}$, $\begin{bmatrix} 1 \\ -1 \\ 1 \end{bmatrix}$, and $\begin{bmatrix} 2 \\ 0 \\ 4 \end{bmatrix}$?

Exercise 1015

Let $f, g \in \mathbb{R}^\mathbb{R}$ be defined by $f : x \mapsto x$ and $g : x \mapsto x^2 - \frac{1}{2}$. Are f and g orthogonal as elements of $C(0, 1)$? Are they orthogonal as elements of $C(0, 2)$?

Exercise 1016

Find a real number c such that $\|v - w\| = c$ for every orthonormal pair $\{v, w\}$ of vectors in $\mathbb{R}^n$, on which the dot product is defined.

Exercise 1017

Let n be a positive integer and let $c_1, \ldots, c_n$ is a list of real numbers. Let $\{v_1, \ldots, v_n\}$ be an orthonormal basis for $\mathbb{R}^n$, let $d = \min\{c_1, \ldots, c_n\}$. For each $1 < i < n$, set $d_i = \sqrt{c_i} - d$ and let $w_i = dv_i$. Let $B \in M_{n \times n}(\mathbb{R})$ be the matrix the columns of which are $d_1, \ldots, d_n$ and let $A = BB^T + dI$. For $1 \leq i \leq n$, show that v_i is an eigenvector of A associated with the eigenvalue c_i.

Exercise 1018

Let $V = C(0, 1)$ on which we have defined the inner product $\langle f, g \rangle = \int_0^1 f(x)g(x)\,dx$, and let $W = \mathbb{R}\{e^x\} \subseteq V$. Find an infinite set of elements of $W^\perp$.

Exercise 1019

Let $V = \mathbb{R}^4$ on which some inner product is defined. Find distinct vectors $v, w, y \in V$ such that $v \perp w$ and $w \perp y$, but not $v \perp y$.

Exercise 1020

Let V be an inner product space over $\mathbb{R}$ and let v and w be vectors in V. Show that $\|v\| = \|w\|$ if and only if $(v + w) \perp (v - w)$.

Exercise 1021

Let $V = C(-1, 1)$ and define an inner product on V by setting $\langle f, g \rangle = \int_{-1}^{1} f(x)g(x)\,dx$. Let W be the subspace of V composed of all even functions. Find $W^{\perp}$.

Exercise 1022

Let n be a positive integer and let $V = \mathcal{M}_{n \times n}(\mathbb{C})$, on which we have an inner product defined by $\langle A, B \rangle = \text{tr}(A^T \overline{B})$. Let W be the subspace of V consisting of all those matrices $A \in V$ satisfying $\text{tr}(A) = 0$. Find $W^{\perp}$.

Exercise 1023

Define an inner product on $\mathbb{R}^2$ with respect to which the vectors $\begin{bmatrix} -1 \\ 2 \end{bmatrix}$ and $\begin{bmatrix} 2 \\ 4 \end{bmatrix}$ are orthogonal.

Exercise 1024

Make use of the Gram–Schmidt process to find an orthonormal basis for the space $\mathbb{R}^3$ together with the dot product, beginning with the initial basis

$$\left\{ \begin{bmatrix} 1 \\ 1 \\ 1 \end{bmatrix}, \begin{bmatrix} 1 \\ -2 \\ 1 \end{bmatrix}, \begin{bmatrix} 1 \\ 2 \\ 3 \end{bmatrix} \right\}.$$

Exercise 1025

Let V be an inner product space and let A be an orthonormal subset of V. Show that A is a maximal orthonormal subset if and only if for every $0_V \neq y \in V$ there exists a $v \in A$ satisfying $\langle v, y \rangle \neq 0$.

Exercise 1026

Let V be an inner product space of finite dimension n over its field of scalars. Show that there exists a subset $\{v_1, \ldots, v_{2n}\}$ of V satisfying the conditions that $\langle v_i, v_j \rangle \leq 0$ for all $1 \leq i \neq j \leq 2n$.

Exercise 1027

Let V be the space of all polynomial functions in $\mathbb{R}^{\mathbb{R}}$ of degree less than 3, with inner product $\langle p, q \rangle = \frac{1}{2} \int_{-1}^{1} p(t)q(t)\,dt$. Find an orthonormal basis $\{p_0, p_1, p_2\}$ of V satisfying $\deg(p_h) = h$ for $h = 0, 1, 2$.

Exercise 1028

Consider the function $\mu : \mathbb{R}^3 \times \mathbb{R}^3 \to \mathbb{R}$ given by

$$\mu : \left(\begin{bmatrix} a \\ b \\ c \end{bmatrix}, \begin{bmatrix} a' \\ b' \\ c' \end{bmatrix} \right) \mapsto 2aa' + ac' + ca' + bb' + cc'.$$

Show that μ is an inner product and find a basis of $\mathbb{R}^3$ orthonormal with respect to μ.

Exercise 1029

Let V be an inner product space over $\mathbb{R}$ and let W be a finitely-generated subspace of V with orthonormal basis $\{w_1, \ldots, w_n\}$. Let $\alpha \in \text{Hom}(V, W)$ be defined by $\alpha : v \mapsto \sum_{i=1}^{n} \langle v, w_i \rangle w_i$. Show that $\alpha(v) - v \in W^\perp$ for all $v \in V$ and that $\|\alpha(v) - v\| < \|w - v\|$ for all $\alpha(v) \neq w \in W$.

Exercise 1030

Let W be the subspace of $\mathbb{R}^4$ spanned by linearly-independent subset

$$\left\{ \begin{bmatrix} 1 \\ 2 \\ 0 \\ 3 \end{bmatrix}, \begin{bmatrix} 4 \\ 0 \\ 5 \\ 8 \end{bmatrix}, \begin{bmatrix} 8 \\ 1 \\ 5 \\ 6 \end{bmatrix} \right\},$$

which is an inner product space with respect to the dot product. Make use of the Gram–Schmidt process to find an orthonormal basis for W.

Exercise 1031

Let n be a positive integer and let $A \in \mathcal{M}_{n \times n}(\mathbb{R})$. If the set of rows of A is orthonormal with respect to the dot product, is the same true for the set of columns of A?

Exercise 1032

Let $n > k$ be positive integers and let $A \in \mathcal{M}_{k \times n}(\mathbb{R})$ satisfy the condition that its set of rows is orthonormal with respect to the dot product. Show that $(A^T A)^2 = A^T A$.

Exercise 1033

Let n be a positive integer and let $A \in \mathcal{M}_{n \times n}(\mathbb{R})$. Show that A is symmetric if and only if for some $k < n$ there exists a matrix $B \in \mathcal{M}_{n \times k}(\mathbb{R})$ and a real number r such that $A = BB^T + rI$ and the columns of B are mutually orthogonal.

Exercise 1034

Let W be the subspace of $\mathbb{R}^4$, which is an inner product space with respect to the dot product, generated by $\left\{ \begin{bmatrix} 2 \\ 1 \\ 0 \\ 0 \end{bmatrix}, \begin{bmatrix} 0 \\ 0 \\ 1 \\ 2 \end{bmatrix}, \begin{bmatrix} 4 \\ 2 \\ 2 \\ 4 \end{bmatrix}, \begin{bmatrix} 1 \\ 1 \\ 1 \\ 1 \end{bmatrix} \right\}$. Find an orthonormal basis for W and an orthonormal basis for $W^\perp$.

Exercise 1035

Consider $\mathbb{R}^4$ as an inner product space with respect to the dot product. Add two

vectors to the set $\left\{ \frac{1}{6} \begin{bmatrix} 1 \\ 1 \\ 3 \\ -5 \end{bmatrix}, \frac{1}{2} \begin{bmatrix} 1 \\ 1 \\ 1 \\ 1 \end{bmatrix} \right\}$ in order to get an orthonormal basis for

this space.

Exercise 1036

Let n be a positive integer and let $V = \mathbb{R}^n$, which is an inner product space with respect to the dot product. Let $\{v_1, \dots, v_n\}$ be an orthonormal basis for V, let $a \in \mathbb{R}$, and let $1 \le h \ne k \le n$. Define vectors $w_1, \dots, w_n$ in V by setting

$$w_i = \begin{cases} \cos(a)v_h - \sin(a)v_k & \text{if } i = h, \\ \sin(a)v_h - \cos(a)v_k & \text{if } i = k, \\ v_i & \text{otherwise.} \end{cases}$$

Is $\{w_1, \dots, w_n\}$ an orthonormal basis for V?

Exercise 1037

Consider $\mathbb{R}^3$ as an inner product space with respect to the dot product. Is there

a $k \in \mathbb{Z}$ such that $\left\{ \begin{bmatrix} 4k \\ 4 \\ -k \end{bmatrix}, \begin{bmatrix} 0 \\ k \\ 4 \end{bmatrix}, \begin{bmatrix} -25 \\ 16k \\ -12k \end{bmatrix} \right\}$ is an orthonormal basis for this

space?

Exercise 1038

Consider $\mathbb{R}^4$ as an inner product space with respect to the dot product. Find an

orthonormal basis for $\mathbb{R} \left\{ \begin{bmatrix} 1 \\ 0 \\ 1 \\ 0 \end{bmatrix}, \begin{bmatrix} 1 \\ 1 \\ 2 \\ 1 \end{bmatrix}, \begin{bmatrix} 0 \\ 1 \\ 1 \\ 2 \end{bmatrix} \right\}$.

Exercise 1039

Define an inner product on $\mathbb{R}^2$ by setting

$$\left\langle \begin{bmatrix} a \\ b \end{bmatrix}, \begin{bmatrix} c \\ d \end{bmatrix} \right\rangle = ac + \frac{1}{2}(ad + bc) + bd.$$

Find an orthonormal basis for this space.

Exercise 1040

Consider $\mathbb{R}^3$ as an inner product space with respect to the dot product. Let a, b, c, d be nonzero real numbers satisfying the conditions that $a^2 + b^2 + c^2 = d^2$

and $ab + ac = bc$. Show that the subset $\left\{\frac{1}{d}\begin{bmatrix} a \\ b \\ c \end{bmatrix}, \frac{1}{d}\begin{bmatrix} b \\ -c \\ a \end{bmatrix}, \frac{1}{d}\begin{bmatrix} c \\ a \\ -b \end{bmatrix}\right\}$ of $\mathbb{R}^3$ is orthonormal.

Exercise 1041
Define an inner product on $\mathbb{R}^3$ by setting

$$\left\langle \begin{bmatrix} a_1 \\ a_2 \\ a_3 \end{bmatrix}, \begin{bmatrix} b_1 \\ b_2 \\ b_3 \end{bmatrix} \right\rangle = a_1b_1 + 2(a_2b_2 + a_3b_3) - (a_1b_2 + a_2b_1) - (a_2b_3 + a_3b_2).$$

Find an orthonormal basis for this space.

Exercise 1042
Let m be a positive integer and let

$$W = \{f \in \mathbb{C}^{\mathbb{Z}} \mid f(i+m) = f(i) \text{ for all } i \in \mathbb{Z}\},$$

which is a subspace of the vector space $\mathbb{C}^{\mathbb{Z}}$ over $\mathbb{C}$. Define a function $\mu :$ $W \times W \to \mathbb{C}$ by setting $\mu : (f, g) \mapsto \sum_{h=0}^{m-1} f(h)\overline{g(h)}$. For each $0 \le j < m$, let $f_j \in W$ be the function defined by

$$f_j(h) = \begin{cases} 1 & \text{if } h \text{ is of the form } j + mi, \text{ for } i \in \mathbb{Z}, \\ 0 & \text{otherwise.} \end{cases}$$

Show that μ is an inner product on W and that, with respect to that product, $\{f_0, \ldots, f_{m-1}\}$ is an orthonormal basis for W.

Exercise 1043
A function $f \subset \mathbb{R}^{\mathbb{R}}$ has *bounded support* if and only if there exist real numbers $a \le b$ such that $f(x) = 0$ for all x not in the interval $[a, b]$ on the real line. Let V be the set of all such functions and define a function $\mu : V \times V \to \mathbb{R}$ by setting $\mu(f, g) = \int_{-\infty}^{\infty} f(x)g(x)\,dx$. For each $k \in \mathbb{Z}$, let $f_k \in V$ be the function defined by

$$f_k : x \mapsto \begin{cases} 1 & \text{if } k \le x \le k+1, \\ 0 & \text{otherwise.} \end{cases}$$

Show that μ is an inner product on V and that the subset $\{f_k \mid k \in \mathbb{Z}\}$ of V is orthonormal with respect to this inner product.

Exercise 1044
Let V be the subspace of $\mathbb{R}^{\mathbb{R}}$ consisting of all infinitely-differentiable functions f which are periodic of period $h > 0$. (In other words, $f(x + h) = f(x)$ for all $x \in \mathbb{R}$.) Define an inner product on V by setting $\langle f, g \rangle = \int_{-h}^{h} f(x)g(x)\,dx$. Let α be the endomorphism of V which assigns to every element of V its derivative. Find α^*.

Exercise 1045
Let V be an inner product space over $\mathbb{R}$ and let $\{v_1, \ldots v_n\}$ be a set of mutually-orthogonal nonzero vectors in V. Let $a_1, \ldots, a_n$ be positive real numbers satisfying $\sum_{i=1}^{n} a_i = 1$, and let $w = \sum_{j=1}^{n} a_i v_i$. Suppose that $w \perp v_i - v_j$ for all $1 \leq i \neq j \leq n$. Then show that $\|w\|^{-2} = \sum_{i=1}^{n} \|v_i\|^{-2}$.

Exercise 1046
Let V be a finitely-generated inner product space and let $\alpha, \beta_1, \beta_2 \in \mathrm{End}(V)$ satisfy $\alpha^* \alpha \beta_1 = \alpha^* \alpha \beta_2$. Show that $\alpha \beta_1 = \alpha \beta_2$.

Exercise 1047
If W and Y are subspaces of a finitely-generated inner product space V, show that $g(W, Y)$ is the maximum of

$$\sup\{d(w, Y) \mid w \in W \text{ and } \|w\| = 1\} \quad \text{and} \quad \sup\{d(y, W) \mid y \in Y \text{ and } \|y\| = 1\}.$$

Exercise 1048
Let W and Y be subspaces of a finite-dimensional inner product space V satisfying $g(W, Y) < 1$. Show that $\dim(W) = \dim(Y)$.

Exercise 1049
Let K be an algebra over a field F on which we have an involution $a \mapsto a^*$ defined. Let c be an element of K satisfying the conditions that $c = c^2$ and that $c + c^* - 1$ is a unit of K. Show that there exists an element $d \in K$ satisfying $d^2 = d = d^*$, $dc = c$, and $cd = d$. Is d necessarily unique?

Exercise 1050
Let V be an inner product space over $\mathbb{C}$ and let α be an endomorphism of V satisfying $\alpha^* = -\alpha$. Show that every eigenvalue of α is purely imaginary.

Selfadjoint Endomorphisms

17

Let V be an inner product space. An endomorphism α of V is *selfadjoint* if and only if $\langle \alpha(v), w \rangle = \langle v, \alpha(w) \rangle$ for all $v, w \in V$. Such endomorphisms always exist since σ_c is selfadjoint for any $c \in \mathbb{R}$. Selfadjoint endomorphisms have important applications in mathematical models in physics. For example, in mathematical models of quantum theory, selfadjoint operators on the state space of a system represent measurements which can be performed on the system. Note that if $\alpha \in \mathrm{End}(V)$ is selfadjoint, then $\langle \alpha(v), v \rangle = \langle v, \alpha(v) \rangle = \overline{\langle \alpha(v), v \rangle}$ and so $\langle \alpha(v), v \rangle \in \mathbb{R}$ for all $v \in V$.

Example Let $V = C(0, 1)$, which is an inner product space over $\mathbb{R}$ in which $\langle f, g \rangle = \int_0^1 f(x)g(x)\,dx$. Then the endomorphism α of V defined by $\alpha(f) : x \mapsto \int_0^1 \cos(x - y)f(y)\,dy$ for all $f \in V$ is selfadjoint.

Proposition 17.1 *Let V be an inner product space. Then:*
(1) *If $\alpha \in \mathrm{End}(V)$ has an adjoint α^*, then $\alpha + \alpha^*$ is selfadjoint;*
(2) *If $\alpha \in \mathrm{End}(V)$ is selfadjoint, so is $c\alpha$ for each $c \in \mathbb{R}$;*
(3) *If $\alpha \in \mathrm{End}(V)$ is selfadjoint, so is α^n for each positive integer n;*
(4) *If $\alpha, \beta \in \mathrm{End}(V)$ are selfadjoint so are $\alpha + \beta$ and $\alpha \bullet \beta$, where $\bullet$ is the Jordan product in $\mathrm{End}(V)$;*
(5) *If $\alpha \in \mathrm{End}(V)$ is selfadjoint and $\beta \in \mathrm{End}(V)$ has an adjoint, then $\beta\alpha\beta^*$ is selfadjoint.*

Proof (1) If $v, w \in V$ then

$$\langle (\alpha + \alpha^*)(v), w \rangle = \langle \alpha(v), w \rangle + \langle \alpha^*(v), w \rangle$$
$$= \langle v, \alpha^*(w) \rangle + \langle v, \alpha(w) \rangle = \langle v, (\alpha + \alpha^*)(w) \rangle.$$

(2) If $v, w \in V$ then $\langle (c\alpha)(v), w \rangle = c\langle \alpha(v), w \rangle = c\langle v, \alpha(w) \rangle = \langle v, (c\alpha)(w) \rangle$.

(3) This follows by an easy induction argument, using Proposition 16.17(3).

(4) The selfadjointness of $\alpha + \beta$ is an immediate consequence of Proposition 16.17(1). Also, recall that $\alpha \bullet \beta = \frac{1}{2}(\alpha\beta + \beta\alpha)$ and so, if $v, w \in V$ then

$$\langle(\alpha \bullet \beta)(v), w\rangle = \frac{1}{2}\langle\alpha\beta(v), w\rangle + \frac{1}{2}\langle\beta\alpha(v), w\rangle$$

$$= \frac{1}{2}\langle v, \beta\alpha(w)\rangle + \frac{1}{2}\langle v, \alpha\beta(w)\rangle = \langle v, (\alpha \bullet \beta)(w)\rangle.$$

(5) By Proposition 16.17, we have $(\beta\alpha\beta^*)^* = \beta^{**}\alpha\beta^* = \beta\alpha\beta^*$.   □

In particular, if α is a selfadjoint endomorphism of an inner product space V, and if $p(X) \in \mathbb{R}[X]$, then $p(\alpha)$ is selfadjoint. The product of selfadjoint endomorphisms of V need not be selfadjoint, as we will see in the example after Proposition 17.4. Thus we see that the set of all selfadjoint endomorphisms of V is a subspace, though not necessarily a subalgebra, of $\mathrm{End}(V)$.

Example Let V be an inner product space over $\mathbb{R}$ and let $\alpha \in \mathrm{End}(V)$ be selfadjoint. Let $a, b \in \mathbb{R}$ satisfy the condition that $a^2 < 4b$. Then, by the previous remark, we know that $\beta = \alpha^2 + a\alpha + b\sigma_1$ is again a selfadjoint endomorphism of V. Moreover, if $0_V \neq v \in V$ then

$$\langle\beta(v), v\rangle = \langle\alpha^2(v), v\rangle + a\langle\alpha(v), v\rangle + b\langle v, v\rangle$$

$$= \langle\alpha(v), \alpha(v)\rangle + a\langle\alpha(v), v\rangle + b\langle v, v\rangle$$

$$= \|\alpha(v)\|^2 + a\langle\alpha(v), v\rangle + b\|v\|^2.$$

By Proposition 15.2, we know that $|\langle\alpha(v), v\rangle| \leq \|\alpha(v)\| \cdot \|v\|$ and so

$$\langle\beta(v), v\rangle \geq \|\alpha(v)\|^2 - |a| \cdot \|\alpha(v)\| \cdot \|v\| + b\|v\|^2$$

$$= \left(\|\alpha(v)\| - \frac{1}{2}|a| \cdot \|v\|\right)^2 + \left(b - \frac{1}{4}a^2\right)\|v\|^2 > 0.$$

Thus $\beta(v) \neq 0_V$ for each $0_V \neq v \in V$, showing that β is monic. In particular, if V is finitely generated, this in fact shows that β is an automorphism of V.

Let V be a finitely-generated inner product space having an orthonormal basis D. If $\alpha \in \mathrm{End}(V)$ and if $\Phi_{DD}(\alpha) = [a_{ij}]$, then we know from Proposition 16.16 that $\Phi_{DD}(\alpha^*) = \Phi_{DD}(\alpha)^H = [\overline{a}_{ij}]^T$. Therefore, if α is selfadjoint we have $a_{ij} = \overline{a}_{ji}$ for all $1 \leq i, j \leq n$. In particular, $a_{ii} = \overline{a}_{ii}$ for all $1 \leq i \leq n$ and so the diagonal entries in $\Phi_{DD}(\alpha)$ belong to $\mathbb{R}$. Matrices A over $\mathbb{C}$ satisfying the condition that $A = A^H$ are known as *Hermitian matrices*. When we are working over $\mathbb{R}$, these are, of course, just the symmetric matrices. It is clear that the sum of Hermitian matrices is again a Hermitian matrix, but the product of Hermitian matrices is not necessarily

Hermitian, just as we have seen that the product of symmetric matrices is not necessarily symmetric. We do note, however, that if a matrix $A \in \mathcal{M}_{n \times n}(\mathbb{C})$ is Hermitian then so is A^2. Indeed, if A and B are Hermitian matrices in $\mathcal{M}_{n \times n}(\mathbb{C})$, then their Jordan product $\frac{1}{2}(AB + BA)$ is a Hermitian matrix and so, in particular, the product of a commuting pair of Hermitian matrices is again Hermitian. Moreover, any matrix $D \in \mathcal{M}_{n \times n}(\mathbb{C})$ can be written in the form $A + iB$, where $A = \frac{1}{2}(D + D^H)$ and $B = -\frac{1}{2}(D - D^H)$ are both Hermitian matrices. If $A \in \mathcal{M}_{n \times n}(\mathbb{C})$ is Hermitian, then so is cA for any $c \in \mathbb{R}$, and so the set of all Hermitian matrices in $\mathcal{M}_{n \times n}(\mathbb{C})$ is a subspace of $\mathcal{M}_{n \times n}(\mathbb{C})$, considered as a vector space over $\mathbb{R}$; indeed, it is a subalgebra of the commutative Jordan $\mathbb{R}$-algebra $\mathcal{M}_{n \times n}(\mathbb{C})^+$. However, this set is not closed under multiplication by complex scalars, and so it is not a vector space over $\mathbb{C}$.

We have already seen that if V is an inner product space and if $\alpha \in \operatorname{End}(V)$ is selfadjoint, then $\langle \alpha(v), v \rangle \in \mathbb{R}$ for all $v \in V$. If V is finitely generated, the reverse is also true, as follows immediately from the following result.

Proposition 17.2 *Let V be an inner product space over $\mathbb{C}$ and let $\alpha \in \operatorname{End}(V)$ have an adjoint. If $\langle \alpha(v), v \rangle \in \mathbb{R}$ for all $v \in V$, then α is selfadjoint.*

Proof For vectors $v, w \in V$, we have

$$\langle \alpha(v + w), v + w \rangle = \langle \alpha(v), v \rangle + \langle \alpha(v), w \rangle + \langle \alpha(w), v \rangle + \langle \alpha(w), w \rangle$$

and since, by assumption, we know that the scalars $\langle \alpha(v + w), v + w \rangle$, $\langle \alpha(v), v \rangle$, and $\langle \alpha(w), w \rangle$ are all real, we see that $\langle \alpha(v), w \rangle + \langle \alpha(w), v \rangle \in \mathbb{R}$ as well. This implies that $\langle \alpha(v), w \rangle + \langle \alpha(w), v \rangle = \langle w, \alpha(v) \rangle + \langle v, \alpha(w) \rangle$ and so $i\langle \alpha(v), w \rangle + i\langle \alpha(w), v \rangle = i\langle w, \alpha(v) \rangle + i\langle v, \alpha(w) \rangle$. Similarly,

$$\langle \alpha(v + iw), v + iw \rangle = \langle \alpha(v), v \rangle - i\langle \alpha(v), w \rangle + i\langle \alpha(w), v \rangle + \langle \alpha(w), w \rangle$$

and so $-i\langle \alpha(v), w \rangle + i\langle \alpha(w), v \rangle \in \mathbb{R}$. This implies that

$$-i\langle \alpha(v), w \rangle + i\langle \alpha(w), v \rangle = i\langle w, \alpha(v) \rangle - i\langle v, \alpha(w) \rangle$$

and so, multiplying by i and adding it to the previous result, we get $2\langle \alpha(v), w \rangle = 2\langle w, \alpha(v) \rangle$, whence $\langle \alpha(v), w \rangle = \langle w, \alpha(v) \rangle$. Therefore, $\alpha = \alpha^*$. $\square$

Proposition 17.3 *Let V be an inner product space and let $\sigma_0 \neq \alpha \in \operatorname{End}(V)$ be selfadjoint. Then there exists a vector $v \in V$ satisfying $\langle \alpha(v), v \rangle \neq 0$.*

Proof First, assume that the field of scalars is $\mathbb{C}$. Then it is easy to check that if $v, w \in V$ then

$$\langle \alpha(v), w \rangle = \frac{1}{4}\left[\langle \alpha(v+w), v+w \rangle - \langle \alpha(v-w), v-w \rangle\right]$$

$$+ \frac{i}{4}\left[\langle \alpha(v+iw), v+iw \rangle - \langle \alpha(v-iw), v-iw \rangle\right].$$

Moreover, each term on the right-hand side of this equality is of the form $\langle \alpha(y), y \rangle$ for some $y \in V$, so if all of these were equal to 0 we would see that $\alpha(v) \perp w$ for all $v, w \in V$, which means that $\alpha(v) = 0_V$ for all $v \in V$, contradicting the hypothesis that $\sigma_0 \neq \alpha$. Thus the desired result must hold.

Now assume that the field of scalars is $\mathbb{R}$. Then for all $v, w \in V$ we have $\langle \alpha(w), v \rangle = \langle w, \alpha(v) \rangle = \langle \alpha(v), w \rangle$ and so

$$\langle \alpha(v), w \rangle = \frac{1}{4}\left[\langle \alpha(v+w), v+w \rangle - \langle \alpha(v-w), v-w \rangle\right].$$

Again, each term on the right-hand side of this equality is of the form $\langle \alpha(y), y \rangle$ for some $y \in V$ so if all of these were all equal to 0 we would have $\alpha(v) \perp w$ for all $v, w \in V$ which, as we have seen in the previous case, leads to a contradiction. $\square$

We now return to a new variant of a question we have already posed: If α is an endomorphism of an inner product space V, when does there exist an orthonormal basis of V composed of eigenvectors of α? If such a basis exists, we say that α is *orthogonally diagonalizable*.

Example Let $V = \mathbb{R}^3$ with the dot product, and let $\alpha \in \text{End}(V)$ be given by
$\alpha: \begin{bmatrix} a \\ b \\ c \end{bmatrix} \mapsto \begin{bmatrix} b \\ a \\ c \end{bmatrix}$. Then $\left\{ \frac{1}{\sqrt{2}}\begin{bmatrix} 1 \\ 1 \\ 0 \end{bmatrix}, \frac{1}{\sqrt{2}}\begin{bmatrix} 1 \\ -1 \\ 0 \end{bmatrix}, \begin{bmatrix} 0 \\ 0 \\ 1 \end{bmatrix} \right\}$ is an orthogonal basis of V composed of eigenvectors of α, so α is orthogonally diagonalizable.

Proposition 17.4 *Let V be an inner product space and let $\alpha \in \text{End}(V)$ be selfadjoint. Then $\text{spec}(\alpha) \subseteq \mathbb{R}$ and eigenvectors of α associated with distinct eigenvalues are orthogonal.*

Proof Let c be an eigenvalue of α and let v be an eigenvector of α associated with c. Then $c\langle v, v \rangle = \langle cv, v \rangle = \langle \alpha(v), v \rangle = \langle v, \alpha(v) \rangle = \langle v, cv \rangle = \overline{c}\langle v, v \rangle$ and so, since $v \neq 0_V$, we see that $c = \overline{c}$, proving that $c \in \mathbb{R}$. Thus we have shown the first assertion.

If c and d are distinct eigenvalues of α associated with eigenvectors v and w, respectively, then $c\langle v, w \rangle = \langle cv, w \rangle = \langle \alpha(v), w \rangle = \langle v, \alpha(w) \rangle = \langle v, dw \rangle = d\langle v, w \rangle$. Since $c \neq d$, this implies that $\langle v, w \rangle = 0$. $\square$

Example The endomorphisms α and β of $\mathbb{C}^2$ which are given by $\alpha : \begin{bmatrix} a \\ b \end{bmatrix} \mapsto \begin{bmatrix} b \\ a \end{bmatrix}$ and $\beta : \begin{bmatrix} a \\ b \end{bmatrix} \mapsto \begin{bmatrix} a \\ -b \end{bmatrix}$ can easily be seen to be selfadjoint. However, $\mathrm{spec}(\beta\alpha) = \{i, -i\}$ and so, by Proposition 17.4, $\beta\alpha$ is not selfadjoint.

We note the following consequence of Proposition 17.4: If the matrix $A \in M_{n \times n}(\mathbb{R})$ is symmetric, then all eigenvalues of A are real, and so the characteristic polynomial of A is completely reducible in $\mathbb{R}[X]$.

By Proposition 17.4, we see that if V is an inner-product space of finite dimension n over $\mathbb{C}$ and if $\alpha \in \mathrm{End}(V)$ is selfadjoint, then the eigenvalues of α can be written uniquely as an n-tuple $(c_1(\alpha), \ldots, c_n(\alpha))$ of real numbers, the entries of which form a nonincreasing sequence. If $\alpha, \beta \in \mathrm{End}(V)$ are selfadjoint, the problem of describing the possible sets of eigenvalues of $\alpha + \beta$ in terms of those of α and β is extremely important in particle physics, and is known as *Weyl's Problem*. Weyl himself showed that if $\alpha, \beta \in \mathrm{End}(V)$, if $1 \leq k \leq i \leq n$, and $1 \leq j \leq n - i + 1$, then $c_{i+j-1}(\alpha) + c_{n-j+1}(\beta) \leq c_i(\alpha + \beta) \leq c_{i-k+1}(\alpha) + c_k(\beta)$ and so, if $j = k = 1$, we have $c_i(\alpha) + c_n(\beta) \leq c_i(\alpha + \beta) \leq c_i(\alpha) + c_1(\beta)$ for each $1 \leq i \leq n$. In particular, $c_1(\alpha + \beta) \leq c_1(\alpha) + c_1(\beta)$ and $c_n(\alpha) + c_n(\beta) \leq c_n(\alpha + \beta)$. Since then, this problem has been extensively studied. A solution in terms of probability measures was given by the Australian mathematicians Anthony H. Dooley and Norman J. Wildberger, together with the Canadian mathematician Joe Repka, in 1993.

Weyl's result has many interesting consequences. We note that if V is an inner product space and if $\alpha \in \mathrm{End}(V)$ is selfadjoint and represented with respect to a given basis by some matrix $A \in M_{n \times n}(\mathbb{C})$ then the diagonal elements of A must also be real. Schur proved that if A is a matrix representing such an endomorphism α having eigenvalues $c_1(\alpha) \geq \cdots \geq c_n(\alpha)$ and diagonal entries $p_1 \geq \cdots \geq p_n$ then $\sum_{j=1}^{k} p_j \leq \sum_{j=1}^{k} c_j(\alpha)$ for all $1 \leq k \leq n - 1$ and $\sum_{j=1}^{n} p_j = \sum_{j=1}^{n} c_j(\alpha)$. The converse was proven by the American mathematician Alfred Horn: if $c_1 \geq \cdots \geq c_n$ and $p_1 \geq \cdots \geq p_n$ are sequences of real numbers satisfying $\sum_{j=1}^{k} p_j \leq \sum_{j=1}^{k} c_j$ for all $1 \leq k \leq n - 1$ and $\sum_{j=1}^{n} p_j = \sum_{j=1}^{n} c_j$ then there exists a selfadjoint endomorphism α of $\mathbb{C}^n$ with eigenvalues $c_1, \ldots, c_n$ and having an orthonormal basis relative to which α is represented by a matrix having diagonal entries $p_1, \ldots, p_n$.

We now turn to the problem of finding the eigenvalues of a selfadjoint endomorphism of a finitely-generated inner product space. This problem arises in many important applications. For example, let Γ be a (nondirected) graph with vertex set $\{1, \ldots, n\}$. We associate to this graph a symmetric matrix, called the *adjacency matrix* $[a_{ij}]$, the entries of which are nonnegative integers, by setting a_{ij} to be the number of edges in Γ connecting vertex i to vertex j. The matrix represents a selfadjoint endomorphism of $\mathbb{R}^n$ with respect to some basis and its spectrum can be used to derive important information about Γ. This technique has important applications in the analysis of computer networks, in the design of error-correcting codes, and in such areas as chemistry, where it is used to make rough estimates of the electron density distribution of molecules. Another example is the following: If

$B = \{v_1, \ldots, v_n\}$ is a set of distinct vectors in $\mathbb{R}^n$, and if $\| \cdot \|$ is a norm on $\mathbb{R}^n$, then the *distance matrix* defined by B is the matrix $[\|v_i - v_j\|]$. This matrix is symmetric and so defines a selfadjoint endomorphism of $\mathbb{R}^n$. Computing the eigenvalues of such matrices has important applications in many areas, including bioinformatics and X-ray crystallography.

Proposition 17.5 *If V is a nontrivial finitely-generated inner product space, then* $\operatorname{spec}(\alpha) \neq \varnothing$ *for any selfadjoint endomorphism α of V.*

Proof Let α be a selfadjoint endomorphism of V. Choose an orthonormal basis $B = \{v_1, \ldots, v_n\}$ for V and let $A = \Phi_{BB}(\alpha)$. Since α is selfadjoint, we know that $A = A^H$. Let $W = \mathbb{C}^n$ on which we have defined the dot product. Then the endomorphism β of W defined by $\beta : w \mapsto Aw$ is selfadjoint. The degree of the characteristic polynomial $|XI - A|$ of β is $n > 0$ and so, by the Fundamental Theorem of Algebra, it has a root $c \in \mathbb{C}$. Thus the matrix $cI - A$ is singular and so there exists a nonzero vector $w \in W$ satisfying $Aw = cw$. In other words, $c \in \operatorname{spec}(\beta)$. By Proposition 17.4, this implies that $c \in \mathbb{R}$ and so $c \in \operatorname{spec}(\alpha)$, even if V is an inner product space over $\mathbb{R}$. $\square$

In particular, we learn from Proposition 17.5 that every symmetric matrix over $\mathbb{R}$ has an eigenvalue in $\mathbb{R}$. Compare this to the example we have already seen of a symmetric matrix in $\mathcal{M}_{2\times 2}(\mathrm{GF}(2))$ having no eigenvalues. Similarly, the symmetric matrix $\begin{bmatrix} 1 & 2 \\ 2 & 3 \end{bmatrix} \in \mathcal{M}_{2\times 2}(\mathbb{Q})$ has no eigenvalues in $\mathbb{Q}$.

Let V be an inner product space finitely generated over $\mathbb{C}$ and let α be a selfadjoint endomorphism of V. We know, by Proposition 17.4, that the eigenvalues of α are all real and that eigenvectors of α associated with distinct eigenvalues are orthogonal. Let us denote the eigenvalues of α by $c_1, \ldots, c_n$ where the indices are so chosen that $c_1 \geq \cdots \geq c_n$. An important result known as the Courant–Fischer Minimax Theorem states that, for each $1 \leq k \leq n$, we have $c_k = \sup\{\inf\{\langle \alpha(w), w\rangle \mid w \in W$ and $\|w\| = 1\}\}$, where the supremum runs over all subspaces W of V having dimension k.

Let us look at this from a different perspective. The function which assigns to each $0_V \neq v \in V$ the scalar $R_\alpha(v) = \langle v, \alpha(v)\rangle \|v\|^{-2}$ is called the *Rayleigh quotient function*. Note that the projection π_v defined in connection with the Gram–Schmidt theorem satisfies the condition that $\pi_v : \alpha(v) \mapsto R_\alpha(v)v$. By what we have already seen, the image D of this function is contained in $\mathbb{R}$. Moreover, if v is an eigenvector of α with associated eigenvalue c, then $R_\alpha(v) = c$, and so $\varnothing \neq \operatorname{spec}(\alpha) \subseteq D$. On the other hand, it is possible to show—though we will not do it here—that D is contained in the closed interval $[c_n, c_1]$ bounded by the largest and the smallest eigenvalues of α, both endpoints of which in fact belong to D. This observation can be used to define the *Rayleigh quotient iterative scheme* to find eigenvalues of a selfadoint endomorphism α:

As an initial guess, choose a normal vector v_0 and let $d_0 = R_\alpha(v_0)$.
For $k = 0, 1, 2, \ldots$ repeat the following steps:
(1) If $\alpha - d_k\sigma_1 \notin \text{Aut}(V)$, then d_k is an eigenvalue of α, and we are done;
(2) Otherwise, $\alpha - d_k\sigma_1 \in \text{Aut}(V)$. Set $y = (\alpha - d_k\sigma_1)^{-1}(v_k)$ and then compute
   $v_{k+1} = \|y\|^{-2}y$ and $d_{k+1} = R_\alpha(v_{k+1})$.

This scheme will indeed produce an eigenvalue of α for all guesses of v_0 except those in a set of measure 0, and when it converges, the convergence is very rapid. Its main disadvantage is the time and effort needed in step (1) of the iteration to decide if $\alpha - d_k\sigma_1$ is an automorphism of V or not (usually, if the matrix representing this endomorphism is nonsingular or not) and, if it is, to compute its inverse; the algorithm is therefore worthwhile only if this can be done without major computational effort.

With kind permission of the Archives of the Mathematisches Forschungsinstitut Oberwolfach (Fischer, Courant); With kind permission of the Science Photo Library (Strutt).

The twentieth-century German mathematicians **Ernst Fischer** and **Richard Courant** studied spaces of functions. Courant, who headed the Mathematics Institute at the University of Göttingen, fled Germany in 1933 and founded a similar institute in New York City, which now bears his name. **John William Strutt, Lord Rayleigh,** was a nineteenth-century British physicist and applied mathematician, who made important contributions to mathematical physics and who won the Nobel prize in 1904 for his discovery of the inert gas argon.

Example Let α be the endomorphism of $\mathbb{R}^3$ represented with respect to the canonical basis by the symmetric matrix $A = \begin{bmatrix} 2 & 1 & 1 \\ 1 & 3 & 1 \\ 1 & 1 & 4 \end{bmatrix}$. Then α is selfadjoint. Choose

$v_0 = \frac{1}{\sqrt{3}} \begin{bmatrix} 1 \\ 1 \\ 1 \end{bmatrix}$. Using the above algorithm, we see that $d_0 = R_\alpha(v_0) = 5$, which is

not an eigenvalue of α. Moreover,

$$v_1 = \begin{bmatrix} 0.3841106399\ldots \\ 0.5121475201\ldots \\ 0.7682212801\ldots \end{bmatrix} \quad \text{and} \quad d_1 = 5.213114754\ldots,$$

$$v_2 = \begin{bmatrix} 0.3971170680\ldots \\ 0.5206615990\ldots \\ 0.7557840528\ldots \end{bmatrix} \quad \text{and} \quad d_2 = 5.214319743\ldots.$$

The actual value of an eigenvalue of α is $5.214319744\ldots$, so we see that convergence was very rapid indeed.

Proposition 17.6 *Let V be a finitely-generated inner product space and let $\alpha \in \mathrm{End}(V)$. If W is a subspace of V invariant under α, then $W^{\perp}$ is invariant under α^*.*

Proof If $w \in W$ and $y \in W^{\perp}$. Then $\alpha(w) \in W$ and so $\langle w, \alpha^*(y) \rangle = \langle \alpha(w), y \rangle = 0$, whence $\alpha^*(y) \in W^{\perp}$. $\square$

If V is a nontrivial inner product space finitely generated over $\mathbb{R}$ and assume that $\alpha \in \mathrm{End}(V)$ is orthogonally diagonalizable. Then there exists an orthonormal basis $B = \{v_1, \ldots, v_n\}$ composed of eigenvectors of α. Thus $\Phi_{BB}(\alpha)$ is a diagonal matrix and so symmetric. In particular, $\Phi_{BB}(\alpha^*) = \Phi_{BB}(\alpha)^T = \Phi_{BB}(\alpha)$, which proves that $\alpha = \alpha^*$ and so α is selfadjoint. The converse of this result follows from the following proposition.

Proposition 17.7 *Let V be a nontrivial finitely-generated inner product space and let $\alpha \in \mathrm{End}(V)$ be selfadjoint. Then α is orthogonally diagonalizable. The converse holds if the field of scalars is $\mathbb{R}$.*

Proof We will prove the result by induction on $n = \dim(V)$. For $n = 1$, we know by Proposition 17.5 that α has an eigenvector $v \in V$, and so $\{v_1\}$ is the desired basis, where $v_1 = \|v\|^{-1}v$. Now assume that $n > 1$ and that the proposition has been established for all spaces of dimension less than n. Pick v_1 as before and let W be the subspace of V generated by $\{v_1\}$. Then $V = W \oplus W^{\perp}$ and, by Proposition 17.6, we know that $W^{\perp}$ is invariant under $\alpha^* = \alpha$. Moreover, $W^{\perp}$ is an inner product space of dimension $n - 1$ and the restriction of α to $W^{\perp}$ is selfadjoint. Therefore, by the induction hypothesis, there exists an orthonormal basis $\{v_2, \ldots, v_n\}$ of $W^{\perp}$ composed of eigenvectors of α. Since v_1 is orthogonal to each of the vectors in this basis, we see that $\{v_1, \ldots, v_n\}$ is an orthonormal basis of V.

Now assume that the field of scalars is $\mathbb{R}$ and that $\alpha \in \mathrm{End}(V)$ is orthogonally diagonalizable. Then there exists an orthonormal basis D of V composed of eigenvectors of α. This means that $\Phi_{DD}(\alpha)$ is a diagonal matrix, which is surely symmetric, and so by Proposition 16.16 we see that α is selfadjoint. $\square$

Example The converse part of Proposition 17.7 is not true if the field of scalars is $\mathbb{C}$. Indeed, consider the endomorphism α of $\mathbb{C}^2$ represented with respect to the canonical basis by the matrix $\begin{bmatrix} 1 & i \\ i & -1 \end{bmatrix}$. The characteristic polynomial of α is X^2, so were it diagonalizable, it would have to be equal to σ_0.

Let V be an inner product space. An endomorphism $\alpha \in \text{End}(V)$ is *positive definite* (resp., *positive semidefinite*) if and only if it is selfadjoint and satisfies the condition that $\langle \alpha(v), v \rangle$ is a positive (resp., nonnegative) real number for all $0_V \neq v \in V$. If there exist $0_V \neq v, w \in V$ satisfying $\langle \alpha(v), v \rangle > 0 > \langle \alpha(w), w \rangle$, then α is *indefinite*.

We see that σ_c is positive definite for any positive real number c. We also note that a positive-definite endomorphism must be monic since if $\alpha(v) = 0_V$ implies that $\langle \alpha(v), v \rangle = 0$ and so $v = 0_V$. Therefore, every positive-definite endomorphism of a finitely-generated inner product space is in fact an automorphism. Positive definite endomorphisms have important applications in optimization and linear programming.[1]

Example Let $D = \{z \in \mathbb{C} \mid |z| < 1\}$. If $z_1, \ldots, z_n$ are distinct complex numbers in D, and if we are given complex numbers $w_1, \ldots, w_n$ in D, one can ask if there exists an analytic function $f : D \rightarrow D$ satisfying $f(z_i) = w_i$ for all $1 \leq i \leq n$. The *Nevanlinna–Pick Interpolation Theorem* states that such a function exists if and only if the matrix $[a_{ij}]$ in which $a_{ij} = (1 - w_i \overline{w_j})(1 - z_i \overline{z_j})^{-1}$ for all $1 \leq i, j \leq n$, represents a positive-definite endomorphism of $\mathbb{C}^n$. This theorem has been generalized considerably in many directions.

Rolf Nevanlinna was a twentieth-century Finnish mathematician who worked mostly in analysis. **Georg Pick** was a twentieth-century Austrian earth geometer, who was a good friend of Einstein.

Note that if α is positive definite and if $0_V \neq v \in V$ then $0 < \langle \alpha(v), v \rangle = \|\alpha(v)\| \cdot \|v\| \cos(t)$, where t is the angle between $\alpha(v)$ and v, showing that $0 \leq t < \frac{\pi}{2}$.

Example Let $V = \mathbb{R}^n$ on which we have the dot product defined, and let B be the canonical basis. An endomorphism α of V is positive definite if and only if $A = \Phi_{BB}(\alpha)$ is a symmetric matrix satisfying the condition that $v^T A v > 0$ for all nonzero vectors $v \in V$. Such matrices have nice properties. For example, it can be shown that if A is of this form then the Gauss–Seidel method applied to an equation $AX = w$ will converge to the unique solution v, for any initial guess v_0 chosen. If α is positive definite, then the norm on V defined by $\|v\|_A = \sqrt{v^T A v}$ is called an *elliptic norm*. Any norm on V can be reasonably approximated by an elliptic norm, a fact of importance in numerical analysis.

---

[1] Systems of linear equations defined by positive-definite endomorphisms of $\mathbb{R}^n$ first appear in Gauss' work on least-squares approximation, which we will consider in a later chapter.

Example As an immediate consequence of the observation in the previous example, we see that the endomorphism α of $\mathbb{R}^2$ defined by $\alpha : \begin{bmatrix} a \\ b \end{bmatrix} \mapsto \begin{bmatrix} a+b \\ b \end{bmatrix}$ satisfies the condition that $\langle \alpha(v), v \rangle \geq 0$ for all $0_V \neq v \in V$, but is not selfadjoint and so is not positive definite since $\Phi_{BB}(\alpha) = \begin{bmatrix} 1 & 1 \\ 0 & 1 \end{bmatrix}$ is not symmetric.

Example Even if $\sigma_0 \neq \alpha \in \text{End}(V)$ is selfadjoint, it may be the case that neither α nor $-\alpha$ is positive definite. For example, if $\alpha \in \text{End}(\mathbb{R}^2)$ is defined by $\alpha : \begin{bmatrix} a \\ b \end{bmatrix} \mapsto \begin{bmatrix} -a \\ b \end{bmatrix}$, then $\alpha \left(\begin{bmatrix} 1 \\ 0 \end{bmatrix} \right) \cdot \begin{bmatrix} 1 \\ 0 \end{bmatrix} = -1 = (-\alpha) \left(\begin{bmatrix} 0 \\ 1 \end{bmatrix} \right) \cdot \begin{bmatrix} 0 \\ 1 \end{bmatrix}$.

Example The endomorphism $\alpha : \begin{bmatrix} a \\ b \end{bmatrix} \mapsto \begin{bmatrix} a+b \\ a+b \end{bmatrix}$ of $\mathbb{R}^2$ is selfadjoint and, for any $v = \begin{bmatrix} a \\ b \end{bmatrix}$, we check that $\langle \alpha(v), v \rangle = (a+b)^2 \geq 0$, so α is positive semidefinite. On the other and, if $v = \begin{bmatrix} 1 \\ -1 \end{bmatrix}$, then $\langle \alpha(v), v \rangle = 0$, so α is not positive definite. Since α is represented with respect to the canonical basis by the matrix $\begin{bmatrix} 1 & 1 \\ 1 & 1 \end{bmatrix}$, this also shows that in order for an endomorphism to be positive definite, it is not sufficient that it be represented by a matrix all of the entries of which are positive.

Let V an inner product space. If α, $\beta \in \text{End}(V)$ are selfadjoint, then $\alpha - \beta$ is also selfadjoint. We write $\alpha > \beta$ whenever $\alpha - \beta$ is positive definite. Thus, α is positive definite if and only if $\alpha > \sigma_0$. We write $\alpha \geq \beta$ if and only if $\alpha > \beta$ or $\alpha = \beta$. We claim this is a partial-order relation on the set of all selfadjoint endomorphisms of V. Indeed, it is sure that $\alpha \geq \alpha$ for all such endomorphisms α. Suppose that α_1, α_2, and α_3 are selfadjoint endomorphisms of V satisfying $\alpha_1 \geq \alpha_2 \geq \alpha_3$. If $\alpha_1 = \alpha_2$ or $\alpha_2 = \alpha_3$ then it is clear that $\alpha_1 \geq \alpha_3$. Let us therefore assume that $\alpha_1 > \alpha_2 > \alpha_3$. Then for all $v \in V$ we see that $\langle (\alpha_1 - \alpha_3)(v), v \rangle = \langle \alpha_1(v) - \alpha_2(v) + \alpha_2(v) - \alpha_3(v), v \rangle = \langle \alpha_1(v) - \alpha_2(v), v \rangle + \langle \alpha_2(v) - \alpha_3(v), v \rangle > 0$ and so $\alpha_1 > \alpha_3$. Finally, assume that $\alpha_1 \geq \alpha_2$ and $\alpha_2 \geq \alpha_1$ but $\alpha_1 \neq \alpha_2$. Then $\alpha_1 > \alpha_2 > \alpha_1$ and so, as we have seen, $\alpha_1 > \alpha_1$, which is a contradiction. Thus we have a partial order on the set of all selfadjoint endomorphisms of V, called the *Loewner partial order*.

The Czech mathematician **Karl Loewner** emigrated to the United States in 1933. His research concentrated in complex function theory and spaces of functions.

Example Let V be a finite-dimensional inner product space. Inequalities of the form $\sum_{i=1}^{n} a_i \alpha_i > \sigma_0$, where the α_i are in $\text{End}(V)$ and the a_i are scalars, play an important part in control theory, and have been studied extensively.

Proposition 17.8 *Let V be an inner product space and let $\alpha \in \text{End}(V)$ be an endomorphism for which α^* exists. Then α is positive definite if and only if the function $\mu : (v, w) \mapsto \langle \alpha(v), w \rangle$ from $V \times V$ to the field of scalars is also an inner product.*

Proof First, let us assume that α is a positive-definite endomorphism of V. If $v_1, v_2, w \in V$ then $\mu(v_1 + v_2, w) = \langle \alpha(v_1 + v_2), w \rangle = \langle \alpha(v_1), w \rangle + \langle \alpha(v_2), w \rangle = \mu(v_1, w) + \mu(v_2, w)$ and, similarly, we show that $\mu(cv, w) = c\mu(v, w)$ for all scalars c. We also see that $\mu(v, w) = \langle \alpha(v), w \rangle = \langle v, \alpha^*(w) \rangle = \langle v, \alpha(w) \rangle = \overline{\langle \alpha(w), v \rangle} = \overline{\mu(w, v)}$. If $0_V \neq v \in V$ then, by the assumption of positive definiteness, we see that $\mu(v, v) = \langle \alpha(v), v \rangle$ is a positive real number, and it is clear that $\mu(0_V, 0_V) = 0$. Thus μ is an inner product on V.

Conversely, assume that μ is an inner product on V. Then for all $v, w \in V$ we have $\langle v, \alpha^*(w) \rangle = \langle \alpha(v), w \rangle = \mu(v, w) = \overline{\mu(w, v)} = \overline{\langle \alpha(w), v \rangle} = \langle v, \alpha(w) \rangle$ and so $\alpha(w) = \alpha^*(w)$ for all $w \in V$, proving that α is selfadjoint. Moreover, for all $v \in V$ we have $\langle \alpha(v), v \rangle = \mu(v, v)$ for all $0_V \neq v \in V$ and so α is positive definite. $\square$

Proposition 17.9 *Let V be an inner product space, with a given inner product $(v, w) \mapsto \langle v, w \rangle$, and let μ be another inner product defined on V. Then there exists a unique positive-definite endomorphism α of V satisfying the condition that $\mu(v, w) = \langle \alpha(v), w \rangle$ for all $v, w \in V$.*

Proof Fix a vector $w \in V$. The function $v \mapsto \mu(v, w)$ belongs to $D(V)$ and so there exists a unique vector $y_w \in V$ satisfying $\mu(v, w) = \langle v, y_w \rangle$ for all $v \in V$. Define a function $\alpha : V \to V$ by $\alpha : w \mapsto y_w$. Then $\langle \alpha(v), w \rangle = \overline{\langle w, \alpha(v) \rangle} = \overline{\mu(w, v)} = \mu(v, w)$ for all $v, w \in V$. We claim that $\alpha \in \text{End}(V)$. Indeed, if $w_1, w_2 \in V$ then for all $y \in V$ we have

$$\langle \alpha(w_1 + w_2), y \rangle = \mu(w_1 + w_2, y) = \mu(w_1, y) + \mu(w_2, y)$$
$$= \langle \alpha(w_1), y \rangle + \langle \alpha(w_2), y \rangle = \langle \alpha(w_1) + \alpha(w_2), y \rangle$$

and so $\alpha(w_1 + w_2) = \alpha(w_1) + \alpha(w_2)$. Similarly, we can show that $\alpha(cw) = c\alpha(w)$ for all $w \in V$ and all scalars c. Thus we see that α is indeed an endomorphism of V satisfying the condition $\mu(v, w) = \langle \alpha(v), w \rangle$ for all $v, w \in V$, and so it is positive definite.

Finally, α has to be unique since if $\mu(v, w) = \langle \beta(v), w \rangle$ for all $v, w \in V$, then $\langle (\alpha - \beta)(v), w \rangle = \langle \alpha(v) - \beta(v), w \rangle = \langle \alpha(v), w \rangle - \langle \beta(v), w \rangle = 0$ for all $v, w \in V$, which implies that $(\alpha - \beta)(v) = 0_V$ for all $v \in V$, showing that $\alpha = \beta$. $\square$

> **Proposition 17.10** *Let V be a finitely-generated inner product space and let $\alpha \in \text{End}(V)$. Then α is positive definite if and only if there exists an automorphism β of V satisfying $\alpha = \beta^*\beta$.*

Proof Assume that there exists an automorphism β of V satisfying $\alpha = \beta^*\beta$. Then, as previously noted, α is selfadjoint. Moreover, for all $0_V \neq v \in V$ we have $\langle \alpha(v), v \rangle = \langle \beta^*\beta(v), v \rangle = \langle \beta(v), \beta^{**}(v) \rangle = \langle \beta(v), \beta(v) \rangle > 0$ since β is an automorphism and hence $\beta(v) \neq 0_V$. Therefore, α is positive definite.

Conversely, assume that α is a positive-definite endomorphism of V. Then the function $\mu : (v, w) \mapsto \langle \alpha(v), w \rangle$ is an inner product on V. Let $\{v_1, \ldots, v_n\}$ be a basis for V which is orthonormal with respect to the original inner product on V and let $\{w_1, \ldots, w_n\}$ be a basis for V which is orthonormal with respect to μ. By Proposition 6.2, we know that there exists a unique endomorphism β of V satisfying $\beta(w_i) = v_i$ for all $1 \leq i \leq n$. Then β is an epimorphism since its image contains a basis for V and so, since V is finitely-generated, it is an automorphism of V. Therefore, if $v = \sum_{i=1}^n a_i w_i$ and $w = \sum_{j=1}^n b_j w_j$ are vectors in V we see that

$$\langle \alpha(v), w \rangle = \mu(v, w) = \mu\left(\sum_{i=1}^n a_i w_i, \sum_{j=1}^n b_j w_j \right)$$

$$= \sum_{i=1}^n \sum_{j=1}^n a_i \overline{b}_j \mu(w_i, w_j) = \sum_{i=1}^n a_i \overline{b}_i$$

and similarly

$$\langle \beta^*\beta(v), w \rangle = \langle \beta(v), \beta(w) \rangle = \left\langle \beta\left(\sum_{i=1}^n a_i w_i \right), \beta\left(\sum_{j=1}^n b_j w_j \right) \right\rangle$$

$$= \sum_{i=1}^n \sum_{j=1}^n a_i \overline{b}_j \langle w_i, w_j \rangle = \sum_{i=1}^n a_i \overline{b}_i,$$

and so we see that $\langle \beta^*\beta(v), w \rangle = \langle \alpha(v), w \rangle$ for all $v, w \in V$, which shows that $\alpha = \beta^*\beta$. $\qquad \square$

From Proposition 17.10 we know that if $A = [a_{ij}] \in \mathcal{M}_{n \times n}(\mathbb{C})$ is a matrix representing a positive-definite endomorphism of $\mathbb{C}^n$, namely if it is a Hermitian matrix satisfying the condition that $v \cdot Av > 0$ for all nonzero vectors $v \in \mathbb{C}^n$, then there exists a nonsingular matrix B such that $A = B^H B$. Indeed, we can choose B to be upper triangular, so that it is a form of LU-decomposition, though it takes only half as many arithmetic operations to perform. This decomposition is known as a *Cholesky decomposition* of A. This decomposition need not be unique. Cholesky decompositions are widely used in building economic and financial models. Because of this wide usage, there are many algorithms available to efficiently calculate Cholesky decompositions of general matrices or of matrices in special forms. Indeed, one of the computational advantages of the Cholesky decomposition is that it is numerically stable, even with no pivoting. On the other hand, if you change even one element of A you have to recompute the Cholesky decomposition of the new matrix from scratch.

With kind permission of the Collections École polytechnique (SABIX).

Major **André-Louis Cholesky** was a cartographer in the French army, who used this method in connection with the mapping of the island of Crete before World War I. It had previously been used by other cartographers, including Myrick H. Doolittle, of the computing division of the US Coast and Geodetic Survey, in 1878. A mathematical formulation had been given earlier by Toeplitz.

The following algorithm calculates a Cholesky decomposition for real symmetric matrices.

For $k = 1, \ldots, n$ perform the following steps:
(1) For each $1 \le i < k$ define $b_{ik} = b_{ii}^{-1}[a_{ik} - \sum_{j=1}^{i-1} b_{ji}b_{jk}]$;
(2) Set $b_{kk} = \sqrt{a_{kk} - \sum_{j=1}^{k-1} b_{jk}^2}$;
(3) For each $k < i \le n$ set $b_{ik} = 0$.

Note that if the matrix A did not satisfy $v \cdot Av > 0$ for all nonzero vectors v, the algorithm would hang up at some stage, trying to take the square root of a negative number. Indeed, attempting a Cholesky decomposition is often used as a test to see whether a given matrix represents a positive-definite endomorphism or not.

Example Let $A = \begin{bmatrix} 5 & 2 & 3 \\ 2 & 1 & 1 \\ 3 & 1 & 4 \end{bmatrix} \in \mathcal{M}_{3\times3}(\mathbb{R})$. This is a symmetric matrix satisfying the condition that $v \cdot Av > 0$ for all nonzero vectors $v \in \mathbb{R}^3$ and having a Cholesky decomposition $B^T B$, where $B = \frac{1}{5} \begin{bmatrix} 5\sqrt{5} & 2\sqrt{5} & 3\sqrt{5} \\ 0 & \sqrt{5} & -\sqrt{5} \\ 0 & 0 & 5\sqrt{2} \end{bmatrix}$.

Notice that the Proposition 17.10 extends our ongoing analogy between the operation $*$ and the conjugate operation on $\mathbb{C}$, just as the notion of "positive definite" is the analog of positivity of complex numbers: a complex number z is (real and) positive if and only if there exists a complex number y such that $z = \bar{y}y$.

Cholesky decompositions do not work for Hermitian matrices representing indefinite endomorphisms of $\mathbb{C}^n$. In such cases, one has to make use of other methods, such as the *Bunch–Kaufman algorithm*, which is quite effective for sparse matrices.

Proposition 17.11 *Let V be an inner product space. If $\alpha \in \text{End}(V)$ is positive definite, then every eigenvalue of α is a positive real number. The converse holds if α is orthogonally diagonalizable.*

Proof Assume that α is positive define. By Proposition 17.4, the eigenvalues of α are real numbers. If $c \in \text{spec}(\alpha)$ is an eigenvalue of α associated with an eigenvector v, then $0 < \langle \alpha(v), v \rangle = \langle cv, v \rangle = c\langle v, v \rangle$ and so $c > 0$, since we know that $\langle v, v \rangle > 0$. Conversely, assume that every eigenvalue of α is positive and that there exists an orthonormal basis B of V composed of eigenvectors of α. Let $v = \sum_{i=1}^n a_i v_i$, where $\{v_1, \ldots, v_n\} \subseteq B$. For each $1 \le i \le n$, let c_i be an eigenvalue of α associated with v_i. We can assume that the v_i are arranged in such a manner that $0 < c_1 \le c_2 \le \cdots \le c_n$. Then

$$\langle \alpha(v), v \rangle = \left\langle \sum_{i=1}^n \alpha(a_i v_i), \sum_{j=1}^n a_j v_j \right\rangle = \sum_{i=1}^n \sum_{j=1}^n a_i \overline{a}_j \langle c_i v_i, v_j \rangle$$

$$= \sum_{i=1}^n \sum_{j=1}^n c_i a_i \overline{a}_j \langle v_i, v_j \rangle = \sum_{i=1}^n c_i |a_i|^2 \ge c_1 \sum_{i=1}^n |a_i|^2 > 0,$$

and so α is positive definite. $\square$

From Propositions 17.11 and 17.7, we see that if V is an finitely-generated inner product space over $\mathbb{R}$ or $\mathbb{C}$ and if $\alpha \in \text{End}(V)$ is positive definite, then there exists a basis of V relative to which α is represented by a diagonal matrix in which the entries of the diagonal are positive real numbers. Such a matrix is, of course, nonsingular.

Proposition 17.12 *Let V and W be inner-product spaces finitely-generated over $\mathbb{R}$ and let $\alpha \in \text{Hom}(V, W)$. Then $\|\alpha\| = \sqrt{c}$, where c is the largest eigenvalue of $\alpha^*\alpha \in \text{End}(V)$, and where $\|\alpha\|$ is the norm induced by the respective inner products on V and W.*

Proof If c is an eigenvalue of $\beta = \alpha^*\alpha$ then there exists a nonzero vector v such that $\beta(v) = cv$ and so $c\|v\|^2 = \langle v, cv \rangle = \langle v, \alpha^*\alpha(v) \rangle = \langle \alpha(v), \alpha(v) \rangle \ge 0$, and so $c \ge 0$. By Proposition 17.7, we know that there exists a basis $\{v_1, \ldots, v_n\}$ of V composed of orthonormal eigenvectors of β. For each $1 \le i \le n$, let c_i be an eigenvalue of β associated with v_i. After renumbering, we can assume that $0 \le c_1 \le \cdots \le c_n$. If

$v = \sum_{i=1}^{n} a_i v_i \in V$, then

$$\|\alpha(v)\|^2 = \langle v, \beta(v) \rangle = \langle v, \alpha^*\alpha(v) \rangle = \sum_{j=1}^{n} \sum_{i=1}^{n} a_j c_i a_i \langle v_j, v_i \rangle$$

$$= \sum_{i=1}^{n} c_i a_i^2 \leq c_n \left(\sum_{i=1}^{n} a_i^2 \right) = c_n \|v\|^2$$

and so $\|\alpha(v)\|^2 / \|v\|^2 \leq c_n$. Therefore, by definition of the induced norm, $\|\alpha\| \leq \sqrt{c_n}$. But one easily sees that $\|\alpha(v_n)\|^2 / \|v_n\|^2 = c_n$ and so $\sqrt{c_n} \leq \|\alpha\|$, proving equality. □

Example Let $\alpha : \mathbb{R}^3 \to \mathbb{R}^2$ be the linear transformation defined by $\alpha : v \mapsto Av$, where $A = \begin{bmatrix} 1 & -2 & 1 \\ 3 & 0 & -1 \end{bmatrix}$. Then $\alpha^*\alpha$ is the endomorphism of $\mathbb{R}^3$ given by $v \mapsto$ $\begin{bmatrix} 10 & -2 & -2 \\ -2 & 4 & -2 \\ -2 & -2 & 2 \end{bmatrix} v$. The eigenvalues of this endomorphism are $0 \leq 8 - 2\sqrt{2} \leq$ $8 + 2\sqrt{2}$ and so $\|\alpha\| = \sqrt{8 + 2\sqrt{2}}$, which is approximately equal to 3.291.

Let V and W be inner product spaces. A linear transformation $\alpha : V \to W$ preserves inner products if and only if $\langle v_1, v_2 \rangle = \langle \alpha(v_1), \alpha(v_2) \rangle$ for all $v_1, v_2 \in V$. Notice that any linear transformation which preserves inner products also preserves distances: $\|v_1 - v_2\| = \|\alpha(v_1 - v_2)\| = \|\alpha(v_1) - \alpha(v_2)\|$ for all $v_1, v_2 \in V$. Also, as a direct consequence of the definition, such a linear transformation preserves the angles between vectors. Conversely, we have already noted that from the norm defined by an inner product we can recover the inner product itself, so that any linear transformation $\alpha : V \to W$ satisfying $\|v_1 - v_2\| = \|\alpha(v_1) - \alpha(v_2)\|$ for all v_1, v_2 also preserves inner products. Such a linear transformation is called an *isometry*.

Proposition 17.13 *Let V be an inner product space over $\mathbb{C}$ and let $\alpha \in$ End(V) be an isometry. Then the eigenvalues of α lie on the unit circle $\{z \in \mathbb{C} \mid |z| = 1\}$.*

Proof If c is an eigenvalue of α with associated eigenvector v, then $\|v\|^2 = \|cv\|^2 = |c|^2 \|v\|^2$ and so $|c| = 1$. □

Example Let V be an inner product space over $\mathbb{R}$ and let $0_V \neq y \in V$. This vector y defines an endomorphism α_y of V by setting

$$\alpha_y : v \mapsto -v + 2 \frac{\langle v, y \rangle}{\langle y, y \rangle} y.$$

This endomorphism is an isometry which satisfies $\alpha_y^2 = \sigma_1$, and y is a fixed point of α_y.

Proposition 17.14 *Let V and W be finitely-generated inner product spaces and having equal dimensions. Then the following conditions on a linear transformation $\alpha : V \to W$ are equivalent:*
(1) α is an isometry;
(2) α is an isomorphism which is an isometry;
(3) If $\{v_1, \ldots, v_n\}$ is an orthonormal basis of V then the set $\{\alpha(v_1), \ldots, \alpha(v_n)\}$ is an orthonormal basis of W.

Proof (1) $\Rightarrow$ (2): If $0_V \neq v \in V$ then $\langle v, v \rangle = \langle \alpha(v), \alpha(v) \rangle$, and so $\alpha(v) \neq 0_W$. Thus we see that $\ker(\alpha) = \{0_V\}$ and so α is an isomorphism, since V and W have the same finite dimension.

(2) $\Rightarrow$ (3): If $\{v_1, \ldots, v_n\}$ is an orthonormal basis of V then, since α is an isomorphism, we see that $\{\alpha(v_1), \ldots, \alpha(v_n)\}$ is a basis for W. Moreover, for all $1 \leq i, j \leq n$ we know that

$$\langle \alpha(v_i), \alpha(v_j) \rangle = \langle v_i, v_j \rangle = \begin{cases} 1 & \text{when } i = j, \\ 0 & \text{otherwise,} \end{cases}$$

and so this basis is orthonormal.

(3) $\Rightarrow$ (1): Let $\{v_1, \ldots, v_n\}$ be an orthonormal basis of V. If $v = \sum_{i=1}^n a_i v_i$ and $y = \sum_{j=1}^n b_j v_j$, then $\langle v, y \rangle = \sum_{i=1}^n a_i \overline{b}_i$. Moreover,

$$\langle \alpha(v), \alpha(y) \rangle = \left\langle \alpha \left(\sum_{i=1}^n a_i v_i \right), \alpha \left(\sum_{j=1}^n b_j v_j \right) \right\rangle$$

$$= \sum_{i=1}^n \sum_{j=1}^n a_i \overline{b}_j \langle \alpha(v_i), \alpha(v_j) \rangle = \sum_{i=1}^n a_i \overline{b}_i = \langle v, y \rangle,$$

and this proves (1). $\qquad\square$

In particular, if V and W are finitely-generated inner product spaces having equal dimensions, then every isometry $\alpha : V \to W$ is an isomorphism. If $w_1, w_2 \in W$, then $\langle w_1, w_2 \rangle = \langle \alpha\alpha^{-1}(w_1), \alpha\alpha^{-1}(w_2) \rangle = \langle \alpha^{-1}(w_1), \alpha^{-1}(w_2) \rangle$ and so we see that α^{-1} is also an isometry. Moreover, there is always at least one isometry α from V to W. Just pick orthonormal bases $\{v_1, \ldots, v_n\}$ for V and $\{w_1, \ldots, w_n\}$ for W and define α by $\alpha : \sum_{i=1}^n a_i v_i \mapsto \sum_{i=1}^n a_i w_i$.

Example The endomorphism of $\mathbb{R}^3$ represented with respect to the canonical basis

by $\dfrac{1}{\sqrt{6}} \begin{bmatrix} \sqrt{3} & \sqrt{2} & -1 \\ \sqrt{3} & -\sqrt{2} & 1 \\ 0 & \sqrt{2} & 2 \end{bmatrix}$ is an isometry.

Example Let W be the set of all matrices $A \in \mathcal{M}_{3 \times 3}(\mathbb{R})$ satisfying $A^T = -A$, which is a subspace of $\mathcal{M}_{3 \times 3}(\mathbb{R})$ of dimension 3. Define an inner product on W as follows: if $A, B \in W$ then $\langle A, B \rangle = \frac{1}{2}\operatorname{tr}(AB^T)$. Let $V = \mathbb{R}^3$, which is an inner product space with respect to the dot product. Define a linear transformation $\alpha : V \to W$ by setting $\alpha : \begin{bmatrix} a \\ b \\ c \end{bmatrix} \mapsto \begin{bmatrix} 0 & -c & b \\ c & 0 & -a \\ -b & a & 0 \end{bmatrix}$. If $A = \begin{bmatrix} 0 & -c & b \\ c & 0 & -a \\ -b & a & 0 \end{bmatrix}$

and $B = \begin{bmatrix} 0 & -f & e \\ f & 0 & -d \\ -e & d & 0 \end{bmatrix}$ then $AB^T = \begin{bmatrix} cf + be & -bd & -dc \\ ea & cf + ad & ec \\ -af & fb & be + ad \end{bmatrix}$ and so

we can check that $\langle A, B \rangle = \begin{bmatrix} a \\ b \\ c \end{bmatrix} \cdot \begin{bmatrix} d \\ e \\ f \end{bmatrix}$ and thus α is an isometry, and hence is an isomorphism.

Example Proposition 17.14 is no longer true if we remove the condition that the spaces are finitely generated. Indeed, let $V = C(0, 1)$, on which we have the inner product $\mu(f, g) = \int_0^1 f(x)g(x)x^2\,dx$, and let W be the same space on which we have the inner product $\langle f, g \rangle = \int_0^1 f(x)g(x)\,dx$. Let $\alpha : V \to W$ be the linear transformation defined by $\alpha : f(x) \mapsto xf(x)$. Then $\mu(f, g) = \langle \alpha(f), \alpha(g) \rangle$ and so α is an isometry. But α is not an isomorphism since the function $x \mapsto x^2 + 1$ does not belong to the image of α.

Let us now return to the case of inner product spaces the dimensions of which are not necessarily equal.

> **Proposition 17.15** *Let V and W be inner product spaces finitely-generated over $\mathbb{R}$ and let $\alpha \in \operatorname{Hom}(V, W)$. Then α is an isometry if and only if $\alpha^*\alpha = \sigma_1 \in \operatorname{End}(V)$.*

Proof By Proposition 16.15, α^* exists. If $\alpha^*\alpha = \sigma_1 \in \operatorname{End}(V)$, and if $v_1, v_2 \in V$ then

$$\|v_1 - v_2\|^2 = \langle v_1 - v_2, v_1 - v_2 \rangle = \langle v_1 - v_2, \alpha^*\alpha(v_1 - v_2) \rangle$$
$$= \langle \alpha(v_1 - v_2), \alpha(v_1 - v_2) \rangle = \|\alpha(v_1) - \alpha(v_2)\|^2,$$

and so $\|v_1 - v_2\| = \|\alpha(v_1) - \alpha(v_2)\|$, proving that α is an isometry. Conversely, if α is an isometry and if $v_1, v_2 \in V$ then $\langle \alpha^*\alpha(v_1), v_2 \rangle = \langle \alpha(v_1), \alpha(v_2) \rangle = \langle v_1, v_2 \rangle$. Therefore, by Proposition 16.14, we see that $\alpha^*\alpha(v_1) = v_1$ for all $v_1 \in V$, so $\alpha^*\alpha = \sigma_1 \in \operatorname{End}(V)$. $\square$

Exercises

Exercise 1051
Let $V = \mathbb{C}[X]$ and define an inner product on V by setting

$$\left\langle \sum_{i=0}^{\infty} a_i X^i, \sum_{i=0}^{\infty} b_i X^i \right\rangle = \sum_{i=0}^{\infty} a_i \overline{b_i}.$$

Let α be the endomorphism of V defined by $\alpha : p(X) \mapsto (X+1)p(X)$. Calculate α^*, or show that it does not exist.

Exercise 1052
Let $V = \mathbb{C}[X]$ and define an inner product on V by setting

$$\left\langle \sum_{i=0}^{\infty} a_i X^i, \sum_{i=0}^{\infty} b_i X^i \right\rangle = \sum_{i=0}^{\infty} a_i \overline{b_i}.$$

Let β be the endomorphism of V defined by $\beta : p(X) \mapsto p(X+1)$. Calculate β^*, or show that it does not exist.

Exercise 1053
Let $p > 1$ be an integer, let $G = \mathbb{Z}/(p)$, and let $V = \mathbb{C}^G$, which is an inner product space over $\mathbb{C}$ with inner product defined by $\langle f, g \rangle = \sum_{n \in G} f(n) \overline{g(n)}$. Let α be the endomorphism of V defined by $\alpha(f) : n \mapsto f(n+1) + f(n-1)$. Is α selfadjoint?

Exercise 1054
Let V be a vector space over $\mathbb{R}$. A nonempty subset K of V is *convex* if and only if $cv + (1 - c)w \in K$ whenever $v, w \in K$ and $0 \leq c \leq 1$. Is the set of all selfadjoint endomorphisms of an inner product space Y over $\mathbb{R}$ necessarily a convex subset of the vector space $\mathrm{End}(Y)$?

Exercise 1055
Let V be an inner product space and let α be an endomorphism of V. Is the endomorphism $\alpha^* \alpha - \sigma_1$ of V selfadjoint?

Exercise 1056
Let n be a positive integer and let V be the space of all polynomial functions in $\mathbb{R}^{\mathbb{R}}$ of degree at most n. Define an inner product on V by setting $\langle f, g \rangle = \int_{-1}^{1} f(t)g(t)\,dt$. Let $\alpha \in \mathrm{End}(V)$ be defined by $\alpha(f) : x \mapsto (1 - x^2)f''(x) - 2xf'(x)$. Show that α is selfadjoint.

Exercise 1057
Let V be an inner product space finitely generated over $\mathbb{C}$ and let α be an endomorphism of V satisfying $\alpha\alpha^* = \alpha^2$. Show that α is selfadjoint.

Exercise 1058

Let V be an inner product space finitely generated over $\mathbb{C}$ and let α and β be selfadjoint endomorphisms of V satisfying the condition that $\alpha\beta$ is a projection. Is $\beta\alpha$ necessarily also a projection?

Exercise 1059

Give an example of nonzero Hermitian matrices A and B satisfying $AB = O = BA$, or show that no such matrices exist.

Exercise 1060

Let $A \in \mathcal{M}_{2\times2}(\mathbb{C})$ be Hermitian. Find real numbers w, x, y, and z satisfying $|A| = w^2 - x^2 - y^2 - z^2$.

Exercise 1061

Let $O \neq A \in \mathcal{M}_{3\times3}(\mathbb{C})$ be a Hermitian matrix. Show that $A^k \neq O$ for all positive integers k.

Exercise 1062

Find complex numbers a and b such that $\begin{bmatrix} a & 0 & b \\ 0 & 2a & a \\ i & 1 & a \end{bmatrix} \in \mathcal{M}_{3\times3}(\mathbb{C})$ is a Hermitian matrix.

Exercise 1063

Determine all Hermitian matrices $A \in \mathcal{M}_{5\times5}(\mathbb{C})$ satisfying $A^5 + 2A^3 + 3A = 6I$.

Exercise 1064

A matrix $A \in \mathcal{M}_{n\times n}(\mathbb{C})$ is *anti-Hermitian* if and only if $A^H = -A$. Show that A is anti-Hermitian if and only if iA is Hermitian.

Exercise 1065

If matrices $A, B \in \mathcal{M}_{n\times n}(\mathbb{C})$ are anti-Hermitian, show that the Lie product of A and B is also anti-Hermitian.

Exercise 1066

Let n be a positive integer and let $A \in \mathcal{M}_{n\times n}(\mathbb{C})$. Show that every eigenvalue of $A^H A$ is a positive real number.

Exercise 1067

Let V be an inner product space and let $\alpha \in \text{End}(V)$ be selfadjoint. Show that $\ker(\alpha) = \ker(\alpha^h)$ for all $h \geq 1$.

Exercise 1068

Let V be a nontrivial finitely-generated inner product space and let $\alpha \in \text{End}(V)$ be orthogonally diagonalizable and satisfy the condition that each of its eigenvalues is real. Is α necessarily selfadjoint?

Exercise 1069

Let $\alpha \in \text{End}(\mathbb{R}^3)$ be represented with respect to the canonical basis by the matrix
$\begin{bmatrix} 1 & 2 & 3 \\ 2 & a & 4 \\ 3 & 4 & 5 \end{bmatrix}$. For which values of a is α positive definite?

Exercise 1070

For each complex number z, let α_z be the endomorphism of $\mathbb{C}^3$ represented with respect to the canonical basis by $\begin{bmatrix} 1 & 1 & -1 \\ 1 & 1 & z \\ -1 & \overline{z} & 1 \end{bmatrix}$. Does there exist a z for which this endomorphism is positive definite?

Exercise 1071

Let V be an inner product space and let $\alpha \in \text{End}(V)$ be positive definite. Is α^2 necessarily positive definite?

Exercise 1072

Let α be a positive definite automorphism of an inner product space V. Is α^{-1} necessarily positive definite?

Exercise 1073

Do there exist $a, b, c, d \in \mathbb{R}$ such that the endomorphism of $\mathbb{R}^4$ represented with respect to some basis by the matrix $\begin{bmatrix} 1 & 1 & a & 0 \\ 1 & 1 & 1 & b \\ c & 1 & 1 & 1 \\ 0 & d & 1 & 1 \end{bmatrix}$ is positive definite?

Exercise 1074

Let V be an inner product space finitely generated over $\mathbb{R}$ and let $\alpha \in \text{End}(V)$. Let D be a fixed basis for V and let $A = \Phi_{DD}(\alpha)$. Recall that we can write $A = B + C$, where $B = \frac{1}{2}(A + A^T)$ is symmetric and $C = \frac{1}{2}(A - A^T)$ is skew symmetric. Let $\beta, \gamma \in \text{End}(V)$ satisfy $B = \Phi_{DD}(\beta)$ and $C = \Phi_{DD}(\gamma)$. Show that α is positive definite if and only if γ is positive definite.

Exercise 1075

Let α be a positive semidefinite endomorphism of $\mathbb{R}^n$, represented with respect to the canonical basis $\{v_1, \ldots, v_n\}$ by a symmetric matrix $A = [a_{ij}] \in \mathcal{M}_{n \times n}(\mathbb{R})$. Show that $|a_{ij}| \leq \frac{1}{2}(a_{ii} + a_{jj})$ for all $1 \leq i, j \leq n$.

Exercise 1076

Let U be the set of all vectors $\begin{bmatrix} a_1 \\ \vdots \\ a_6 \end{bmatrix} \in \mathbb{R}^6$ satisfying the condition that

$\begin{bmatrix} a_1 & a_2 & a_2 \\ a_2 & a_4 & a_5 \\ a_3 & a_5 & a_6 \end{bmatrix}$ is positive semidefinite. Is U a convex subset of $\mathbb{R}^6$?

Exercise 1077

Do there exist real numbers a, b, c, d such that the matrix $\begin{bmatrix} 1 & 1 & a & 0 \\ 1 & 1 & 1 & b \\ c & 1 & 1 & 1 \\ 0 & d & 1 & 1 \end{bmatrix}$ is

positive semidefinite?

Exercise 1078
A selfadjoint endomorphism α of $\mathbb{R}^n$ is *almost positive semidefinite* if and only

if $\alpha(v) \cdot v \geq 0$ for all nonzero vectors $v = \begin{bmatrix} a_1 \\ \vdots \\ a_n \end{bmatrix}$ satisfying $\sum_{i=1}^{n} a_i = 0$. Give

an example of an endomorphism of $\mathbb{R}^3$ which is almost positive semidefinite but
not positive semidefinite.

Exercise 1079
Let k and n be positive integers. A symmetric matrix in $\mathcal{M}_{k+n,k+n}(\mathbb{R})$ is
quasidefinite when it is of the form $\begin{bmatrix} -B & A^T \\ A & C \end{bmatrix}$ where B is a matrix representing a positive-definite endomorphism of $\mathbb{R}^k$ with respect to the canonical basis,
and C is a matrix representing a positive-definite endomorphism of $\mathbb{R}^n$ with respect to the canonical basis. Show that a quasidefinite matrix is nonsingular, and
that its inverse is again quasidefinite.

Exercise 1080
Let $V = \mathbb{R}^2$ together with the dot product. Find positive-definite endomorphisms
α and β of V satisfying the condition that their Jordan product is not positive
definite.

Exercise 1081
Let V be an inner product space over $\mathbb{R}$ and let $\alpha \in \text{End}(V)$. Show that α is
positive definite if and only if $\alpha + \alpha^*$ is positive definite.

Exercise 1082
Let V be an inner product space finitely generated over $\mathbb{C}$ and let α and β be
positive-definite endomorphisms of V satisfying $\alpha\beta = \sigma_0$. Is it necessarily true
that $\alpha = \sigma_0$ or $\beta = \sigma_0$?

Exercise 1083

Let $V = \left\{ \begin{bmatrix} a \\ b \\ b \\ c \end{bmatrix} \middle|\, a, b, c \in \mathbb{R} \right\}$ and let $W = \mathbb{R}^3$, both of which together with the

dot product, are inner product spaces of dimension 3 over $\mathbb{R}$. Find an isomorphism $\alpha : V \to W$ which is also an isometry.

Exercise 1084

Let V be an inner product space and let α be an endomorphism of V which is an isometry. Does α also preserve angles between vectors?

Exercise 1085

Let α be a positive-definite endomorphism of a finite-dimensional inner product space V represented with respect to some fixed basis by an $n \times n$ matrix $A = [a_{ij}]$. Show that $|A| \leq \prod_{i=1}^{n} a_{ii}$.

Exercise 1086

Let n be a positive integer and let α be a positive-definite endomorphism of $\mathbb{C}^n$ represented with respect to the canonical basis by the matrix $A = [a_{ij}] \in \mathcal{M}_{n \times n}(\mathbb{C})$. Show that a_{ii} is a positive real number for all $1 \leq i \leq n$.

Exercise 1087

Let $\alpha : \mathbb{R}^3 \to \mathbb{R}^2$ be the linear transformation defined by $\alpha : \begin{bmatrix} a \\ b \\ c \end{bmatrix} \mapsto \begin{bmatrix} b - 2c \\ a + c \end{bmatrix}$.

Calculate $\mathrm{spec}(\alpha\alpha^*)$ and $\mathrm{spec}(\alpha^*\alpha)$.

Exercise 1088

Let n be a positive integer and let $V = \mathbb{R}^n$, on which we have defined the dot product. Let α be a positive-definite endomorphism of V represented with respect to the canonical basis by the matrix $A = [a_{ij}] \in \mathcal{M}_{n \times n}(\mathbb{R})$. Show that $|A| > 0$. Is it necessarily true that $\mathrm{tr}(A) > 0$?

Exercise 1089

Find endomorphisms $\alpha, \beta \in \mathrm{End}(\mathbb{C}^2)$ satisfying $\alpha > \beta$ (in the sense of Loewner) but not $\alpha^2 > \beta^2$.

Exercise 1090

Let V be an inner product space and let $\alpha, \beta \in \mathrm{End}(V)$ be positive definite. Is it necessarily true that $\alpha + \beta > \beta$?

Exercise 1091

Let V and W be inner product spaces finitely generated over $\mathbb{C}$. Let $\alpha \in \mathrm{Hom}(V, W), \theta \in \mathrm{Aut}(V)$, and $\varphi \in \mathrm{Aut}(W)$, where the automorphisms θ and

φ are positive definite. Show that the automorphism $\theta - \alpha^*\varphi\alpha$ of V is positive definite if and only if the automorphism $\varphi - \alpha\theta\alpha^*$ of W is positive definite.

Exercise 1092
Let $\alpha \in \mathrm{End}(\mathbb{R}^n)$ be represented with respect to the canonical basis by a symmetric matrix A. Show that $\langle \alpha(v), v \rangle \geq 0$ for all nonzero vectors $v \in \mathbb{R}^n$ if and only if $\alpha + c\sigma_1$ is positive definite for every positive real number c.

Exercise 1093
Let V be the vector space of all infinitely-differential functions in $\mathbb{R}^\mathbb{R}$, on which we define the inner product $\langle f, g \rangle = \int_0^\pi f(x)g(x)\,dx$. Let W be the subspace of all functions $f \in V$ satisfying $f(0) = f(\pi) = 0$. Show that the endomorphism of W defined by $f \mapsto f''$ is selfadjoint.

Exercise 1094
Let V be a finitely-generated inner product space and let $\alpha \in \mathrm{End}(V)$ be selfadjoint. Show that $\|\alpha(v)\| \leq \|v\|$ for all $v \in V$.

Exercise 1095
Let V be an inner product space finitely generated over $\mathbb{R}$ of dimension greater than 1, and let α be a selfadjoint endomorphism of V. Show that there are eigenvalues $c < d$ of α satisfying $c\|v\|^2 \leq \langle \alpha(v), v \rangle \leq d\|v\|^2$ for all $v \in V$.

Exercise 1096
Let V be an inner product space finitely generated over $\mathbb{R}$ and let $\alpha, \beta \in \mathrm{End}(V)$ be selfadjoint. Assume that the eigenvalues of α all lie in the interval $[a, b]$ on the real line and that the eigenvalues of β all lie in the interval $[c, d]$ on the real line. Show that the eigenvalues of $\alpha + \beta$ all lie in the interval $[a + c, b + d]$ on the real line.

Exercise 1097
Let V be an inner product space finitely generated over $\mathbb{C}$ and let α be a positive-definite selfadjoint automorphism of V. Show that $\langle (\alpha + \alpha^{-1})(v), v \rangle \geq 2\langle v, v \rangle$ for all $v \in V$.

Exercise 1098
Let V be an inner product space finitely generated over $\mathbb{R}$ and let α be a positive-definite selfadjoint automorphism of V. Show that $\langle \alpha^{-1}(v), v \rangle = \max\{2\langle v, w \rangle - \langle \alpha(w), w \rangle \mid w \in W\}$ for all $v \in V$.

Exercise 1099
Let V be a finite-dimensional inner product space over $\mathbb{C}$ and let $\alpha \neq \sigma_1$ positive-definite endomorphism of V. Show that there exists no positive integer p satisfying $\alpha^p = \sigma_1$.

Exercise 1100
Let V be a vector space finitely-generated over $\mathbb{R}$ and let $\alpha \in \text{End}(V)$ be selfadjoint. Show that at least one of the values $\pm\|\alpha\|$ is an eigenvalue of α and any eigenvalue c of α satisfies $-\|\alpha\| \leq c \leq \|\alpha\|$.

Exercise 1101
Let $A = \begin{bmatrix} a & b \\ \bar{b} & c \end{bmatrix} \in \mathcal{M}_{2\times2}(\mathbb{C})$ be Hermitian and let $r \geq s$ be the (necessarily real) eigenvalues of A. Show that $|b| \leq \frac{1}{2}(r - s)$.

Exercise 1102
Let V be a nontrivial finitely-generated inner product space and let α and β be selfadjoint endomorphisms of V satisfying $\alpha\beta = \beta\alpha$. Show that α and β have a common eigenvector.

Exercise 1103
Let n be a positive integer. An endomorphism α of $\mathbb{R}^n$ is *copositive* if and only if it is selfadjoint and satisfies the condition that $\alpha(v) \cdot v$ is a positive real number whenever v is a nonzero vector all components of which are nonnegative. Clearly, positive-definite endomorphisms are copositive. Give an example of a copositive endomorphism which is not positive definite.

Exercise 1104
Is the endomorphism of $\mathbb{C}^3$ represented with respect to the canonical basis by the
matrix $\begin{bmatrix} 4 & 2-i & 1+i \\ 2+i & 3 & 0 \\ 1-i & 0 & 2 \end{bmatrix} \in \mathcal{M}_{3\times3}(\mathbb{C})$ positive definite?

Exercise 1105
Let α be the endomorphism of $\mathbb{R}^2$ defined by setting $\alpha : \begin{bmatrix} a \\ b \end{bmatrix} \mapsto \begin{bmatrix} 2a-b \\ 2b-a \end{bmatrix}$.
Show that α is positive definite by constructing an endomorphism β of $\mathbb{R}^2$ satisfying $\alpha = \beta^*\beta$.

Exercise 1106
Find selfadjoint automorphisms α and β of $\mathbb{R}^2$ satisfying the condition that $\alpha \geq \beta \geq -\alpha$ but $\alpha \not\geq \sqrt{\beta^*\beta}$.

Exercise 1107
Let $\alpha \in \text{End}(\mathbb{R}^n)$ be represented with respect to the canonical basis by a symmetric matrix $A = [a_{ij}]$. Let $\beta \in \text{End}(\mathbb{R}^n)$ be represented by the matrix $B = [e^{a_{ij}}]$. If α is positive semidefinite, is β necessarily positive semidefinite? Is β necessarily positive definite?

Unitary and Normal Endomorphisms 18

Let V be an inner product space. An automorphism of V which is an isometry is called a *unitary automorphism*. It is easy to see that if α and β are unitary automorphisms of V then $\alpha\beta$ and α^{-1} are also unitary automorphisms of V. It is also clear that σ_1 is unitary. Therefore, the set of all unitary automorphisms of V is a group of automorphisms.

Proposition 18.1 *Let V be an inner product space and let $\alpha \in \text{Aut}(V)$ have an adjoint. Then α is unitary if and only if $\alpha^* = \alpha^{-1}$.*

Proof If α is unitary then $\langle \alpha(v), w \rangle = \langle \alpha(v), \alpha\alpha^{-1}(w) \rangle = \langle v, \alpha^{-1}(w) \rangle$ for all $v, w \in V$ and so $\alpha^* = \alpha^{-1}$. Conversely, if $\alpha^* = \alpha^{-1}$ then $\langle \alpha(v), \alpha(w) \rangle = \langle v, \alpha^*\alpha(w) \rangle = \langle v, w \rangle$ for all $v, w \in V$ and so α is unitary. $\qquad\square$

As a direct consequence of Proposition 17.14, we see that if V is an inner product space finitely generated over its field of scalars then for $\alpha \in \text{End}(V)$ the following conditions are equivalent:

(1) α is an isometry;
(2) α is unitary;
(3) α maps an orthonormal basis of V to an orthonormal basis of V.

If V is an inner product space finitely generated over its field of scalars F, and if α is a unitary automorphism of V represented by a matrix $A = [a_{ij}] \in M_{n \times n}(F)$ with respect to a given orthonormal basis, then we see that $A^{-1} = A^H \in M_{n \times n}(F)$. A matrix of this form over F is called a *unitary matrix*. If A is a unitary matrix then so is A^{-1} since $(A^{-1})^H = (A^H)^{-1}$. Also, if A and B are unitary matrices then $(AB)^{-1} = B^{-1}A^{-1} = B^H A^H = (AB)^H$ so AB is also unitary. The converse is false. For example, the matrix $A = \begin{bmatrix} -1 & 1 \\ 0 & 1 \end{bmatrix} \in M_{2 \times 2}(\mathbb{R})$ is not unitary, but $A^2 = I$ is.

Thus we see that the set of unitary matrices in $M_{n \times n}(F)$ define a group of automorphisms of F^n and so an equivalence relation $\sim$ defined by the condition that

J.S. Golan, *The Linear Algebra a Beginning Graduate Student Ought to Know*,
DOI 10.1007/978-94-007-2636-9_18, © Springer Science+Business Media B.V. 2012

$A \sim B$ if and only if there exists a unitary matrix P such that $A = P^{-1}BP$. Matrices equivalent in this sense are *unitarily similar*. As an immediate consequence of the definition, we see that A is unitary if and only if the set of columns (resp., rows) of A is an orthonormal basis of F^n (resp., $\mathcal{M}_{1 \times n}(F)$) endowed with the dot product.

Proposition 18.2 *Let n be a positive integer and let $A = [a_{ij}]$ and $B = [b_{ij}]$ be unitarily-similar matrices in $\mathcal{M}_{n \times n}(\mathbb{C})$. Then*

$$\sum_{i=1}^{n}\sum_{j=1}^{n}|a_{ij}|^2 = \sum_{i=1}^{n}\sum_{j=1}^{n}|b_{ij}|^2.$$

Proof We note that $\sum_{i=1}^{n}\sum_{j=1}^{n}|a_{ij}|^2 = \mathrm{tr}(A^H A)$. If P is a unitary matrix satisfying $B = P^{-1}AP$ then $\mathrm{tr}(B^H B) = \mathrm{tr}(P^{-1}A^H P P^{-1}AP) = \mathrm{tr}(P^{-1}A^H AP) = \mathrm{tr}(A^H AP^{-1}P) = \mathrm{tr}(A^H A)$, and we are done. $\square$

Example If $c, d \in \mathbb{C}$ satisfy the condition that $|c|^2 + |d|^2 = 1$, then the matrix $\begin{bmatrix} c & d \\ -\overline{d} & \overline{c} \end{bmatrix} \in \mathcal{M}_{2 \times 2}(\mathbb{C})$ is unitary. A matrix of this form is known as a *Givens rotation matrix*. More generally, if $n > 3$ then a matrix $A = [a_{ij}] \in \mathcal{M}_{n \times n}(\mathbb{C})$ is a Givens rotation matrix if and only if there exist integers $1 \le h < k \le n$ and nonzero complex numbers c and d satisfying $|c|^2 + |d|^2 = 1$ such that

$$a_{ij} = \begin{cases} c & \text{if } i = j \in \{h, k\}, \\ 1 & \text{if } i = j \notin \{h, k\}, \\ d & \text{if } i = h \text{ and } j = k, \\ -\overline{d} & \text{if } i = k \text{ and } j = h, \\ 0 & \text{otherwise.} \end{cases}$$

These matrices play important roles in numerical algorithms.

James Wallace Givens, a former assistant to von Neumann and considered one of the fathers of the twentieth-century American numerical analysis, made major contributions to numerical matrix computation.

Example The matrix $A = \frac{1}{2}\begin{bmatrix} 1-i & 1+i \\ 1+i & 1-i \end{bmatrix} \in \mathcal{M}_{2 \times 2}(\mathbb{C})$ is unitary. This matrix has important applications in the modeling of quantum computing, where it is often denoted by $\sqrt{NOT}$, since $A^2 = \begin{bmatrix} 0 & 1 \\ 1 & 0 \end{bmatrix}$ represents the negation operator in this context.

Example It is easy to show that $A_b = \begin{bmatrix} \sqrt{1+b^2} & bi \\ -bi & \sqrt{1+b^2} \end{bmatrix} \in M_{2\times2}(\mathbb{C})$ satisfies $A_b A_b^T = I$ for any real number b, but, except for the case of $b = 0$, it is not unitary.

Unitarily-similar matrices are surely similar, but the converse is not true.

Example The matrices $\begin{bmatrix} 3 & 1 \\ -2 & 0 \end{bmatrix}$ and $\begin{bmatrix} 1 & 1 \\ 0 & 2 \end{bmatrix}$ in $M_{2\times2}(\mathbb{R})$ are similar but not unitarily similar, as we can see from Proposition 18.2.

Proposition 18.3 (Schur's Theorem) *If n is a positive integer, then every matrix in $M_{n\times n}(\mathbb{C})$ is unitarily similar to an upper-triangular matrix.*

Proof We will proceed by induction on n. For $n = 1$, the result is trivial since every 1×1 matrix is upper triangular. Assume now that $n > 1$ and that the result has been established for $M_{(n-1)\times(n-1)}(\mathbb{C})$. Let $A = [a_{ij}] \in M_{n\times n}(\mathbb{C})$. Since we are working over $\mathbb{C}$, we know that the characteristic polynomial of A is completely reducible, and so A has an eigenvalue, call it c_1. Corresponding to that eigenvalue, we have a normal eigenvector $v_1 = \begin{bmatrix} d_1 \\ \vdots \\ d_n \end{bmatrix}$ in which we can assume that $d_1 \in \mathbb{R}$.

We now are able to construct a basis $\{v_1, \ldots, v_n\}$ for $\mathbb{C}^n$ to which we can apply the Gram–Schmidt procedure, and thus assume that it is in fact an orthonormal basis (the vector v_1 does not change, since it was assumed to be normal to begin with). The matrix P_1, the columns of which are these vectors, is therefore unitary. Now set $A_1 = P_1^{-1}AP_1$. It is easy to see that the first column of A_1 is of the form $\begin{bmatrix} c_1 \\ 0 \\ \vdots \\ 0 \end{bmatrix}$

so we can write A_1 in block form as $\begin{bmatrix} c_1 & x \\ O & A_2 \end{bmatrix}$, where $A_2 \in M_{(n-1)\times(n-1)}(\mathbb{C})$. By the induction hypothesis, there is a unitary matrix $Q \in M_{(n-1)\times(n-1)}(\mathbb{C})$ such that $Q^{-1}A_2Q$ is an upper-triangular matrix. Now set $P_2 = \begin{bmatrix} 1 & O \\ O & Q \end{bmatrix}$. Then P_2 is a unitary matrix in $M_{n\times n}(\mathbb{C})$ and $P_2^{-1}P_1^{-1}AP_1P_2 = \begin{bmatrix} c_1 & y \\ O & Q^{-1}A_2Q \end{bmatrix}$ is an upper triangular matrix in $M_{n\times n}(\mathbb{C})$. Since P_1P_2 is again unitary, we are done. □

If we are working over $\mathbb{R}$, then a matrix A representing a unitary automorphism of $\mathbb{R}^n$ satisfies $A^{-1} = A^T$. Such a matrix is called an *orthogonal matrix*. It is clear that the matrix I is orthogonal and that A^{-1} is orthogonal whenever A is orthogonal. If A and B are orthogonal matrices then $(AB)^{-1} = B^{-1}A^{-1} = B^T A^T = (AB)^T$ and so AB is also orthogonal. As an immediate consequence of the definition, we see that A is orthogonal if and only if the set of columns (resp., rows) of A is an orthonormal basis of $\mathbb{R}^n$ (resp., $\mathcal{M}_{1 \times n}(\mathbb{R})$) endowed with the dot product. It is also clear that A is orthogonal if and only if A^T is orthogonal.

Example Permutation matrices, which we considered earlier, are clearly orthogonal.

Example The matrices $\begin{bmatrix} \cos(t) & \sin(t) \\ -\sin(t) & \cos(t) \end{bmatrix}$ and $\begin{bmatrix} \cos(t) & \sin(t) \\ \sin(t) & -\cos(t) \end{bmatrix}$ are orthogonal for every $t \in \mathbb{R}$, and one can show that these are the only orthogonal matrices in $\mathcal{M}_{2 \times 2}(\mathbb{R})$. Indeed, suppose that the matrix $\begin{bmatrix} a_{11} & a_{12} \\ a_{21} & a_{22} \end{bmatrix} \in \mathcal{M}_{2 \times 2}(\mathbb{R})$ is orthogonal. Then $a_{11}^2 + a_{12}^2 = 1 = a_{21}^2 + a_{22}^2$ so $-1 \leq a_{11} \leq 1$. Hence there exists a real number t such that $a_{11} = \cos(t)$. Then $a_{12}^2 = 1 - a_{11}^2 = 1 - \cos^2(t) = \sin^2(t)$ and so $a_{12} = \pm \sin(t)$. Also, $a_{11} = \cos(-t)$ and $\sin(-t) = -\sin(t)$. Thus, replacing t by $-t$ if necessary we can assume that $a_{11} = \cos(t)$ and $a_{12} = \sin(t)$. Similarly, there exists an angle s such that $a_{22} = \cos(s)$ and $a_{21} = \sin(s)$. Matrices of the first type are just Givens rotation matrices; matrices of the second type are known as *Jacobi reflection matrices*.

Since $0 = a_{11}a_{21} + a_{12}a_{22} = \cos(t)\sin(s) + \sin(t)\cos(s) = \sin(t + s)$, we see that $t + s = 0$ or $t + s = \pi$. If $t + s = 0$, we obtain $A = \begin{bmatrix} \cos(t) & \sin(t) \\ -\sin(t) & \cos(t) \end{bmatrix}$. If $t + s = \pi$, then $s = \pi - t$ and so $A = \begin{bmatrix} \cos(t) & \sin(t) \\ \sin(t) & -\cos(t) \end{bmatrix}$ since $\sin(t) = \sin(\pi - t)$ and $-\cos(t) = \cos(\pi - t)$.

One can also show that every orthogonal matrix in $\mathcal{M}_{3 \times 3}(\mathbb{R})$ is similar to a matrix of the form $\begin{bmatrix} \cos(t) & \sin(t) & 0 \\ -\sin(t) & \cos(t) & 0 \\ 0 & 0 & \pm 1 \end{bmatrix}$ for some $t \in \mathbb{R}$. More generally, if $n > 2$ then every orthogonal matrix in $\mathcal{M}_{n \times n}(\mathbb{R})$ is similar to a matrix in block form $[D_{ij}]$, where $D_{ij} = O$ if $i \neq j$ and D_{ii} is either 1, -1, or a 2×2 matrix of the form $\begin{bmatrix} \cos(t) & \sin(t) \\ -\sin(t) & \cos(t) \end{bmatrix}$.

Example Let n be a positive integer. If $0 \leq c \leq 1$ is a real number, the matrix $\begin{bmatrix} (\sqrt{c})I & (-\sqrt{1-c})I \\ (\sqrt{1-c})I & (\sqrt{c})I \end{bmatrix} \in \mathcal{M}_{2n \times 2n}(\mathbb{R})$ is orthogonal, where I denotes the identity matrix in $\mathcal{M}_{n \times n}(\mathbb{R})$.

Example Let n be a positive integer and let $V = \mathbb{R}^n$, on which we have defined the dot product. If $v \in V$ is a normal vector, then the matrix $A = I - 2(v \wedge v)$ is a

Householder matrix. These matrices are clearly symmetric. Moreover, if $A = I - 2(v \wedge v)$ then $A^T A = A^2 = (I - 2[v \wedge v])^2 = I - 4v(v^T v)v^T + 4v(v^T v)v^T = I$ and so A is orthogonal. Householder matrices have important uses in numerical analysis. We should also mention that if $u \neq v$ are vectors in V satisfying $\|u\| = \|v\|$, then the vector $w = \|v - u\|^{-1}(v - u)$ defines a Householder matrix $A = I - 2(w \wedge w)$ satisfying $Au = Av$. Since a Householder matrix is totally determined by one vector, it is easy to store in a computer. One of the important uses of Householder matrices is to compute QR-decompositions of matrices in a manner far more stable numerically than via the use of the Gram–Schmidt method.

Alston Householder, a twentieth-century American mathematician, was among the pioneer researchers of the numerical analysis of matrices using computers, who developed many of the basic algorithms used in this field.

The complex analog of Householder matrices are matrices of the form $I - 2ww^H$, where $w \in \mathbb{C}^n$. Such matrices are Hermitian and unitary and, too, have an important role in numerical computation.

Example A general method for the construction of orthogonal matrices, due to the contemporary American mathematician George W. Soules, is given as follows: Let $n > 1$ be an integer and let $w_1 \in \mathbb{R}^n$ be a normal vector all of the entries of which are all positive. Let $1 \leq k < n$ and write $w_1 = \begin{bmatrix} u \\ v \end{bmatrix}$, where $u \in \mathbb{R}^k$ and $v \in \mathbb{R}^{n-k}$. Set $a = \frac{\|v\|}{\|u\|}$ and $w_2 = \begin{bmatrix} au \\ -a^{-1}v \end{bmatrix}$. Then it is easy to see that w_2 is normal and orthogonal to w_1. Moreover, by further partitioning the vectors au and $-a^{-1}v$, we can eventually construct a mutually-orthogonal normal vectors $w_1, w_2, \ldots, w_n$. The matrix with these vectors as columns is then orthogonal.

Notice that if F is either $\mathbb{R}$ or $\mathbb{C}$, and if $A \in \mathcal{M}_{n \times n}(F)$ is a unitary matrix the columns of which are $v_1, \ldots, v_n$, then the identity $AA^H = I$ implies that $\{v_1, \ldots, v_n\}$ is an orthonormal set of vectors in F^n, on which we have the dot product, and hence it is a basis for this space. Conversely, if $\{v_1, \ldots, v_n\}$ is an orthonormal basis of F^n then the matrix the columns of which are these vectors is unitary. Similarly, a matrix in $\mathcal{M}_{n \times n}(\mathbb{R})$ is orthogonal if and only if the set of its columns forms an orthonormal basis for $\mathbb{R}^n$ with the dot product. Another way of putting this is that a matrix in $\mathcal{M}_{n \times n}(\mathbb{R})$ the columns of which are $v_1, \ldots, v_n$ is orthogonal if and only if $\sum_{i=1}^{n} v_i \wedge v_i = I$.

Proposition 18.4 *Let* V *be an inner product space of finite dimension* n *over* $\mathbb{R}$. *Let* α *be a unitary automorphism of* V, *which is represented by a matrix* $A \in M_{n \times n}(\mathbb{R})$ *with respect to a given orthonormal basis of* V. *Then* $|A| = \pm 1$.

Proof We know that if α is represented by $A = [a_{ij}]$ with respect to the given basis, then α^* is represented by A^T. From Proposition 18.1, we deduce that $AA^T = I$ and so $|A|^2 = |A| \cdot |A^T| = |I| = 1$, which in turn implies that $|A| = \pm 1$. □

Example The converse of Proposition 18.4 is false, even for matrices the columns of which are orthogonal. Thus, the matrix $\begin{bmatrix} 0.25 & 0 \\ 0 & 4 \end{bmatrix}$ has determinant 1, but does not represent a unitary automorphism of $\mathbb{R}^2$.

The orthogonal matrices in $M_{n \times n}(\mathbb{R})$ having determinant equal to 1 are known as the *special orthogonal matrices*, and the set of all such matrices is denoted by SO(n). This subset of $M_{n \times n}(\mathbb{R})$ is clearly closed under taking products as well as taking inverses, since if $A \in$ SO(n) then $|A^{-1}| = |A^T| = |A| = 1$. If $A \in M_{n \times n}(\mathbb{R})$ is a special orthogonal matrix, where n is an odd integer, then $1 \in$ spec(A). To see this, we note that $|A - 1I| = |A - I| = |A - AA^T| = |A| \cdot |I - A^T| = |I - A^T| = |I - A|$ and, since n is odd, $|I - A| = (-1)^n |A - I|$. Thus we must have $|A - I| = 0$, and so $1 \in$ spec(A).

Example We have already noted that the only orthogonal matrices in $M_{2 \times 2}(\mathbb{R})$ are of the form $\begin{bmatrix} \cos(t) & \sin(t) \\ -\sin(t) & \cos(t) \end{bmatrix}$ or $\begin{bmatrix} \cos(t) & \sin(t) \\ \sin(t) & -\cos(t) \end{bmatrix}$ for some $t \in \mathbb{R}$. Matrices of the first type are special, whereas matrices of the second type are not.

Example The matrix $\begin{bmatrix} 0 & -1 & 0 & 0 & 0 \\ 1 & 0 & 0 & 0 & 0 \\ 0 & 0 & -1 & 0 & 0 \\ 0 & 0 & 0 & 1 & 0 \\ 0 & 0 & 0 & 0 & -1 \end{bmatrix} \in M_{5 \times 5}(\mathbb{R})$ is special orthogonal.

Let V be an inner product space. An endomorphism $\alpha \in$ End(V) is *normal* if and only if α^* exists and satisfies $\alpha^* \alpha = \alpha \alpha^*$. From this definition, it is clear that α is normal if and only if α^* is normal. Clearly, selfadjoint endomorphisms of V are normal, as are unitary automorphisms.

Example If $a, b \in \mathbb{R}$ satisfy $b \neq 0$ and $a^2 + b^2 \neq 1$, then the automorphism α of $\mathbb{R}^2$ defined by $\begin{bmatrix} c \\ d \end{bmatrix} \mapsto \begin{bmatrix} ac + bd \\ ad - bc \end{bmatrix}$ is normal but neither unitary nor selfadjoint.

Example If $0 \neq a, b \in \mathbb{R}$ then the automorphism α of $\mathbb{C}^2$ defined by

$$\begin{bmatrix} c \\ d \end{bmatrix} \mapsto \begin{bmatrix} c + id \\ c - id \end{bmatrix}$$

is normal but neither unitary and nor selfadjoint.

Proposition 18.5 *Let V be an inner product space. An endomorphism $\alpha \in \mathrm{End}(V)$ for which α^* exists is normal if and only if $\|\alpha(v)\| = \|\alpha^*(v)\|$ for all $v \in V$.*

Proof If α is normal and $v \in V$, then $\|\alpha(v)\|^2 = \langle \alpha(v), \alpha(v) \rangle = \langle v, \alpha^*\alpha(v) \rangle = \langle v, \alpha\alpha^*(v) \rangle = \langle v, \alpha^{**}\alpha^*(v) \rangle = \langle \alpha^*(v), \alpha^*(v) \rangle = \|\alpha^*(v)\|^2$ and so $\|\alpha(v)\| = \|\alpha^*(v)\|$. Conversely, assume that this condition holds. Then for each $v \in V$ we have

$$\langle (\alpha\alpha^* - \alpha^*\alpha)(v), v \rangle = \langle \alpha\alpha^*(v), v \rangle - \langle \alpha^*\alpha(v), v \rangle$$
$$= \langle \alpha^*(v), \alpha^*(v) \rangle - \langle \alpha(v), \alpha(v) \rangle = 0.$$

But $\alpha\alpha^* - \alpha^*\alpha$ is selfadjoint and so, by Proposition 17.3, we see that $\alpha\alpha^* - \alpha^*\alpha = \sigma_0$, and so $\alpha\alpha^* = \alpha^*\alpha$. $\quad\square$

As a consequence of Proposition 18.5 we see that if α is a normal endomorphism of an inner product space V and if $v \in V$ then $v \in \ker(\alpha) \Leftrightarrow \|\alpha(v)\| = 0 \Leftrightarrow \|\alpha^*(v)\| = 0 \Leftrightarrow v \in \ker(\alpha^*)$ and so $\ker(\alpha) = \ker(\alpha^*)$.

We now take a short look at the extensive theory of eigenvalues of normal endomorphisms of inner product spaces. We will restrict our attention to finite-dimensional spaces, since the theory for infinite-dimensional spaces requires additional topological assumptions.

With kind permission of the American Mathematical Society.

The study of eigenvalues of normal and selfadjoint endomorphisms of inner product spaces was developed simultaneously by the American mathematician **Marshall Stone** and by John von Neumann, inspired by problems in quantum theory.

Proposition 18.6 *Let V be an inner product space and let $\alpha \in \mathrm{End}(V)$ be normal. Then every eigenvector of α is also an eigenvector of α^* and if c is an eigenvalue of α then $\bar{c}$ is an eigenvalue of α^*.*

Proof If $v \in V$ then, as we have noted, $\|\alpha(v)\| = \|\alpha^*(v)\|$. For a scalar c, we see that

$$(\alpha - c\sigma_1)^*(\alpha - c\sigma_1) = (\alpha^* - \bar{c}\sigma_1)(\alpha - c\sigma_1) = (\alpha - c\sigma_1)(\alpha^* - \bar{c}\sigma_1)$$

$$= (\alpha - c\sigma_1)(\alpha - c\sigma_1)^*,$$

and so $\alpha - c\sigma_1$ is also normal. Thus, $\|(\alpha - c\sigma_1)(v)\| = \|(\alpha^* - \bar{c}\sigma_1)(v)\|$ for $v \in V$ and so, in particular, we see that $v \in \ker(\alpha - c\sigma_1)$ if and only if $v \in \ker(\alpha^* - \bar{c}\sigma_1)$. In other words, v is an eigenvector of α associated with the eigenvalue c if and only if it is an eigenvector of α^* associated with the eigenvalue $\bar{c}$.   □

Since $\alpha^{**} = \alpha$ for any endomorphism α of V, we see from Proposition 18.6 that if α is normal then a scalar c is an eigenvalue of α if and only if $\bar{c}$ is an eigenvalue of α^*.

Another interesting consequence of Proposition 18.6 is the following: Let V be a finitely-generated inner product space and let $\alpha \in \mathrm{Aut}(V)$ be unitary. Then α is surely normal. If $c \in \mathrm{spec}(\alpha)$ then $c \neq 0$ since α is an automorphism. If v is an eigenvector associated with c then $v = (\alpha^*\alpha)(v) = \alpha^*(cv) = c\alpha^*(v)$ and so $\alpha^*(v) = c^{-1}v$. This shows that c^{-1} is an eigenvalue of α^* and hence, by Proposition 18.6, $c^{-1} \in \mathrm{spec}(\alpha)$.

Example In Proposition 17.5, we saw that if α is a selfadjoint endomorphism of an inner product space V finitely generated over $\mathbb{R}$, then $\mathrm{spec}(\alpha) \neq \varnothing$. This is not necessarily true for normal endomorphisms of inner product spaces which are not selfadjoint. For example, let $V = \mathbb{R}^2$ together with the dot product, and if $\alpha \in \mathrm{End}(V)$ is defined by $\alpha : \begin{bmatrix} a \\ b \end{bmatrix} \mapsto \begin{bmatrix} -b \\ a \end{bmatrix}$, then we have already seen that $\mathrm{spec}(\alpha) = \varnothing$. One can easily check that α is normal but not selfadjoint.

Proposition 18.7 *Let V be an inner product space finitely generated over $\mathbb{C}$ and let $\alpha \in \mathrm{End}(V)$. Then α is normal if and only if it is orthogonally diagonalizable.*

Proof Assume that α is normal. We will proceed by induction on $n = \dim(V)$. First, assume that $n = 1$. Since we are working over $\mathbb{C}$, we know that $\mathrm{spec}(\alpha) \neq \varnothing$ and so there exists a normal eigenvector v_1 of α. Then $V = \mathbb{C}v_1$ and we are done. Now assume that $n > 1$ and that the result has been proven for subspaces of dimension $n - 1$. Again, there exists a normal eigenvector v_1 of α. Set $W = \mathbb{C}v_1$. The subspace W of V is invariant under α, and so, by Proposition 18.6, it is also invariant under α^*. Therefore, $W^\perp$ is invariant under $\alpha^{**} = \alpha$. The restriction of α to $W^\perp$ is a normal endomorphism, the adjoint of which is the restriction of α^* to $W^\perp$. By induction, we know that there exists an orthonormal basis $\{v_2, \ldots, v_n\}$ composed of eigenvectors of α, and so $\{v_1, \ldots, v_n\}$ is the basis of V we are seeking.

Conversely, assume that there exists an orthonormal basis of $B = \{v_1, \ldots, v_n\}$ composed of eigenvectors of α. Then $\Phi_{BB}(\alpha) = [a_{ij}]$ is a diagonal matrix satisfying the condition that each a_{ii} is an eigenvalue of α. Moreover, $\Phi_{BB}(\alpha^*) = \Phi_{BB}(\alpha)^H$ and this too is a diagonal matrix. Since diagonal matrices commute, we see that $\alpha\alpha^* = \alpha^*\alpha$, and so α is normal. $\square$

Note that Proposition 18.7 does not imply that if V is an inner product space finitely generated over $\mathbb{C}$ and if $\alpha \in \mathrm{End}(V)$ is normal, then every basis B of V composed of eigenvectors of α is necessarily orthonormal or that its elements are even necessarily mutually orthogonal, merely that one such basis exists.

Example Let α be the endomorphism of $\mathbb{C}^4$ represented with respect to the canon-

ical basis by the matrix $A = \begin{bmatrix} 1 & 2 & 0 & 0 \\ -2 & 1 & 0 & 0 \\ 0 & 0 & 3 & -1 \\ 0 & 0 & 1 & 3 \end{bmatrix}$. One easily checks that $AA^H = A^H A$, and so α is a normal automorphism of $\mathbb{C}^4$. The characteristic polynomial of A is

$$X^4 - 8X^3 + 27X^3 - 50X + 50 = (X^2 - 6X + 10)(X^2 - 2X + 5),$$

and so $\mathrm{spec}(\alpha) = \{3 \pm i, 1 \pm 2i\}$. The set

$$\left\{ \frac{1}{\sqrt{2}} \begin{bmatrix} -i \\ 1 \\ 0 \\ 0 \end{bmatrix}, \frac{1}{\sqrt{2}} \begin{bmatrix} i \\ 1 \\ 0 \\ 0 \end{bmatrix}, \frac{1}{\sqrt{2}} \begin{bmatrix} 0 \\ 0 \\ 1 \\ -i \end{bmatrix}, \frac{1}{\sqrt{2}} \begin{bmatrix} 0 \\ 0 \\ 1 \\ i \end{bmatrix} \right\}$$

is an orthonormal basis for $\mathbb{C}^4$ composed of eigenvectors of α.

Proposition 18.8 *Let V be a finitely-generated inner product space. Then the following conditions on a projection $\alpha \in \mathrm{End}(V)$ are equivalent:*
(1) *α is normal;*
(2) *α is selfadjoint;*
(3) *$\ker(\alpha) = \mathrm{im}(\alpha)^{\perp}$.*

Proof (1) $\Rightarrow$ (2): From (1) we know that $\|\alpha(v)\| = \|\alpha^*(v)\|$ for all $v \in V$. In particular, $\alpha(v) = 0_V$ if and only if $\alpha^*(v) = 0_V$ so that $\ker(\alpha) = \ker(\alpha^*)$. If $v \in V$ and $w = v - \alpha(v)$ then $\alpha(w) = \alpha(v) - \alpha^2(v) = \alpha(v) - \alpha(v) = 0_V$ and so $\alpha^*(w) = 0_V$. Therefore, $\alpha^*(v) = \alpha^*\alpha(v)$ for all $v \in V$, whence $\alpha^* = \alpha^*\alpha$. This implies that $\alpha = \alpha^{**} = (\alpha^*\alpha)^* = \alpha^*\alpha^{**} = \alpha^*\alpha = \alpha^*$, which proves (2).
(2) $\Rightarrow$ (3): If $v, w \in V$ then, from (2), we see that $\langle \alpha(v), w \rangle = \langle v, \alpha(w) \rangle$. In particular, if $v \in \ker(\alpha)$ then $\langle v, \alpha(w) \rangle = 0$ for all $w \in V$, which is to say that $v \in \mathrm{im}(\alpha)^{\perp}$. Conversely, if $v \in \mathrm{im}(\alpha)^{\perp}$ then $\langle v, \alpha(w) \rangle = 0$ for all $w \in V$, which

implies that $\alpha(v)$ is orthogonal to every element of V. Therefore, $\alpha(v) = 0_V$ and so $v \in \ker(\alpha)$. This proves (3).

(3) $\Rightarrow$ (1): Let $v, w \in V$. Since α is a projection, we have $v - \alpha(v) \in \ker(\alpha)$. It is also clear that $\alpha(w) \in \text{im}(\alpha)$. Therefore,

$$0 = \langle v - \alpha(v), \alpha(w) \rangle = \langle v, \alpha(w) \rangle - \langle \alpha(v), \alpha(w) \rangle = \langle v, \alpha(w) \rangle - \langle v, \alpha^*\alpha(w) \rangle,$$

and since this is true for all $v, w \in V$, we have $\alpha = \alpha^*\alpha$. This implies that $\alpha = \alpha^*$ is selfadjoint and therefore surely normal, proving (1). $\square$

We note that if V is a finitely-generated inner product space and if $\alpha \in \text{End}(V)$ is normal, then, by Propositions 16.5, 16.7 and 18.5, we have $V = \ker(\alpha) \oplus \text{im}(\alpha)$, and, in particular, $\{\text{im}(\alpha), \ker(\alpha)\}$ is an independent set of subspaces of V. Moreover, $v \perp v'$ for all $v \in \ker(\alpha)$ and $v' \in \text{im}(\alpha)$. While the direct-sum decomposition is valid for any projection, it is the normality which ensures the orthogonality.

Proposition 18.9 *Let V be a finitely-generated inner product space. Let $W_1, \ldots, W_n$ be subspaces of V and, for each $1 \leq i \leq n$, let α_i be the projection of V onto the subspace W_i coming from the decomposition $V = W_i \oplus W_i^\perp$. Then the following conditions are equivalent:*

(1) *$V = \bigoplus_{i=1}^n W_i$ and $W_h^\perp = \bigoplus_{j \neq h} W_j$ for all $1 \leq h \leq n$;*
(2) *$\alpha_1 + \cdots + \alpha_n = \sigma_1$ and $\alpha_i \alpha_j = \sigma_0$ for all $i \neq j$;*
(3) *If B_i is an orthonormal basis of W_i for each i, then $B = \bigcup_{i=1}^n B_i$ is an orthonormal basis of V.*

Proof This has essentially already been established when we talked about the decomposition of a vector space into a direct sum of subspaces. $\square$

Proposition 18.10 *Let F be either $\mathbb{R}$ or $\mathbb{C}$ and let V be a finitely-generated inner product space over F. If $p(X) \in F[X]$ and if α is a normal endomorphism of V, then $p(\alpha)$ is a normal endomorphism of V.*

Proof If $p(X) = \sum_{i=0}^n a_i X^i$. Then $p(\alpha) = \sum_{i=0}^n a_i \alpha^i$ and $p(\alpha)^* = \sum_{i=0}^n \overline{a}_i (\alpha^*)^i$. Since $\alpha\alpha^* = \alpha^*\alpha$, it follows from the definition of the product that $p(\alpha)p(\alpha)^* = p(\alpha)^* p(\alpha)$. Therefore, $p(\alpha)$ is a normal endomorphism of V. $\square$

Proposition 18.11 *Let V be a finitely-generated inner product space and let α be a normal endomorphism of V. If the minimal polynomial of α is completely reducible, then it does not have multiple roots.*

Proof Let $p(X)$ be the minimal polynomial of α, which we assume is completely reducible. Assume that there exists a scalar c and a polynomial $q(X)$ such that $p(X) = (X - c)^2 q(X)$. Since $p(\alpha) = \sigma_0$, we have $(\alpha - c\sigma_1)^2 q(\alpha) = \sigma_0$ and so $\ker((\alpha - c\sigma_1)^2 q(\alpha)) = V$. By Proposition 18.10, we know that $\beta = \alpha - c\sigma_1$ is a normal endomorphism of V. Let $v \in V$ and let $w = q(\alpha)(v)$. Then $\beta^2(w) = 0_V$ and so $\beta(w) \in \mathrm{im}(\beta) \cap \ker(\beta) = \{0_V\}$. Thus we see that $\beta q(\alpha)(v) = 0_V$ for all $v \in V$ and hence α annihilates the polynomial $(X - c)q(X)$, contradicting the minimality of $p(X)$. $\square$

Proposition 18.12 (Spectral Decomposition Theorem) *Let V be an inner product space finitely generated over $\mathbb{C}$ and let α be a normal endomorphism of V. Then there exist scalars $c_1, \ldots, c_n$ and projections $\alpha_1, \ldots, \alpha_n$ of V satisfying:*

(1) $\alpha = c_1 \alpha_1 + \cdots + c_n \alpha_n$;

(2) $\sigma_1 = \alpha_1 + \cdots + \alpha_n$;

(3) $\alpha_h \alpha_j = \sigma_0$ *for all* $h \neq j$.

Moreover, these c_j and α_j are unique. The c_j are precisely the distinct eigenvalues of α and each α_j is the projection of V onto the eigenspace W_j associated with c_j coming from the decomposition $V = W_j \oplus W_j^\perp$.

Proof Let $p(X)$ be the minimal polynomial of α, which we will write in the form $p(X) = \prod_{i=1}^{n}(X - c_i)$, where the c_i are complex numbers which, by Proposition 18.11, are distinct. For each $1 \leq j \leq n$, let $p_j(X)$ be the jth Lagrange interpolation polynomial determined by the c_i.

Let $f(X)$ be a polynomial of degree at most $n - 1$. Then the polynomial $f(X) - \sum_{i=1}^{n} f(c_i)p_i(X)$ is of degree at most $n - 1$ and has n distinct roots $c_1, \ldots, c_n$. Thus it must be the 0-polynomial and so $f(X) = \sum_{i=1}^{n} f(c_i)p_i(X)$. In particular, we see that $1 = \sum_{i=1}^{n} p_i(X)$ and $X = \sum_{i=1}^{n} c_i p_i(X)$. Set $\alpha_j = p_j(\alpha)$. Then $\sigma_1 = \sum_{i=1}^{n} \alpha_i$ and $\alpha = \sum_{i=1}^{n} c_i \alpha_i$. Note that $\alpha_j \neq \sigma_0$ since $\alpha_j = p_j(\alpha)$ and the degree of $p_j(X)$ is less than the degree of the minimal polynomial of α. Moreover, if $h \neq j$ then there exists a polynomial $u(X) \in \mathbb{C}[X]$ satisfying $\alpha_h \alpha_j = u(\alpha)p(\alpha) = u(\alpha)\sigma_0 = \sigma_0$. Thus we see that for all $1 \leq j \leq n$ we have $\alpha_j = \alpha_j \sigma_1 = \sum_{i=1}^{n} \alpha_j \alpha_i = \alpha_j^2$ and so each α_j is a projection. Thus we see that $\{\mathrm{im}(\alpha_j) \mid 1 \leq j \leq n\}$ is an independent set of subspaces of V.

Since the minimal polynomial and the characteristic polynomial of α have the same roots, we know that $\mathrm{spec}(\alpha) = \{c_1, \ldots, c_n\}$. To show that $W_h = \mathrm{im}(\alpha_h)$, we have to prove that a vector v belongs to $\mathrm{im}(\alpha_h)$ if and only if $\alpha(v) = c_h v$. Indeed, if $\alpha(v) = c_h v$ then $c_h[\sum_{j=1}^{n} \alpha_j(v)] = c_h v = \alpha(v) = \sum_{j=1}^{n}(c_j \alpha_j)(v)$ and so $\sum_{j=1}^{n}[(c_h - c_j)\alpha_j](v) = 0_V$. Thus, for all $j \neq h$, we have $\alpha_j(v) = 0_V$ and so $v = \alpha_h(v) \in \mathrm{im}(\alpha_h)$.

Finally, we note that α_h is the projection coming from the decomposition $V = W_j \oplus W_j^\perp$ since α_h is a polynomial in α and hence normal and so the result follows from the remark after Proposition 18.8. $\square$

Note that we could have deduced Proposition 18.12 directly from Proposition 18.7. What is important in the above proof is the explicit construction of the projection maps as polynomials in α.

If $\alpha = \sum_{i=1}^{n} c_i \alpha_i$ is as in Proposition 18.12, then $\alpha^k = (\sum_{i=1}^{n} c_i \alpha_i)^k = \sum_{i=1}^{n} c_i^k \alpha_i$ for any positive integer k, and from this we see that if $p(X) \in \mathbb{C}[X]$ then $p(\alpha) = \sum_{i=1}^{n} p(c_i) \alpha_i$.

Proposition 18.13 *Let V be an inner product space finitely generated over $\mathbb{C}$. A normal endomorphism α of V is positive definite if and only if each of its eigenvalues is positive.*

Proof If α is positive definite then, by Proposition 17.11, each of its eigenvalues is positive. Conversely, assume each of the eigenvalues of α is positive. By Proposition 18.12, we write $\alpha = \sum_{i=1}^{n} c_i \alpha_i$, where the c_i are the eigenvalues of α and the α_i are projections in $\text{End}(V)$ satisfying $\alpha_i \alpha_j = \sigma_0$ for $i \neq j$. If $0_V \neq v \in V$ then $\langle \alpha(v), v \rangle = \sum_{i=1}^{n} \sum_{j=1}^{n} c_i \langle \alpha_i(v), \alpha_j(v) \rangle = \sum_{i=1}^{n} c_i \|\alpha_i(v)\|^2 > 0$ and so α is positive definite. $\square$

Example Let $V = \mathbb{R}^3$. For each $a \in \mathbb{R}$, let $\alpha_a \in \text{End}(V)$ be the normal endomorphism of V represented with respect to the canonical basis by the matrix
$\begin{bmatrix} 1 & a & a \\ a & 1 & a \\ a & a & 1 \end{bmatrix}$. Then $\text{spec}(\alpha) = \{2a + 1, 1 - a\}$ and so, by Proposition 18.13, α is positive definite precisely when $-1 < 2a < 2$.

As a consequence of Proposition 18.13 and the comments before it, we see that if α is a positive-definite endomorphism of a finitely-generated inner product space V over $\mathbb{C}$ then there exists an endomorphism $\sqrt{\alpha}$ of V satisfying $(\sqrt{\alpha})^2 = \alpha$. This endomorphism is defined by $\sqrt{\alpha} = \sum_{i=1}^{n} (\sqrt{c_i}) \alpha_i$, where the c_i are the eigenvalues of α, and where the α_i are defined as in Proposition 18.12. In particular, if β is an automorphism of V then, by Proposition 17.10, we can talk about $\sqrt{\beta^* \beta}$, which is also positive definite by Proposition 18.13.

Proposition 18.14 *Let V be an inner product space finitely generated over $\mathbb{C}$ and let $\alpha \in \text{Aut}(V)$. Then there exists a unique positive-definite automorphism θ of V and a unique unitary automorphism ψ of V satisfying $\alpha = \psi\theta$.*

Proof By Proposition 17.10, we know that the automorphism $\alpha^* \alpha$ of V is positive definite and so we can set $\theta = \sqrt{\alpha^* \alpha}$. Let $\varphi = \theta \alpha^{-1}$. Then $\varphi^* = (\alpha^{-1})^* \theta^* = (\alpha^*)^{-1} \theta$ so $\varphi^* \varphi = (\alpha^*)^{-1} \theta \theta \alpha^{-1} = (\alpha^*)^{-1} \alpha^* \alpha \alpha^{-1} = \sigma_1$, proving that φ is unitary by Proposition 18.1, and hence belongs to $\text{Aut}(V)$. If we now define $\psi = \varphi^{-1}$, we see that $\alpha = \psi\theta$. Moreover, we note that $\theta \in \mathrm{Aut}(V)$ since $\theta = \varphi\alpha$. To prove uniqueness, assume that $\psi\theta = \psi'\theta'$, where ψ and ψ' are unitary automorphisms of V and where θ and θ' are positive-definite automorphisms of V. Then $\psi^2 = \psi\theta^*\theta\psi = \psi'(\theta')^*\theta'\psi' = (\psi')^2$. Since ψ is positive definite, this implies that $\psi = \psi'$ and so, since ψ is an automorphism, we have $\theta = \psi^{-1}\psi\theta = \psi^{-1}\psi\theta' = \theta'$. $\square$

The representation of an automorphism α of an inner product space finitely generated over $\mathbb{C}$ in the form given in Proposition 18.14 is sometimes called the *polar decomposition* of α.[1] If we move over to matrices, we see that the *polar decomposition* of a nonsingular matrix $A \in \mathcal{M}_{n \times n}(\mathbb{C})$ is of the form $A = UM$, where U is a unitary matrix and M is a positive-definite Hermitian matrix. Similarly, there exists a unitary matrix U' and a Hermitian matrix M' satisfying $A^H = U'M'$ and so $A = M'(U')^H$, where $(U')^H$ is again unitary. In the case we are working over $\mathbb{R}$, the matrix U is orthogonal, and M is symmetric and positive definite. Because polar decompositions are important in applications, several iterative algorithms exist to compute them.

Example If a and b are nonzero real numbers, then the polar decomposition of the matrix $\begin{bmatrix} a & -b \\ b & a \end{bmatrix}$ is $\begin{bmatrix} \cos(\theta) & -\sin(\theta) \\ \sin(\theta) & \cos(\theta) \end{bmatrix} \begin{bmatrix} r & 0 \\ 0 & r \end{bmatrix}$, where $\theta = \arctan(\frac{b}{a})$ and $r = \sqrt{a^2 + b^2}$.

Proposition 18.15 (Singular Value Decomposition Theorem) *Let V and W be inner product spaces of finite dimensions k and n, respectively, and let $\alpha \in \mathrm{Hom}(V, W)$. Then there exists an integer $t \leq \min\{k, n\}$, together with positive real numbers $c_1 \geq c_2 \geq \cdots \geq c_t$ and with orthonormal bases $\{v_1, \dots, v_k\}$ of V and $\{w_1, \dots w_n\}$ of W satisfying*

$$\alpha(v_i) = \begin{cases} c_i w_i & \text{if } 1 \leq i \leq t, \\ 0_W & \text{otherwise} \end{cases}$$

and

$$\alpha^*(w_i) = \begin{cases} c_i v_i & \text{if } 1 \leq i \leq t, \\ 0_V & \text{otherwise.} \end{cases}$$

Proof If α is the 0-map, then the result is immediate, so assume that is not the case. We note that $\beta = \alpha^*\alpha$ is a selfadjoint endomorphism of V and so, by Proposition 17.7, it is orthogonally diagonalizable. Hence V has an orthonormal basis $\{v_1, \dots, v_k\}$ composed of eigenvectors of β, where each v_i is associated with an eigenvalue b_i. By Proposition 17.10, we know that each b_i belongs

---

[1] Polar decompositions were first studied by the French engineer **Léon Autonne** at the beginning of the twentieth century.

to $\mathbb{R}$. Moreover, for each i we note that $b_i = b_i\langle v_i, v_i\rangle = \langle b_i v_i, v_i\rangle = \langle \beta(v_i), v_i\rangle = \langle \alpha(v_i), \alpha(v_i)\rangle \geq 0$. Indeed, renumbering if necessary, we can assume that there exists an integer $t \leq k$ such that $b_1 \geq b_2 \geq \cdots \geq b_t > 0$ while $b_{t+1} = \cdots = b_k = 0$. For each $1 \leq i \leq t$, set $c_i = \sqrt{b_i}$ and let $w_i = c_i^{-1}\alpha(v_i) \in W$. If $i \neq j$ then
$\langle w_i, w_j\rangle = (c_i c_j)^{-1}\langle \alpha(v_i), \alpha(v_j)\rangle = (c_i c_j)^{-1}\langle \beta(v_i), v_j\rangle = (c_i c_j)^{-1} b_i\langle v_i, v_j\rangle = 0$
while, for each $1 \leq i \leq t$, we have $\langle w_i, w_i\rangle = c_i^{-2}\langle \alpha(v_i), \alpha(v_i)\rangle = c_i^{-2}\langle \beta(v_i), v_i\rangle = c_i^{-2}\langle b_i v_i, v_i\rangle = \langle v_i, v_i\rangle = 1$. Thus we see that the set $\{w_1, \ldots, w_t\}$ is orthonormal. Moreover, for each $1 \leq i \leq t$ we have $\|\alpha(v_i)\|^2 = b_i$ so $\|\alpha(v_i)\| = c_i$ and $\alpha^*(w_i) = \alpha^*(c_i^{-1}\alpha(v_i)) = c_i^{-1}\alpha^*\alpha(v_i) = c_i^{-1}\beta(v_i) = c_i^{-1}b_i v_i = c_i v_i$. For $t + 1 \leq i \leq k$ we have $\alpha^*\alpha(v_i) = \beta(v_i) = 0_V$ and so $0 = \langle \beta(v_i), v_i\rangle = \langle \alpha(v_i), \alpha(v_i)\rangle$, which implies that $\alpha(v_i) = 0_W$. Thus $v_i \in \ker(\alpha)$ for each $t + 1 \leq i \leq k$.

We are therefore left with the matter of defining $w_{t+1}, \ldots, w_n$ in the case $t < n$. By Proposition 16.18, we know that $\ker(\alpha^*) = \operatorname{im}(\alpha)^\perp$ and so, if we pick an orthonormal basis $\{w_{i+1}, \ldots, w_n\}$ for $\ker(\alpha^*)$ we see that $\{w_1, \ldots, w_n\}$ is an orthonormal basis for W having the desired properties. $\square$

The first version of the Singular Value Decomposition Theorem was proven by the nineteenth-century Italian mathematician **Eugenio Beltrami**; it was subsequently extended by many others, including Camille Jordan and Sylvester. Schmidt extended this theorem to infinite-dimensional spaces. Effective algorithms for computation of singular value decompositions were developed by the twentieth-century American computer scientist **Gene H. Golub**, along with William Kahan.

The scalars $c_1 \geq c_2 \geq \cdots \geq c_t$ given in the Proposition 18.15 are called the *singular values* of the linear transformation α. The number c_1/c_t, called the *spectral condition number*, is used as a measure of the numerical instability of the matrix representing $\alpha^*\alpha \in \operatorname{End}(V)$ with respect to the given basis.

If we consider the special case of a linear transformation $\alpha : \mathbb{C}^k \to \mathbb{C}^n$ represented with respect to the canonical bases by a matrix $A \in \mathcal{M}_{n \times k}(\mathbb{C})$, the Singular Value Decomposition Theorem says that there exist unitary matrices $P \in \mathcal{M}_{n \times n}(\mathbb{C})$ and $Q \in \mathcal{M}_{k \times k}(\mathbb{C})$ such that A can be written as $P\begin{bmatrix} D & O \\ O & O \end{bmatrix}Q^H$, where $D \in \mathcal{M}_{t \times t}(\mathbb{R})$ is a diagonal matrix having the singular values of α on the diagonal. These singular values are precisely the square roots of the eigenvalues of $A^H A$. The columns of Q form an orthonormal basis for $\mathbb{C}^k$ consisting of eigenvectors of $A^H A$, and the columns of P form an orthonormal basis for $\mathbb{C}^n$.

If $\alpha : \mathbb{R}^k \to \mathbb{R}^n$ then, of course, the matrices P and Q are orthogonal and
$$A = P\begin{bmatrix} D & O \\ O & O \end{bmatrix}Q^T.$$

Example The matrix $A = \frac{1}{10} \begin{bmatrix} 20 & 20 & -20 & 20 \\ 1 & 17 & 1 & -17 \\ 18 & 6 & 18 & -6 \end{bmatrix}$ can be written as a product

$P \begin{bmatrix} 4 & 0 & 0 & 0 \\ 0 & 3 & 0 & 0 \\ 0 & 0 & 2 & 0 \end{bmatrix} Q$, where

$$P = \frac{1}{5} \begin{bmatrix} 5 & 0 & 0 \\ 0 & 3 & -4 \\ 0 & 4 & 3 \end{bmatrix} \quad \text{and} \quad Q = \frac{1}{2} \begin{bmatrix} 1 & 1 & -1 & 1 \\ 1 & 1 & 1 & -1 \\ 1 & -1 & 1 & 1 \\ 1 & -1 & -1 & -1 \end{bmatrix}$$

are orthogonal and where the singular values of A are $4, 3, 2$.

Singular value decompositions have many applications, and play important roles in the mathematics of optimization, data compression, population genetics, and image processing. They are especially useful since accurate and relatively-efficient algorithms for computing these decompositions are readily available in many common linear-algebra software packages. In particular, in many applications one needs to compute the singular value decomposition of a product of a large number of matrices (often over 1,000) and there exist algorithms to do that without having to multiply out the matrices explicitly.

Exercises

Exercise 1108
Let $A \in M_{n \times n}(\mathbb{C})$ be similar to a unitary matrix. Is A^{-1} necessarily similar to A^H?

Exercise 1109
Let n be a positive integer and let $A \in M_{n \times n}(\mathbb{C})$ be a nonsingular matrix having a singular value decomposition $A = PDQ^H$, where P and Q are unitary matrices and $D = \begin{bmatrix} c_1 & & O \\ & \ddots & \\ O & & c_n \end{bmatrix}$ is a diagonal matrix with $c_1 \geq \cdots \geq c_n$. If B is a singular matrix, show that $\|A - B\| \geq c_n$, where $\| \cdot \|$ denotes the spectral norm.

Exercise 1110
Let $n > 1$ be an integer and let V be the subspace of $\mathbb{C}[X]$ consisting of all polynomials of degree at most n. Let $0 \neq c \in \mathbb{C}$ and let α be the endomorphism of V defined by $\alpha : p(X) \mapsto p(X + c)$. Is it possible to define an inner product on V relative to which α is normal?

Exercise 1111

Let $a, b, c \in \mathbb{C}$. Find the set of all triples (x, y, z) of complex numbers satisfying

the condition that the matrix $\begin{bmatrix} a & x & y \\ 0 & b & z \\ 0 & 0 & c \end{bmatrix}$ represents a normal endomorphism

of $\mathbb{C}^3$, endowed with the dot product, with respect to the canonical basis.

Exercise 1112

Show that any Givens rotation matrix in $\mathcal{M}_{2\times2}(\mathbb{R})$ can be written as the product of two Jacobi reflection matrices.

Exercise 1113

Let n be a positive integer. A matrix $A \in \mathcal{M}_{n\times n}(\mathbb{C})$ is *normal* if and only if $A^H A = AA^H$. Show that every normal upper-triangular matrix is a diagonal matrix.

Exercise 1114

Let n be a positive integer and let $A \in \mathcal{M}_{n\times n}(\mathbb{R})$. Then A is normal if and only if $A^T A = AA^T$. If A is normal, is e^A normal? Is the converse of this statement true?

Exercise 1115

Let $V = \mathbb{R}^2$, together with the dot product. Show that a matrix in $\mathcal{M}_{2\times2}(\mathbb{R})$ is of the form $\Phi_{BB}(\alpha)$ for some normal endomorphism α of V which is not selfadjoint if and only if it is of the form $\begin{bmatrix} a & -b \\ b & a \end{bmatrix}$ for real numbers a and $b \neq 0$.

Exercise 1116

Let V be an inner product space finitely generated over $\mathbb{R}$ and let S be the set of all isometries V. Is S an $\mathbb{R}$-subalgebra of $\text{End}(V)$?

Exercise 1117

Let n be a positive integer and let $V = \mathbb{C}^n$ on which we have the dot product. If $\alpha \in \text{End}(V)$, let $G(\alpha) = \{\langle \alpha(v), v \rangle \mid \|v\| = 1\}$. For the special case $n = 2$, find $G(\alpha)$ and $G(\beta)$, where α is represented with respect to the canonical basis by

the matrix $\begin{bmatrix} 1 & 0 \\ 0 & 0 \end{bmatrix}$ and β is represented with respect to the canonical basis by

the matrix $\begin{bmatrix} 0 & 2 \\ 0 & 0 \end{bmatrix}$.

Exercise 1118

Let $V = \mathbb{R}^3$ on which we have the dot product, and let W be the space of all polynomial functions in $\mathbb{R}^{\mathbb{R}}$ of degree at most 2, on which we define the inner product $\langle f, g \rangle = \int_0^1 f(x)g(x)\,dx$. Let $\alpha \in \text{Hom}(V, W)$ be defined by

$$\alpha\left(\begin{bmatrix} a \\ b \\ c \end{bmatrix}\right) : x \mapsto 1 + \frac{b}{2} + \frac{c}{6} + (b - c)x + cx^2.$$

Is this linear transformation an isometry?

Exercise 1119

Let V be an inner product space and let α be an endomorphism of V satisfying the condition that $\alpha^*\alpha = \sigma_0$. Show that $\alpha = \sigma_0$.

Exercise 1120

Let $V = \mathbb{R}^3$ with the dot product, and let α be the automorphism of V defined by

$$\alpha : \begin{bmatrix} a \\ b \\ c \end{bmatrix} \mapsto \begin{bmatrix} -c \\ -b \\ -a \end{bmatrix}. \text{ Is } \alpha \text{ unitary?}$$

Exercise 1121

Is the matrix $\begin{bmatrix} 0 & 0 & 0 & i \\ 0 & 0 & 1 & 0 \\ 0 & 1 & 0 & 0 \\ i & 0 & 0 & 0 \end{bmatrix} \in \mathcal{M}_{4\times 4}(\mathbb{C})$ unitary?

Exercise 1122

Find a real number a satisfying the condition that the matrix

$$a \begin{bmatrix} -9 + 8i & -10 - 4i & -16 - 18i \\ -2 - 24i & 1 + 12i & -10 - 4i \\ 4 - 10i & -2 - 24i & -9 + 8i \end{bmatrix} \in \mathcal{M}_{3\times 3}(\mathbb{C})$$

is unitary.

Exercise 1123

Find a real number a satisfying the condition that the matrix

$$\frac{1}{24} \begin{bmatrix} 12 & 6 - 12i & 12 + 6i & 6 - 6i \\ 6 + 12i & a & 5i & 3 + i \\ 12 - 6i & -5i & a & 1 - 3i \\ 6 + 6i & 3 - i & 1 + 3i & -22 \end{bmatrix} \in \mathcal{M}_{4\times 4}(\mathbb{C})$$

is unitary.

Exercise 1124

Given a real number a, check if the matrix

$$\begin{bmatrix} -\sin^2(a) + i\cos^2(a) & (1 + i)\sin(a)\cos(a) \\ (1 + i)\sin(a)\cos(a) & -\cos^2(a) + i\sin^2(a) \end{bmatrix} \in \mathcal{M}_{2\times 2}(\mathbb{C})$$

is unitary.

Exercise 1125

Find all possible triples a, b, c of real numbers, if any exist, such that the matrix
$\frac{1}{3} \begin{bmatrix} 1 & -2 & 2 \\ 2 & -1 & 2 \\ a & b & c \end{bmatrix}$ is orthogonal.

Exercise 1126

For which $a \in \mathbb{R}$ is $\frac{1}{1+2a^2} \begin{bmatrix} 1 & -2a & 2a^2 \\ 2a & 1-2a^2 & 2a \\ 2a^2 & 2a & 1 \end{bmatrix} \in \mathcal{M}_{3 \times 3}(\mathbb{R})$ an orthogonal

matrix?

Exercise 1127

Is the matrix $\frac{1}{2} \begin{bmatrix} 1 & 1 & 1 & 1 \\ 1 & 1 & -1 & -1 \\ 1 & -1 & 1 & -1 \\ 1 & -1 & -1 & 1 \end{bmatrix} \in \mathcal{M}_{4 \times 4}(\mathbb{R})$ orthogonal?

Exercise 1128

If $v = \begin{bmatrix} c \\ d \end{bmatrix} \in \mathbb{R}^2$, show that there exists an orthogonal matrix $A \in \mathcal{M}_{2 \times 2}(\mathbb{R})$ and

a real number b satisfying the condition that $Av = \begin{bmatrix} b \\ 0 \end{bmatrix}$.

Exercise 1129

Let a and b be real numbers, not both equal to 0. Show that the matrix

$$\frac{1}{a^2 + ab + b^2} \begin{bmatrix} ab & a(a+b) & b(a+b) \\ a(a+b) & -b(a+b) & ab \\ b(a+b) & ab & -a(a+b) \end{bmatrix} \in \mathcal{M}_{3 \times 3}(\mathbb{R})$$

is orthogonal.

Exercise 1130

Find all $a \in \mathbb{R}$ such that the matrix $\begin{bmatrix} 2a & -2a & a \\ -2a & -a & 2a \\ a & 2a & 2a \end{bmatrix} \in \mathcal{M}_{3 \times 3}(\mathbb{R})$ is orthog-

onal.

Exercise 1131

Let $A = \begin{bmatrix} 4 & -1 & 1 \\ -1 & 4 & -1 \\ 1 & -1 & 4 \end{bmatrix} \in \mathcal{M}_{3 \times 3}(\mathbb{R})$. Find an orthogonal matrix P such that

$P^T A P$ is a diagonal matrix.

Exercise 1132

Let $A = \begin{bmatrix} 1 & 0 & 0 & 0 \\ 0 & 0 & 1 & 0 \\ 0 & 1 & 0 & 0 \\ 0 & 0 & 0 & 1 \end{bmatrix} \in \mathcal{M}_{4 \times 4}(\mathbb{R})$. Find an orthogonal matrix P such that

$P^T A P$ is a diagonal matrix.

Exercise 1133

Let n be a positive integer and let A and B be orthogonal matrices in $\mathcal{M}_{n \times n}(\mathbb{R})$ satisfying $|A| + |B| = 0$. Show that $|A + B| = 0$.

Exercise 1134

Let $A = \frac{1}{2} \begin{bmatrix} 1 & -1 & 2\sqrt{2} \\ 2\sqrt{2} & 2\sqrt{2} & 0 \\ -1 & 1 & 2\sqrt{2} \end{bmatrix} \in \mathcal{M}_{3 \times 3}(\mathbb{R})$. Find an infinite number of pairs

(P, Q) of orthogonal matrices such that PAQ is a diagonal matrix.

Exercise 1135

Find an $a \in \mathbb{R}$ such that the matrix $\begin{bmatrix} a & -\frac{4}{5} & 0 \\ \frac{4}{5} & a & 0 \\ 0 & 0 & 1 \end{bmatrix} \in \mathcal{M}_{3 \times 3}(\mathbb{R})$ is orthogonal.

Exercise 1136

Let $A, B \in \mathcal{M}_{k \times n}(\mathbb{R})$ be matrices such that the columns of each form orthonormal bases for the same subspace W of $\mathbb{R}^k$. Show that $AA^T = BB^T$.

Exercise 1137

Let $A, B \in \mathcal{M}_{n \times n}(\mathbb{R})$ be orthogonal matrices. Is the matrix $\begin{bmatrix} A & O \\ O & B \end{bmatrix} \in \mathcal{M}_{2n \times 2n}(\mathbb{R})$ necessarily orthogonal?

Exercise 1138

Let n be a positive integer and let $A \in \mathcal{M}_{n \times n}(\mathbb{R})$ be a skew-symmetric matrix. Show that $(A - I)^{-1}(A + I)$ is an orthogonal matrix which does not have 1 as an eigenvalue.

Exercise 1139

Find two distinct functions $f_1, f_2 : \mathbb{R}^4 \setminus \left\{ \begin{bmatrix} 0 \\ 0 \\ 0 \\ 0 \end{bmatrix} \right\} \to \mathbb{R}$ satisfying the condition

that

$$f_i\left(\begin{bmatrix} a \\ b \\ c \\ d \end{bmatrix}\right)\begin{bmatrix} a^2+b^2-c^2-d^2 & 2(bc-da) & 2(bd-ca) \\ 2(bc-da) & a^2+b^2-c^2-d^2 & 2(cd-ba) \\ 2(bd-ca) & 2(cd-ba) & a^2+b^2-c^2-d^2 \end{bmatrix}$$

is always an orthogonal matrix.

Exercise 1140

Let n be a positive integer and let $A, B \in \mathcal{M}_{n \times n}(\mathbb{R})$ satisfy $A^2 + B^2 = I$. Is the matrix $\begin{bmatrix} A & -B \\ B & A \end{bmatrix} \in \mathcal{M}_{2n \times 2n}(\mathbb{R})$ necessarily orthogonal?

Exercise 1141

Let $O \neq A \in \mathcal{M}_{3 \times 3}(\mathbb{C})$ be a matrix satisfying $\mathrm{adj}(A) = A^H$. Show that A is a unitary matrix having determinant 1.

Exercise 1142

Let n be a positive integer and let α be the endomorphism of $\mathbb{C}^n$ defined by $\alpha : v \mapsto iv$. Is α normal?

Exercise 1143

Let V be an inner product space and let $\alpha, \beta \in \mathrm{End}(V)$ be normal. Is $\beta\alpha$ necessarily normal?

Exercise 1144

Let V be an inner product space finitely-generated over $\mathbb{C}$ and let $\alpha \in \mathrm{End}(V)$ satisfy the condition that every eigenvector of $\beta = \alpha + \alpha^*$ is also an eigenvector of $\gamma = \alpha - \alpha^*$. Prove that α is normal.

Exercise 1145

Let α be the endomorphism of $\mathbb{C}^2$ represented with respect to the canonical basis by the matrix $A = \begin{bmatrix} 2 & i \\ i & 2 \end{bmatrix}$. Is α normal?

Exercise 1146

Let V be an inner product space over $\mathbb{C}$ and let $\alpha \in \mathrm{End}(V)$ be normal. If $c \in \mathbb{C}$, is the endomorphism $\alpha - c\sigma_1$ necessarily normal?

Exercise 1147

Let V be an inner product space finitely generated over $\mathbb{C}$ and let $\sigma_0 \neq \alpha \in \mathrm{End}(V)$ be normal. Show that α is not nilpotent.

Exercise 1148

Let $a, b \in \mathbb{R}$ let $\alpha \in \text{End}(\mathbb{R}^3)$ be represented with respect to the canonical basis by $\begin{bmatrix} a & -2 & b \\ b & a & -2 \\ -2 & 3 & a \end{bmatrix}$. For which values of a and b is this endomorphism normal?

Exercise 1149

Let $\alpha \in \text{End}(\mathbb{R}^3)$ be represented with respect to the canonical basis by the matrix $\frac{1}{3}\begin{bmatrix} 14 & 2 & 14 \\ 2 & -1 & -16 \\ 14 & -16 & 5 \end{bmatrix}$. Show that α is selfadjoint and find an orthonormal basis of $\mathbb{R}^3$ composed of eigenvectors of α.

Exercise 1150

Let α be the endomorphism of $\mathbb{C}^3$ represented with respect to the canonical basis by the matrix $\begin{bmatrix} 6 & -2 & 3 \\ 3 & 6 & -2 \\ -2 & 3 & 6 \end{bmatrix}$. Show that α is normal and find an orthonormal basis of $\mathbb{C}^3$ composed of eigenvectors of α.

Exercise 1151

Let V be an inner product space and let $\sigma_0 \neq \alpha \in \text{End}(V)$ be a normal projection. Show that $\|\alpha(v)\| \leq \|v\|$ for all $v \in V$, with equality whenever $v \in \text{im}(\alpha)$. Give an example where this does not hold for α which is not normal.

Exercise 1152

Let n be a positive integer and let F be any field. A matrix $A \in \mathcal{M}_{n \times n}(F)$ is *antiorthogonal* if and only if $A^{-1} = -A^T$. Give an example of an antiorthogonal matrix in $\mathcal{M}_{2 \times 2}(\text{GF}(3))$.

Exercise 1153

Let $\alpha \in \text{End}(\mathbb{R}^4)$ be defined by $\alpha : \begin{bmatrix} a_1 \\ a_2 \\ a_3 \\ a_4 \end{bmatrix} \mapsto \begin{bmatrix} a_1 \\ a_2 \\ a_3 + a_4 \\ a_4 - a_3 \end{bmatrix}$. Show that α is normal but not selfadjoint.

Exercise 1154

Let $\alpha : \mathbb{R}^2 \to \mathbb{R}^3$ be the linear transformation represented with respect to the canonical bases by the matrix $A = \begin{bmatrix} 1 & 1 \\ 2 & 2 \\ 2 & 2 \end{bmatrix}$. Find the singular values of α.

Exercise 1155

Let n be a positive integer, let F be a field, and let I be the identity matrix in $\mathcal{M}_{n\times n}(F)$. A matrix $A \in \mathcal{M}_{2n\times 2n}(F)$ is *symplectic* if and only if

$$A\begin{bmatrix} O & I \\ -I & O \end{bmatrix} A^T = \begin{bmatrix} O & I \\ -I & O \end{bmatrix}.$$

If $B, C \in \mathcal{M}_{n\times n}(\mathbb{R})$, show that $B + iC \in \mathcal{M}_{n\times n}(\mathbb{C})$ is unitary if and only if the matrix $\begin{bmatrix} B & -C \\ C & B \end{bmatrix} \in \mathcal{M}_{2n\times 2n}(\mathbb{R})$ is symplectic.

Exercise 1156

For which $c, d \in \mathbb{C}$ is the matrix $\begin{bmatrix} 0 & -1/\sqrt{2} & d \\ c & 1/2 & i/2 \\ i/\sqrt{2} & -i/2 & 1/2 \end{bmatrix}$ Hermitian? For which values of c and d is it unitary?

Exercise 1157

A polynomial $p(X) \in \mathbb{C}[X]$ of degree $n \geq 0$ is a *reciprocal polynomial* if and only if $p(X) = \pm X^n p(X^{-1})$. Show that characteristic polynomials of orthogonal matrices are reciprocal and that the set of all reciprocal polynomials, together with the 0-polynomial, forms a subalgebra of $\mathbb{C}[X]$.

Exercise 1158

(*Cayley representation*) For any real number t, with $\cos(t) \neq -1$, find a skew-symmetric matrix $A \in \mathcal{M}_{2\times 2}(\mathbb{R})$ satisfying

$$\begin{bmatrix} \cos(t) & \sin(t) \\ -\sin(t) & \cos(t) \end{bmatrix} = (I - A)(I + A)^{-1}.$$

Exercise 1159

Let V be a vector space finitely generated over $\mathbb{C}$ and let α be an automorphism of V having polar decomposition $\alpha = \psi\theta$, where ψ is unitary and θ is positive definite. Show that α is normal if and only if $\alpha = \theta\psi$.

Moore–Penrose Pseudoinverses **19**

Let V and W be inner product spaces, and let $\alpha : V \to W$ be a linear transformation. We know that there exists a linear transformation $\beta : W \to V$ satisfying the condition that $\beta\alpha$ is the identity function on V and $\alpha\beta$ is the identity function on W if and only if α is an isomorphism; in this case, $\beta = \alpha^{-1}$. If both spaces are finitely generated, we also know that such an isomorphism can exist only when $\dim(V) = \dim(W)$. If α is not an isomorphism, it is possible to weaken the notion of the inverse of a function. Given a linear transformation $\alpha : V \to W$, we say that a linear transformation $\beta : W \to V$ is a *Moore–Penrose pseudoinverse* of α if and only if the following conditions are satisfied:

(1) $\alpha\beta\alpha = \alpha$ and $\beta\alpha\beta = \beta$;
(2) The endomorphisms $\beta\alpha \in \mathrm{End}(V)$ and $\alpha\beta \in \mathrm{End}(W)$ are selfadjoint.

With kind permission of the Archives of the Mathematisches Forschungsinstitut Oberwolfach (Penrose).

Eliakim Hastings Moore developed this construction in 1922, but it did not receive much attention at the time; it was rediscovered independently in 1955 by **Sir Roger Penrose**, a contemporary British applied mathematician, best known for his collaboration with the physicist Stephen Hawking.

Example The two parts of condition (2) in the definition of the Moore–Penrose pseudoinverse are independent. To see this, consider the linear transformation $\alpha : \mathbb{R}^3 \to \mathbb{R}^2$ defined by $\alpha : v \mapsto \begin{bmatrix} 1 & 2 & 3 \\ 0 & 1 & 0 \end{bmatrix} v$. For any $c, d \in \mathbb{R}$, let $\beta : \mathbb{R}^2 \to \mathbb{R}^3$ be the linear transformation defined by $\beta : w \mapsto \begin{bmatrix} 1 - 3c & -2 - 3d \\ 0 & 1 \\ c & d \end{bmatrix}$. Then one can check that $\alpha\beta\alpha = \alpha$ and $\beta\alpha\beta = \beta$ and that $\alpha\beta = \sigma_1$ in $\mathrm{End}(\mathbb{R}^2)$. On the other

J.S. Golan, *The Linear Algebra a Beginning Graduate Student Ought to Know*,   441
DOI 10.1007/978-94-007-2636-9_19, © Springer Science+Business Media B.V. 2012

hand, $\beta\alpha : v \mapsto \begin{bmatrix} 1-3c & -6c-3d & 3-9c \\ 0 & 1 & 0 \\ c & 2c+d & 3c \end{bmatrix} v$, and it is easy enough to choose
c and d so that this matrix is not symmetric and hence $\beta\alpha$ is not selfadjoint.

We will denote the Moore–Penrose pseudoinverse of α by α^+. Of course, in order
to justify this notation we have to show that β exists and is unique, which we will
do for the case that V and W are finitely generated. We will begin with uniqueness.

Proposition 19.1 *Let V and W be inner product spaces and let $\alpha : V \to W$
be a linear transformation. If α has a Moore–Penrose pseudoinverse, it must
be unique.*

Proof Suppose that $\beta, \gamma \in \mathrm{Hom}(W, V)$ are Moore–Penrose pseudoinverses of α.
Then $\beta = \beta\alpha\beta = (\beta\alpha)^*\beta = \alpha^*\beta^*\beta = (\alpha\gamma\alpha)^*\beta^*\beta = (\gamma\alpha)^*\alpha^*\beta^*\beta = \gamma\alpha\alpha^*\beta^*\beta = $
$\gamma\alpha(\beta\alpha)^*\beta = \gamma\alpha\beta\alpha\beta = \gamma\alpha\beta = \gamma\alpha\gamma\alpha\beta = \gamma(\alpha\gamma)^*\alpha\beta = \gamma\gamma^*\alpha^*\alpha\beta = \gamma\gamma^*\alpha^*(\alpha\beta)^*$
$= \gamma\gamma^*(\alpha\beta\alpha)^* = \gamma\gamma^*\alpha^* = \gamma(\alpha\gamma)^* = \gamma\alpha\gamma = \gamma$ and so we have proven unique-
ness.   □

In particular, if $\alpha : V \to W$ is an isomorphism, then, by Proposition 19.1, we
have $\alpha^+ = \alpha^{-1}$. If α is the 0-function then so is α^+.

Proposition 19.2 *Let V and W be finitely-generated inner product spaces
and let $\alpha : V \to W$ be a linear transformation.*
(1) *If α is a monomorphism, then α^+ exists and equals $(\alpha^*\alpha)^{-1}\alpha^*$. Moreover,
   $\alpha^+\alpha$ is the identity function on V;*
(2) *If α is an epimorphism, then α^+ exists and equals $\alpha^*(\alpha\alpha^*)^{-1}$. Moreover,
   $\alpha\alpha^+$ is the identity function on W.*

Proof (1) From Proposition 16.20, we see that if α is a monomorphism then
$\alpha^*\alpha \in \mathrm{Aut}(V)$, and so $(\alpha^*\alpha)^{-1}$ exists. Set $\beta = (\alpha^*\alpha)^{-1}\alpha^*$. Then $\beta\alpha$ is the identity function on V, and so $\beta\alpha$ is a selfadjoint endomorphism of V which satis-
fies $\alpha\beta\alpha = \alpha$ and $\beta\alpha\beta = \beta$. Finally, $(\alpha\beta)^* = [\alpha(\alpha^*\alpha)^{-1}\alpha^*]^* = \alpha[(\alpha^*\alpha)^{-1}]^*\alpha^* = $
$\alpha[(\alpha^*\alpha)^*]^{-1}\alpha^* = \alpha(\alpha^*\alpha)^{-1}\alpha^* = \alpha\beta$, and so $\alpha\beta$ is also selfadjoint. Thus $\beta = \alpha^+$.
   (2) From Proposition 16.20, we see that if α is an epimorphism then $\alpha\alpha^* \in$
$\mathrm{Aut}(W)$ and so $(\alpha\alpha^*)^{-1}$ exists. As in (1), we see that $\alpha^*(\alpha\alpha^*)^{-1} = \alpha^+$.   □

Example Let $\alpha : \mathbb{R}^2 \to \mathbb{R}^3$ be the linear transformation represented with respect to
the canonical bases by the matrix $\begin{bmatrix} 1 & 2 \\ -1 & 3 \\ 2 & 4 \end{bmatrix}$. This is a monomorphism and so, by

Proposition 19.2, α^+ exists and is represented with respect to the canonical bases
by the matrix $\frac{1}{25}\begin{bmatrix} 3 & -10 & 6 \\ 1 & 5 & 2 \end{bmatrix}$.

Proposition 19.3 *Let V and W be finitely-generated inner product spaces.
Then every $\alpha \in \mathrm{Hom}(V, W)$ has a Moore–Penrose pseudoinverse $\alpha^+ \in \mathrm{Hom}(W, V)$.*

Proof Let $Y = \mathrm{im}(\alpha)$, and write $\alpha = \mu\beta$, where $\beta : V \to Y$ is an epimorphism
given by $\beta : v \mapsto \alpha(v)$, and $\mu : Y \to W$ is the inclusion monomorphism. By Proposition 19.2, we know that $\beta^+ \in \mathrm{Hom}(Y, V)$ and $\mu^+ \in \mathrm{Hom}(W, Y)$ exist and satisfy
the conditions that $\beta\beta^+$ and $\mu^+\mu$ are equal to the identity function on Y. Therefore,
we see that $(\mu\beta)(\beta^+\mu^+)(\mu\beta) = \mu\beta$ and $(\beta^+\mu^+)(\mu\beta)(\beta^+\mu^+) = \beta^+\mu^+$ and we see
that $(\beta^+\mu^+)(\mu\beta) = \beta^+\beta$ and $(\mu\beta)(\beta^+\mu^+) = \mu\mu^+$ are selfadjoint. Thus α^+ exists
and equals $\beta^+\mu^+$.   □

As an immediate consequence of this, we note that if α is an endomorphism
of a finitely-generated inner produce space V then, by Proposition 6.11, we see
that $\mathrm{rk}(\alpha\alpha^+) \le \mathrm{rk}(\alpha)$ and $\mathrm{rk}(\alpha) = \mathrm{rk}(\alpha\alpha^+\alpha) \le \mathrm{rk}(\alpha\alpha^+)$ and so $\mathrm{rk}(\alpha\alpha^+) = \mathrm{rk}(\alpha)$.
Similarly, $\mathrm{rk}(\alpha^+\alpha) = \mathrm{rk}(\alpha)$.

If F is either $\mathbb{R}$ or $\mathbb{C}$, and if we are given a linear transformation $\alpha : F^k \to F^n$
which is represented with respect to the canonical bases by the matrix $A = [a_{ij}]$,
then we will denote the matrix representing α^+ with respect to these bases by A^+.
Thus the matrix A^+ has the following properties:
(1) $AA^+A = A$ and $A^+AA^+ = A^+$;
(2) The matrices AA^+ and A^+A are symmetric (in the case $F = \mathbb{R}$) or Hermitian
   (in the case $F = \mathbb{C}$).

Example Let $A = \begin{bmatrix} 1 & -1 & 2 \\ 2 & 1 & -2 \\ 3 & 0 & 0 \end{bmatrix} \in \mathcal{M}_{3\times3}(\mathbb{R})$. This matrix is clearly singular and

hence A^{-1} does not exist. However, we can check that $A^+ = \frac{1}{45}\begin{bmatrix} 5 & 5 & 10 \\ -5 & 4 & -1 \\ 10 & -8 & 2 \end{bmatrix}$.

For nonsingular square matrices of the same size A and B, we know that
$(AB)^{-1} = B^{-1}A^{-1}$. A similar equality does not hold for the Moore–Penrose pseudoinverse, as the following example shows.

Example If $A = \begin{bmatrix} 2 & 6 \\ 1 & 3 \end{bmatrix}$ and $B = \begin{bmatrix} 1 & 2 \\ 2 & 4 \end{bmatrix}$ in $\mathcal{M}_{2\times2}(\mathbb{R})$, then $AB = \begin{bmatrix} 14 & 28 \\ 7 & 14 \end{bmatrix}$.

Then $A^+ = \frac{1}{50}\begin{bmatrix} 2 & 1 \\ 6 & 3 \end{bmatrix}$ and $B^+ = \frac{1}{25}\begin{bmatrix} 1 & 2 \\ 2 & 4 \end{bmatrix}$ so $B^+A^+ = \frac{7}{1250}\begin{bmatrix} 2 & 1 \\ 4 & 2 \end{bmatrix}$, while

$(AB)^+ = \frac{1}{175}\begin{bmatrix} 2 & 1 \\ 4 & 2 \end{bmatrix}$.

The Singular Value Decomposition Theorem can be used to compute pseudoinverses. This is important since, as we have remarked previously, there exist several relatively efficient and stable numerical algorithms for computing such decompositions.

Example Let $\alpha : \mathbb{R}^k \to \mathbb{R}^n$ be a monomorphism represented with respect to the canonical bases by a matrix A. By Proposition 19.2, we have $A^+ = (A^TA)^{-1}A^T$. By Proposition 18.15, we set $A = PEQ^T$, where $P \in \mathcal{M}_{k\times k}(\mathbb{R})$ and $Q \in \mathcal{M}_{k\times n}(\mathbb{R})$ are orthogonal matrices and $E \in \mathcal{M}_{k\times n}(\mathbb{R})$ is of the form $\begin{bmatrix} D & O \\ O & O \end{bmatrix}$ for a diagonal matrix $D \in \mathcal{M}_{t\times t}(\mathbb{R})$, the diagonal entries of which are nonzero. Then

$$A^+ = \left(A^TA\right)^{-1}A^T = \left(QE^TP^TPEQ^T\right)^{-1}QE^TP^T$$

$$= \left(QE^TEQ^T\right)^{-1}QE^TP^T = Q\begin{bmatrix} D^{-2} & O \\ O & O \end{bmatrix}Q^TQE^TP^T$$

$$= Q\begin{bmatrix} D^{-1} & O \\ O & O \end{bmatrix}P^T.$$

Example If $A(t) = \begin{bmatrix} 1 & 0 \\ 0 & t \end{bmatrix}$ for all real numbers t then we see that $A(t)^+ = \begin{bmatrix} 1 & 0 \\ 0 & t^{-1} \end{bmatrix}$ when $t \neq 0$, but is equal to $\begin{bmatrix} 1 & 0 \\ 0 & 0 \end{bmatrix}$ for $t = 0$. Thus we see that not only is $\lim_{t\to0} A(t)$ not equal to $A(0)$, but indeed that the value of $A(t)$ moves farther and farther away from $A(0)$ as t approaches 0.

Thus we see that the Moore–Penrose pseudoinverse is not computationally stable. This means that one has to be very careful in actual applications. Because of the importance and utility of Moore–Penrose pseudoinverses, there exists a considerable literature on techniques for computing A^+ or A^+A, given a matrix A. One of the methods used in practice for computing the Moore–Penrose pseudoinverse over $\mathbb{R}$ is a recursive one, known as *Greville's method*, which is based on the following result: If $A \in \mathcal{M}_{k\times n}(\mathbb{R})$, and if we write $A = [\,B\ v\,]$, where $B \in \mathcal{M}_{k\times(n-1)}(\mathbb{R})$, then

$$A^+ = \begin{bmatrix} B^+(I - v \wedge w) \\ w \end{bmatrix}, \text{ where}$$

$$w = \begin{cases} (\|(I - BB^+)v\|)^{-2}(I - BB^+)v & \text{if } \|(I - BB^+)v\| \neq 0, \\ (1 + \|B^+v\|^2)^{-1}(B^+)^T B^+ v & \text{otherwise.} \end{cases}$$

© Mrs Greville (Greville); © Adi Ben-Israel (Ben-Israel)

In the 1970s, the American mathematician **Thomas N.E. Greville** and the American/Israeli mathematician **Adi Ben-Israel** popularized and reinvigorated the use of the Moore–Penrose pseudoinverse as a computational tool.

Another technique is to break A up into blocks, if possible. Indeed, if $A = \begin{bmatrix} A_{11} & A_{12} \\ A_{21} & A_{22} \end{bmatrix}$, where A_{11} is a nonsingular square matrix the rank of which equals the rank of A, then, by *Zlobec's formula*, we have $A^+ = \begin{bmatrix} A_{11} & A_{12} \end{bmatrix}^* B^* \begin{bmatrix} A_{11} \\ A_{21} \end{bmatrix}^*$, where $B = \left(\begin{bmatrix} A_{11} & A_{12} \end{bmatrix} A^* \begin{bmatrix} A_{11} \\ A_{21} \end{bmatrix} \right)^{-1}$.

One can also use convergence methods to compute the Moore–Penrose pseudoinverse of a matrix. If $A \in \mathcal{M}_{k \times n}(\mathbb{C})$ then, by Proposition 17.4, we know that the eigenvalues of A^*A are real. Let c be the largest such eigenvalue and pick a real number b satisfying $0 < bc < 2$. For each integer $p \geq 2$, define the sequence $Y_0, Y_1, \dots$ of matrices in $\mathcal{M}_{n \times k}(\mathbb{C})$ as follows:

(1) $Y_0 = bA^*$;

(2) If $k \geq 0$ and Y_k has already been defined, set $T_k = I - Y_k A$ and set $Y_{k+1} = Y_k + \sum_{i=1}^{p-1} T_k^i Y_k$. Then the sequence $Y_0, Y_1, \dots$ converges to A^+.

Another method is the following: if $A \in \mathcal{M}_{n \times n}(\mathbb{R})$ is an arbitrary symmetric matrix we can define matrices $A_0, A_1, \dots$ by setting $A_0 = A$ and $A_{k+1} = [I + (I - A_k)(I + A_k)^{-1}]A_k$ for all $k \geq 0$. Also, we can define real numbers $c_0, c_1, \dots$ by setting $c_0 = 1$ and $c_{i+1} = 2c_i + 1$ for each $i \geq 0$. Then the *Kovarik algorithm* states that if none of the numbers $-c_i^{-1}$ is an eigenvalue of A, the sequence $A_0, A_1, \dots$ converges to A^+A.

Let F be either $\mathbb{R}$ or $\mathbb{C}$, let k and n be positive integers, and let $A \in \mathcal{M}_{k \times n}(F)$. We now look at what the matrix A^+ says about a solution (if any) to a system of linear equations of the form $AX = w$, where $w \in F^k$. First of all, we note that in general the following proposition holds.

Proposition 19.4 *Let F be either $\mathbb{R}$ or $\mathbb{C}$, let k and n be positive integers, let $A \in M_{k \times n}(F)$, and let $w \in F^k$. The system of linear equations $AX = w$ has a solution if and only if $(AA^+)w = w$.*

Proof If there is a vector $v \in F^n$ satisfying $Av = w$ then $(AA^+)w = (AA^+)(Av) = (AA^+A)v = Av = w$. Conversely, if $(AA^+)w = w$ then $A(A^+w) = w$ and so $Av = w$, where $v = A^+w$. $\qquad \square$

We also note that, in the situation above, if $y \in F^n$ then $A(I - A^+A)y = \begin{bmatrix} 0 \\ \vdots \\ 0 \end{bmatrix}$,

and so we also see that $A^+w + (I - A^+A)y$ is also a solution to the system $AX = w$, assuming that the system has any solutions at all. Conversely, any solution to this system is of the form $A^+w + u$, where $Au = \begin{bmatrix} 0 \\ \vdots \\ 0 \end{bmatrix}$, and so $(I - A^+A)u = u$.

Proposition 19.5 *Let F be either $\mathbb{R}$ or $\mathbb{C}$, let k and n be positive integers, and let $A \in M_{k \times n}(F)$. Let $w \in F^k$. If the system $AX = w$ has a solution then in the set of all solutions to this system of linear equations there is precisely one having a minimal norm, and it is A^+w.*

Proof If u is a solution to this system, then we have already seen that it is of the form $A^+w + (I - A^+A)y$. But we note that

$$\begin{aligned}
\langle A^+w, (I - A^+A)y \rangle &= \langle A^+AA^+w, (I - A^+A)y \rangle \\
&= \langle A^+w, (A^+A)(I - A^+A)y \rangle \\
&= \langle A^+w, (A^+A - A^+AA^+A)y \rangle \\
&= \left\langle A^+w, \begin{bmatrix} 0 \\ \vdots \\ 0 \end{bmatrix} \right\rangle = 0
\end{aligned}$$

so

$$\begin{aligned}
\|u\|^2 = \langle u, u \rangle &= \langle A^+w + (I - A^+A)y, A^+w + (I - A^+A)y \rangle \\
&= \langle A^+w, A^+w \rangle + \langle (I - A^+A)y, (I - A^+A)y \rangle \\
&= \|A^+w\|^2 + \|(I - A^+A)y\|^2,
\end{aligned}$$

which implies that $\|u\| \geq \|A^+w\|$. $\qquad \square$

Example Let $A = \begin{bmatrix} 2 & -1 & 1 \\ 1 & 1 & 1 \end{bmatrix}$ and $w = \begin{bmatrix} 1 \\ 2 \end{bmatrix}$. Then $A^+ = \frac{1}{14} \begin{bmatrix} 4 & 2 \\ -5 & 8 \\ 1 & 4 \end{bmatrix}$ and so

the solution to the system $AX = w$ having minimal norm is $A^+ w = \frac{1}{14} \begin{bmatrix} 8 \\ 11 \\ 9 \end{bmatrix}$. Its

norm is $\frac{1}{14}\sqrt{266}$.

But what happens if the system $AX = w$ does not have a solution? Suppose that F is either $\mathbb{R}$ or $\mathbb{C}$ and that $A \in \mathcal{M}_{k \times n}(F)$, and $w \in F^k$, where k and n be positive integers. Then the system $(A^+ A)X = A^+ w$ always has a solution, namely $A^+ w$, and, by Proposition 19.5, this is in fact the solution of minimal norm of this equation, which is the *best approximation* to a solution of $AX = w$.

Example Consider the system of linear equations $AX = w$, where

$$A = \begin{bmatrix} 2 & -4 & 5 \\ 6 & 0 & 3 \\ 2 & -4 & 5 \\ 6 & 0 & 3 \end{bmatrix} \quad \text{and} \quad w = \begin{bmatrix} 1 \\ 3 \\ -1 \\ 3 \end{bmatrix}.$$

Then

$$A^+ = \frac{1}{72} \begin{bmatrix} -2 & 6 & -2 & 6 \\ -5 & 3 & -5 & 3 \\ 4 & 0 & 4 & 0 \end{bmatrix} \quad \text{and} \quad A^+ w = \frac{1}{4} \begin{bmatrix} 2 \\ 1 \\ 0 \end{bmatrix}.$$

In order to emphasize the use of Proposition 19.5, we briefly consider the *least squares method*, which is an important tool in many areas of applied mathematics and statistics. This method was developed at the beginning of the nineteenth century by Gauss and Legendre and, independently, by the American mathematical pioneer Robert Adrain. Suppose that we have before us the results of several observations, which, depending on values $t_1, \ldots, t_n$ of a real parameter, give us real values $c_1, \ldots, c_n$. Our theory tells us that the set of points $\{(t_i, c_i) \mid 1 \le i \le n\}$ in the Euclidean plane should lie on a straight line. However, because of measuring and/or computational errors, this does not quite work out. So we want to find the equation of the line in the plane which best fits our observed data. In other words, we want to find a solution of minimal norm to the system of linear equations $\{X_1 + t_i X_2 = c_i \mid 1 \le i \le n\}$, which can be written as

$$\begin{bmatrix} 1 & t_1 \\ 1 & t_2 \\ \vdots & \vdots \\ 1 & t_n \end{bmatrix} \begin{bmatrix} X_1 \\ X_2 \end{bmatrix} = \begin{bmatrix} c_1 \\ c_2 \\ \vdots \\ c_n \end{bmatrix}.$$

As we have seen, the solution of minimal norm, if it exists, is $\begin{bmatrix} 1 & t_1 \\ 1 & t_2 \\ \vdots & \vdots \\ 1 & t_n \end{bmatrix}^{+} \begin{bmatrix} c_1 \\ c_2 \\ \vdots \\ c_n \end{bmatrix}$.

Otherwise, this is the best approximation to the solution the system.

With kind permission of the University of Pennsylvania Libraries.

Irish-born **Robert Adrain** emigrated to the United States in 1798. He published his own mathematics journal, but his work received no international attention at the time.

Example To find the equation of the line in the Euclidean plane which best fits the set of points $\{(1, 3), (2, 7), (3, 8), (4, 11)\}$, we calculate

$$\begin{bmatrix} 1 & 1 \\ 1 & 2 \\ 1 & 3 \\ 1 & 4 \end{bmatrix}^{+} \begin{bmatrix} 3 \\ 7 \\ 8 \\ 11 \end{bmatrix} = \left(\frac{1}{20} \begin{bmatrix} 20 & 10 & 0 & -10 \\ -6 & -2 & 2 & 6 \end{bmatrix} \right) \begin{bmatrix} 3 \\ 7 \\ 8 \\ 11 \end{bmatrix} = \frac{1}{2} \begin{bmatrix} 2 \\ 5 \end{bmatrix}$$

so the line we want is given by $\{(t, 1 + \frac{5}{2}t) \mid t \in \mathbb{R}\}$.

We can use the same method to find the best fit of any polynomial of a higher degree to a set of points. For example, if we wish to find a parabola which best fits the set of points $\{(t_i, c_i) \mid 1 \leq i \leq n\}$ in the Euclidean plane, we have to find a best approximation to a solution of the system of linear equations $\{X_1 + t_i X_2 + t_i^2 = c_i \mid$

$1 \leq i \leq n\}$, which we know is $\begin{bmatrix} 1 & t_1 & t_1^2 \\ 1 & t_2 & t_2^2 \\ \vdots & \vdots & \vdots \\ 1 & t_n & t_n^2 \end{bmatrix}^{+} \begin{bmatrix} c_1 \\ c_2 \\ \vdots \\ c_n \end{bmatrix}$.

Example To find the equation of the parabola in the Euclidean plane which best fits the set of points $\{(1, 3), (2, 7), (3, 8), (4, 11)\}$, we calculate

$$\begin{bmatrix} 1 & 1 & 1 \\ 1 & 2 & 4 \\ 1 & 3 & 9 \\ 1 & 4 & 16 \end{bmatrix}^{+} \begin{bmatrix} 3 \\ 7 \\ 8 \\ 11 \end{bmatrix} = \left(\frac{1}{20} \begin{bmatrix} 45 & -15 & -25 & 15 \\ -31 & 23 & 27 & -19 \\ 5 & -5 & -5 & 5 \end{bmatrix} \right) \begin{bmatrix} 3 \\ 7 \\ 8 \\ 11 \end{bmatrix}$$

$$= \frac{1}{4} \begin{bmatrix} -1 \\ 15 \\ -1 \end{bmatrix},$$

and so the parabola we want is given by $\{(t, -\frac{1}{4} + \frac{15}{4}t - \frac{1}{4}t^2) \mid t \in \mathbb{R}\}$.

Needless to say, we can also consider a much more general context. Suppose that W is a finitely-generated subspace of $\mathbb{R}^A$. Given a set of observations $\{(t_i, c_i) \mid 1 \leq i \leq n\} \subseteq A \times \mathbb{R}$, we want to find the function $g \in W$ which best approximates these observations.

To do this, we pick a basis $\{f_1, \ldots, f_k\}$ for W. Then we want to find a best approximation to a solution of the system of linear equations

$$\{X_1 f_1(t_i) + \cdots + X_k f_k(t_i) = c_i \mid 1 \leq i \leq n\},$$

which can be written as $\begin{bmatrix} f_1(t_1) & \cdots & f_k(t_1) \\ \vdots & \ddots & \vdots \\ f_1(t_n) & \cdots & f_k(t_n) \end{bmatrix} \begin{bmatrix} X_1 \\ \vdots \\ X_k \end{bmatrix} = \begin{bmatrix} c_1 \\ \vdots \\ c_n \end{bmatrix}$. As we have seen,

this is $\begin{bmatrix} f_1(t_1) & \cdots & f_k(t_1) \\ \vdots & \ddots & \vdots \\ f_1(t_n) & \cdots & f_k(t_n) \end{bmatrix}^{+} \begin{bmatrix} c_1 \\ \vdots \\ c_n \end{bmatrix}$.

Least-squares approximations are often used to find best-fit solutions to very large systems of linear equations of the form $AX = w$ which, in theory, have an exact solution but in practice that solution cannot be found because of errors in measurement of the data and computational errors. Indeed, Gauss developed this method for finding solutions to the very large systems of linear equations which resulted from laying down a triangulation grid for a geodetic survey of the state of Hanover he conducted in 1818. In 1978, the American National Geodetic Survey used it to solve a system of over 2.5 million linear equations in 400,000 unknowns which resulted from the updating of the triangulation grid for the continental United States.

The constructions presented in this chapter can be generalized considerably. Indeed, if $(K, \bullet)$ is an associative unital algebra over a field F on which we have defined an involution $a \mapsto a^*$, then an element a of K has a Moore–Penrose pseudoinverse b if and only if the following conditions are satisfied:
(1) $a \bullet b \bullet a = a$ and $b \bullet a \bullet b = b$;
(2) $(b \bullet a)^* = b \bullet a$ and $(a \bullet b)^* = a \bullet b$.

Proposition 19.1 can easily be modified to show that such a pseudoinverse, if it exists, is unique. Pseudoinverses of this sort show up in the study of C^*-algebras, or, more generally, associative unital algebras $(K, \bullet)$ that satisfy the *Gelfand–Naimark property*, namely that $e + a^* \bullet a$ is a unit of K for each $a \in K$, where e is the multiplicative identity of K. In such algebras, it is possible to show that if $a \in K$ satisfies the condition that there exists an element $b \in K$ satisfying $a \bullet b \bullet a = a$ and $b \bullet a \bullet b = b$, then a has a Moore–Penrose pseudoinverse.

With kind permission of the Archives of the Mathematisches Forschungsinstitut Oberwolfach (Gelfand); With kind permission of the American Mathematical Society (Naimark).

Israil Moisseevich Gelfand was a twentieth-century Russian mathematician who emigrated to the United States. He worked in many areas of analysis and mathematical biology. **Mark Aronovich Naimark** was a twentieth-century Ukrainian mathematician who worked primarily in functional analysis.

Finally, one should note that the Moore–Penrose pseudoinverse is just one of many "pseudoinverses" in the mathematical literature, each designed for a fairly specific purpose. The first of these was introduced by Fredholm in 1903 to deal with integral operators. Others are based on specific situations which arise in algebra or analysis, or which are used to implement specific computational methods.

Example Let V be a vector space finitely generated over a field F and let $\alpha \in \mathrm{End}(V)$. Let $k = \inf\{0 < h \in \mathbb{N} \mid \mathrm{rk}(\alpha^h) = \mathrm{rk}(\alpha^{h+1})\}$. Then the *Drazin pseudoinverse* of α is the endomorphism β of V satisfying $\alpha^{k+1}\beta = \alpha^k$, $\beta\alpha\beta = \beta$, and $\alpha\beta = \beta\alpha$. If such a β exists, it is necessarily unique. It is immediate that if $\alpha \in \mathrm{Aut}(V)$ then $k = 1$ and $\beta = \alpha^{-1}$. If α is nilpotent then its Drazin pseudoinverse is σ_0. Drazin pseudoinverses have important applications in differential equations and in mathematical economics.

Exercises

Exercise 1160
Let V and W be finitely-generated inner product spaces and let $\alpha \in \mathrm{Hom}(V, W)$. Let $\beta \in \mathrm{Hom}(W, V)$ satisfy $\alpha\beta\alpha = \alpha$. Show that $\mathrm{rk}(\beta) \geq \mathrm{rk}(\alpha)$, with equality holding if and only if $\beta\alpha\beta = \beta$.

Exercise 1161
Let $A = \begin{bmatrix} 1 & 1 \\ -1 & 1 \\ 2 & 3 \end{bmatrix} \in \mathcal{M}_{3\times2}(\mathbb{R})$. Calculate A^+.

Exercise 1162
Let $A = [5\ 0\ 0] \in \mathcal{M}_{1\times3}(\mathbb{Q})$. Calculate A^+.

Exercise 1163
Let $A = [a_1 \ldots a_n] \in \mathcal{M}_{1\times n}(\mathbb{C})$, where n is a positive integer. Show that $A^+ = (AA^H)^{-1}A^H$.

Exercise 1164

Let $A = \begin{bmatrix} 2 & 2 & 0 \\ 1 & 2 & 1 \\ 1 & 2 & 1 \end{bmatrix} \in \mathcal{M}_{3\times3}(\mathbb{R})$. Calculate A^+.

Exercise 1165

Let V and W be finitely-generated inner product spaces, and let $\alpha \in \text{Hom}(V, W)$. For any nonzero scalar c, show that $(c\alpha)^+ = \frac{1}{c}\alpha^+$.

Exercise 1166

Let n be a positive integer and let $A \in \mathcal{M}_{n\times n}(\mathbb{R})$ be a diagonal matrix. Calculate A^+.

Exercise 1167

Let V and W be finitely-generated inner product spaces, and let $\alpha \in \text{Hom}(V, W)$. Show that $(\alpha^*)^+ = (\alpha^+)^*$.

Exercise 1168

Let $V = \mathbb{R}^2$, which is endowed with the dot product and let $\alpha : V \to \mathbb{R}$ be the linear functional defined by $\alpha : \begin{bmatrix} a \\ b \end{bmatrix} \mapsto a$. Let $\beta : \mathbb{R} \to V$ be the linear transformation defined by $\beta : a \mapsto \begin{bmatrix} a \\ a \end{bmatrix}$. Show that $(\alpha\beta)^+ \neq \beta^+\alpha^+$.

Exercise 1169

Let n be a positive integer and let $A = [a_{ij}] \in \mathcal{M}_{n\times n}(\mathbb{R})$ be the matrix all entries of which are equal to 1. Show that $A^+ = n^{-2}A$.

Exercise 1170

Let $A \in \mathcal{M}_{k\times n}(\mathbb{R})$ be a matrix of the form $\begin{bmatrix} C & O \\ O & O \end{bmatrix}$, where C is a $t \times t$ nonsingular diagonal matrix. Show that $A^+ = \begin{bmatrix} D & O \\ O & O \end{bmatrix}$, where $D = C^{-1}$.

Exercise 1171

Let $V = \mathbb{R}^n$ on which we have the dot product defined, and let $\alpha \in \text{End}(V)$ satisfy the condition that $\text{ker}(\alpha) = \text{im}(\alpha)^\perp$. Show that the restriction β of α to $\text{im}(\alpha)$ is an automorphism of $\text{im}(\alpha)$ and that the restriction of α^+ to $\text{im}(\alpha)$ equals β^{-1}.

Exercise 1172

Let n be a positive integer and let $A, B \in \mathcal{M}_{n\times n}(\mathbb{R})$ be matrices satisfying the conditions $ABA = A$, $BAB = B$, and $A^2 = A$. Is it necessarily true that $B^2 = B$?

Exercise 1173

Let h, k, m, and n be positive integers, let $A \in \mathcal{M}_{h \times k}(\mathbb{R})$, let $B \in \mathcal{M}_{m \times n}(\mathbb{R})$, and let $C \in \mathcal{M}_{h \times n}(\mathbb{R})$. Show that there exists a matrix $X \in \mathcal{M}_{k \times m}(\mathbb{R})$ satisfying $AXB = C$ if and only if $AA^+CB^+B = C$.

Exercise 1174

Let k and n be positive integers and let $B \in \mathcal{M}_{k \times k}(\mathbb{R})$ and $C \in \mathcal{M}_{n \times n}(\mathbb{R})$ be orthogonal matrices. For $A \in \mathcal{M}_{k \times n}(\mathbb{R})$, show that $(BAC)^+ = C^T A^+ B^T$.

Exercise 1175

Let k and n be positive integers and let $A \in \mathcal{M}_{k \times k}(\mathbb{R})$ and $B \in \mathcal{M}_{k \times n}(\mathbb{R})$. Let $C \in \mathcal{M}_{n \times n}(\mathbb{R})$ be nonsingular. Prove that

$$\begin{bmatrix} A & AB \\ O & C \end{bmatrix}^+ = \begin{bmatrix} A^+ & -A^+ABC^{-1} \\ O & C^{-1} \end{bmatrix}.$$

Bilinear Transformations and Forms

Let V, W, and Y be vector spaces over a field F. We say that a function $f : V \times W \to Y$ is a *bilinear transformation* if and only if the function $v \mapsto f(v, w_0)$ belongs to $\mathrm{Hom}(V, Y)$ for any given vector $w_0 \in W$ and the function $w \mapsto f(v_0, w)$ belongs to $\mathrm{Hom}(W, Y)$ for any given vector $v_0 \in V$. The set of all bilinear transformations from $V \times W$ to Y will be denoted by $\mathrm{Bil}(V \times W, Y)$. If $f, g \in \mathrm{Bil}(V \times W, Y)$ and if $c \in F$ then $f + g$ and cf also belong to $\mathrm{Bil}(V \times W, Y)$, and so $\mathrm{Bil}(V \times W, Y)$ is a subspace of the vector space $Y^{V \times W}$ over F. Also, any bilinear transformation $f : V \times W \to Y$ defines a bilinear transformation $f^{\mathrm{op}} : W \times V \to Y$, called the *opposite transformation* of f, by setting $f^{\mathrm{op}} : (w, v) \mapsto f(v, w)$. It is clear that the function

$$(\,)^{\mathrm{op}} : \mathrm{Bil}(V \times W, Y) \to \mathrm{Bil}(W \times V, Y)$$

is an isomorphism of vector spaces. We say that a bilinear transformation $f \in \mathrm{Bil}(V \times V, Y)$ is *symmetric* if and only if $f = f^{\mathrm{op}}$. It is *skew symmetric* if and only if $f = -f^{\mathrm{op}}$.

In particular, if we consider a single vector space V over a field F, then we note that $f \in \mathrm{Bil}(V \times V, V)$ if and only if the operation $\bullet$ on V defined by $v \bullet w = f(v, w)$ turns V into an F-algebra. This algebra is commutative if and only if f is symmetric.

Example Let n be a positive integer and let $V = \mathbb{R}^n$, on which we have the dot product defined. A classical problem in geometry is to ask if there exists a bilinear transformation $f \in \mathrm{Bil}(V \times V, V)$ satisfying the condition that $\|f(v, w)\| = \|v\| \cdot \|w\|$ for all $v, w \in V$. Euler showed that such a transformation exists for the case $n = 4$. At the end of the nineteenth century, Hurwitz showed that such transformations exist only when $n = 1, 2, 4$, or 8.

J.S. Golan, *The Linear Algebra a Beginning Graduate Student Ought to Know*, DOI 10.1007/978-94-007-2636-9_20, © Springer Science+Business Media B.V. 2012

Adolph Hurwitz was a nineteenth-century German mathematician who taught both Hilbert and Minkowski.

Example Let F be a field and let k, n, and t be positive integers. Set $V = \mathcal{M}_{k \times n}(F)$, $W = \mathcal{M}_{t \times n}(F)$, and $Y = \mathcal{M}_{k \times t}(F)$. Then there exists a bilinear transformation $V \times W \to Y$ defined by $(A, B) \mapsto AB^T$. In particular, we have a bilinear transformation $F^n \times F^n \to \mathcal{M}_{n \times n}(F)$ given by $(v, w) \mapsto v \wedge w$. More generally, if V, W, and Y are as mentioned, every matrix $C \in \mathcal{M}_{n \times n}(F)$ defines a bilinear transformation $V \times W \to Y$ by setting $(A, B) \mapsto ACB^T$.

Example For vector spaces V and W over a field F, the function $\mathrm{Hom}(V, W) \times V \to W$ given by $(\alpha, v) \mapsto \alpha(v)$ is a bilinear transformation.

Let V, W, and Y be vector spaces over a field F. The image of a bilinear transformation $f \in \mathrm{Bill}(V \times W, Y)$ is not necessarily a subspace of Y, as the following example shows.

Example Consider the bilinear transformation $f : \mathbb{R}^2 \times \mathbb{R}^2 \to \mathcal{M}_{2 \times 2}(\mathbb{R})$ defined by $f : (v, w) \mapsto v \wedge w$. The image of f contains $\begin{bmatrix} 0 & 1 \\ 0 & 0 \end{bmatrix}$ and $\begin{bmatrix} 0 & 0 \\ 1 & 0 \end{bmatrix}$, but is not a subspace since $\begin{bmatrix} 0 & 1 \\ 1 & 0 \end{bmatrix} \notin \mathrm{im}(f)$.

As with linear transformations, bilinear transformations are totally determined by their behavior on bases. That is to say, let V and W be vector spaces over a field F, and let $B = \{v_i \mid i \in \Omega\}$ and $D = \{w_j \mid j \in \Lambda\}$ be bases of V and W, respectively. Let Y be a vector space over F and let $f_0 : B \times D \to Y$ be a function. Then there exists a unique bilinear transformation $f \in \mathrm{Bill}(V \times W, Y)$ satisfying $f(v_i, w_j) = f_0(v_i, w_j)$ for all i and j, namely the function defined by $f : (\sum_{i \in \Omega} a_i v_i, \sum_{j \in \Lambda} b_j w_j) \mapsto \sum_{i \in \Omega} \sum_{j \in \Lambda} a_i b_j f_0(v_i, w_j)$. In the case that $V = W = Y$, we have already noted this fact in Proposition 5.5.

Proposition 20.1 *If V, W, and Y are vector spaces over a field F, then* $\mathrm{Bill}(V \times W, Y)$ *is isomorphic to* $\mathrm{Hom}(V, \mathrm{Hom}(W, Y))$.

Proof Define a function $\theta : \mathrm{Bill}(V \times W, Y) \to \mathrm{Hom}(V, \mathrm{Hom}(W, Y))$ as follows: given a bilinear transformation $f \in \mathrm{Bill}(V \times W, Y)$ and a vector $v \in V$, then

$\theta(f)(v) : w \mapsto f(v, w)$. It is straightforward to check that indeed $\theta(f)(v) \in$ Hom(W, Y) for all $f \in$ Bill$(V \times W, Y)$ and all $v \in V$. Moreover, $\theta(f)(v_1 + v_2) = \theta(f)(v_1) + \theta(f)(v_2)$ and $\theta(f)(cv) = c\theta(f)(v)$ for all $v, v_1, v_2 \in V$ and all $c \in F$, so $\theta(f) \in$ Hom$(V, \text{Hom}(W, Y))$ for all $f \in$ Bill$(V \times W, Y)$. Finally, $\theta(f + g) = \theta(f) + \theta(g)$ and $\theta(cf) = c\theta(f)$ for all $f, g \in$ Bill$(V \times W, Y)$ and all $c \in F$, and so we have shown that θ is a linear transformation.

It is also possible to define a function

$$\varphi : \text{Hom}\big(V, \text{Hom}(W, Y)\big) \to \text{Bill}(V \times W, Y)$$

by setting $\varphi(\alpha) : (v, w) \mapsto \alpha(v)(w)$ for all $v \in V$ and $w \in W$, and again it is easy to show that this is a linear transformation. If $\alpha \in$ Hom$(V, \text{Hom}(W, Y))$ and $v \in V$, then $\theta\varphi(\alpha)(v) : w \mapsto \varphi(\alpha)(v)(w) = \alpha(v)(w)$ and so $\theta\varphi(\alpha)(v) = \alpha(v)$ for all $v \in V$. Thus $\theta\varphi(\alpha) = \alpha$ for all $\alpha \in$ Hom$(V, \text{Hom}(W, Y))$, and so $\theta\varphi$ is the identity function on Hom$(V, \text{Hom}(W, Y))$. Conversely, if $f \in$ Bill$(V \times W, Y)$ then

$$\varphi\theta(f) : (v, w) \mapsto \theta(f)(v)(w) = f(v, w)$$

for all $v \in V$ and $w \in W$ and so $\varphi\theta(f) = f$ for all $f \in$ Bill$(V \times W, Y)$, proving that $\varphi\theta$ is the identity function on Bill$(V \times W, Y)$. Thus we have established that θ is an isomorphism, with $\theta^{-1} = \varphi$. $\qquad\square$

Let V and W be vector spaces over a field F. A bilinear transformation $f : V \times W \to F$ is called a *bilinear form*. We will denote the set of all such bilinear forms by Bill$(V \times W)$, instead of Bill$(V \times W, F)$. By what we have seen above, Bill$(V \times W)$ is a subspace of $F^{V \times W}$ which is isomorphic to Hom$(V, D(W))$. If V and W are vector spaces over a field F, then a bilinear form $f \in$ Bill$(V \times W)$ is *nondegenerate* if and only if for each $0_V \neq v \in V$ there exists a $w \in W$ satisfying $f(v, w) \neq 0$ and for each $0_W \neq w \in W$ there exists a $v \in V$ satisfying $f(v, w) \neq 0$.

With kind permission of the Archives of the Mathematisches Forschungsinstitut Oberwolfach.

Mathematicians at the beginning of the nineteenth century, such as Gauss and Jacobi, preferred to state their results in terms of bilinear forms rather than in terms of matrices. Sylvester contributed greatly to the theory of bilinear forms, as did the influential nineteenth-century German mathematician **Karl Weierstrass**.

Example If V is an inner product space over $\mathbb{R}$, then the function $(v, w) \mapsto \langle v, w \rangle$ belongs to Bill$(V \times V)$. This is not true, of course, if our field of scalars is $\mathbb{C}$.

Example If F is a field and $V = F^n$ for some positive integer n, then the function $(v, w) \mapsto v \odot w$ belongs to Bill$(F^n \times F^n)$. This function is particularly useful in the case $F = \text{GF}(2)$. Indeed, if $v \in \text{GF}(2)^n$, then

$$v \odot v = \begin{cases} 0 & \text{if an even number of entries in } v \text{ are equal to } 1, \\ 1 & \text{if an odd number of entries in } v \text{ are equal to } 1. \end{cases}$$

This value is known as the *parity* of v.

More generally, let A be a finite set and let V be the collection of all subsets of A. Define a function $f : V \times V \to \mathrm{GF}(2)$ by setting

$$f(A, B) = \begin{cases} 0 & \text{if } A \cap B \text{ has an even number of elements,} \\ 1 & \text{if } A \cap B \text{ has an odd number of elements.} \end{cases}$$

Then $f \in \mathrm{Bill}(V \times V)$.

Example If V is a vector space over a field F, we have a nondegenerate bilinear form in $\mathrm{Bill}(D(V) \times V)$ given by $(\delta, v) \mapsto \delta(v)$. Similarly, if $\delta_1, \delta_2 \in D(V)$, we have a bilinear form in $\mathrm{Bill}(V \times V)$ given by $(v, w) \mapsto \delta_1(v)\delta_2(w)$, which is nondegenerate if δ_1 and δ_2 are not the 0-functional.

Example If F is a field and if k and n are positive integers, then each matrix $A \in M_{k \times n}(F)$ defines a bilinear form in $\mathrm{Bill}(F^k \times F^n)$ by $(v, w) \mapsto v \odot Aw$.

Example If F is a field of characteristic not equal to 2 and of V is a vector space over F, then any $f \in \mathrm{Bill}(V \times V)$ can be written as a sum of a symmetric bilinear form and a skew-symmetric bilinear form, namely $f = f_1 + f_2$, where $f_1 : (v, w) \mapsto \frac{1}{2}[f(v, w) + f(w, v)]$ and $f_2 : (v, w) \mapsto \frac{1}{2}[f(v, w) - f(w, v)]$. Moreover, this representation is unique, for if $f = g_1 + g_2$, where g_1 is symmetric and g_2 is skew symmetric, then for each $(v, w) \in V \times V$ we have $f(v, w) + f(w, v) = 2g_1(v, w)$ and $f(v, w) - f(w, v) = 2g_2(v, w)$, from which we deduce that $g_i = f_i$ for $i = 1, 2$.

If V and W are vector spaces over F of finite dimension k and n, respectively, then any bilinear form on $V \times W$ can be represented as in the previous example. Indeed, if we fix bases $B = \{v_1, \ldots, v_k\}$ for V and $D = \{w_1, \ldots, w_n\}$ for W, then for any $f \in \mathrm{Bill}(V \times W)$ we define the matrix $T_{BD}(f) = [f(v_i, w_j)] \in M_{k \times n}(F)$ and check that if $v = \sum_{i=1}^{k} a_i v_i$ and $w = \sum_{j=1}^{n} b_j w_j$, then $f(v, w) = v \odot T_{BD}(f)w$. Indeed, for fixed B and D, the function $f \mapsto T_{BD}(f)$ is an isomorphism from $\mathrm{Bill}(V \times W)$ to $M_{k \times n}(F)$.

Example Let F be a field and let $V = F^2$. Consider the bases $B = \left\{ \begin{bmatrix} 1 \\ 0 \end{bmatrix}, \begin{bmatrix} 0 \\ 1 \end{bmatrix} \right\}$ and $D = \left\{ \begin{bmatrix} 1 \\ -1 \end{bmatrix}, \begin{bmatrix} 1 \\ 1 \end{bmatrix} \right\}$ of V. If $f \in \mathrm{Bill}(V \times V)$ is given by $f : \left(\begin{bmatrix} a \\ b \end{bmatrix}, \begin{bmatrix} c \\ d \end{bmatrix} \right) \mapsto (a + b)(c + d)$, then it is easy to verify that $T_{BB}(f) = \begin{bmatrix} 1 & 1 \\ 1 & 1 \end{bmatrix}$ and $T_{DD}(f) = \begin{bmatrix} 0 & 0 \\ 0 & 4 \end{bmatrix}$.

Proposition 20.2 *Let V and W be vector spaces finitely generated over a field F and having bases $B = \{v_1, \ldots, v_k\}$ and $D = \{w_1, \ldots, w_n\}$, respectively. Let $C = \{x_1, \ldots, x_k\}$ and $E = \{y_1, \ldots, y_n\}$ also be bases for V and W, respectively, and let $P = [p_{ir}] \in M_{k \times k}(F)$ and $Q = [q_{js}] \in M_{n \times n}(F)$ be nonsingular matrices satisfying*

$$\begin{bmatrix} x_1 \\ \vdots \\ x_k \end{bmatrix} = P \begin{bmatrix} v_1 \\ \vdots \\ v_k \end{bmatrix} \quad and \quad \begin{bmatrix} y_1 \\ \vdots \\ y_n \end{bmatrix} = Q \begin{bmatrix} w_1 \\ \vdots \\ w_n \end{bmatrix}.$$

Then for $f \in \mathrm{Bill}(V \times W)$ we see that $T_{CD}(f) = PT_{BD}(f)Q^T$.

Proof As a direct consequence of the definitions, we see that

$$f(x_i, y_j) = \left(\sum_{r=1}^{k} p_{ir} v_r, \sum_{s=1}^{n} q_{js} w_s \right) = \sum_{r=1}^{k} \sum_{s=1}^{n} p_{ir} f(v_r, w_s) q_{js},$$

and this is precisely the (i, j)th-entry of $PT_{BD}(f)Q^T$.   □

In particular, we see that if $f \in \mathrm{Bill}(V \times V)$, where V is a vector space of finite dimension n over a field F, and if B and D are bases of V, then there exists a nonsingular matrix $P \in M_{n \times n}(F)$ satisfying $T_{DD}(f) = PT_{BB}(f)P^T$. In general, matrices A and C in $M_{n \times n}(F)$ are *congruent* if and only if there exists a nonsingular matrix $P \in M_{n \times n}(F)$ satisfying $C = PAP^T$. Congruence is easily checked to be an equivalence relation on $M_{n \times n}(F)$, which joins the relations of equivalence and similarity, that we have already defined. Congruent matrices clearly have the same rank, so that the rank of a matrix of the form $T_{BB}(f)$ depends only on f and not on the choice of basis B. Therefore, we call this the *rank* of the bilinear form f. Thus, for example, the bilinear forms in $\mathrm{Bill}(V \times V)$ of rank 1 are precisely those of the form $(v, w) \mapsto \alpha(v)\beta(w)$, where $\alpha, \beta \in D(V)$.

A matrix congruent to a symmetric matrix is again symmetric. Indeed, if $A \in M_{n \times n}(F)$ is symmetric, then for any nonsingular matrix P we have $(PAP^T)^T = P^{TT}A^T P^T = PAP^T$.

Example The matrix $A = \begin{bmatrix} 1 & -6 & -6 \\ -6 & 40 & 39 \\ -6 & 39 & 39 \end{bmatrix} \in M_{3 \times 3}(\mathbb{R})$ is congruent to I, since

$PAP^T = I$, where $P = \begin{bmatrix} 1 & 0 & 0 \\ 3 & \frac{1}{2} & 0 \\ \sqrt{3} & -\frac{\sqrt{3}}{2} & \frac{2\sqrt{3}}{3} \end{bmatrix}$.

As was the case with inner products, we can define orthogonality with respect to an arbitrary bilinear form. This concept has important applications when we are working over fields other than $\mathbb{R}$ or $\mathbb{C}$, and especially in areas such as algebraic coding theory, where all of the work is done over finite fields. Let V be a vector space over a field F and let $f \in \mathrm{Bill}(V \times V)$. Vectors $v, w \in V$ are f-*orthogonal* if and only if $f(v, w) = 0$. In this case, we will write $v \perp_f w$. (One has to be careful here, it may be true that $v \perp_f w$ but false that $w \perp_f v$; this will not happen, of course, if f is symmetric.) If A is a nonempty subset of V, then we can talk about the *right* f-*orthogonal complement* of A to be the set $A^{\perp_f} = \{w \in V \mid v \perp_f w \text{ for all } v \in A\}$. Complements of this form may behave very differently than complements defined by inner products, as the following example shows.

Example Let $F = \mathrm{GF}(2)$ and let $V = F^4$. Define $f \in \mathrm{Bill}(V \times V)$ by setting

$f(v, w) = v \odot w$. Then $W = \left\{ \begin{bmatrix} 0 \\ 0 \\ 0 \\ 0 \end{bmatrix}, \begin{bmatrix} 1 \\ 1 \\ 1 \\ 1 \end{bmatrix}, \begin{bmatrix} 1 \\ 0 \\ 1 \\ 0 \end{bmatrix}, \begin{bmatrix} 0 \\ 1 \\ 0 \\ 1 \end{bmatrix} \right\}$ is a subspace of V

which satisfies $W^{\perp_f} = W$.

We note that $V^{\perp_f}$ is trivial if and only if for any $0_V \neq w \in V$ there exists a vector $v \in V$ satisfying $f(v, w) \neq 0$. This condition is not a consequence of our definitions, and we must explicitly state it when we need it. It holds, of course, if f is nondegenerate.

Example Let $V = \mathbb{R} \left\{ \begin{bmatrix} 1 \\ 1 \\ 1 \\ 1 \end{bmatrix}, \begin{bmatrix} 1 \\ 1 \\ -1 \\ -1 \end{bmatrix} \right\} \subseteq \mathbb{R}^4$. If $f \in \mathrm{Bill}(V \times V)$ is defined by

$f : \left(\begin{bmatrix} a_1 \\ a_2 \\ a_3 \\ a_4 \end{bmatrix}, \begin{bmatrix} b_1 \\ b_2 \\ b_3 \\ b_4 \end{bmatrix} \right) \mapsto a_1 b_1 + a_2 b_2 + a_3 b_3 - a_4 b_4$, then $\begin{bmatrix} 0 \\ 0 \\ 1 \\ 1 \end{bmatrix} \in V^{\perp_f}$ and, indeed,

$V^{\perp_f} = \mathbb{R} \left\{ \begin{bmatrix} 1 \\ -1 \\ 0 \\ 0 \end{bmatrix}, \begin{bmatrix} 0 \\ 0 \\ 1 \\ 1 \end{bmatrix} \right\}$.

Proposition 20.3 *Let V be a vector space over a field F, and let $f \in \mathrm{Bill}(V \times V)$. If A is a nonempty subset of V then:*
(1) $A^{\perp_f}$ *is a subspace of V;*
(2) $A^{\perp_f} = (FA)^{\perp_f}$;
(3) *If $A \subseteq B$ then $B^{\perp_f} \subseteq A^{\perp_f}$.*
Moreover, if $\{A_i \mid i \in \Omega\}$ is a collection of nonempty subsets of V, then
$(\bigcup_{i \in \Omega} A_i)^{\perp_f} = \bigcap_{i \in \Omega} A_i^{\perp_f}$.

Proof The proof of (1)–(3) is an immediate consequence of the definitions. To prove the last statement, we note that if $w \in V$, then $w \in (\bigcup_{i \in \Omega} A_i)^{\perp_f}$ if and only if for each $i \in \Omega$ and each $v \in A_i$ we have $f(v, w) = 0$. This is true if and only if $w \in A_i^{\perp_f}$ for each $i \in \Omega$, namely if and only if $w \in \bigcap_{i \in \Omega} A_i^{\perp_f}$. $\qquad\square$

Proposition 20.4 *Let V be a vector space finitely generated over a field F and let $f \in \text{Bill}(V \times V)$ satisfy the condition that $V^{\perp_f}$ is trivial. Then each subspace W of V satisfies the following conditions:*
(1) *If $\delta \in D(W)$ there exists a $v \in V$ such that $\delta(w) = f(v, w)$ for all $w \in W$;*
(2) *$\dim(W) + \dim(W^{\perp_f}) = \dim(V)$.*

Proof (1) Every vector $v \in V$ defines a linear functional $\delta_v \in D(V)$ by setting $\delta_v : y \mapsto f(y, v)$. Moreover, the function $v \mapsto \delta_v$ from V to $D(V)$ is a linear transformation, which is a monomorphism as a result of the condition that $V^{\perp_f}$ is trivial. But $\dim(V) = \dim(D(V))$ since V is finitely generated, and hence this is an isomorphism. Now let $\delta \in D(W)$ and let Y be a complement of W in V. Then the function from V to F given by $w + y \mapsto \delta(w)$ belongs to $D(V)$ and so there exists a vector $v \in V$ such that it equals δ_v. In particular, $\delta(w) = f(v, w)$ for all $w \in W$, proving (1).

(2) The function from V to $D(W)$ which assigns to each $v \in V$ the restriction of δ_v to W is a linear transformation which, by (1), is an epimorphism. The kernel of this epimorphism consists of all vectors $v \in V$ satisfying $f(w, v) = 0$ for all $w \in W$, and that is precisely $W^{\perp_f}$. Therefore, by Proposition 6.10, we have (2). $\qquad\square$

In particular, we see from Proposition 20.4 that a necessary and sufficient condition for us to have $V = W \oplus W^{\perp_f}$ is that W and $W^{\perp_f}$ be disjoint.

Proposition 20.5 *Let V and W be vector spaces finitely generated over a field F and let $f \in \text{Bill}(V \times W)$ be a bilinear form which is not the 0-function. Then there exist bases $\{v_1, \dots, v_k\}$ and $\{w_1, \dots, w_n\}$ of V and W, respectively, and there exists a positive integer $1 \le t \le \min\{k, n\}$ such that*

$$f(v_i, w_j) = \begin{cases} 1 & \text{if } i = j \le t, \\ 0 & \text{otherwise.} \end{cases}$$

Proof Since f is not the 0-function, there exist vectors $v_1 \in V$ and $y_1 \in W$ such that $f(v_1, y_1) \ne 0$. Therefore, if we set $w_1 = f(v_1, y_1)^{-1} y_1$, we have $f(v_1, w_1) = 1$. Let $V_1 = Fv_1$ and $W_1 = Fw_1$. If we set $W_2 = \{w \in W \mid f(v_1, w) = 0\}$, then $W_1 \cap W_2 = \{0_W\}$ since $cw_1 \notin W_2$ for all $0 \ne c \in F$. We claim that $W = W_1 \oplus W_2$. Indeed, if

$w \in W$ and if $c = f(v_1, w)$ then we see that

$$f(v_1, w - cw_1) = f(v_1, w) - cf(v_1, w_1) = c - c = 0$$

and so $w - cw_1 \in W_2$, which proves the claim. In a similar way, we have $V = V_1 \oplus V_2$, where $V_2 = \{v \in V \mid f(v, w_1) = 0\}$. Thus we see that $f(v, w) = 0$ whenever $(v, w) \in [V_1 \times W_2] \cup [V_2 \times W_1]$.

By passing to the oppose form if necessary, we can assume without loss of generality that $k \leq n$. If $k = 1$, we choose $\{v_1\}$ as a basis for V and $\{w_1, \ldots, w_n\}$ as a basis for W, where $\{w_2, \ldots, w_n\}$ is an arbitrary basis for W_2. This proves the proposition, with $t = 1$. Now assume that $k > 1$ (which implies $n > 1$) and that the proposition has been proven whenever $\dim(V) < k$. In particular, we will look at the restriction of f to $V_2 \times W_2$. By the induction hypothesis, there exist bases $\{v_2, \ldots, v_k\}$ of V_2 and $\{w_2, \ldots, w_n\}$ of W_2 such that

$$f(v_i, w_j) = \begin{cases} 1 & \text{if } 2 \leq i = j \leq t, \\ 0 & \text{otherwise.} \end{cases}$$

Then $\{v_1, \ldots, v_k\}$ and $\{w_1, \ldots, w_n\}$ are the bases we want. $\square$

We see that if V and W are vector spaces finitely generated over a field F and if $f \in \text{Bill}(V \times W)$, then Proposition 20.5 says that there exist bases of V and W with respect to which f is represented by a matrix of the form $\begin{bmatrix} I & O \\ O & O \end{bmatrix}$.

We will be particularly interested in symmetric bilinear forms. As an immediate consequence of the definition, we see that if V is a vector space finitely generated over a field F and if B is a given basis for V, then a bilinear form $f \in \text{Bill}(V \times V)$ is symmetric if and only if the matrix $T_{BB}(f)$ is symmetric. Moreover, every symmetric matrix is $T_{BB}(f)$ for some symmetric bilinear form $f \in \text{Bill}(V \times V)$.

Example Let B be the canonical basis of $\mathbb{R}^3$ and let $A = \begin{bmatrix} 1 & -5 & 3 \\ -5 & 1 & 7 \\ 3 & 7 & 4 \end{bmatrix}$. Then $A = T_{BB}(f)$, where $f \in \text{Bill}(\mathbb{R}^3 \times \mathbb{R}^3)$ is defined by

$$f\left(\begin{bmatrix} a_1 \\ a_2 \\ a_3 \end{bmatrix}, \begin{bmatrix} b_1 \\ b_2 \\ b_3 \end{bmatrix}\right) = a_1 b_1 + a_2 b_2 - 5(a_1 b_2 + a_2 b_1)$$

$$+ 3(a_1 b_3 + a_3 b_1) + 7(a_2 b_3 + a_3 b_2) + 4a_3 b_3.$$

Proposition 20.6 *Let F be a set of characteristic other than 2 and let V be a vector space finitely-generated over F. Let $f \in \text{Bill}(V \times V)$ be symmetric. Then there exists a basis $B = \{v_1, \ldots, v_n\}$ of V such that $T_{BB}(f)$ is a diagonal matrix.*

Proof The proposition is trivially true if f is the 0-function, and so we can assume that is not the case. We will proceed by induction on $n = \dim(V)$. For $n = 1$, the result is again immediate, and so we can assume that $n > 1$ and that the result has been established for all spaces having dimension less than n. We first claim is that there exists a vector $v \in V$ satisfying $f(v, v) \neq 0$. Indeed, assume that this is not the case. Then if v and w are arbitrary vectors in V we have

$$0 = f(v + w, v + w) = f(v, v) + 2f(v, w) + f(w, w) = 2f(v, w)$$

and since the characteristic of F is not 2, this implies that $f(v, w) = 0$, contradicting our assumption that f is not the 0-function. Hence we can select a vector $v_1 \in V$ satisfying $f(v_1, v_1) \neq 0$.

Let $V_1 = Fv_1$ and let $V_2 = V_1^{\perp_f}$. From the definition of V_1 it is clear that V_1 and V_2 are disjoint, and from Proposition 20.3 it follows that $V = V_1 \oplus V_2$. In particular, $\dim(V_2) = n - 1$ and so, by the induction hypothesis, there exists a basis $C = \{v_2, \ldots, v_n\}$ of V_2, such that, if f_2 is the restriction of f to V_2, then $T_{CC}(f_2)$ is a diagonal matrix. Since $f(v_1, v_i) = 0$ for all $2 \leq i \leq n$, it follows that $B = \{v_1, \ldots, v_n\}$ does indeed give us the desired result. □

Thus we see that every symmetric matrix over a field of characteristic other than 2 is congruent to a diagonal matrix.

Proposition 20.7 *Let V be a vector space finitely-generated over $\mathbb{C}$ and let $f \in \mathrm{Bill}(V \times V)$ be a symmetric bilinear form of rank r. Then there exists a basis $B = \{v_1, \ldots, v_n\}$ of V satisfying the following conditions:*

(1) $T_{BB}(f)$ is a diagonal matrix;

(2) $f(v_i, v_i) = \begin{cases} 1 & \text{if } 1 \leq i \leq r, \\ 0 & \text{otherwise.} \end{cases}$

Proof By Proposition 20.6, we know that there is a basis $B = \{v_1, \ldots, v_n\}$ of V satisfying the condition that $T_{BB}(f)$ is a diagonal matrix. This matrix is of rank r and so, renumbering the basis elements if necessary, we can assume that $f(v_i, v_i) \neq 0$ when and only when $1 \leq i \leq r$. For each $1 \leq i \leq r$, define $c_i = f(v_i, v_i)^{-1/2} \in \mathbb{C}$, and replace v_i by $c_i v_i$ to get a basis satisfying (2) as well. □

Let V be a vector space finitely-generated over a field F of characteristic other than 2 and let $f \in \mathrm{Bill}(V \times V)$ be a bilinear form. The function $q : V \to F$ defined by $q : v \mapsto f(v, v)$ is called the *quadratic form* defined by f. Note that if $a \in F$ and $v \in V$ then $q(av) = f(av, av) = a^2 f(v, v) = a^2 q(v)$. Moreover, if $f \in \mathrm{Bill}(V \times V)$ and if $g \in \mathrm{Bill}(V \times V)$ is the symmetric bilinear form $g : (v, w) \mapsto \frac{1}{2}[f(v, w) + f(w, v)]$ then the quadratic forms defined by f and g are the same. Therefore, without loss of generality, we will always assume that all quadratic forms over such fields are defined by symmetric bilinear forms. We further see that different symmetric bilinear forms define different quadratic forms, since, for any

$v, w \in V$, we have $f(v, w) = \frac{1}{2}[q(v + w) - q(v) - q(w)]$. The classification of quadratic forms is of great importance in analytic geometry and in number theory.

With kind permission of the Archives of the Mathematisches Forschungsinstitut Oberwolfach (Witt).

The theory of quadratic forms over $\mathbb{R}$ was developed by Gauss and his student Eisenstein, and the need to study such forms was one of the factors which led to the development of determinant theory. Their work was extended to quadratic forms over $\mathbb{C}$ by the nineteenth-century British mathematician **Henry Smith**. The fundamental development in the theory of symmetric bilinear forms on vector spaces over fields of characteristic other than 2 is due to the twentieth-century German mathematician **Ernst Witt**.

Let V be a vector space over $\mathbb{R}$. A quadratic form $q : V \to \mathbb{R}$ is *positive* if and only if $q(v) > 0$ for all $0_V \neq v \in V$. If $q : V \to \mathbb{R}$ is a positive quadratic form defined by a symmetric bilinear form $f \in \text{Bill}(V \times V)$, then f must be nondegenerate. Indeed, if $0_V \neq v \in V$ then $f(v, v) \neq 0$.

Example If V is the vector space of all polynomial functions from $\mathbb{R}$ to itself, then we have a symmetric bilinear form from $V \times V$ to $\mathbb{R}$ defined by $(f, g) \mapsto \int_0^1 f(t)g(t) \, dt$, which in turn defines the positive quadratic form $f \mapsto \int_0^1 f(t)^2 \, dt$.

Example Let $V = \mathbb{R}^4$ and let $f \in \text{Bill}(V \times V)$ be the symmetric bilinear form

$$\left(\begin{bmatrix} a_1 \\ a_2 \\ a_3 \\ a_4 \end{bmatrix}, \begin{bmatrix} b_1 \\ b_2 \\ b_3 \\ b_4 \end{bmatrix} \right) \mapsto a_1b_1 + a_2b_2 + a_3b_3 - a_4b_4, \text{ which lies at the center of}$$

Minkowski's mathematical formulation of Einstein's relativity theory. The quadratic

form defined by this bilinear form is $\begin{bmatrix} a_1 \\ a_2 \\ a_3 \\ a_4 \end{bmatrix} \mapsto a_1^2 + a_2^2 + a_3^2 - a_4^2$. A similar symmet-

ric bilinear form is the *Lorentz form* $\left(\begin{bmatrix} a_1 \\ a_2 \\ a_3 \\ a_4 \end{bmatrix}, \begin{bmatrix} b_1 \\ b_2 \\ b_3 \\ b_4 \end{bmatrix} \right) \mapsto a_1b_1 + a_2b_2 + a_3b_3 -$

$c^2 a_4 b_4$, where c is the speed of light. The quadratic form defined by this bilinear

form is $\begin{bmatrix} a_1 \\ a_2 \\ a_3 \\ a_4 \end{bmatrix} \mapsto a_1^2 + a_2^2 + a_3^2 - c^2 a_4^2$.

With kind permission of the Museum Boerhaave Leiden.

The Dutch physicist **Hendrick Antoon Lorentz**, the first to conceive of the notion of the electron, won a Nobel prize in 1902. His work formed a basis for much of Einstein's theory.

A more general result, also based on the work of Lorentz and Minkowski, gives a fascinating "reversal" of the Cauchy–Schwarz–Bunyakovski inequality. Let n be a positive integer and consider the subset (not subspace) U of $\mathbb{R}^{n+1}$ consisting of all vectors of the form $\begin{bmatrix} a \\ v \end{bmatrix}$, where a is a nonnegative real number and $v \in \mathbb{R}^n$ satisfies $\|v\| \leq a$. For $u = \begin{bmatrix} a \\ v \end{bmatrix}$ and $y = \begin{bmatrix} b \\ w \end{bmatrix}$ in U, let us define $u \boxdot y$ to be $ab - v \cdot w$. By our assumption on U, we note at $u \boxdot u \geq 0$ for every $u \in U$. Then one can show that $u \boxdot y \geq [\sqrt{u \boxdot u}][\sqrt{y \boxdot y}]$. This inequality is often known as the *lightcone inequality* because of its applications in physics.

Example If V is an inner product space over $\mathbb{R}$, then we have already noted that the function $f : (v, w) \mapsto \langle v, w \rangle$ is a symmetric bilinear form. The quadratic form defined by f is given by $v \mapsto \|v\|^2$. This quadratic form is surely positive. The converse is also true. If $f \in \mathrm{Bill}(V \times V)$ is a symmetric bilinear form defining a positive quadratic form, then f is an inner product on V, in the sense of Chap. 15.

By Proposition 20.7, we see that if V is a vector space finitely generated over $\mathbb{C}$ and if $f \in \mathrm{Bill}(V \times V)$ is symmetric and has rank r, we can find a basis $\{v_1, \ldots, v_n\}$ of V such that the quadratic form q defined by f is given by $q : \sum_{i=1}^n a_i v_i \mapsto \sum_{i=1}^r a_i^2$.

Example Let F be either $\mathbb{R}$ or $\mathbb{C}$. Let n be a positive integer and let $A \in \mathcal{M}_{n \times n}(F)$ be symmetric. Let $f \in \mathrm{Bill}(F^n \times F^n)$ be the symmetric bilinear form given by $f : (v, w) = v^T A w$, and let q be the quadratic form defined by f. The set $\{q(v) \mid \|v\| = 1\}$ (here the norm is the one defined by the dot product on F^n) is called the *numerical range* of the matrix A. In the case $F = \mathbb{C}$, this is always a bounded convex subset which contains all of the eigenvalues of A. For the special case $n = 2$, this set is an ellipse with its foci at the eigenvalues of A, assuming that they are distinct, or a circle with center at the sole eigenvalue of A, assuming that A has only one eigenvalue of multiplicity 2. For $n > 2$, the characterization of the numerical range is much more complicated.

Proposition 20.8 *Let n be a positive integer and let $A \in \mathcal{M}_{n \times n}(\mathbb{R})$ be symmetric. Let $f \in \text{Bill}(\mathbb{R}^n \times \mathbb{R}^n)$ be the symmetric bilinear form given by $f : (v, w) \mapsto v^T A w$, and let q be the quadratic form defined by f. Let $c_1 \geq c_2 \geq \cdots \geq c_n$ be the eigenvalues of A. Then the numerical range of A lies in the closed interval $[c_n, c_1]$. Moreover, both endpoints of this interval belong to the numerical range of A.*

Proof By Proposition 17.7, we know that there exists an orthonormal basis $B = \{v_1, \dots, v_n\}$ of V consisting of eigenvectors of A. Moreover, if $v \in V$ then $v = \sum_{i=1}^{n} \langle v, v_i \rangle v_i$ by Proposition 17.9, and so $1 = \|v\|^2 = \langle v, v \rangle = \sum_{i=1}^{n} \langle v, v_i \rangle^2$. We also see that $Av = \sum_{i=1}^{n} \langle v, v_i \rangle A(v_i) = \sum_{i=1}^{n} c_i \langle v, v_i \rangle v_i$. Thus $v^T A v = \langle v, Av \rangle = \sum_{i=1}^{n} c_i \langle v, v_i \rangle^2$. But $c_1 = c_1 (\sum_{i=1}^{n} \langle v, v_i \rangle^2) \geq \sum_{i=1}^{n} c_i \langle v, v_i \rangle^2 \geq c_n (\sum_{i=1}^{n} \langle v, v_i \rangle^2) = c_n$. Therefore, the numerical range of A lies in the closed interval $[c_n, c_1]$.

If v is a normal eigenvector of A corresponding to c_n, then $v^T A v = \langle v, Av \rangle = \langle v, c_n v \rangle = c_n \langle v, v \rangle = c_n$, and similarly for the case of an eigenvector of A satisfying $\|v\| = 1$ and corresponding to c_1. $\qquad \square$

In order to see the geometric significance of quadratic forms, let us recall that a *general quadratic equation* in three unknowns over $\mathbb{R}$ is one of the form

$$\left(a_{11} X_1^2 + a_{22} X_2^2 + a_{33} X_3^2\right) + 2(a_{12} X_1 X_2 + a_{13} X_1 X_3 + a_{23} X_2 X_3)$$
$$+ b_1 X_1 + b_2 X_2 + b_3 X_3 + c = 0$$

in which not all of the a_{ij} are equal to 0. Such an equation can be written in the form $f(v, v) + w \cdot v + c = 0$, where $f \in \text{Bill}(\mathbb{R}^3, \mathbb{R}^3)$ is the symmetric bilinear form defined with respect to the canonical basis by the matrix $A = \begin{bmatrix} a_{11} & a_{12} & a_{13} \\ a_{12} & a_{22} & a_{23} \\ a_{13} & a_{23} & a_{22} \end{bmatrix}$, where $w = \begin{bmatrix} b_1 \\ b_2 \\ b_3 \end{bmatrix}$, and where $v = \begin{bmatrix} X_1 \\ X_2 \\ X_3 \end{bmatrix}$. The graph of such an equation is a *quadratic surface*. The various quadratic surfaces in $\mathbb{R}^3$ can then be classified by considering congruence classes of the matrices A, a task very important in analytic geometry.

We will now return to the general case of bilinear transformations. Let F be a field, let V and W be vector spaces over F, and let $G = F^{(V \times W)}$. Then G is a subspace of $F^{V \times W}$ having a basis $\{g_{v,w} \mid (v, w) \in V \times W\}$, where

$$g_{v,w} : (v', w') \mapsto \begin{cases} 1 & \text{if } (v', w') = (v, w), \\ 0 & \text{otherwise.} \end{cases}$$

Let H be the subspace of G generated by all functions of the form

$$g_{v_1 + v_2, w} - g_{v_1, w} - g_{v_2, w}, \quad g_{v, w_1 + w_2} - g_{v, w_1} - g_{v, w_2}, \quad g_{av, w} - a g_{v, w},$$

$$\text{or} \quad g_{v, aw} - a g_{v, w}$$

for all $v, v_1, v_2 \in V$, $w, w_1, w_2 \in W$, and $a \in F$. Let us pick a complement of H in G, and call it $V \otimes W$. By Proposition 7.8, we know that $V \otimes W$ is unique up to isomorphism. Let α be the projection of G with image $V \otimes W$ coming from the decomposition $G = H \oplus (V \otimes W)$ and, for all $v \in V$ and $w \in W$, denote $\alpha(g_{v,w})$ by $v \otimes w$. Then $B = \{v \otimes w \mid (v, w) \in V \times W\}$ is a generating set for $V \otimes W$. It is important to emphasize that the elements of $V \otimes W$ are linear combinations of elements of B. In quantum physics, elements of $V \otimes W \smallsetminus B$, for suitable spaces V and W, are known as *entangled tensors* and these have important physical interpretations. Elements of B are known as *simple tensors*.

If $v_1, v_2 \in V$ and $w \in W$, then

$$[v_1 + v_2] \otimes w - (v_1 \otimes w) - (v_2 \otimes w) = \alpha(g_{v_1+v_2,w} - g_{v_1,w} - g_{v_2,w}) = 0_G$$

and so $[v_1 + v_2] \otimes w = (v_1 \otimes w) + (v_2 \otimes w)$. Similarly, if $v \in V$ and $w_1, w_2 \in W$ then $v \otimes [w_1 + w_2] = v \otimes w_1 + v \otimes w_2$. We also see that if $v \in V$, $w \in W$ and $c \in F$, then $cv \otimes w = c(v \otimes w) = v \otimes cw$. The vector space $V \otimes W$ is called the *tensor product* of V and W.

With kind permission of the Archives of the Mathematisches Forschungsinstitut Oberwolfach (Chevalley).

There are many equivalent definitions of the tensor product. The definition given here is due to the twentieth-century French mathematician **Claude Chevalley**. The notion of a tensor was first introduced in differential calculus by the nineteenth-century Italian mathematicians **Gregorio Ricci-Curbastro** and **Tullio Levi-Civita** and became a central tool in relativity theory.

From the definition of the tensor product, we see that the function t_{VW} from $V \times W$ to $V \otimes W$ given by $(v, w) \mapsto v \otimes w$ is a bilinear transformation. This transformation has a very special significance, due to the following theorem, which allows us to move from bilinear transformations to linear transformations.

Proposition 20.9 *Let V, W, and Y be vector spaces over a field F. For each bilinear transformation $f \in \mathrm{Bill}(V \times W, Y)$ there exists a unique linear transformation $\alpha \in \mathrm{Hom}(V \otimes W, Y)$ satisfying $f = \alpha t_{VW}$.*

Proof Given $f \in \mathrm{Bill}(V \times W, Y)$, there exists a linear transformation $\beta \in \mathrm{Hom}(G, Y)$ defined on the elements of a basis of the space G defined above, given by the condition that $\beta : g_{v,w} \mapsto f(v, w)$. Since f is a bilinear transformation, $H \subseteq \ker(\beta)$ and so we can define the linear transformation $\alpha \in \mathrm{Hom}(V \otimes W, Y)$ by setting

$\alpha : \sum_{i=1}^{n} a_i [v_i \otimes w_i] \mapsto \sum_{i=1}^{n} a_i f(v_i, w_i)$. This function is well-defined since if $\sum_{i=1}^{n} a_i [v_i \otimes w_i] = \sum_{i=1}^{n} b_i [v_i \otimes w_i]$ in $V \otimes W$ then $\sum_{i=1}^{n} (a_i - b_i) g_{v_i, w_i} \in H \subseteq \ker(\beta)$. Therefore, we see that $\alpha(\sum_{i=1}^{n} a_i [v_i \otimes w_i]) = \alpha(\sum_{i=1}^{n} b_i [v_i \otimes w_i])$. Clearly, α is a linear transformation and satisfies $f = \alpha t_{VW}$.

We are left to prove uniqueness. Suppose that $\gamma \in \mathrm{Hom}(V \otimes W, Y)$ satisfies $f = \gamma t_{VW}$. In particular, $\alpha(v \otimes w) = \gamma(v \otimes w)$ for all $(v, w) \in V \times W$. That is to say, α and γ act identically on a generating set for $V \otimes W$ and so, in particular, on a basis for $V \otimes W$ contained in this generating set. Therefore, by Proposition 6.2, it follows that $\alpha = \gamma$. $\square$

The following proposition is very important, and is often used as a basis for the definition of the tensor product.

Proposition 20.10 *If V, W, and Y are vector spaces over a field F, then the vector spaces $\mathrm{Hom}(V \otimes W, Y)$ and $\mathrm{Hom}(V, \mathrm{Hom}(W, Y))$ are isomorphic.*

Proof The function $\mathrm{Hom}(V \otimes W, Y) \to \mathrm{Bill}(V \times W, Y)$ defined by $\beta \mapsto \beta t_{VW}$ is clearly a linear transformation, and from Proposition 20.8 it follows that this is an isomorphism. Therefore, the result follows from Proposition 20.1. $\square$

Example Let V and W be vector spaces over a field F and let $\delta_1 \in D(V)$ and $\delta_2 \in D(W)$ be linear functionals. Then there exists a bilinear form in $\mathrm{Bill}(V \times W)$ defined by $(v, w) \mapsto \delta_1(v)\delta_2(w)$. From Proposition 20.9, it follows that there exists a linear functional $\delta_1 \otimes \delta_2 \in D(V \otimes W)$ satisfying $\delta_1 \otimes \delta_2 : \sum_{i=1}^{n} a_i [v_i \otimes w_i] \mapsto \sum_{i=1}^{n} a_i \delta_1(v_i)\delta_2(w_i)$.

Example More generally, let V and W be vector spaces over a field F, let α be an endomorphism of V, and let β be an endomorphism of W. The function from $V \times W$ to $V \otimes W$ defined by

$$(v, w) \mapsto \alpha(v) \otimes \beta(w)$$

is a bilinear transformation and so defines an endomorphism $\alpha \otimes \beta$ of $V \otimes W$ satisfying $\alpha \otimes \beta : \sum_{i=1}^{n} a_i [v_i \otimes w_i] \mapsto \sum_{i=1}^{n} a_i [\alpha(v_i) \otimes \beta(w_i)]$.

By Proposition 5.13, we know that if $V \oplus W$ is a vector space finitely-generated over a field F, then $\dim(V \oplus W) = \dim(V) + \dim(W)$. We now prove the "multiplicative" analog of this assertion for tensor products.

Proposition 20.11 *Let V and W be vector spaces finitely generated over a field F. Then $V \otimes W$ is also finitely generated, and $\dim(V \otimes W) = \dim(V)\dim(W)$.*

Proof Let us choose bases $\{v_1, \ldots, v_k\}$ of V and $\{w_1, \ldots, w_n\}$ of W. Then for any $v = \sum_{i=1}^{k} a_i v_i \in V$ and $w = \sum_{j=1}^{n} b_j w_j \in W$, we see that $v \otimes w = \sum_{i=1}^{k} \sum_{j=1}^{n} a_i b_j (v_i \otimes w_j)$. Thus we see that $\{v_i \otimes w_j \mid 1 \le i \le k \text{ and } 1 \le j \le n\}$ is a generating set for $V \otimes W$, showing that $V \otimes W$ is finitely-generated. Moreover, by Proposition 20.10 and Proposition 14.8, we see that the dimension of $V \otimes W$ is equal to the dimension of $D(V \otimes W)$ and hence to the dimension of $\mathrm{Hom}(V, D(W))$, and this is equal to the dimension of $\mathrm{Hom}(V, W)$, which is precisely $\dim(V) \dim(W)$. $\square$

In particular, we see that in the context of Proposition 20.10, the set $\{v_i \otimes w_j \mid 1 \le i \le k \text{ and } 1 \le j \le n\}$ is in fact a basis of $V \otimes W$.

Example Let F be a field and let k and n be positive integers. Then, by Proposition 20.11, we know that $\dim(F^k \otimes F^n) = kn = \dim(\mathcal{M}_{k \times n}(F))$, and so the vector spaces $F^k \otimes F^n$ and $\mathcal{M}_{k \times n}(F)$ are isomorphic. Indeed, if we choose bases $\{v_1, \ldots, v_k\}$ of V and $\{w_1, \ldots, w_n\}$ of W, then the function $v_i \otimes w_j \mapsto v_i \wedge w_j$ extends to an isomorphism between these two spaces.

Example Let F be a field, let n be a positive integer, and let V be a vector space finitely generated over F and having a basis $\{v_1, \ldots, v_k\}$. The dimension of the vector space $\mathcal{M}_{n \times n}(V)$ over F is $n^2 k$. Consider the bilinear transformation $f : \mathcal{M}_{n \times n}(F) \times V \to \mathcal{M}_{n \times n}(V)$ defined by $([a_{ij}], v) \mapsto [a_{ij}v]$. By Proposition 20.9, we know that this bilinear transformation defines a linear transformation $\alpha : \mathcal{M}_{n \times n}(F) \otimes V \to \mathcal{M}_{n \times n}(V)$ and it is clear that this is an epimorphism. But, by Proposition 20.10, we see that the dimension of $\mathcal{M}_{n \times n}(F) \otimes V$ is also equal to $n^2 k$ and so α must be an isomorphism.

Example Let F be a field and let k, n, s, and t be positive integers. Let $f : \mathcal{M}_{k \times n}(F) \times \mathcal{M}_{s \times t}(F) \to \mathcal{M}_{ks \times nt}(F)$ be the function defined by

$$f : (A, B) \mapsto \begin{bmatrix} a_{11}B & \cdots & a_{1n}B \\ \vdots & \ddots & \vdots \\ a_{k1}B & \cdots & a_{kn}B \end{bmatrix}.$$

This is a bilinear transformation of vector spaces over F and so, by Proposition 20.8, it defines a linear transformation $\alpha : \mathcal{M}_{k \times n}(F) \otimes \mathcal{M}_{s \times t}(F) \to \mathcal{M}_{ks \times nt}(F)$ which, again, can be shown to be an isomorphism. In the literature, it is usual to write $A \otimes B$ instead of $f(A, B)$. This matrix is called the *Kronecker product* of the matrices A and B. Kronecker products are very important in matrix theory and its applications. It is easy to see that for all such matrices A and B we have $(A \otimes B)^T = A^T \otimes B^T$. Moreover, if $k = n$ and $s = t$ and if A and B are nonsingular, then $A \otimes B$ is nonsingular, and $(A \otimes B)^{-1} = A^{-1} \otimes B^{-1}$. We also note that if A and B are symmetric then so is $A \otimes B$. Furthermore, Cholesky or QR-factorizations of $A \otimes B$ come immediately from the corresponding factorizations of A and B.

As an example of the use of Kronecker products, we note the following result, established in the 1970s by the American mathematicians Michael Gauger and Christopher Byrnes: Let F be a field, let n be a positive integer, and let $A, B \in \mathcal{M}_{n \times n}(F)$. Let I be the multiplicative identity of $\mathcal{M}_{n \times n}(F)$. Then the matrices A and B are similar if and only if they have the same characteristic polynomial and the $n^2 \times n^2$ matrices $A \otimes I - I \otimes A$, $B \otimes I - I \otimes B$, and $A \otimes I - I \otimes B$ all have the same rank.

Because of the utility of Kronecker products, one can raise the following problem: Given positive integers k, n, s, and t, and given $C \in \mathcal{M}_{ks \times nt}(\mathbb{R})$, find matrices $A \in \mathcal{M}_{k \times n}(\mathbb{R})$ and $B \in \mathcal{M}_{s \times t}(\mathbb{R})$ such that $\|C - A \otimes B\|$ is minimal. Several algorithms have been developed for finding a solution to this problem, the first by the American computer scientists Charles Van Loan and Nikos Pitsianis.

Let V, V', W, and W' be vector spaces over a field F. If $\alpha \in \operatorname{Hom}(V, V')$ and $\beta \in \operatorname{Hom}(W, W')$ then we have a bilinear transformation $V \times W \to V' \otimes W'$ defined by $(v, w) \mapsto \alpha(v) \otimes \beta(w)$ and so, by Proposition 20.8, there exists a linear transformation from $V \otimes W$ to $V' \otimes W'$ satisfying $v \otimes w \mapsto \alpha(v) \otimes \beta(w)$. We will denote this linear transformation by $\alpha \otimes \beta$.

Proposition 20.12 *Let V, V', W, and W' be vector spaces finitely generated over a field F. Any element of the space $\operatorname{Hom}(V \otimes W, V' \otimes W')$ is of the form $\sum_{i=1}^{n} \alpha_i \otimes \beta_i$, where $\alpha_i \in \operatorname{Hom}(V, V')$ and $\beta_i \in \operatorname{Hom}(W, W')$ for each $1 \leq i \leq n$.*

Proof The function $(\alpha, \beta) \mapsto \alpha \otimes \beta$ from $\operatorname{Hom}(V, V') \times \operatorname{Hom}(W, W')$ to $\operatorname{Hom}(V \otimes W, V' \otimes W')$ is bilinear and so defines a linear transformation $\varphi : \operatorname{Hom}(V, V') \otimes \operatorname{Hom}(W, W') \to \operatorname{Hom}(V \otimes W, V' \otimes W')$. We are done if we can show that φ is an isomorphism. By Propositions 8.1 and 20.11, we know that

$$
\begin{aligned}
\dim\big(\operatorname{Hom}(V, V') \otimes \operatorname{Hom}(W, W')\big) &= \dim\big(\operatorname{Hom}(V, V')\big) \dim\big(\operatorname{Hom}(W, W')\big) \\
&= \dim(V) \dim(V') \dim(W) \dim(W') \\
&= \dim(V \otimes W) \dim(V' \otimes W') \\
&= \dim\big(\operatorname{Hom}(V \otimes W, V' \otimes W')\big),
\end{aligned}
$$

and so it suffices to prove that φ is a monomorphism.

Indeed, assume that $\sum_{i=1}^{n} \alpha_i \otimes \beta_i \in \ker(\varphi)$, where the set $\{\beta_1, \ldots, \beta_n\}$ is linearly independent, and where none of the α_i is the 0-function. Then $\sum_{i=1}^{n} \alpha_i(v) \otimes \beta_i(w) = 0_{V' \otimes W'}$ for all $v \in V$ and all $w \in W$. Pick $v \in V$ satisfying $\alpha_1(v) \neq 0_{V'}$. By renumbering if necessary, we can assume that $\{\alpha_1(v), \ldots, \alpha_k(v)\}$ is a maximal linearly-independent subset of $\{\alpha_1(v), \ldots, \alpha_n(v)\}$. Therefore, for each $k < h \leq n$ there exists a scalar b_{hj}, not all of them being equal to 0, such that $\alpha_h(v) = \sum_{j=1}^{k} b_{hj} \alpha_j(v)$ and so

$$0_{V'\otimes W'} = \sum_{i=1}^{k} \alpha_i(v) \otimes \beta_i(w) + \sum_{h=k+1}^{m} \left(\sum_{j=1}^{k} b_{hj}\alpha_j(v) \right) \otimes \beta_h(w)$$

$$= \sum_{i=1}^{k} \alpha_i(v) \otimes \beta_i(w) + \sum_{j=1}^{k} \alpha_j(v) \otimes \left(\sum_{h=k+1}^{m} b_{hj}\beta_h(w) \right)$$

$$= \sum_{i=1}^{k} \alpha_i(v) \otimes \left(\beta_i(w) + \sum_{h=k+1}^{m} b_{hj}\beta_h(w) \right).$$

Since the set $\{\alpha_1(v), \ldots, \alpha_k(v)\}$ is linearly independent, we must have $\beta_i(w) + \sum_{h=k+1}^{m} b_{hj}\beta_h(w) = 0_{W'}$ for all $1 \le i \le k$ and all $w \in W$. Hence $\beta_i + \sum_{h=k+1}^{m} b_{hj}\beta_h$ is the 0-function for all $1 \le i \le k$, contradicting the assumption that the set $\{\beta_1, \ldots, \beta_n\}$ is linearly independent. We therefore conclude that $\ker(\varphi)$ is trivial, which is what we needed to prove. $\qquad\square$

Proposition 20.13 *If U, V, and W are vector spaces over a field F, then*

$$U \otimes (V \otimes W) \cong (U \otimes V) \otimes W.$$

Proof The bilinear transformation $U \times (V \otimes W) \to (U \otimes V) \otimes W$ defined by $(u, v \otimes w) \mapsto (u \otimes v) \otimes w$ induces a linear transformation $\alpha : U \otimes (V \otimes W) \to (U \otimes V) \otimes W$ which satisfies $u \otimes (v \otimes w) \mapsto (u \otimes v) \otimes w$. Similarly, we have a linear transformation $\beta : (U \otimes V) \otimes W \to U \otimes (V \otimes W)$ which satisfies $(u \otimes v) \otimes w \mapsto u \otimes (v \otimes w)$. Since $\alpha\beta$ and $\beta\alpha$ are clearly the respective identity maps, we see that α must be the isomorphism we seek. $\qquad\square$

Proposition 20.14 *If V and W are vector spaces over a field F, then $V \otimes W \cong W \otimes V$.*

Proof The bilinear transformation $V \times W \to W \otimes V$ defined by $(v, w) \mapsto w \otimes v$ induces a linear transformation α from $V \otimes W$ to $W \otimes V$ satisfying $\alpha : v \otimes w \mapsto w \otimes v$. Similarly, there exists a linear transformation $\beta : W \otimes V \to V \otimes W$ satisfying $\beta : w \otimes v \mapsto v \otimes w$. Since $\alpha\beta$ and $\beta\alpha$ are clearly the respective identity maps, we see that α must be the isomorphism we seek. $\qquad\square$

Finally, let us briefly mention two algebras built on the notion of the tensor product. The study of these algebras is beyond the scope of this book. However, the reader should be aware of them and will find it fruitful to explore them further. In what ensues, V is an arbitrary vector space over a field F.

(I) For each nonnegative integer k, we define the vector space $V^{\otimes k}$ over F by setting $V^{\otimes 0} = V$ and $V^{\otimes k} = V^{\otimes(k-1)} \otimes V$ if $k > 0$. Let $T(V) = \coprod_{k=0}^{\infty} V^{\otimes k}$. We can define a product $\bullet$ on $T(V)$ by setting $(v_1 \otimes \cdots \otimes v_k) \bullet (v_{k+1} \otimes \cdots \otimes v_m) = v_1 \otimes \cdots \otimes v_{k+m}$ for all $v_1, \ldots, v_{k+m} \in V$ and extend linearly. This is an F-algebra, known as the *tensor algebra* of V over F. The tensor algebra has several important properties, one of which is that if K is any algebra over F then any linear transformation $\alpha : V \to K$ can be uniquely extended to a homomorphism of F-algebras from $T(V)$ to K. Moreover, if W is a vector space over F then any linear transformation $\alpha : V \to W$ can be uniquely extended to a homomorphism of F-algebras from $T(V)$ to $T(W)$. (In the language of category theory, this says that $T(\cdot)$ is a functor from the category of vector spaces over F to the category of F-algebras.)

(II) Let Y be the subspace of $V \otimes V$ generated by $\{v \otimes v \mid v \in V\}$. Then a complement of Y in $V \otimes V$ is called an *exterior square* of V and is denoted by $V \wedge V$. This space is unique up to isomorphism. If α is the projection of $V \otimes V$ with image $V \wedge V$ and kernel Y, denote $\alpha(v \otimes w)$ by $v \wedge w$. Since $(v+w) \otimes (v+w) = v \otimes v + v \otimes w + w \otimes v + w \otimes w$ for all $v, w \in V$, we see that $v \wedge w = -w \wedge v$ for all $v, w \in V$. Therefore, if V is finitely-generated over F with basis $\{v_1, \ldots, v_n\}$, we see that $\{v_i \wedge v_j \mid 1 \le i < j \le n\}$ is a basis for $V \wedge V$, and hence $\dim(V \wedge V) = \binom{n}{2} = \frac{1}{2}n(n-1)$. This construction can be iterated to more than two factors. If $k > 0$ is an integer, we can consider the subspace Y of $V^{\otimes k}$ generated by all expressions of the form $v_1 \otimes \cdots \otimes v_k$ in which $v_i = v_j$ for some $i \ne j$. A complement of Y is denoted by $\bigwedge^k V$ and is called the *kth exterior power* of V. If V has finite dimension n, then $\dim(\bigwedge^k V) = \binom{n}{k}$. In particular, we note that $\bigwedge^k V$ is trivial when $k > n$. The subspace $\bigwedge(V) = \coprod_{k=0}^{n}(\bigwedge^k V)$ of $T(V)$ is known as the *exterior algebra* of V, and has important applications in geometry and cohomology theory. One can show that if $(K, \bullet)$ is a unital F-algebra and if $\alpha : V \to K$ is a linear transformation satisfying the condition that $\alpha(v) \bullet \alpha(v) = 0_K$ for all $v \in V$, then α can be uniquely extended to a homomorphism of unital F-algebras from $\bigwedge(V)$ to K.

Exercises

Exercise 1176
Let n be a positive integer and let V be the space of all polynomials in $\mathbb{C}[X]$ of degree at most n. For $p(X) = \sum a_i X^i$ and $q(X) = \sum b_i X^i$ in V, we define the nth *Bézout matrix* $\mathrm{Bez}_n(f, g) \in \mathcal{M}_{n \times x}(\mathbb{C})$ defined by f and g as follows: $\mathrm{Bez}_n(f, g) = [c_{ij}]$, where $c_{ij} = \sum_{k=1}^{m(i,j)}[a_{j+k-1}b_{i-k} - a_{i-k}b_{j+k-1}]$ and $m(i, j) = \min\{i, n+1-j\}$. Show that the function $\mathrm{Bez}_n : V \times V \to \mathcal{M}_{n \times x}(\mathbb{C})$ is a bilinear transformation satisfying the conditions that $\mathrm{Bez}_n(f, f) = O$ for all $f \in V$. If $n = \max\{\deg(f), \deg(g)\}$, show that $\mathrm{Bez}_n(f, g)$ is nonsingular if and only if f and g have no common roots.

Étienne Bézout was an eighteenth-century French mathematician.

Exercise 1177

Find $a \in \mathbb{R}$ such that the matrices $\begin{bmatrix} 1 & a & a \\ 0 & 1 & a \\ 0 & 0 & 1 \end{bmatrix}$ and $\begin{bmatrix} 1 & -1 & -1 \\ -1 & 1 & -1 \\ -1 & -1 & 1 \end{bmatrix}$ define the same bilinear form in $\text{Bill}(\mathbb{R}^3, \mathbb{R}^3)$.

Exercise 1178

Let $V = \mathbb{Q}^\mathbb{Q}$. Is the function from $V \times V$ to $\mathbb{Q}$ given by $(f, g) \mapsto (f + g)(\frac{1}{2}) \cdot (f - g)(2)$ a bilinear form?

Exercise 1179

Let F be a field and let $u : \mathbb{N} \times \mathbb{N} \to F$ be an arbitrary function. Is the function $f_u : F[X] \times F[X] \to F[X]$ defined by

$$f_u : \left(\sum_{i=0}^{\infty} a_i X^i, \sum_{j=0}^{\infty} b_j X^j \right) \mapsto \sum_{k=0}^{\infty} \left(\sum_{i+j=k} u(i, j) a_i b_j X^k \right)$$

a bilinear transformation?

Exercise 1180

Let B be the canonical basis for the vector space $V = \mathbb{R}^2$. Find a bilinear form $f \in \text{Bill}(V \times V)$ satisfying the condition $T_{BB}(f) = \begin{bmatrix} 2 & 2 \\ 4 & -1 \end{bmatrix}$.

Exercise 1181

Let B be the canonical basis for $\mathbb{R}^3$. Find $T_{BB}(f)$, where $f : \mathbb{R}^3 \times \mathbb{R}^3 \to \mathbb{R}$ is the bilinear form defined by

$$f : \left(\begin{bmatrix} a \\ b \\ c \end{bmatrix}, \begin{bmatrix} a' \\ b' \\ c' \end{bmatrix} \right) \mapsto aa' + 2bc' + cc' + 2cb' - ab' + bb' - ba'.$$

Exercise 1182

Let $f : \mathbb{R}^3 \times \mathbb{R}^3 \to \mathbb{R}$ be the bilinear form defined by

$$f : (v, w) \mapsto v \cdot \begin{bmatrix} 0 & 1 & 0 \\ 1 & 0 & 2 \\ 0 & 1 & 1 \end{bmatrix} w.$$

Find the matrix representing f with respect to the basis $\left\{ \begin{bmatrix} 1 \\ 0 \\ 0 \end{bmatrix}, \begin{bmatrix} 0 \\ 1 \\ 1 \end{bmatrix}, \begin{bmatrix} 1 \\ 0 \\ 1 \end{bmatrix} \right\}$

of $\mathbb{R}^3$.

Exercise 1183

Let V and W be vector spaces over a field F and let $\alpha \in \mathrm{Hom}(V, W)$. For each $g \in \mathrm{Bill}(W \times W)$, let us define the bilinear form $g_\alpha \in \mathrm{Bill}(V \times V)$ by setting $g_\alpha : (v, v') \mapsto g(\alpha(v), \alpha(v'))$. Is the function $g \mapsto g_\alpha$ a linear transformation?

Exercise 1184

Let F be a field of characteristic other than 2 and let V be a vector space over F. Let $f \in \mathrm{Bill}(V \times V)$. Show that $f(v, v) \neq 0$ for all $0_V \neq v \in V$ if and only if for every nontrivial subspace W of V and for every $0_V \neq w \in W$ there exists a vector $w' \in W$ satisfying $f(w, w') \neq 0$.

Exercise 1185

Show that if V and W are vector spaces finitely generated over a field F of unequal dimensions, then there is no nondegenerate $f \in \mathrm{Bill}(V \times W)$.

Exercise 1186

Let F be a field of characteristic 0 and let the bilinear form $f \in \mathrm{Bill}(F^3 \times F^3)$ be defined by $f : \left(\begin{bmatrix} a \\ b \\ c \end{bmatrix}, \begin{bmatrix} a' \\ b' \\ c' \end{bmatrix} \right) \mapsto aa' + bb' - cc'$. Is there a nontrivial subspace W of V satisfying $f(w, w') = 0$ for all $w, w' \in W$?

Exercise 1187

Let $f \in \mathrm{Bill}(\mathbb{R}^4 \times \mathbb{R}^4)$ be defined by $f : (v, w) \mapsto v \cdot (Aw)$, where

$$A = \begin{bmatrix} 1 & 0 & 0 & 0 \\ 0 & 1 & 0 & 0 \\ 0 & 0 & 1 & 0 \\ 0 & 0 & 0 & -1 \end{bmatrix}.$$

Find a basis $\{v_1, v_2, v_3, v_4\}$ of $\mathbb{R}^4$ satisfying the condition that $f(v_i, v_i) = 0$ for all $1 \leq i \leq 4$.

Exercise 1188

Let $f : \mathcal{M}_{n \times n}(F) \times \mathcal{M}_{n \times n}(F) \to F$ be the function defined by $f : (A, B) \mapsto \mathrm{tr}(AB)$, where F is a field and n is a positive integer. Is f a bilinear form? Is f symmetric?

Exercise 1189

Let n be a positive integer and let $f : \mathcal{M}_{n \times n}(\mathbb{C}) \times \mathcal{M}_{n \times n}(\mathbb{C}) \to \mathbb{C}$ be the function defined by $f : (A, B) \mapsto n \cdot \mathrm{tr}(AB) - \mathrm{tr}(A)\,\mathrm{tr}(B)$. Show that f is a symmetric bilinear form.

Exercise 1190

Let V be a vector space over a field F and let $f \in \mathrm{Bill}(V \times V)$ be a symmetric bilinear form. Let $Y = F \times V$ and define an operation $\bullet$ on Y by setting

$$\begin{bmatrix} a \\ v \end{bmatrix} \bullet \begin{bmatrix} b \\ w \end{bmatrix} = \begin{bmatrix} ab + f(v, w) \\ aw + bv \end{bmatrix} \quad \text{for all } a, b \in F \text{ and all } v, w \in V.$$

Show that $(Y, \bullet)$ is a Jordan algebra.

Exercise 1191

Let V be a vector space over $\mathbb{Q}$. Is the function $V \times V \to V$ defined by $(v, v') \mapsto v + v'$ a bilinear transformation?

Exercise 1192

Are the matrices $\begin{bmatrix} 0 & 0 & 0 \\ 1 & 1 & 0 \\ 1 & 1 & 1 \end{bmatrix}$ and $\frac{1}{25} \begin{bmatrix} 25 & -5 & 35 \\ 0 & -3 & 21 \\ 0 & -4 & 28 \end{bmatrix}$ in $\mathcal{M}_{3 \times 3}(\mathbb{R})$ congruent?

Exercise 1193

Find an upper-triangular matrix in $\mathcal{M}_{3 \times 3}(\mathbb{R})$ congruent to $\begin{bmatrix} 1 & 0 & -2 \\ -1 & 1 & 0 \\ 0 & -2 & 4 \end{bmatrix}$ or show that there is no such matrix.

Exercise 1194

Let F be a field and let n be a positive integer. A matrix $A = [a_{ij}] \in \mathcal{M}_{n \times n}(F)$ is an *upper Hessenberg matrix* if and only if $a_{ij} = 0$ whenever $i - j \geq 2$. Is every matrix in $\mathcal{M}_{n \times n}(\mathbb{R})$ necessarily congruent to an upper Hessenberg matrix?

© Brigitte Bossert.

Karl Hessenberg was a twentieth-century German engineer.

Exercise 1195

Let F be a field. Show that every upper triangular matrix in $\mathcal{M}_{3\times3}(F)$ is congruent to a lower triangular matrix.

Exercise 1196

Let n be a positive integer and let A be a nonsingular symmetric matrix in $\mathcal{M}_{n\times n}(\mathbb{C})$. Show that A is congruent to A^{-1}.

Exercise 1197

Find a matrix $P \in \mathcal{M}_{3\times3}(\mathbb{R})$ such that the matrix $P \begin{bmatrix} 2 & 1 & 3 \\ 1 & 0 & 1 \\ 3 & 1 & 3 \end{bmatrix} P^T$ is diagonal.

Exercise 1198

Let $A = \begin{bmatrix} 1 & 1 & 1 & 1 \\ 1 & 1 & 1 & 1 \\ 1 & 1 & 1 & 1 \\ 1 & 1 & 1 & 1 \end{bmatrix} \in \mathcal{M}_{4\times4}(\mathbb{R})$. Find a nonsingular matrix $P \in \mathcal{M}_{4\times4}(\mathbb{R})$ such that PAP^T is diagonal.

Exercise 1199

Find a diagonal matrix in $\mathcal{M}_{4\times4}(\mathbb{R})$ congruent to the matrix

$$\begin{bmatrix} 1 & 2 & 3 & 2 \\ 2 & 3 & 5 & 8 \\ 3 & 5 & 8 & 10 \\ 2 & 8 & 10 & -8 \end{bmatrix}.$$

Exercise 1200

Is the matrix $\begin{bmatrix} 1 & i & 1+i \\ i & 0 & 2-i \\ 1+i & 2-i & 10+2i \end{bmatrix} \in \mathcal{M}_{3\times3}(\mathbb{C})$ congruent to I?

Exercise 1201

Let n be a positive integer and let α be a positive-definite endomorphism of $\mathbb{R}^n$ represented with respect to the canonical basis by the matrix A. If A' is a matrix congruent to A, does it too represent a positive-definite endomorphism of $\mathbb{R}^n$ with respect to the canonical basis?

Exercise 1202

Let V be a vector space finitely generated over be a field F of characteristic other than 2. If $f \in \text{Bill}(V \times V)$ is symmetric and not the 0-function, show that there exists a vector $v \in V$ satisfying $f(v, v) \neq 0$.

Exercise 1203

Let V be a vector space finitely generated over be a field F of characteristic other than 2. If $f \in \mathrm{Bill}(V \times V)$, show that $f(v, v) = 0$ for all $v \in V$ if and only if $f(v, w) = -f(w, v)$ for all $v, w \in V$.

Exercise 1204

Let n be a positive integer, let F be a field, and let $A \in \mathcal{M}_{n \times n}(F)$. Show that there exists a symmetric matrix $B \in \mathcal{M}_{n \times n}(F)$ satisfying $v \cdot Av = v \cdot Bv$ for all $v \in F^n$.

Exercise 1205

Find a bilinear form $f \in \mathrm{Bill}\left(\mathbb{R}^3 \times \left[\begin{array}{c} 1 \\ 3 \\ -1 \\ 3 \end{array} \right] \mathbb{R}^3 \right)$ which defines the quadratic

form $\left[\begin{array}{c} a \\ b \\ c \end{array} \right] \mapsto a^2 - 2ab + 4ac - 2bc + 2c^2$.

Exercise 1206

Let $f \in \mathrm{Bill}(\mathbb{R}^3, \mathbb{R}^3)$ be the symmetric bilinear form defined by the matrix $\left[\begin{array}{ccc} -3 & 1 & 0 \\ 1 & -6 & 1 \\ 0 & 1 & 7 \end{array} \right]$. Find the quadratic form defined by f.

Exercise 1207

Let $f \in \mathrm{Bill}(\mathbb{R}^3, \mathbb{R}^3)$ be the symmetric bilinear form defined by the matrix $\left[\begin{array}{ccc} 2 & -1 & 5 \\ -1 & 0 & \frac{1}{3} \\ 5 & \frac{1}{3} & -3 \end{array} \right]$. Find the quadratic form defined by f.

Exercise 1208

Find a symmetric bilinear form $f \in \mathrm{Bill}(\mathbb{R}^3, \mathbb{R}^3)$ which defines the quadratic

form $\left[\begin{array}{c} a \\ b \\ c \end{array} \right] \mapsto 2ab + 4ac + 6bc$.

Exercise 1209

Let F be a field of characteristic other than 2, and let V be a vector space over F. Let $q : V \to F$ be a function satisfying the condition that $q(v + w) + q(v - w) = 2q(v) + 2q(w)$ for all $v, w \in V$. Show that the function $f : V \times V \to F$ defined by $f : (v, w) \mapsto \frac{1}{4}[q(v + w) - q(v - w)]$ is a symmetric bilinear form.

Exercise 1210
Let V be a vector space over a field F of characteristic other than 2, and let $f \in \mathrm{Bill}(V \times V)$ be a symmetric bilinear form which defines a quadratic form $q : V \to F$. Show that

$$q(u + v + w) = q(u + v) + q(u + w) + q(v + w) - q(u) - q(v) - q(w)$$

for all $u, v, w \in V$.

Exercise 1211
Let V be a vector space over a field F. Show that $V \cong F \otimes V$.

Exercise 1212
Let V and W be vector spaces over a field F. Let $x \in V \otimes W$ be written in the form $x = \sum_{i=1}^{n} v_i \otimes w_i$, where n is minimal in the sense that there is no way to express x in the form $\sum_{i=1}^{k} v_i' \otimes w_i'$ for any $k < n$. Show that $\{v_1, \ldots, v_n\}$ is a linearly-independent subset of V and that $\{w_1, \ldots, w_n\}$ is a linearly-independent subset of W.

Exercise 1213
Let K be a field containing F as a subfield. If V is a vector space over F, show that $K \otimes V$ is a vector space over K.

Exercise 1214
Let V be a vector space of finite dimension n over a field F and let Y be the subspace of $V \otimes V$ generated by all elements of the form $v \otimes v' - v' \otimes v$, where $v, v' \in V$. Find the dimension of Y.

Exercise 1215
Let V and W be finite dimensional vector spaces over a field F. Let $v, v' \in V$ and $w, w' \in W$ be vectors satisfying the condition $v \otimes w = v' \otimes w'$ and this is not the identity element of $V \otimes W$ with respect to addition. Show that there exists a scalar $c \in F$ such that $v = cv'$ and $w' = cw$.

Exercise 1216
Let F be a field and, for all $A, B \in \mathcal{M}_{2 \times 2}(F)$, denote the Kronecker product of A and B by $A \otimes B$. If $\{H_1, \ldots, H_4\}$ is the canonical basis for $\mathcal{M}_{2 \times 2}(F)$, is $\{H_i \otimes H_j \mid 1 \leq i, j \leq 4\}$ a basis for $\mathcal{M}_{4 \times 4}(F)$.

Exercise 1217
Find the numerical range of the quadratic form $q : \mathbb{R}^2 \to \mathbb{R}$ defined by $q : v \mapsto$
$$v^T \begin{bmatrix} 1 & 0 \\ 0 & 0 \end{bmatrix} v.$$

Exercise 1218
Let n be a positive integer and let F be a field. If $A \in \mathcal{M}_{n \times n}(F)$ is a magic matrix, is the same true for $A \otimes A \in \mathcal{M}_{2n \times 2n}(F)$?

Exercise 1219

Let F be a field and let k and n be positive integers. If matrices $A \in \mathcal{M}_{k \times k}(F)$ and $B \in \mathcal{M}_{n \times n}(F)$ have eigenvalues a and b, respectively, show that ab is an eigenvalue of $A \otimes B$.

Exercise 1220

Let F be a field and let k and n be positive integers. If matrices $A \in \mathcal{M}_{k \times k}(F)$ and $B \in \mathcal{M}_{n \times n}(F)$ have eigenvalues a and b respectively, find a matrix $C \in \mathcal{M}_{kn \times kn}(F)$ with eigenvalue $a + b$.

Exercise 1221

Let F be a field of characteristic other than 2 and let V be a vector space over F. Find the minimal polynomial of the endomorphism α of $V \otimes V$ defined by $\alpha : \sum_{i=1}^{n} a_i(v_i \otimes w_i) \mapsto \sum_{i=1}^{n} a_i(w_i \otimes v_i)$.

Exercise 1222

Let F be a field, let k, n, s, and t be positive integers, and consider matrices $A \in \mathcal{M}_{k \times n}(F)$ and $B \in \mathcal{M}_{s \times t}(F)$. Is the rank of $A \otimes B$ necessarily equal to the product of the ranks of A and B?

Exercise 1223

Let V, V', W, W' be vector spaces over a field F and let $\alpha : V \to V'$ and $\beta : W \to W'$ be monic linear transformations. Let $\alpha \otimes \beta$ be the linear transformation from $V \otimes V'$ to $W \otimes W'$ defined by $\alpha \otimes \beta : \sum_{i=1}^{n} a_i(v_i \otimes v_i') \mapsto \sum_{i=1}^{n} a_i[\alpha(v_i) \otimes \beta(v_i')]$. Is $\alpha \otimes \beta$ monic?

Exercise 1224

Let F be a field and let $(K, \bullet)$ and $(L, *)$ be F-algebras. Define an operation $\diamond$ on $V \otimes W$ by setting $(v \otimes w) \diamond (v' \otimes w') = (v \bullet v') \otimes (w * w')$ for all $v, v' \in K$ and $w, w' \in L$. Is $(K \otimes L, \diamond)$ an F-algebra?

Exercise 1225

Let $V = \mathbb{R}^2$ and let $W = V \otimes V$. If $w \in W$ is normal, do there necessarily exist normal vectors $v, v' \in V$ such that $w = v \otimes v'$?

Exercise 1226

Let V be a vector space over $\mathbb{R}$. Show that the complexification of V is isomorphic to $\mathbb{C} \otimes V$.

Exercise 1227

Let V be an inner product space over $\mathbb{R}$ having a basis $\{v_i \mid i \in \Omega\}$ and let W be an inner product space over $\mathbb{R}$ having a basis $\{w_j \mid j \in \Lambda\}$ Define a function $\mu : (V \otimes W) \times (V \otimes W) \to \mathbb{R}$ by setting

$$\mu : \left(\sum_{i \in \Omega} \sum_{j \in \Lambda} a_{ij}(v_i \otimes w_j), \sum_{i \in \Omega} \sum_{j \in \Lambda} b_{ij}(v_i' \otimes w_j') \right)$$

$$\mapsto \sum_{i \in \Omega} \sum_{j \in \Lambda} a_{ij} b_{ij} \left[\langle v_i, v_i' \rangle + \langle w_j, w_j' \rangle \right].$$

Is μ an inner product on $V \otimes W$?

Exercise 1228
Let V be a vector space over a field F and let $\alpha \in \text{End}(V)$. Is the function $V \wedge V \to V \wedge V$ defined by $\sum_{i=1}^{n} c_i(v_i \wedge w_i) \mapsto \sum_{i=1}^{n} c_i(\alpha(v_i) \wedge \alpha(w_i))$ a linear transformation?

Exercise 1229
Let n be a positive integer and let $A, B \in \mathcal{M}_{n \times n}(\mathbb{R})$ be orthogonal matrices. Is their Kronecker product $A \otimes B$ an orthogonal matrix?

Exercise 1230
Let n be a positive integer and let $A, B \in \mathcal{M}_{n \times n}(\mathbb{R})$ be permutation matrices. Is their Kronecker product $A \otimes B$ a permutation matrix?

Exercise 1231
Let k and n be positive integers and let F be a field. Let $A \in \mathcal{M}_{k \times k}(F)$ and let $B \in \mathcal{M}_{n \times n}(F)$. Is it necessarily true that $\text{tr}(A \otimes B) = \text{tr}(A)\text{tr}(B)$?

Exercise 1232
Let k and n be positive integers and let F be a field. For $A \in \mathcal{M}_{k \times k}(F)$ and $B \in \mathcal{M}_{n \times n}(F)$, find a matrix $C \in \mathcal{M}_{kn \times kn}(F)$ such that $e^C = e^A \otimes e^B$.

Exercise 1233

The matrix $\begin{bmatrix} 1 & 0 & 0 & 0 \\ 0 & 0 & 1 & 0 \\ 0 & 1 & 0 & 0 \\ 0 & 0 & 0 & 1 \end{bmatrix}$ plays an important part in quantum information the-

ory. Write this matrix as a sum of Kronecker products of $\begin{bmatrix} 1 & 0 \\ 0 & 1 \end{bmatrix}$ and the three Pauli matrices.

Summary of Notation

$\mathrm{Re}(z)$, 7

$\mathrm{im}(z)$, 7

$\bar{z}$, 8

$\mathbb{Q}(\sqrt{p})$, 8

$\mathbb{Z}/(p)$, 8

$\mathrm{GF}(p)$, 8

$\prod_{i \in \Omega} V_i$, 23

$\mathcal{M}_{k \times n}(V)$, 24

V^{Ω}, 23

χ_B, 24

$\coprod_{i \in \Omega} V_i$, 26

$V^{(\Omega)}$, 26

Fv, 26

$C(a, b)$, 26

FD, 27

$\mathrm{affh}(D)$, 29

$\sum_{i \in \Omega} W_i$, 30

K^-, 42

$v \times w$, 42

K^+, 43

$\deg(f)$, 44

$F[X]$, 44

$F[g(X)]$, 45

$\Phi_q(X)$, 48

$\mu(d)$, 48

$F[X_1, \ldots, X_n]$, 50

$\dim(V)$, 71

$U \oplus W$, 74

$\bigoplus_{i \in \Omega} W_i$, 74

$\mathrm{gr}(f)$, 91

$\mathrm{Hom}(V, W)$, 92

$\ker(\alpha)$, 94

$\mathrm{im}(\alpha)$, 94

A^T, 95

$\mathrm{Aff}(V, W)$, 96

$\cong$, 97

$\mathrm{rk}(\alpha)$, 98

$\mathrm{null}(\alpha)$, 98

V/W, 110

$\mathrm{End}(V)$, 113

σ_c, 113

$\mathrm{Aut}(V)$, 115

ε_{hk}, 116

$\varepsilon_{h;c}$, 116

$\varepsilon_{hk;c}$, 116

Φ_{BD}, 133

$v \odot w$, 136

$v \wedge w$, 136

E_{hk}, 153

$E_{h;c}$, 153

$E_{hk;c}$, 153

S_n, 224

$\mathrm{sgn}(\pi)$, 224

$|A|$, 225

$\mathrm{adj}(A)$, 235

$\mathrm{spec}(\alpha)$, 255

$\rho(A)$, 258

$\mathrm{comp}(p)$, 265

$A \sim B$, 266

$\mathrm{Ann}(v)$, 272

$m_A(X)$, 274

$F[\alpha]v_0$, 297

$\mathrm{LR}(V)$, 298

$D(V)$, 317

$\mathrm{tr}(A)$, 318

$\langle v, w \rangle$, 333

J.S. Golan, *The Linear Algebra a Beginning Graduate Student Ought to Know*,
DOI 10.1007/978-94-007-2636-9, © Springer Science+Business Media B.V. 2012

Index to Thumbnail Photos

B

A

B

J.S. Golan, *The Linear Algebra a Beginning Graduate Student Ought to Know*,
DOI 10.1007/978-94-007-2636-9, © Springer Science+Business Media B.V. 2012

M

Cyrus C. Macduffee, 266, © Mathematical Association of America 2011; all rights reserved

Colin Maclaurin, 237, Maclaurin, C., A Treatise of Fluxions, T.W. and T. Ruddimans, 1742

Andrei Andreyevich Markov, 151, source unknown

James Clerk Maxwell, 22, courtesy of Special collections, fine Arts Library, Harvard University

William Henry Metzler, 354, source unknown

John Milnor, 72, Archives of the Mathematisches Forschungsinstitut Oberwolfach

Hermann Minkowski, 340, ETH-Bibliothek Zurich, Image Archive

Cleve Moler, 380, courtesy of The MathWorks, Inc.

Eliakim Hastings Moore, 9, courtesy of the American Mathematical Society

N

Mark Aronovich Naimark, 450, courtesy of the American Mathematical Society

Alexander Ivanovich Nekrasov, 206, source unknown

Rolf Nevanlinna, 403, source unknown

Emmy Noether, 40, Bryn Mawr College Library, Special Collections

O

Andrew Odlyzko, 301, with kind permission of Andrew Odlyzko

Alexander Markovich Ostrowski, 47, Archives of the Mathematisches Forschungsinstitut Oberwolfach

P

Luca Pacioli, 299, source unknown

Henri Padé, 239, source unknown

Giuseppe Peano, 67, Department of Mathematics, University of Torino, Italy, www.peano2008.unito.it

Benjamin Peirce, 40, Harvard University Archives, HUP

Charles Sanders Peirce, 40, Harvard University Archives, HUP

Roger Penrose, 441, Archives of the Mathematisches Forschungsinstitut Oberwolfach

Oskar Perron, 271, Archives of the Mathematisches Forschungsinstitut Oberwolfach

Johann Pfaff, 228, source unknown

Georg Pick, 403, source unknown

Salvadore Pincherle, 89, Archives of the Mathematisches Forschungsinstitut Oberwolfach

Jules Henri Poincaré, 189, The University of Chicago Yerkes Observatory, courtesy AIP Emilio Segre Visual Archives, Physics Today Collection and Tenn Collection

R

Gregorio Ricci-Curbastro, 465, source unknown

Frigyes Riesz, 383, source unknown

Heinz Rutishauser, 379, with kind permission of Walter Gander

S

Yousef Saad, 375, with kind permission of Y. Saad

Jean Claude Saint-Venant, 22, source unknown

Heinrich Scherk, 227, Archives of the Mathematisches Forschungsinstitut Oberwolfach

Erhardt Schmidt, 26, Archives of the Mathematisches Forschungsinstitut Oberwolfach

Isaac Jacob Schoenberg, 56, Archives of the Mathematisches Forschungsinstitut Oberwolfach

Otto Schreier, 334, source unknown

Martin Schultz, 375, with kind permission of Martin Schultz

Issai Schur, 161, Archives of the Mathematisches Forschungsinstitut Oberwolfach

Hermann Schwarz, 337, ETH-Bibliothek Zurich, Image Archive

Takazaku Seki Kowa, 191, source unknown

Kenjiro Shoda, 320, source unknown

Thomas Simpson, 187, source unknown

Henry Smith, 462, source unknown

Sergei Lvovich Sobolev, 359, Archives of the Mathematisches Forschungsinstitut Oberwolfach

Richard V. Southwell, 207, Neville Miles, Imperial College London

Ernst Steinitz, 6, Archives of the Mathematisches Forschungsinstitut Oberwolfach

G. W. Stewart, 380, with kind permission of Eric de Sturler

Eduard Stiefel, 301, Archives of the Mathematisches Forschungsinstitut Oberwolfach

Marshall Stone, 425, courtesy of the American Mathematical Society

Volker Strassen, 166, with kind permission of Volker Strassen

John William Strutt (Lord Rayleigh), 401, Science Photo Library

Charles-François Sturm, 256, source unknown

Peter Swinnerton-Dyer, 169, with kind permission of Sir Peter Swinnerton-Dyer

James Joseph Sylvester, 24, source unknown

T

Henry Taber, 318, Clarke University Archives

William Guthrie Tait, 66, courtesy of Special collections, Fine Arts Library, Harvard University

John Tate, 280, Archives of the Mathematisches Forschungsinstitut Oberwolfach

Olga Taussky-Todd, 349, Archives of the Mathematisches Forschungsinstitut Oberwolfach

Angus Taylor, 72, with kind permission of UC Berkeley

William Thurston, 271, Archives of the Mathematisches Forschungsinstitut Oberwolfach

Otto Toeplitz, 203, Archives of the Mathematisches Forschungsinstitut Oberwolfach

Alan Turing, 169, National Portrait Gallery

Index